Nitrogen Fixation in Agriculture, Forestry, Ecology, and the Environment

Nitrogen Fixation: Origins, Applications, and Research Progress

VOLUME 4

Nitrogen Fixation in Agriculture, Forestry, Ecology, and the Environment

Edited by

Dietrich Werner
Philipps-University,
Marburg, Germany

and

William E. Newton
Department of Biochemistry,
Virginia Polytechnic Institute and State University,
Blacksburg, Virginia, U.S.A.

Springer

A C.I.P. Catalogue record for this book is available from the Library of Congress.

ISBN-13 978-90-481-6897-2 (PB)
ISBN-13 978-1-4020-3544-9 (e-book)

Published by Springer,
P.O. Box 17, 3300 AA Dordrecht, The Netherlands.

www.springeronline.com

Background figure caption:

"A seed crop of clover (*Trifolium hirtum*) in flower near Moora, Western Australia. Photograph courtesy of Mike Davies, Senior Technical Officer, Pasture Research Group of Agriculture WA and reproduced with permission."

Vol. 4-specific figure caption:

"First year soybean cultivation from a field trial in Puerto Rico with both inoculated and non-inoculated rows. Photograph courtesy of R. Stewart Smith, Nitragin Company, USA, and reproduced with permission."

Printed on acid-free paper

Printed in the Netherlands

TABLE OF CONTENTS

SERIES PREFACE

Nitrogen Fixation: Origins, Applications, and Research Progress

Nitrogen fixation, along with photosynthesis as the energy supplier, is the basis of all life on Earth (and maybe elsewhere too!). Nitrogen fixation provides the basic component, fixed nitrogen as ammonia, of two major groups of macromolecules, namely nucleic acids and proteins. Fixed nitrogen is required for the N-containing heterocycles (or bases) that constitute the essential coding entities of deoxyribonucleic acids (DNA) and ribonucleic acids (RNA), which are responsible for the high-fidelity storage and transfer of genetic information, respectively. It is also required for the amino-acid residues of the proteins, which are encoded by the DNA and that actually do the work in living cells. At the turn of the millennium, it seemed to me that now was as good a time as any (and maybe better than most) to look back, particularly over the last 100 years or so, and ponder just what had been achieved. What is the state of our knowledge of nitrogen fixation, both biological and abiological? How has this knowledge been used and what are its impacts on humanity?

In an attempt to answer these questions and to capture the essence of our current knowledge, I devised a seven-volume series, which was designed to cover all aspects of nitrogen-fixation research. I then approached my long-time contact at Kluwer Academic Publishers, Ad Plaizier, with the idea. I had worked with Ad for many years on the publication of the Proceedings of most of the International Congresses on Nitrogen Fixation. My personal belief is that congresses, symposia, and workshops must not be closed shops and that those of us unable to attend should have access to the material presented. My solution is to capture the material in print in the form of proceedings. So it was quite natural for me to turn to the printed word for this detailed review of nitrogen fixation. Ad's immediate affirmation of the project encouraged me to share my initial design with many of my current co-editors and, with their assistance, to develop the detailed contents of each of the seven volumes and to enlist prospective authors for each chapter.

There are many ways in which the subject matter could be divided. Our decision was to break it down as follows: nitrogenases, commercial processes, and relevant chemical models; genetics and regulation; genomes and genomics; associative, endophytic, and cyanobacterial systems; actinorhizal associations; leguminous symbioses; and agriculture, forestry, ecology, and the environment. I feel very fortunate to have been able to recruit some outstanding researchers as co-editors for this project. My co-editors were Mike Dilworth, Claudine Elmerich, John Gallon, Euan James, Werner Klipp, Bernd Masepohl, Rafael Palacios, Katharina Pawlowski, Ray Richards, Barry Smith, Janet Sprent, and Dietrich Werner. They worked very hard and ably and were most willing to keep the volumes moving along reasonably close to our initial timetable. All have been a pleasure to work with and I thank them all for their support and unflagging interest.

Nitrogen-fixation research and its application to agriculture have been ongoing for many centuries – from even before it was recognized as nitrogen fixation. The Romans developed the crop-rotation system over 2000 years ago for maintaining and improving soil fertility with nitrogen-fixing legumes as an integral component. Even though crop rotation and the use of legumes was practiced widely but intermittently since then, it wasn't until 1800 years later that insight came as to how legumes produced their beneficial effect. Now, we know that bacteria are harbored within nodules on the legumes' roots and that they are responsible for fixing N_2 and providing these plants with much of the fixed nitrogen required for healthy growth. Because some of the fixed nitrogen remains in the unharvested parts of the crop, its release to the soil by mineralization of the residue explains the follow-up beneficial impact of legumes. With this realization, and over the next 100 years or so, commercial inoculants, which ensured successful bacterial nodulation of legume crops, became available. Then, in the early 1900's, abiological sources of fixed nitrogen were developed, most notable of these was the Haber-Bosch process. Because fixed nitrogen is almost always the limiting nutrient in agriculture, the resulting massive increase in synthetic fixed-nitrogen available for fertilizer has enabled the enormous increase in food production over the second half of the 20th century, particularly when coupled with the new "green revolution" crop varieties. Never before in human history has the global population enjoyed such a substantial supply of food.

Unfortunately, this bright shiny coin has a slightly tarnished side! The abundance of nitrogen fertilizer has removed the necessity to plant forage legumes and to return animal manures to fields to replenish their fertility. The result is a continuing loss of soil organic matter, which decreases the soil's tilth, its water-holding capacity, and its ability to support microbial populations. Nowadays, farms do not operate as self-contained recycling units for crop nutrients; fertilizers are trucked in and meat and food crops are trucked out. And if it's not recycled, how do we dispose of all of the animal waste, which is rich in fixed nitrogen, coming from feedlots, broiler houses, and pig farms? And what is the environmental impact of its disposal? This problem is compounded by inappropriate agricultural practice in many countries, where the plentiful supply of cheap commercial nitrogen fertilizer, plus farm subsidies, has encouraged high (and increasing) application rates. In these circumstances, only about half (at best) of the applied nitrogen reaches the crop plant for which it was intended; the rest leaches and "runs off" into streams, rivers, lakes, and finally into coastal waters. The resulting eutrophication can be detrimental to marine life. If it encroaches on drinking-water supplies, a human health hazard is possible. Furthermore, oxidation of urea and ammonium fertilizers to nitrate progressively acidifies the soil – a major problem in many agricultural areas of the world. A related problem is the emission of nitrogen oxides (NO_x) from the soil by the action of microorganisms on the applied fertilizer and, if fertilizer is surface broadcast, a large proportion may be volatilized and lost as ammonia. For urea in rice paddies, an extreme example, as much as 50% is volatilized and lost to the atmosphere. And what goes up must come down; in the case of fertilizer nitrogen, it returns to Earth in the rain, often acidic in nature. This uncontrolled deposition has unpredictable environmental effects, especially in pristine environments like forests, and may also affect biodiversity.

Some of these problems may be overcome by more efficient use of the applied fertilizer nitrogen. A tried and tested approach (that should be used more often) is to ensure that a balanced supply of nutrients (and not simply applying more and more) is applied at the right time (maybe in several separate applications) and in the correct place (under the soil surface and not broadcast). An entirely different approach that could slow the loss of fertilizer nitrogen is through the use of nitrification inhibitors, which would slow the rate of conversion of the applied ammonia into nitrate, and so decrease its loss through leaching. A third approach to ameliorating the problems outlined above is through the expanded use of biological nitrogen fixation. It's not likely that we shall soon have plants, which are capable of fixing N_2 without associated microbes, available for agricultural use. But the discovery of N_2-fixing endophytes within the tissues of our major crops, like rice, maize, and sugarcane, and their obvious benefit to the crop, shows that real progress is being made. Moreover, with new techniques and experimental approaches, such as those provided by the advent of genomics, we have reasons to renew our belief that both bacteria and plants may be engineered to improve biological nitrogen fixation, possibly through developing new symbiotic systems involving the major cereal and tuber crops.

In the meantime, the major impact might be through agricultural sustainability involving the wider use of legumes, reintroduction of crop-rotation cycles, and incorporation of crop residues into the soil. But even these practices will have to be performed judiciously because, if legumes are used only as cover crops and are not used for grazing, their growth could impact the amount of cultivatable land available for food crops. Even so, the dietary preferences of developed countries (who eats beans when steak is available?) and current agricultural practices make it unlikely that the fixed-nitrogen input by rhizobia in agricultural soils will change much in the near-term future. A significant positive input could accrue, however, from matching rhizobial strains more judiciously with their host legumes and from introducing "new" legume species, particularly into currently marginal land. In the longer term, it may be possible to engineer crops in general, but cereals in particular, to use the applied fertilizer more efficiently. That would be a giant step the right direction. We shall have to wait and see what the ingenuity of mankind can do when "the chips are down" as they will be sometime in the future as food security becomes a priority for many nations. At the moment, there is no doubt that commercially synthesized fertilizer nitrogen will continue to provide the key component for the protein required by the next generation or two.

So, even as we continue the discussion about the benefits, drawbacks, and likely outcomes of each of these approaches, including our hopes and fears for the future, the time has arrived to close this effort to delineate what we know about nitrogen fixation and what we have achieved with that knowledge. It now remains for me to thank personally all the authors for their interest and commitment to this project. Their efforts, massaged gently by the editorial team, have produced an indispensable reference work. The content is my responsibility and I apologize upfront for any omissions and oversights. Even so, I remain confident that these volumes will serve well the many scientists researching nitrogen fixation and related fields, students considering the nitrogen-fixation challenge, and administrators wanting to either become acquainted with or remain current in this field. I also acknowledge the many scientists who were not direct contributors to this series of books, but whose contributions to the field are documented in their pages. It would be remiss of me not to acknowledge also the patience and assistance of the several members of the Kluwer staff who have assisted me along the way. Since my initial dealings with Ad Plaizier, I have had the pleasure of working with Arno Flier, Jacco Flipsen, Frans van Dunne, and Claire van Heukelom; all of whom provided encouragement and good advice – and there were times when I needed both!

It took more years than I care to remember from the first planning discussions with Ad Plaizier to the completion of the first volumes in this series. Although the editorial team shared some fun times and a sense of achievement as volumes were completed, we also had our darker moments. Two members of our editorial team died during this period. Both Werner Klipp (1953-2002) and John Gallon (1944-2003) had been working on Volume II of the series, *Genetics and Regulation of Nitrogen-Fixing Bacteria*, and that volume is dedicated to their memory. Other major contributors to the field were also lost in this time period: Barbara Burgess, whose influence reached beyond the nitrogenase arena into the field of iron-sulfur cluster biochemistry; Johanna Döbereiner, who was the discoverer and acknowledged leader in nitrogen-fixing associations with grasses; Lu Jiaxi, whose "string bag" model of the FeMo-cofactor prosthetic group of Mo-nitrogenase might well describe its mode of action; Nikolai L'vov, who was involved with the early studies of molybdenum-containing cofactors; Dick Miller, whose work produced new insights into MgATP binding to nitrogenase; Richard Pau, who influenced our understanding of alternative nitrogenases and how molybdenum is taken up and transported; and Dieter Sellmann, who was a synthetic inorganic chemistry with a deep interest in how N_2 is activated on metal sites. I hope these volumes will in some way help both preserve their scientific contributions and reflect their enthusiasm for science. I remember them all fondly.

Only the reactions and interest of you, the reader, will determine if we have been successful in capturing the essence and excitement of the many sterling achievements and exciting discoveries in the research and application efforts of our predecessors and current colleagues over the past 150 years or so. I sincerely hope you enjoy reading these volumes as much as I've enjoyed producing them.

William E. Newton
Blacksburg, February 2004

PREFACE

Nitrogen Fixation in Agriculture, Forestry, Ecology and the Environment

Most grant applications that involve basic research on biological nitrogen fixation (BNF) emphasize the economic importance of this basic biological process. But just what is the economic value of BNF? In several chapters of this volume, the authors report their estimations of the amount of nitrogen fixed biologically. For example, the amount of nitrogen fixed each year by the legume-rhizobia symbiosis is around 70 million tonnes. If we assume that at least 70% of this amount is utilised in agricultural and agroforestry systems and so replaces commercial N-fertiliser, we can estimate that around 50 million tonnes of fixed nitrogen are made available to support agricultural production. In Europe, the price of fertiliser-N to the farmer at the beginning of 2004 was 650€ per tonne, which is equivalent to US$ 800. Putting these two figures together indicates just how enormous the credit made by legume-based BNF to agriculture is. Of course, this price varies significantly in other regions of the world and is closely linked to the prices of oil and gas. This fertiliser-oil linkage is the reason that funding for both the basic and applied aspects of BNF research has increased in periods with high oil and gas prices and decreased when energy prices are moderate.

Because of their importance in worldwide agriculture, the volume starts with surveys of the current uses and benefits, plus the future potential, of legumes in agricultures around the World. In Chapter 1, production and BNF of many tropical legumes, as well as fodder and green-manure legumes, such as *Mucuna*, are described to illustrate their potential. The "Phaseomics" consortium and its research targets are also summarized here. This consortium involves forty-six research groups working with *Phaseolus* sp. and is offered as a model for future collaborative research projects focussed on a single crop. Chapter 1 also suggests that, in addition to the already established legume crops, such *as Phaseolus, Vigna, Arachis, Cicer, Cajanus and Mucuna* species, many other tropical legumes may become regionally important crops, depending on when more research funding becomes available to develop breeding programmes that include both crop protection and crop nutrition by BNF.

Soybean is one of the four most important crops worldwide and it is the number-one oil and protein supplier for animal and human nutrition. The 2002 world production of 180 million tons exceeded the combined production of all other grain legumes, such as groundnuts, beans, pigeon peas and chickpeas. Production and BNF in soybeans in the four major growing regions for this crop are described in Chapters 2-5, with Chapter 2 considering North America, Chapter 3 covering South America, Chapter 4 describing the situation in India, and Chapter 5 giving an overview of the history and current soybean cultivation in China, the original home of the soybean. The regional aspects are especially emphasized in this extensive coverage to illustrate the importance of both plant breeding (and consequent plant production) and their concurrent adaptation to regional priorities, where world market and trade regulations are becoming increasingly important. The areas planted with soybeans are 57,000 km^2 in India, 83,000 km^2 in China, 290,000 km^2 in South America, and 300,000 km^2 in North America. All other areas add only about 70,000 km^2, making a global total of about 800,000 km^2. This area planted to soybean is more than the combined land area of France and the UK. Other crop legumes are planted on about 700,000 km^2, which together with the soybean area, adds up to 1.5 million km^2 of the Earth's land area planted to cultivated legumes. With such a large land area dedicated to their cultivation, it is not surprising that considerable research emphasis and, in many countries strict regulatory control, is given to the preparation, production, and application of rhizobial inoculants for legumes. This is one of the few areas where research on symbiotic nitrogen fixation has directly achieved a level of economic relevance (Chapter 11). The rhizobial strain-selection programmes for soybeans and common beans in Brazil are successful examples of continuous efforts to promote inoculation with the appropriate strains and the correct techniques.

The pasture and grassland areas on the five continents total about 30 million km^2, but we do not have an accurate estimate of what percentage is covered with legume species. If we assume 10% coverage, they would occupy an area of 3 million km^2. When we add the area covered with nodulated legume trees and shrubs (Chapter 7), which is larger than previously assumed, especially in Africa, and the more than 160 species of nitrogen-fixing trees and shrubs associated with actinorhiza (Chapter 8), we can make the cautious estimate that about 10 million km^2 of the land area of this planet are covered with symbiotic nitrogen-fixing species. This is more than the area of the United States of America, again highlighting the importance of these symbiotic species to humankind.

Among the important influences on symbiotic nitrogen fixation with these crop and related species are soil-stress factors. The major stress factors, soil water, soil pH, soil temperature, and nutrient limitations, have been studied in detail and are described in Chapter 6. The various combinations of these factors, together with the large number of host plant-microsymbiont permutations, make it impossible still to predict precisely the amount of nitrogen that will be fixed by a single plant species in any region of the World. For example, more than forty such factors, which affect nitrogen fixation by actinorhizal plants in the field, are listed in figure 3 of Chapter 8. Another fascinating research area involves the interaction of nitrogen-fixing symbioses with arbuscular mycorrhizae (AM) to give a tripartite association (Chapter 10). The AM is probably the oldest symbiosis of higher plants with microbes. In some phases of the tripartite interaction, the legume-rhizobia symbiosis and the AM symbiosis not only share the same plant host, but also share signal-transduction pathways (see also M. Parniske *et al.*, (2002) *Nature*).

A second BNF-research area of agricultural relevance and potential involves the endophytic associations of diazotrophs with cereals and other grasses. These associations appear less formalised than the rhizobia-legume symbiosis and, unfortunately, we still do have not enough field data to quantify their economic impact. Attempts to develop such data are often frustrated by the very complicated situation that exists. For example, many of these close endophytic associations of nitrogen-fixing microbes with grasses include microbial species that cannot be cultivated directly with the result that their presence can only be detected by indirect methods, such as *in situ nifH* mRNA expression (Chapter 9 and see also volume 5 of this series, *Associative and Endophytic Nitrogen-fixing Bacteria and Cyanobacterial Associations*).

The volume closes with consideration of the other major processes in the Nitrogen Cycle and their relationship to nitrogen fixation. The economic impact of free-living nitrogen-fixing species is probably negligible because, before the nitrogen fixed by the bacterial cells has a chance to reach the plant, a large portion of it is likely lost and returned to the atmosphere as N_2 by the combined processes of nitrification and denitrification. However, the ecological importance of the complete Nitrogen Cycle for the the quality of soils and soil functions can hardly be overestimated. Therefore, the final chapters of this volume cover these two important processes; nitrification is covered in Chapter 12 and denitrification in Chapter 13. Both of these integral parts of the Nitrogen Cycle have microorganisms as their essential components. The progress made in understanding the physiology, biochemistry, and molecular genetics that undergird these two processes is as equally impressive as that made in the field of nitrogen fixation as described in the seven volumes of this series.

We could not close this volume without thanking everybody involved with its production. We are especially grateful to all the authors of this volume for contributing their experience, considerable effort, and time (over and above their regular work load) to produce interesting, up-to-date, and extensively referenced chapters. We also wish to give special thanks to Mrs. Lucette Claudet in Marburg for her dedicated work on this volume.

D. Werner wants to thank the EU for the International Development Project ICA 4-CT-2001-10057, related to several chapters of this volume.

Dietrich Werner
Marburg, January 2005

William E. Newton
Blackburg, January 2005

LIST OF CONTRIBUTORS

Rosario AZCÓN
Departemento Microbiologia del
Suelo y Systemas Simbioticos,
Estacion Experimental del
Zaidin, CSIC,
C/ Profesor Albareda 1,
Apdo 419, E-18008 Granada,
Spain.
Email: razcon@eez.csic.es

Concepción AZCÓN-AGUILAR
Departemento Microbiologia del
Suelo y Systemas Simbioticos,
Estacion Experimental del
Zaidin, CSIC,
C/ Profesor Albareda 1,
Apdo 419, E-18008 Granada,
Spain.
Email: cazcon@eez.csic.es

José-Miguel BAREA
Departemento Microbiologia del
Suelo y Systemas Simbioticos,
Estacion Experimental del
Zaidin, CSIC,
C/ Profesor Albareda 1,
Apdo 419, E-18008 Granada,
Spain.
Email:
JoseMiguel.Barea@eez.csic.es

Eberhard BOCK
Fachbereich Biologie
der Universität Hamburg,
Institut für Allgemeine Botanik
Mikrobiologie, Ohnhorststr. 18,
D-22609 Hamburg, Germany.
Email: spieck@mikrobiologie.uni-
hamburg.de

Rubens José CAMPO
EMBRAPA-CNPSO Soja
Cx. Postal 231,
86001-970, Londrina PR, Brazil.
Email: rjcampo@cnpso.embrapa.br

María Jesús DELGADO IGEÑO
Estación Experimental del
Zaidín, CSIC,
C/ Profesor Albareda 1,
E-18008 Granada, Spain.
Email: mdelgado@eez.csic.es

Claudia FIENCKE
Institut für Bodenkunde,
Universität Hamburg,
Allende-Platz 2,
D-20146 Hamburg, Germany.
Email: c.fiencke@ifb.uni-
hamburg.de

Julio C. FRANCHINI
EMBRAPA-CNPSO Soja,
Cx. Postal 231,
86001-970, Londrina PR, Brazil.
Email:
franchin@soja.cnpso.embrapa.br

Peter H. GRAHAM
University of Minnesota,
Dept. of Soil, Water, and
Climate,
1991 Upper Buford Circle,
St Paul, MN 55108, USA.
Email: pgraham@soils.umn.edu

Mariangela HUNGRIA
EMBRAPA–CNPSO Soja
Cx. Postal 231,
86001-970, Londrina PR, Brazil.
Email: hungria@cnpso.embrapa.br

Thomas HUREK
Allgemeine Mikrobiologie
Fachbereich 2 Biologie/Chemie,
Universität Bremen,
Postfach 33 04 40,
D-28334 Bremen, Germany
Email: thurek@uni-bremen.de

M. F. LOUREIRO
UFMT/FAMEV,
Av. Fernando Correa s/n
Campus Universitário,
78000-900 Cuiabá, MT, Brazil.
Email: loureiromf@uol.com.br

S. K. MAHNA
Department of Botany, Maharshi
Dayanand Saraswati University,
K-22, Gandhi-Nagar, Naka
Madar, Ajmer - 305001, India.
Email: mahnask@hotmail.com

Ileda C. MENDES
Embrapa Cerrados.
Cx. Postal 08223, 73301-970
Planaltina, DF, Brazil.
Email:
mendesi@cpac.embrapa.br

William E. NEWTON
Department of Biochemistry,
Virginia Polytechnic Institute and
State University,
Blacksburg, VA 24061, USA.
Email: wenewton@vt.edu

Steven G. PUEPPKE
Associate Dean for Research,
College of Agricultural,
Consumer and Environmental
Sciences, 211 Mumford Hall,
University of Illinois,
Urbana, IL 61801, USA
Email : pueppke@uiuc.edu

Barbara REINHOLD-HUREK
Allgemeine Mikrobiologie
Fachbereich 2 Biologie/Chemie,
Universität Bremen,
Postfach 33 04 40,
D-28334 Bremen, Germany.
Email: breinhold@uni-
bremen.de

David J. RICHARDSON
Centre for Metalloprotein
Spectroscopy and Biology,
School of Biological Sciences,
University of East Anglia
Norwich NR4 7TJ, U.K.
Email: d.richardson@uea.ac.uk

D.N. RODRIGUEZ-NAVARRO
Centro de Investigación y
Formación Agraria "Las Torres
y Tomejil", Apartado Oficial,
E-41200-Alcalá del Río, Sevilla,
Spain
Email:
dulcenombre.rodriguez@untade
andalucia.es

Ricardo RUSSO
Institute for Forestry and Natural
Resources, EARTH University,
P.O. Box 4442-1000, Costa Rica
Email: r-russo@earth.ac.cr

José Enrique RUIZ SAINZ
Departamento de Microbiologia
Facultad de Biologia,
Universidade de Sevilla
Apartado 1095, 41080 Sevilla
Spain.
Email: rsainz@us.es

Michael J. SADOWSKY
Dept. of Microbiology and Soil Science, University of Minnesota, Borlaug Hall, 1991 Upper Buford Circle, St. Paul, MN 55113, USA.
Email: Sadowsky@umn.edu

Rob J.M. van SPANNING
Department of Molecular Cell Physiology, Faculty of EarTh and Life Sciences, Free University, De Boelelaan 1087, 1081 HV Amsterdam, The Netherlands.
Email: spanning@bio.vu.nl

Eva SPIECK
Institut für Allgemeine Botanik, Abteilung Mikrobiologie, Universität Hamburg, Ohnhorststr. 18, D-22609 Hamburg, Germany.
Email: spieck@mikrobiologie.uni-hamburg.de

Janet I. SPRENT
32 Birkhill Avenue, Wormit, Fife DD6 8PW, U.K.
Email: JISprent@aol.com

Jane E. THOMAS-OATES
Department of Chemistry, University of York, Heslington, York, YO10 5DD, U.K.
Email: jeto1@york.ac.uk

J. M. VINARDELL
Departamento de Microbiologia Facultad de Biologia, Universidade de Sevilla Apartado 1095, E-41080, Sevilla Spain.
Email: jvinar@us.es

Dietrich WERNER
FG Zellbiologie und Angewandte Botanik, Fachbereich Biologie, Philipps-Universität Marburg, Karl-von-Frisch-Strasse, D-35032 Marburg, Germany.
E-mail: werner@staff.uni-marburg.de

J. C. ZHOU
Department of Microbiology Huazhong Agricultural University, Shi Zi Shan Street, P.O. Box 430070, Wuhan People's Republic of China.
Email: zjc42926@public.wh.hb.cn

Chapter 1

PRODUCTION AND BIOLOGICAL NITROGEN FIXATION OF TROPICAL LEGUMES

D. WERNER

Fachbereich Biologie, Fachgebiet Zellbiologie und Angewandte Botanik, Philipps-Universität Marburg, D-35032 Marburg, Germany

1. INTRODUCTION

More than 60% of grain-legume production is soybeans (Table 1). Therefore, this crop is covered by several chapters in this volume as a model for several other grain legumes in which our knowledge is much less developed due to the limited areas of production and, therefore, lower economic impact. However, for the people in many areas, especially in South America and Africa, other grain legumes are vital

Table 1. World Production (FAO Yearbook, 2001)

	Million tons
Soybeans	176.6
Groundnuts	35.1
Dry beans (*Phaseolus* and *Vigna*)	16.8
Dry peas	10.5
Chickpeas	6.1
Dry faba beans	3.7
Lentils	3.1
Green beans	4.7
Green peas	7.1

both for the diets of the population and for the production by small farmers. In terms of sustainable agriculture, both fodder legumes and legumes used as green

D. Werner and W. E. Newton (eds.), Nitrogen Fixation in Agriculture, Forestry, Ecology, and the Environment, 1-13.

manure are equally important. As examples of recent progress in understanding production and biological nitrogen fixation (BNF) in these legumes, the following species will be discussed: beans (*Phaseolus* sp. and *Vigna*), chickpea (*Cicer arietinum*), pigeon pea (*Cajanus cajan*), peanuts (*Arachis* sp.), *Mucuna* and other tropical legumes.

The protein, lipid, and carbohydrate contents of grain legumes are summarized in Table 2. The protein content of most species is in the range of 20-30% in dry seeds. Only *Psophocarpus tetragonolobus* with 33%, *Glycine max* with *ca.* 34%, and *Lupinus mutabilis* with 48% are significantly beyond this range. The lipid content is very low with only 1-2% in *Vigna*, *Phaseolus*, *Pisum*, *Cajanus*, *Lens*, and *Dolichos* species. Three genera with a protein content greater than 25% have high lipid concentrations; these are 16% in *Psophocarpus tetragonolobus*, 18% in *Glycine max*, and up to 48% in *Arachis hypogaea* (peanuts).

Table 2. Protein, lipid and carbohydrate content (%) in dry seeds of grain legumes. Modified from Souci et al., *1994.*

Crop-species	*Protein*	*Lipid*	*Carbohydrate*
Arachis hypogaea	**25.3**	**48.1**	8.3
Cajanus cajan	20.2	1.4	47.0
Canavalia ensiformis	25.5	2.5	50.0
Cicer arietinum	19.8	3.4	41.2
Dolichos lablab	22.0	1.5	50.0
Glycine max	**33.7**	**18.1**	6.3
Lens culinaris	23.5	1.4	52.0
Phaseolus acutifolius	24.5	1.5	65.5
Phaseolus lunatus	20.6	1.4	45.0
Phaseolus vulgaris	21.3	1.6	40.1
Pisum sativum	22.9	1.4	41.2
Psophocarpus tetragonolobus	**33.1**	**16.2**	30.8
Vicia faba	23.0	2.0	55.0
Vigna aconitifolia	23.6	1.1	56.5
Vigna angularis	20.7	1.4	56.4
Vigna mungo	23.1	1.2	41.5
Vigna radiata	24.0	1.1	43.6
Vigna subterranea	19.0	7.0	54.0
Vigna umbellata	21.5	0.3	60.9
Vigna unguiculata	23.5	1.4	41.7

2. *PHASEOLUS* SP. AND *VIGNA* SP. (BEANS)

More than 200 species of the genus *Phaseolus* have been described (Smartt, 1988). The most important species economically are *Phaseolus vulgaris* var. nanus, *Phaseolus coccineus*, *Phaseolus lunatus*, *Phaseolus acutifolius* and *Phaseolus semierectus*. Common names for *Phaseolus vulgaris* are dwarf beans, bush beans and dry beans (in spanish "frielos secos"). The gene center for *Phaseolus* species is in Central and South America and for *Vigna* in Asia.

The original wild form of *Phaseolus vulgaris* is *Phaseolus vulgaris* ssp. *Aborigineus*, which is found in moist valleys in middle America. The genus *Phaseolus* has a chromosome number of 2N = 22. The main gene banks are at the CIAT in Cali (Columbia), in Gatersleben (Germany), and in the Wavilov-Institute in St Petersburg (Russia). A large number of genetic variations are known that impact yield, seed colour, and early ripening. Specific morphological and physiological characters are a shallow root system, self pollination, and light-independent development. Species can develop normally under shading conditions, making this genus suitable for agroforestry systems.

The DNA content of *Phaseolus*, *Vigna*, and *Psophocarpus* species are summarized in Table 3. It ranges from 590 units in *Phaseolus vulgaris* up to 1110 in *Vigna aconitifolia*. Desired characteristics of new bean cultivars have been formulated by Broughton *et al.* (2003) and the targets and genetical components are summarized in Table 4. They include α-amylase-inhibitors, specific sugar synthesis, and *Rhizobium* nodulation characteristics.

Table 3. DNA content of Phaseolus, Vigna and Psophocarpus species. Modified from Bennett and Leitch (2001) and Smartt (1990).

Vernacular name	*Latin name (sub-family - Papilionoideae)*	*IC DNA content*
Common beans	*Phaseolus vulgaris* L.	590
Lima beans	*Phaseolus lunatus* L.	690
Scarlet runner beans	*Phaseolus coccineus* L.	670
Tepary beans	*Phaseolus acutifolius* A. Gray	740
Long beans/cowpeas	*Vigna unguiculatu* (L.) Walp.	590
Mung beans (green gram)	*Vigna radiata* (L.) Wilczek	520
Urd (black gram)	*Vigna mungo* (L.) Hepper	540
Adzuki beans	*Vigna angularis* (Willd.) Ohwi and Ohashi	540
Rice beans	*Vigna umbellata* (Jacq.) Ohwi and Ohashi	570
Moth beans	*Vigna aconitifolia* (Jacq.) Maréchal	1110
Winged beans	*Psophocarpus tetragonolobus* (L.) DC	780

Several other targets and plant components have been proposed by the Phaseomics Consortium, which has 43 groups worldwide (Table 5). These targets include classical physiological traits, such as nutrients and soil factors, and also new molecular genetical techniques to facilitate and improve breeding programmes. At the end, as in classical breeding, the goal is stable or adapted and productive new cultivars.

It is interesting to compare these new goals with those of an FAO/IAEA coordinated research programme that ran from 1986 to 1991 (Bliss and Hardarson,

Table 4. Targets and genes for bean breeding. Modified from Broughton et al. *(2003).*

Problem	*Target*	*Genetic component*
Anti-nutritional factors	α-Amylase inhibitors, arcelins, lectins, phenolics, tannins, phytates, trypsin inhibitors	Often single genes
Flatulence	Raffinose, stachyose, verbascose	Genotypic variation
Hard-to-cook	Cotyledonary middle lamella	Genotypic variation
Low %Ndfa		Single, complex loci
Low protein seed levels	Phaseolin and APA gene families	*fin* locus
Plant type	Determinancy genes	*st* locus
Pod shatter	Pod string	Many
Poor nodulation	Legume–*Rhizobium* Interactions	COK-4 (protein kinase), B4 cluster, and others
Sensitivity to *Colletotrichum*	Several major loci and QTLs for resistance	Single, complex locus
Seeds low in S-amino acids	Phaseolin	Genotypic variation
Susceptibility to seed-boring insects	Arcelin-phytohem-agglutinin-α, amylase inhibitor (APA)	Single, complex locus

1993). The main topics then were methodology to study nitrogen fixation in the field, the use of genotypic variation in different bean breeding lines and cultivars, and the classical eco-physiology of nitrogen fixation. With nine different bean lines, the amount of N_2 fixed varied between 18 and 36 $kg.ha^{-1}$ (kilograms per hectare) after 56 days (Kipe-Nolt and Giller, 1993). In another experimental series, the amount of N_2 fixed within 60 days varied between 18 and 50 $kg.ha^{-1}$ (Kipe-Nolt *et al.*, 1993). Beans are nodulated by several species and genera of rhizobia, as summarized in Table 6. Bean species originating in Latin America harbour mainly *Rhizobium etli* strains in their nodules, whereas in Europe and in Africa, other *Rhizobium* species are present (Giller, 2001).

The protein, lipid, and carbohydrate contents of seven *Vigna* species are summarized in Table 2. They have an important role in agriculture in many Asian and African countries. The lipid content is rather low (1-2%) with the exception of *Vigna subterranean*, which has a lipid concentration of 7% in dry seeds. *Vigna mungo* (black gram), *Vigna radiata* (green gram), and *Vigna unguiculata* (cowpea) are used as either seeds or as vegetables as green beans. The results of the Phaseomics Consortium, together with the methodologies used and developed, can be transferred in the future to several *Vigna* species with agricultural relevance.

Table 5. Phaseomics Consortium and research targets

M. Aguilar (Argentina)	Effects of soil stresses on nodulation
S. Broughton, F. De Lima (Australia)	Development of a standard insect screening system for beans
Gary Cobon (Australia)	Proteomic analyses of beans
C. Atkins, P. Smith (Australia)	Assimilation of fixed N
H. Irving, M. Kelly (Australia)	Signal transduction in host plants in response to Nod factors
E. Luyten, C. Snoeck, J. Michiels, J. Vanderleyden (Belgium)	Isolation and characterisation of differentially expressed genes following inoculation with *R. etli* CNPAF512; Micro-array analysis of differently expressed *P. vulgaris* genes
N. Terryn, M. Van Montagu (Belgium)	Genetic transformation of *Phaseolus vulgaris* and *P. acutifolius*
J.-P. Baudoin, A. Maquet (Belgium)	Inter-specific hybridisation among *Phaseolus* species
G. Volckaert (Belgium)	Genome sequencing
S.-M. Tsai, D.H. Moon, A. Vettore, A.G. Simpson (Brazil)	Phytic acid and aluminium tolerance
M. Melotto, L.E.A. Camargo (Brazil)	Bean EST project – BEST
A. Pedrosa, M. Guerra (Brazil)	Cytogenetic-based physical map of *P. vulgaris*
H. Antoun, S. Laberge (Canada)	Novel genes induced by PGPR, mycorrhizae and cold stress
K. Bett, B. Tar'an, A. Vandenberg, P. Balasubramanian (Canada)	Frost tolerance Generation and analysis of ESTs Development of SSR and SNP markers for *P. vulgaris* and *Phaseolus* genetic maps
K.P. Pauls, T.E. Michaels (Canada)	Development of molecular markers in *P. vulgaris*
S. Beebe, M. Blair, J. Tohme (Colombia)	Bean technology at CIAT
J. Macas, V. Nasinec (Czech Republic)	Analysis of repetitive sequences
J. Dolezel (Czech Republic)	Physical and cytogenetic mapping
J.-J. Drevon (France)	Tolerance of symbiotic nitrogen fixation to phosphorus deficiencies
T. Langin, V. Geffroy (France)	Anthracnose and beans
E. Samain, H. Driguez (France)	Synthesis of sulphated pentamers Sythesis of acetyl-fucosylated pentamers) Availability of Nod-factors

W. Streit (Germany)	Vitamins and minerals
R. Papa (Italy)	*Phaseolus* Italian network; Population genetics, molecular biology and plant breeding
R. Bollini, B. Campion, L. Lioi, A. R. Piergiovanni, F. Sparvoli (Italy)	Bean storage proteins
F. Shah (Malaysia)	Floral and pod development; Pest and fungal resistance
G. Hernández, M. Lará (Mexico)	Functional genomics of symbiosis
F. Sánchez, C. Quinto (Mexico)	Nodule organogenesis; Nod-factors and signal transduction
A. Covarrubias, J. Acosta (Mexico)	Molecular and cellular bases of the plant adaptive responses to water deficit
J.J. Peña Cabriales (Mexico)	Involvement of trehalose in drought tolerance
J. Simpson, L. Herrera-Estrella (Mexico)	Development of efficient transformation systems for beans Sequencing of ESTs
J.-P. Nap (The Netherlands)	Gene and genomic analysis and implementation of agricultural production schemes
E.C. Schröder (Puerto Rico)	Improvement of the *P. vulgaris-Rhizobium* symbiosis in Puerto Rico
C. Gehring, G. Bradley (South Africa)	Stress responses – roles of natriuretic peptides
A.M. De Ron, M. Santalla (Spain)	Germplasm, cropping systems, bean evolution, dry bean breeding, scarlet bean breeding and snap bean breeding
B. Broughton, X. Perret (Switzerland)	Rhizobial determinants of effective nodulation of beans
R. Strasser (Switzerland)	Functional behavioral patterns of living plants
P. Gepts (USA)	Evolutionary genomics of beans
B. Haselkorn (USA)	Acetyl CoA carboxylase in *P. vulgaris*
J. Kelly (USA)	Bean breeding
P. McClean (USA)	Genomics, especially ESTs
T. McDermott, D. Bergey (USA)	Phosphorous metabolism in *Rhizobium tropici*-bean symbiosis
R. Musser (USA)	Functional genomic analyses of bean defense responses to insect attack
S. Mackenzie (USA)	Genomics and transcriptomics
D. Noel (USA)	Determinants of infection and bean nodule development
E. Triplett (USA)	Genomic micro-array hybridisation

Table 6. Rhizobium species and strains tested in Phaseolus vulgaris-plant assays. Modified from Martinez-Romero (2003)

	Nodulation	N_2 fixation
Rhizobium etli	+	+
Rhizobium tropici	+	+
R. leguminosarum bv. phaseoli	+	+
R. leguminosarum bv. trifolii	+	+
R. mongolense	+	+
R. gallicum	+	+
R. giardinii	+	–/+
Sinorhizobium fredii	+	+ or –
Sinorhizobium meliloti	+	+ or –
Sinorhizobium terangae ORS51	+	–
Sinorhizobium arboris HAMBI 1396	+	+
Sinorhizobium kostiense HAMBI 1476	+	–/+
Sinorhizobium sp. BR816	+	+
Sinorhizobium sp. NGR234	+	+
Mesorhizobium loti	+/–	–
Mesorhizobium huakuii	+/–	+
Bradyrhizobium japonicum	+	–
Bradyrhizobium sp. *(Phaseolus lunatus)*	+	+
Bradyrhizobium elkani	+	–
Bradyrhizobium sp. (Lespedeza) B038	+	–
Azorhizobium caulinodans	+/–	–

3. *ARACHIS HYPOGAEA* (GROUNDNUT, PEANUT)

Cultivated peanuts originated from wild peanuts (*Arachis* spp.) with hotspots of species richness in Mato Grosso and in Campo Grande in Brazil, where several species, such as *Arachis archeri, A. setinervosa, A. marginata, A. hatschabchii, A. helodes, A. magna, A. gracilis, A. cruziana, A. pietrarellii, A. williamsii* and *A. monticola*, are now under threat of extinction (Jarvis *et al.*, 2003). *Arachis hypogaea* was directly domesticated from *Arachis monticola* with pods splitting open underground and having smaller seeds (Sauer, 1993). Within the species *Arachis hypogaea*, several subspecies, such as *Arachis hypogaea* ssp. *fastigiata*, ssp. *nambyquarae*, ssp. *oleifera*, ssp. *sylvestris*, ssp. *communis*, ssp. *gigantea*, ssp. *hirsuta*, ssp. *macrocarpa* and ssp. *Stenocarpa*, have been described (USDA-ARS, 2000; Wiersema and León, 1999). From a core collection of 1700 groundnut accessions, 900 belong to the subspecies *fastigiata* and about 800 to the subspecies *hypogaea*, based on 16 morphological and 32 agronomic characteristics (Upadhyaya, 2003). The molecular diversity in *Arachis hypogaea* germplasm is rather low compared to wild species of *Arachis* sp. However, new RAPD data revealed four distinct genetic groups within 26 accessions (He and Prakash, 1997). Compared to soybeans, regeneration and transformation systems (using

Agrobacterium and the particle gun) are less developed in peanuts (Chandra and Pental, 2003).

Worldwide about 22 million ha are cultivated with groundnuts, with about 8 million in India, 8 million in Sub-Saharan Africa and 4 million in China. About 70% of the global groundnut production is in the areas of the semi-arid tropics. The international research mandate for peanuts is at the ICRISAT (International Crops Research Institute for the Semi-Arid Tropics) in Hyderabad, India (http://www.cgiar.org/areas/ground.htm). Although a major producer of soybeans, agronomic research in the United States of America also includes work with crop rotation systems, including peanuts with cotton and soybean and corn (Jordan *et al.*, 2002). Only 50% Ndfa for groundnut was found in field studies in Vietnam (Hiep *et al.*, 2002). In a separate field experiment, three micronutrients, boron, zinc and molybdenum, increased pod yield of groundnuts by 20-46%, indicating that the classical limitations of groundnut production are still present (Tripathy *et al.*, 1999). Also, physical factors, such as soil temperature, can increase source-sink economy in peanuts (Awal *et al.*, 2003).

4. *CICER ARIETINUM* (CHICKPEA)

World production of chickpea is around 6 million tons per year (Table 1), but in 1994, production was 8 million tons per year. The main production areas are India and Pakistan. The seed production is *ca.* 0.8 tons per hectare.

Chickpea seeds have already been found in archaeological sites from 7400 BC. Larger seeds with some evidence for domestication have been found from 6500 BC. Chickpeas, as peas and lentils, are therefore old annual legumes from the ancient neolithic period (Zohary and Hopf, 1993). Common names for chickpeas are bengal gram in India, garbanzo in Latin America, Hamaz in the Arab world, Nohud in Turkey, and Shimbra in Ethopia (Muehlbauer and Tullu, 1997). The center of their origin is probably the Caucasus and the north of Persia. Average annual yields are *ca.* 600 $kg.ha^{-1}$ in Africa, 500-900 $kg.ha^{-1}$ in India, and 1600 $kg.ha^{-1}$ in North and Central America. The germplasm of chickpea is maintained at the two international centers, ICRISAT in India and ICARDA in Syria, and also in the USDA-ARS Regional Plant Introduction Station at Pullman, U.S.A. Also in the U.S., yield ranges are very large, with around 1000-1800 $kg.ha^{-1}$. About 80% of the US production is exported to India, Spain, and Mexico (http://extension.oregonstate.edu). The two main fungal diseases of chickpeas are *Fusarium oxysporum* sp. *Ciceris*, which causes the plant to wilt, and *Ascochyta rubiei*, which causes the Ascochyta blight with brown spots on leaves, stemps, pods and also on the seeds. *Pythium* sp., *Rhizoctonia* sp., and *Sclerotinia* sp. are other fungal diseases of chickpeas.

In different areas of India, the amount of N_2 fixation, assessed by the ^{15}N natural abundance methods, varied between 58% Ndfa up to 86% Ndfa, with an average of 78% Ndfa (Aslam *et al.*, 2003). The nitrogen fixed was from 90 to 180 $kg.ha^{-1}$. By using ^{15}N labelling, a fixation rate of around 90 kg N per ha was found in field-grown chickpeas in Australia (Khan *et al.*, 2003). There is a large genomic heterogeneity of rhizosphere strains that nodulate chickpeas. Out of 30 isolates from chickpea nodules, five genomic species were identified; these included *Mezorhizobium ciceri* and *Rhizobium mediterraneum* (Nour *et al.,* 1995).

Plant genotypes of chickpea have been successfully selected for adaptation to soils with a high salinity and high pH, with plants nodulating even at pH 9.0 to 9.2 (Rao *et al.,* 2002). Chickpea has, like peanut, a low level of intraspecific genetic diversity. Therefore, wild relatives from 43 other species of the genus *Cicer* are being used to incorporate specific morphological and physiological traits, especially resistance to abiotic and biotic stresses. This programme provides a set of very interesting future tasks for both classical and molecular plant breeders (Croser *et al.,* 2003).

5. *CAJANUS CAJAN* (PIGEON PEA)

Pigeon peas are nodulated by slow-growing rhizobia, which cluster into at least four different groups. One of these groups is closely related to *Bradyrhizobium elkanii* (Ramsubhag *et al.,* 2002). The fixation rate of *Cajanus cajan* is estimated at 90 kg N per ha, which is in the middle of the range of fixation data for related systems (Table 7) (Gathumbi *et al.,* 2002).

Table 7. Average N_2 fixation of legumes, estimated with the ^{15}N natural abundance and other methods. Data from Werner (1992), Gathumbi et al. *(2002), Hauser and Nolte (2002), Samba* et al. *(2002) and Aslam* et al. *(2003)*

Species	*Fixation (kg N . ha^{-1} per season)*
Arachis hypogaea	100
Cajanus cajan	91
Calliandra calothyrsus	24
Cicer arietinum	135
Crotolaria grahamiana	142
Crotolaria achroleuca	83
Glycine max	100
Lens culinaris	80
Lupinus sp.	150
Macroptilium atropurpureum	64
Medicago sativa	250
Mucuna pruriens	130
Pisum sp.	150
Tephrosia vogelii	100
Trifolium sp.	250
Vicia faba	200

Cajanus cajan is probably of African origin. It grows up to 4 m high and the pods include up to 8 seeds with *ca.* 20% protein. It is used as green seeds together with the pods, mainly in India. The yields are between 500 and 2000 $kg.ha^{-1}$, which is mainly due to the large size of the plant. World production, which is statistically evaluated, is around 2 million tons per year, but the local production may be much larger. On the plant side, RAPD-PCR methods allow discrimination between *Cajanus cajan* from 18 other legume species, including *Phaseolus vulgaris, P. coccineus, P. lunatus, Glycine max, Vicia faba, Vigna radiata, V. mungo, V. aconitifolia, Pisum sativum, Cicer arietinum, Lathyrus sativus, L. latifolius, Lens culinaris, Lupinus angustifolius, L. luteus, L. albus, Medicago sativa*, and *Onobrychis viciifolia* (Weder, 2002).

6. *MUCUNA PRURIENS* (VELVET BEAN) AND OTHER LEGUMES

Common names for *Mucuna pruriens* are "Nescafé", mauritius bean, itchy bean, krame, picapica, chiporro, frijol, and buffalobean. Scientific synonyms are *Mucuna aterrima, M. atropurpurea, M. cochinchinensis, M. cyanosperma, M. deeringiana, M. esquirolii, M. prunita, M. utilis, Stizolobium aterrinum*, and *Negretia pruriens*. The values of the genetic similarity in *Mucuna* species range from 0.68 to 1.00, indicating the broad genetic base of this genus and the potential for improvements by breeding techniques (Capo-chichi *et al.*, 2003). Seed production is in the range of 650-1300 $kg.ha^{-1}$ and is supported by biological nitrogen fixation from 87-171 $kg.ha^{-1}$ (Hauser and Nolte, 2002).

Agriculturally, the most important use of *Mucuna pruriens* is as a cover crop and green manure. However, due to an unusual large number of secondary plant metabolites in the seeds, extracts from either the seeds or seed powder have documented medicinal properties, which include anabolic, androgenic, anthelmintic, anti-Parkinson, antipyretic, antispasmodic, diuretic, spermatogenic, and teratogenic effects. These effects are due to the presence of several alkaloids, saponins, and sterols, plus a high concentration (7-10%) of L-dopa. Also very important is the high concentration of serotonin. The most important medical effect of *Mucuna* seed components is the slowing down of the progression of Parkinson's symptoms, the reduction of blood-sugar levels, and the stimulation of growth hormones by testosterone levels (Grover *et al.*, 2002). Probably, *Mucuna* seeds should be put on the doping list by the International Olympic Committee. Chemical constituents from *Mucuna* sp. also have nematocidal activity (Demuner *et al.*, 2003; Morris and Walker, 2002). Phytochemical separation of the active components indicated that the isoflavonoid, prunetin, was the most active component, causing a 70% mortality of *Meloidogyne incognita*.

The use of *Mucuna pruriens* as green manure and in crop rotation with maize has proved to be especially successful. A 50-100% yield increase, compared to non-fertilized control plots, has been found in West Africa with *Mucuna* varieties *jaspaeda, utilis*, and *cochinchinensis* (Hauser and Nolte, 2002). These data were confirmed in Mexico, using the system *Zea mays-Mucuna pruriens-Zea mays*, when a 50% yield increase was found. More than a 100% increase was found with the *Mucuna pruriens-Zea mays* system (Eilitta *et al.*, 2003a). A comparison of the effect of the white- and mottled-seeded with the black-seeded *Mucuna* varieties indicated superiority of the white and mottled varieties (Eilitta *et al.,* 2003b).

In principal, any number of legumes that produce a large plant biomass could be used as green manure, including the 18 tropical legume species listed in Table 8. Using legumes, such as *Mucuna* sp., as a cover crop has another important effect; when planted directly after the maize harvest, it out-competes weeds (http://www.forages.css.orst.edu). Furthermore, the use of *Mucuna* species in crop rotation influences habitat conservation in forest areas by increasing the quality of the soil and, thereby, reducing the pressure on forests in Middle and South America as well as in Africa (http://www.peregrinefund.org/maya_project).

Table 8. Tropical legumes as cover crops and green manure. Modified from http://www.unu.edu/unupress, http://www.iirr.org/saem/page147-152.htm, Werner (1992), and Wang et al.*(2002).*

Scientific name	Common name
Arachis prostate	Huban clover
Cajanus cajan	Pigeon pea
Calopogonium mucunoides	Calopo
Centrosema pubescens	Centro
Cicer arietinum	Chickpea
Crotalaria juncea	San hemp
Desmodium uncinatum	Spanish clover
Dolichos lablab	Lablab bean
Glycine wighrii	Glycine (perennial soybean)
Medicago sativa	Alfalfa
Mucuna utilis	Mucuna
Phaseolus aureus	Green gram
Phaseolus lathyroides	Phasey bean
Psophocarpus palustris	Psophocarpus
Pueraria phaseoloides	Kudzu
Sesbania rostrata	Sesbania
Trifolium sp.	Clover
Vigna sinensis	Cowpea
Voandzeia subterranea	Bambara groundnut

In areas with significant sheep, goat or cattle production, the use of legumes as green manure competes with the demand of their use as fodder legumes, but the actual use by farmers and livestock keepers in Africa is still rather low (Sumberg, 2002). The use of non-native legume species can, of course, also be a problem if they become weeds, *e.g.*, as demonstrated for *Crotalaria zanzibarica* in Taiwan, which has spread within 70 years to a large number of localities on this island (Wu *et al.,* 2003).

ACKNOWLEDGEMENTS

The author thanks the European Union for support in the INCO DEV project "Soybean BNF Mycorrhization for Production in South Asia" ICA4-CT-2001-10057.

REFERENCES

Aslam, M., Mahmood, I. A., Peoples, M. B., Schwenke, G. D., and Herridge, D. F. (2003). Contribution of chickpea nitrogen fixation to increased wheat production and soil organic fertility in rain-fed cropping. *Biol. Fertil. Soils, 38*, 59-64.

Awal, M. A., Ikeda, T., and Itoh, R. (2003). The effect of soil temperature on source-sink economy in peanut (*Arachis hypogaea*). *Environ. Exp. Bot., 50*, 41-50.

Bennett, M. D., and Leitch, I. J. (2001). Plant DNA C-values. Retrieved September 1, 2001, from http://www.rbgkew.org.uk/cvalues/homepage.html.

Bliss, F. A., and Hardarson, G. (Eds.) (1993). *Enhancement of biological nitrogen fixation of common bean in Latin America*. Dordrecht, The Netherlands: Kluwer Academic Publishers.

Broughton, W. J., Hernández G., Blair, M., Beebe, S., Gepts, P., and Vanderleyden J. (2003). Beans (*Phaseolus* spp.) – model food legumes. *Plant Soil, 252*, 55-128.

Capo-chichi, Weaver, B., and Morton, C. M. (2003). The use of molecules markers to study genetic diversity in *Mucuna*. *Trop. Subtrol. Agroecosyst., 1*, 309-318.

Chandra, A., and Pental, D. (2003). Regeneration and genetic transformation of grain legumes: An overview. *Curr. Sci., 84*, 381-387.

Croser, J. S., Ahmad, F., Clarke, H. J., and Siddique, K. H. M. (2003). Utilisation of wild *Cicer* in chickpea improvement – progress, constraints, and prospects. *Aust. J. Agric. R., 54*, 429-444.

Demuner, A. J., Barbosa, L. C. D., do Nascimento, J. C., Vieira, J. J., and dos Santos, M. A. (2003). Isolation and nematocidal activity evaluation of chemical constituents from *Mucuna cinerea* against *Meloidogyne incognita* and *Heterodera glycines*. *Quimica Nova, 26*, 335-339.

Eilitta, M., Sollenberger, L. E., Littell, R. C., and Harrington, L. W. (2003a). On-farm experiments with maize-mucuna systems in the Los Tuxtlas region of Veracruz, Mexico. I. *Mucuna* biomass and maize grain yield. *Exp. Agric., 39*, 5-17.

Eilitta, M., Sollenberger, L. E., Littell, R. C., and Harrington, L. W. (2003b). On-farm experiments with maize-mucuna systems in the Los Tuxtlas region of Veracruz, southern Mexico. II. *Mucuna* variety evaluation and subsequent maize grain yield. *Exp. Agric., 39*, 19-27.

Gathumbi, S. M., Cadisch, G., and Giller, K. E. (2002). N-15 natural abundance as a tool for assessing N-2-fixation of herbaceous, shrub and tree legumes in improved fallows. *Soil Biol. Biochem., 34*, 1059-1071.

Giller, K. E. (2001). *Nitrogen fixation in tropical cropping systems*. 2nd Ed. Wallingford, U.K.: CABI Publishing.

Grover, J. K., Yadav, S., and Vats, V. (2002). Medicinal plants of India with anti-diabetic potential. *J. Ethnopharmacol., 81*, 81-100.

Hauser, S., and Nolte, C. (2002). Biomass production and N fixation in five *Mucuna pruriens* varieties and their effect on maize yields in the forest zone of Cameroon. *J. Plant Nutr. Soil Sci., 165*, 101-109.

He, G., and Prakash, S. (1997). Identification of polymorphic DNA markers in cultivated groundnut (*Arachis hypogaea* L.). *Euphytica, 97*, 143-149.

Hiep, N. H., Diep, C. N., and Herridge, D. (2002). Nitrogen fixation of soybean and groundnut in the Mekong Delta, Vietnam. In D. Herridge (Ed.), *Inoculants and nitrogen fixation of legumes in Vietnam*. ACIAR Proceedings.

FAO (2002). Production. *Yearbook 2001*. Vol. 55. Rome, Italy: Food and Agriculture Organization of the United Nations.

Jarvis, A., Ferguson, M. E., Williams, D. E., Guarino, L., Jones, P. G., Stalker, H. T., *et al.* (2003). Biogeography of wild *Arachis*: Assessing conservation status and setting future priorities. *Crop Sci., 43*, 1100-1108.

Jordan, D. L., Bailey, J. E., Barnes, J. S., Bogle, C. R., Bullen, S. G., Brown, A. B., Edmisten, K. L., Dunphy, E. J., and Johnson, P. D. (2002). *Agron. J., 94*, 1289-1294.

Kahn, D. F., Peoples, M. B., Schwenke, G. D., Felton, W. L., Chen, D. L., and Herridge, D. F. (2003). Effects of below-ground nitrogen on N balances of field-grown fababean, chickpea, and barley. *Aust. J. Agric. Res., 54*, 333-340.

Kipe-Nolt, J. A., and Giller, K. E. (1993). A field evaluation using the ^{15}N isotope dilution method of lines of *Phaseolus vulgaris* L. bred for increased nitrogen fixation. *Plant Soil, 152*, 107-114.

Kipe-Nolt, J. A., Vargas, H., and Giller, K. E. (1993). Nitrogen fixation in breeding lines of *Phaseolus vulgaris* L. *Plant Soil, 152*, 103-106.

Martinez-Romero, E. (2003). Diversity of *Rhizobium-Phaseolus vulgaris* symbiosis: Overview and perspectives. *Plant Soil, 252*, 11-23.

Morris, J. B., and Walker, J. T. (2002). Non-traditional legumes as potential soil amendments for nematode control. *J. Nematol., 34*, 358-361.

Muehlbauer, I. J., and Tullu, A. (1997). *Cicer arietinum* L. Purdue University: *NewCROP FactSHEET.* Retrieved from http://www.hort.purdue.edu/newcrop/cropfactsheets/Chickpea.html.

Nour, S. M., Cleyet-Marel, J. C., Normand, P., and Fernandez, M. P. (1995). Genomic heterogeneity of strains nodulating chickpeas (*Cicer arietinum* L.) and description of *Rhizobium mediterraneum* sp. nov. *Intl. J. Syst. Bacteriol., 45*, 640-648.

Ramsubhag, A., Umaharan, P., and Donawa, A. (2002). Partial 16S rRNA gene sequence diversity and numerical taxonomy of slow growing pigeonpea (*Cajanus cajan* L. Millsp) nodulating rhizobia. *FEMS Microbiol. Lett., 216*, 139-144.

Rao, D. L. N., Giller, K. E., Yeo, A. R., and Flowers, T. J. (2002). The effects of salinity and sodicity upon nodulation and nitrogen fixation in chickpea (*Cicer arietinum*). *Ann. Bot., 89*, 563-570.

Samba, R. T., Sylla, S. N., Neyra, M., Gueye, M., Dreyfus, B., and Ndoye, I. (2002). Biological nitrogen fixation in *Crotalaria* species estimated using the ^{15}N isotope dilution method. *African J. Biotechnol., 1*, 17-22.

Sauer, J. D. (1993) *Historical geography of crop plants – a select roster.* Boca Raton, FL: CRC Press.

Smartt, J. (1988). *Grain legumes. Evolution and genetic resources.* Cambridge, U.K.: Cambridge University Press.

Souci, S. W., Fachmann, W., and Kraut, H. (1994). *Die Zusammensetzung der Lebensmittel. Nährwert-Tabellen 1986/87.* 5. Aufl. Stuttgart: Medpharm Scientific Publishers.

Sumberg, J. (2002). The logic of fodder legumes in Africa. *Food Policy, 27*, 285-300.

Tripathyi, S. K., Patra, A. K., and Samui, S. C. (1999). Effect of micronutriens on nodulation, growth, yield and nutrient uptake by groundnut (*Arachis hypogaea* L.). *Indian J. Plant Physiol., 4*, 207-209.

Upadhyaya, H. D. (2003). Phenotypic diversity in groundnut (*Arachis hypogaea* L.) core collection assessed by morphological and agronomical evaluations. *Gen. Resources Crop Evol., 50*, 539-550.

USDA, ARS (2000). *National genetic resources program.* Germplasm resources information network (GRIN). Beltsville, MD: National Germplasm Resources Laboratory (http://www.ars-grin.gov.)

Wang, K. H., Sipes, B. S., and Schmitt, D. P. (2002). *Crotalaria* as a cover crop for nematode management: A review. *Nematropica, 32*, 35-57.

Weder, J. K. P. (2002). Identification of food and feed legumes by RAPD-PCR. *Food Sci. Technol., 35*, 504-511.

Werner, D. (1992). *Symbiosis of plants and microbes.* London, U.K.: Chapman and Hall.

Wiersema, J. H., and León, B. (1999). *World economic plants: A standard reference.* Boca Raton, FL: CRC Press.

Wu, S. H., Chaw, S. M., and Rejmanek, M. (2003). Naturalized Fabaceae (Leguminosae) species in Taiwan: The first approximation. *Bot. Bull. Academia Sinica, 44*, 59-66.

Zohary, D., and Hopf, M. (1993). Domestication of plants in the old World – The origin and spread of cultivated plants in West Asia, Europe, and the Nile Valley. Oxford, U.K.: Clarendon Press.

Chapter 2

NITROGEN FIXATION BY SOYBEAN IN NORTH AMERICA

S. G. PUEPPKE

College of Agricultural, Consumer, and Environmental Sciences, University of Illinois, Urbana, IL 61821, USA

1. SOYBEAN: PATHWAYS TO NORTH AMERICA AND ESTABLISHMENT AS A CROP

The soybean, *Glycine max* (L.) Merr., is one of the world's most important and versatile crop species. Currently grown on 76 million hectares of land on five continents (FAO, 2003), soybeans are a staple part of the human diet, particularly in Asia, where they are consumed both as a vegetable and as a variety of processed food products. Large quantities of whole soybeans, soybean oil, and soybean protein products also are traded on world markets.

A member of the legume subfamily Papilionoideae, *G. max* is known only as a cultivated species and has never been found growing in the wild (Hymowitz, 1970; 1990). Its weedy uncultivated ancestor, *Glycine soja* Sieb and Zucc., is widely distributed in Asia, where it grows in fields and open areas. Soybean was domesticated from *G. soja* by farmers in northeastern China, who recognized its value as a food crop. The precise time frame is unknown, but soybean was likely brought into cultivation more than three millennia ago, during the Shang Dynasty (*ca.* 1700-1100 B.C.). It had spread to adjacent parts of China and Korea by the first century A.D. and then on to other parts of Asia, where it became an increasingly important component of the human diet.

An English seaman, Samuel Bowen, introduced soybean to North America in the decade just prior to the American Revolution (Hymowitz and Harlan, 1983). He had obtained a quantity of seeds from China *via* England, and he gave them to an acquaintance named Henry Yonge. Yonge was the Surveyor General of the Colony of Georgia, and beginning in 1765, he grew soybeans on his plantation near Savannah. Bowen had some of the harvested seeds processed into soy sauce and

D. Werner and W. E. Newton (eds.), Nitrogen Fixation in Agriculture, Forestry, Ecology, and the Environment, 15-23.

noodles and began experimenting with other manufacturing methods, but the undertaking ended with his death in 1777.

Soybeans were reintroduced to North America on a number of occasions after Bowen, and we know that American botanists and horticulturists planted and studied them during the first half of the nineteenth century (Hymowitz, 1990). It was the year 1851, though, that marked the rebirth of investigations into the potential of soybean as a North American crop plant. This time, the seeds came from Japan *via* San Francisco, where Dr. Benjamin Franklin Edwards had been pressed into service to examine a group of Japanese fishermen for contagious diseases (Hymowitz, 1990). The physician found the fishermen to be healthy and, in gratitude, they presented him with a package of soybean seeds. Dr. Edwards took this unusual gift with him when he returned home to Alton, Illinois. Fortuitously, this small city lies within a vast region of farmland that is well suited to soybean production. Edwards presented the seeds to a horticulturist friend, who immediately planted them and began to distribute succeeding crops of seeds through a network of horticulturists. Within just a few years, a loosely organized group of farmers and gardeners had evaluated soybean across the U. S. Midwest as a potential forage crop for animals. As was common then, they communicated their experiences with the new crop by means of notices and articles in the popular agricultural press.

The ability of legumes to fix nitrogen symbiotically had not yet been established and, if these early American soybean growers noticed root nodules, they could not have been aware of their agronomic significance. Symbiotic nitrogen fixation was initially documented by the experiments of Hellriegel and Willfarth in 1888 (Fred *et al.,* 1932) and the soybean root-nodule organism, *Bradyrhizobium japonicum*, was first isolated 7 years later (Kirchner, 1895). These scientific discoveries in Europe coincided with yet another introduction of soybean into the U.S., one that would firmly establish the species as a major agricultural crop.

This time, agricultural scientists deliberately brought the crop into the country, investigated its agronomic potential, and collaborated with other researchers to assure the success of the endeavor (Hymowitz, 1990). The first experiments were conducted by scientists from the New Jersey Agricultural Experiment Station at Rutgers College, who had brought seeds from Europe in 1878, multiplied them in the field, and soon distributed germplasm to agricultural experiment stations in other states. Seeds were also imported directly from Asia and, by the final decades of the nineteenth century, soybeans had become the subject of serious ongoing investigations to determine their suitability as an American crop.

The artificial inoculant industry dates back to 1895, when Nobbe and Hiltner were first granted patents on the use of pure cultures of rhizobia for legume seed treatments (Fred *et al.,* 1932). Even prior to this time, scientists had begun to determine the feasibility of ensuring nodulation and nitrogen fixation by transfer of small amounts of soil from fields in which soybeans had been grown to new fields. Cottrell and colleagues (1900) in Kansas and Lane (1900) in New Jersey were among the first to conduct such experiments with soybean in the United States and, as early as 1904, farmers had begun to offer "infected soil" for sale (Windish, 1981). Inoculation came into common practice during the early part of the twentieth century (Figure 1) and, as soybean began to establish itself commercially, individual states started to produce inoculants for farmers. In 1928 alone, Missouri and Wisconsin distributed 38,522 and 14,793 cultures, respectively, to soybean farmers (Fred *et al.*, 1932).

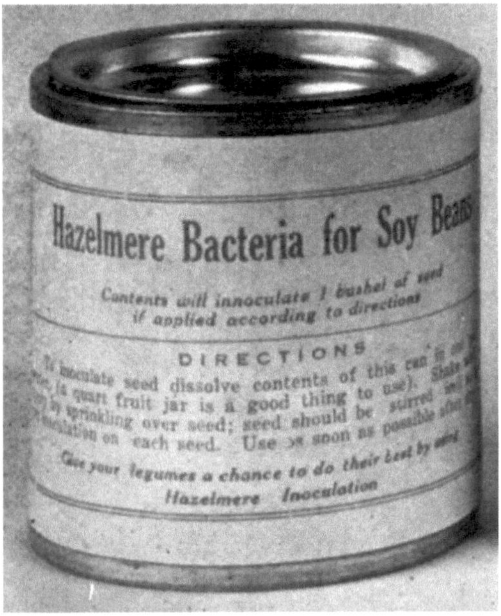

Figure 1. An early soybean inoculant produced by the Hazelmere Bacteria Company in Bowling Green, Indiana. Adapted from Fred et al., 1932.

2. SOYBEAN PRODUCTION IN NORTH AMERICA

Official records of soybean production in the United States did not appear until 1924 (Barnhart, 1954). Previously and for some time thereafter, the crop was grown on a relatively small scale, mostly as either a forage legume or a green manure to enhance soil fertility. Production initially was centered in the eastern part of the country, such that the leading states in 1919 were North Carolina, Kentucky, Mississippi, Virginia, and Alabama. Soybean production areas moved west rapidly over the ensuing five years, such that, by 1924, the leading states were Illinois, Indiana, Tennessee, North Carolina, and Missouri (Windish, 1981).

Soybean first became a significant American crop during World War II, when traditional sources of vegetable oils were disrupted by the conflict. Production of soybean for oil exceeded that for hay in 1941 (Barnhart, 1954) and, by 1942, the United States overtook Manchuria as the world's leading producer of the crop (Windish, 1981). The latter change is illustrated in Figure 2, which documents the growth and geographical redistribution of soybean production in the United States from 1927 to 2000. Soybean production was in its infancy in 1927 and scattered throughout the Midwest and southeast. By 1961, when soybean had become a major crop and was planted on 10.9 million hectares (FAO, 2003), production was concentrated in three Midwestern states; Illinois, Indiana, and Iowa. Smaller but significant acreages were also scattered in the south central and southeastern regions and in Minnesota, a state originally assumed to be too cold to sustain growth of crop (Cottrell *et al.*, 1900). Huge increases in soybean production are evident in the data from 2000, which show the strengthening and continued dominance of Illinois, Indiana, and Iowa. U. S. soybeans covered an area of 29.3 million hectares in 2000, when production exceeded 75 million metric tons for the first time (FAO, 2003).

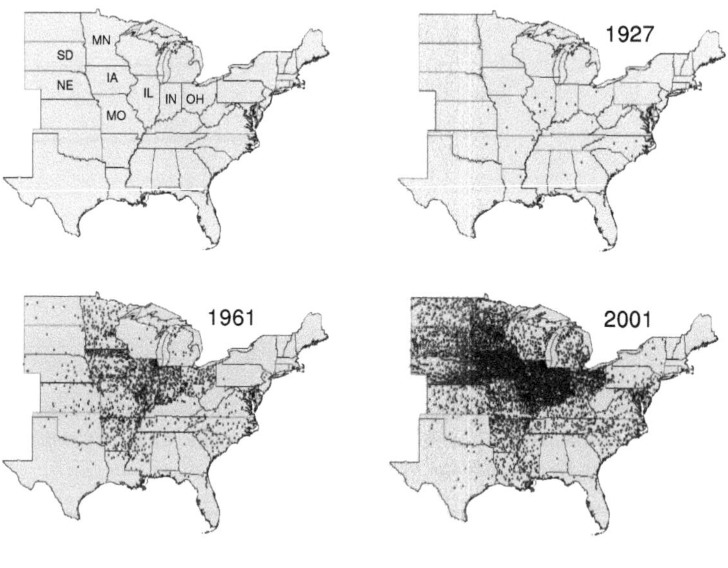

Figure 2. Changes in the magnitude and geographical distribution of United States soybean production from 1927 to 2001.

One dot is equivalent to 9,000 metric tons. The upper left hand map identifies the principal soybean-growing states of the Midwest: IA = Iowa, IL = Illinois, IN = Indiana, MN = Minnesota, MO = Missouri, NE = Nebraska, OH = Ohio, SD = South Dakota. Data Source: National Agricultural Statistics Service, United States Department of Agriculture.

Mexico and Canada produce relatively minor amounts of soybeans in comparison to the United States (Figure 3). Plantings in Mexico have fluctuated significantly over the past 40 years, and production has declined over the past decade. Soybean oil is the most important plant oil produced in Mexico, but domestic soybean production does not meet domestic needs. As recently as 1996-1997, imports from the United States accounted for nearly half of the soybeans crushed for oil in Mexico (Juarez, 1998).

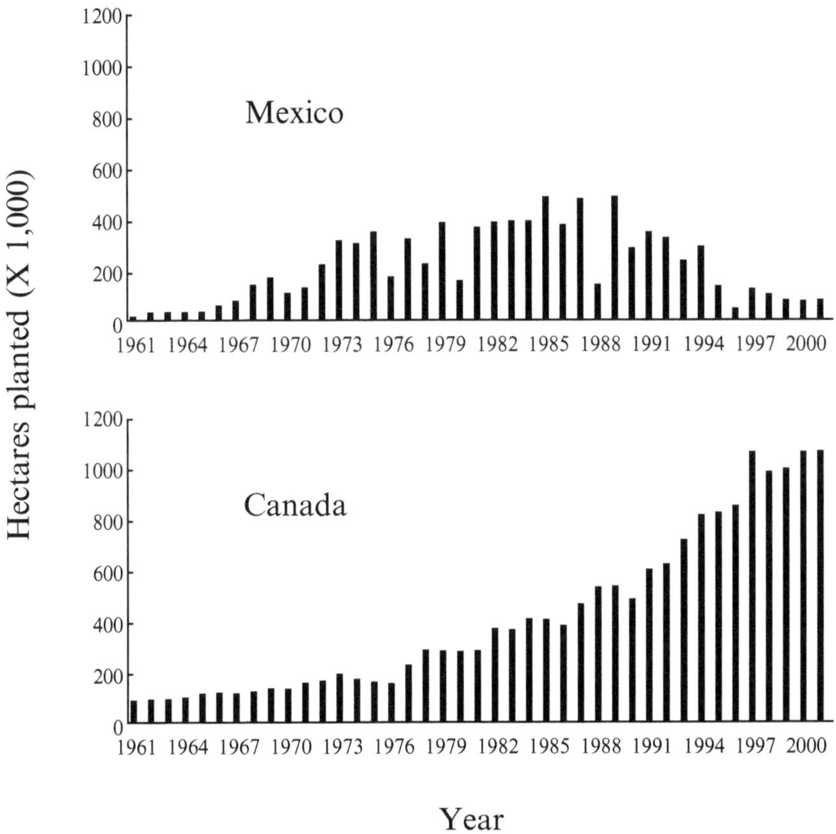

Figure 3. Soybean production in Mexico and Canada from 1961 to 2001.
Data Source: FAOSTAT Agricultural Database of the United Nations Food and Agricultural Organization.

Soybean was grown in Canada as early as 1939, when a total of 3,960 hectares was planted (Anonymous, 1939) and test fields were sown in Western Canada as early as 1941 (Rennie and Dubetz, 1984). But it was only during the past 25 years, as early maturing varieties were developed, that Canada became a significant soybean producer. The area planted to soybeans in Canada increased from 85,800 hectares in 1961 to 279,200 hectares in 1981 to 1.1 million hectares in 2001 (FAO, 2003). About 83% of Canada's current production areas lie in Ontario, primarily in the southern counties of the province along the north shore of the Great Lakes (Ontario Ministry of Agriculture and Food, 2003). Together, Mexican and Canadian soybean acreages are equivalent to less than 5% of the U. S. total.

3. MAJOR SOYBEAN CROPPING SYSTEMS

Soybean is normally grown in sequence with other crops throughout the major production areas of the American Midwest (Hoeft *et al.,* 2000). Crop rotations of this sort are common agricultural practices and they were in place prior to widespread introduction of soybean in the middle of the twentieth century. The appearance of soybean at this time coincided with and was facilitated by another fundamental change, the replacement of animal traction with machinery. Before widespread mechanization, the rotation cycle usually encompassed four years and included oats, two years of a legume, such as either alfalfa or clover, a year of corn, and then oats again. This sequence ensured adequate feed for farm animals as well as an abundant supply of fixed nitrogen, which was not yet available as commercial fertilizer.

Prior to World War II, it was possible to simply replace one of the forage legumes with soybeans for use as hay (Barnhart, 1954). With the advent of vast soybean acreages and the diminished need for animal feed, the rotation was simplified to today's typical corn-soybean sequence. About 80% of the total cropland in the Midwest is currently managed in this way (Hoeft *et al.,* 2000). The widespread acceptance of this practice is evident from the near identity of total corn and total soybean acreages in any given area and growing season. In 2001, for example, Illinois farmers planted 4.4 million hectares of corn and 4.3 million hectares of corn (Illinois Agricultural Statistics Service, 2002). In some of the more southern areas of the Midwest, corn often is replaced with either wheat or sorghum, but the two-year rotation with soybean is maintained (Hoeft *et al.,* 2000). Thus, regardless of the identity of the second crop and as a rule of thumb, farmers in the American Midwest generally plant soybean every other year.

4. BIOLOGICAL NITROGEN FIXATION BY SOYBEAN IN NORTH AMERICA

Biological nitrogen fixation has been assessed by a variety of protocols, each with its own inherent advantages and shortcomings (Bremner, 1977; Burris, 1974; Chalk, 1985; Danso, 1986; Danso *et al.,* 1993; Hardarson and Danso, 1993; Knowles, 1981). ^{15}N-based techniques have become the method of choice for legumes growing in the field (Unkovich and Pate, 2000), in part because they allow the calculation of nitrogen fixation as kg N/ha (Hardarson and Danso, 1993).

Two different ^{15}N-based protocols have been devised (Bergerson, 1980; Hardarson, 1990; Peoples *et al.,* 1989). The natural abundance method takes advantage of the fact that the heavy isotope is relatively more prevalent in the soil than in the atmosphere. Tissues from plants with direct access to N_2 from the atmosphere, thus, have lower ^{15}N enrichment than do plants that derive nitrogen exclusively from the soil, and the resulting differences in isotope abundances can be quantified. Isotope dilution-based techniques rely on enrichment of the soil supporting the growth of nitrogen-fixing legumes and nonfixing reference plants - often non-nodulating mutants - with ^{15}N-labeled fertilizers. Because the fixing plants capture gaseous N_2 from the air, the ratio of ^{15}N to ^{14}N in their tissues decreases relative to that in the non-fixing plants and this difference allows the percentage of nitrogen derived from the atmosphere to be calculated.

The literature on global nitrogen fixation is filled with uncertainty (Burns and Hardy, 1975; Hardy and Havelka, 1975; LaRue and Patterson, 1981; Smil, 2001; Sprent, 1986). This situation is partly due to differences of opinion about the most appropriate application of techniques to measure nitrogen fixation - and the interpretation of experimental results. But it also reflects a whole series of uncontrolled variables that are known to influence nitrogen fixation in the field. These include environmental factors, such as moisture, temperature, soil type, and nutrient status (especially that of mineral nitrogen), as well as plant genotype, adequacy and efficiency of nodulation (Brockwell and Bottomley, 1995), and a whole variety of agronomic practices.

LaRue and Patterson (1981) summarized a series of early experiments to determine nitrogen fixation by soybean and concluded that "the upper limit for average fixation by soybeans in American agriculture is 75 kg N/ha." This estimate is based on the assumption that soybeans growing in U.S. Midwestern fields accumulate a total of 150 kg N/ha and that 50% of the total is supplied by nitrogen fixation. This estimate agrees well with data from nine sets of field experiments, which were conducted in the 1970s and early 1980s. Three were based on isotope-dilution methods and were conducted on typical silty loam soils in the Midwest. The amounts of fixed nitrogen recorded in these experiments were reasonably uniform, given the observed variations in growth conditions, and ranged from 43-146 kg N/ha in eastern Nebraska (Deibert *et al.,* 1979) to 76-152 kg N/ha in south-central Minnesota (Ham and Cardwell, 1978) to 103 kg N/ha in east-central Illinois (Johnson *et al.,* 1975).

Unkovich and Pate (2000) revisited the issue of nitrogen fixation by legumes in the field and have summarized both a vast amount of experimental data as well as recent refinements in our understanding of measurement techniques. They conclude that average nitrogen fixation by dry land soybean supplies 100 kg N/ha on a worldwide basis and, if root biomass is included in the calculations, the amount increases to 142 kg N/ha. These figures are greater and likely more accurate than earlier estimations, but they remain subject to the considerable variability and uncertainty inherent in making generalizations about a crop cultivated under diverse agronomic regimes and across wide geographical areas. The practice of irrigation illustrates just one of these uncertainties. When soybean is grown under irrigation, levels of nitrogen fixation are elevated to 175 kg N/ha - with values approaching 250 kg N/ha if roots are included (Peoples *et al.,* 1995). Just one factor, assured optimal moisture can, therefore, nearly double nitrogen fixation in comparison to average field conditions.

5. PERSPECTIVES

If we accept the most recent figure of 142 kg N/ha, then we can estimate that, for 2001, soybeans growing in North America captured about 4.3 million metric tons of nitrogen from the atmosphere and cycled it into plant organic matter. This is a significant number but, as has been pointed out (Graham and Vance, 2000; Smil, 1997), the world's agriculture is relying less and less on biologically fixed nitrogen and more and more on synthetic fertilizer. Given the immense energy required to synthesize anhydrous ammonia (Smil, 2001), this substitution is likely not sustainable over time.

Dry mass accumulation - plant growth - is often viewed as a principle driver for biological nitrogen fixation, but it would be erroneous to conclude that, as soybean yields have increased in recent decades, nitrogen fixation has just simply increased in a corresponding fashion (Unkovich and Pate, 2000). Fertilizer use is rising in cropping systems, such as the corn-soybean rotation of the U. S. Midwest, and it is well known that nodulation and nitrogen fixation are directly inhibited by combined nitrogen (Streeter, 1988). New soybean varieties are being released and we continue to know very little about rhizobia in cropped soils. These complexities both confound our understanding of biological nitrogen fixation in soybean and offer fertile ground for future investigations.

ACKNOWLEDGEMENTS

I thank Geetanjali Tandon and the National Soybean Research Laboratory for assistance with Figure 2.

REFERENCES

Anonymous (1939). Apportionment of areas, agricultural production and numbers of livestock in various countries. *International yearbook of agricultural statistics 1941-42 to 1945-46* (Vol. III.). Rome, Italy: Villa Borghese.

Barnhart, F. (1954). *Soybeans*. Cape Girardeau, Missouri: Missouri Printing and Stationery Co.

Bergerson, F. J. (1980). *Methods for evaluating biological nitrogen fixation*. Chichester, U.K.: Wiley and Sons.

Bremner, J. M. (1977). Use of nitrogen-tracer techniques for research on nitrogen fixation. In A. Ayanaba and P. J. Dart (Eds.), *Biological nitrogen fixing farming systems of the tropics* (pp. 335-352). Chichester, U.K.: Wiley and Sons.

Brockwell, J., and Bottomley, P. J. (1995). Recent advances in inoculant technology and perspectives for the future. *Soil Biol. Biochem., 27*, 683-697.

Burns, R. C., and Hardy, R. W. F. (1975). *Nitrogen fixation in bacteria and higher plants*. New York, NY: Springer.

Burris, R. H. (1974). Methodology. In A. Quispel (Ed.), *The biology of nitrogen fixation* (pp. 9-33). Amsterdam, The Netherlands: North-Holland Publishing Co.

Chalk, P. M. (1985). Estimation of N_2 fixation by isotope dilution: An appraisal of techniques involving ^{15}N enrichment and their application. *Soil Biol. Biochem., 17*, 389-410.

Cottrell, H. M., Otis, D. H., and Haney, J. G. (1900). Soil inoculation for soy beans. *Kansas State Agricultural College Experiment Station, Bulletin, 96*, 97-116.

Danso, S. K. A. (1986). Estimation of N_2-fixation by isotope dilution: An appraisal of techniques involving ^{15}N enrichment and their application. *Soil Biol. Biochem., 18*, 243-244.

Danso, S. K. A., Hardarson, G., and Zapata, F. (1993). Misconceptions and practical problems in the use of ^{15}N soil enrichment techniques for estimating N_2 fixation. *Plant Soil, 152*, 25-52.

Deibert, E. J., Bijeriego, M., and Olson R. A. (1979). Utilization of ^{15}N fertilizer by nodulating and non-nodulating soybean isolines. *Agron. J., 71*, 717-723.

FAO (2003). United Nations Food and Agricultural Organization, *FAOSTAT Agricultural Database*. See: http://apps.fao.org/cgi-bin/nph-db.pl?subset=agriculture. Downloaded January 10, 2003.

Fred, E. B., Baldwin, I. L., and McCoy, E. (1932). *Root nodule bacteria and leguminous plants.* Madison, WI: University of Wisconsin Press.

Graham, P. H., and Vance, C. P. (2000). Nitrogen fixation in perspective: An overview of research and extension needs. *Field Crops Res., 65*, 93-106.

Ham, G. E., and Caldwell, A. C. (1978). Fertilizer placement effects on soybean yield, N_2 fixation, and ^{32}P uptake. *Agron. J., 70*, 779-783.

Hardarson, G. (1990). *Use of nuclear techniques in studies of soil-plant relationships.* Training Course Series No. 2. Vienna, Austria: International Atomic Energy Agency.

Hardarson, G., and Danso, S. K. (1993). Methods for measuring biological nitrogen fixation in grain legumes. *Plant Soil, 152*, 19-23.

Hardy, R. W. F., and Havelka, U. D. (1975). Nitrogen fixation research: A key to world food? *Science, 188*, 633-643.

Hoeft, R. G., Nafziger, E. D., Johnson, R. R., and Aldrich, S. R. (2000). *Modern corn and soybean.* Champaign, IL: MCSP Publications.

Hymowitz, T. (1970). On the domestication of the soybean. *Econ. Bot., 24*, 408-421.

Hymowitz, T. (1990). Soybeans: The success story. In J. Janick and J. Simon (Eds.), *Advances in new crops* (pp. 159-163). Portland, OR: Timber Press.

Hymowitz, T., and Harlan, J. R. (1983). Introduction of soybean to North America by Samuel Bowen in 1765. *Econ. Bot., 37*, 371-379.

Illinois Agricultural Statistics Service (2002). See: www.agstats.state.il.us/annual/2002.

Johnson, J. W., Welch, L. F., and Kurtz, L. T. (1975). Environmental implications of N fixation by soybeans. *J. Environ. Qual., 4*, 303-306.

Juarez, B. (1998). *Soy leads U. S. oilseed exports to Mexico.* United States Department of Agriculture, Foreign Agricultural Service. See http://www.fas.usda.gov/info/agexporter/1998/soyleads.htm.

Kirchner, O. (1895). Die wurzelknöllchen der sojabohne. *Beitr. Biol. Pflanz., 7*, 213-223.

Knowles, R. (1981). The measurement of nitrogen fixation. In A. H. Gibson and W. E. Newton (Eds.), *Current perspectives in nitrogen fixation* (pp. 327-333). Amsterdam, The Netherlands: Elsevier Press.

Lane, C. B. (1900). *An experiment in inoculating soy bean.* Twentieth Annual Report. New Jersey Agricultural Experiment Station. Camden, NJ: Sinickson Chew and Co.

LaRue, T. A., and Patterson, T. G. (1981). How much nitrogen do legumes fix? *Adv. Agron., 34*, 15-38.

Ontario Ministry of Agriculture and Food (2003). *Soybeans – area and production, Ontario by county, 2001.* See: www.gov.on.ca/omafra/english/stats/crops/ctysoybeans01.htm.

Peoples, M. B., Faizah, A. W., Rerkasem, B., and Herridge, D. F. (1989). *Methods for evaluating nitrogen fixation by nodulated legumes in the field.* Canberra, Australia: Australian Centre for International Agricultural Research.

Peoples, M. B., Gault, R.R., Lean, B., Sykes, J. D., and Brockwell, J. (1995). Nitrogen fixation by soybean in commercial irrigated crops of central and southern New South Wales. *Soil Biol. Biochem., 27*, 553-561.

Rennie, R. J., and Dubetz, S. (1984). Multistrain vs. single strain *Rhizobium japonicum* inoculants for early maturing (00 and 000) soybean cultivars: N_2 fixation quantified by ^{15}N isotope dilution. *Agron. J., 76*, 498-502.

Smil, V. (1997). Some unorthodox perspectives on agricultural biodiversity - the case for legume culture. *Agric. Ecosyst. Environ., 62*, 135-144.

Smil, V. (2001). *Enriching the earth.* Cambridge, MA: MIT Press.

Sprent, J. I. (1986). Benefits of *Rhizobium* to agriculture. *Trends Biol., 4*, 124-129.

Streeter, J. (1988). Inhibition of legume nodule formation and N_2 fixation by nitrate. *Crit. Rev. Plant Sci., 7*, 1-22.

Unkovich, M. J., and Pate, J. S. (2000). An appraisal of recent field measurements of symbiotic N_2 fixation by annual legumes. *Field Crops Res., 65*, 211-228.

Windish, L. G. (1981). *The soybean pioneers.* Henry, IL: M and H Printing.

Chapter 3

THE IMPORTANCE OF NITROGEN FIXATION TO SOYBEAN CROPPING IN SOUTH AMERICA

M. HUNGRIA[1], J. C. FRANCHINI[1], R. J. CAMPO[1]
AND P. H. GRAHAM[2]
[1]*Embrapa Soja, Cx. Postal 231, 86001-970, Londrina, PR, Brazil.*
[2]*University of Minnesota, Department of Soil, Water, and Climate, 1991 Upper Buford Circle, St Paul, MN 55108, USA.*

1. INTRODUCTION

Although soybean (*Glycine max* L. Merr.) is a relatively recent crop introduction to South America, the region currently accounts for about 45% of world soybean production. This paper provides a brief overview of both soybean production in this region and some of the problems that have been overcome in reaching current production and yield levels.

2. TAXONOMY, ORIGINS, AND IMPORTANCE OF SOYBEAN

Hymowitz and Newell (1980) divide the genus *Glycine* into the subgenera, *Glycine* and *Soja*. The former genus includes seven wild perennial species, most of which are indigenous to Australia (Brown *et al.*, 1985), whereas the latter includes the cultivated soybean, *Glycine max* (L.) Merr., and its annual wild counterpart, *G. soja* Sieb. and Zucc. *G. soja*, with twice the nucleotide diversity of *G. max* (Cregan *et al.*, 2002) occurs throughout northern, north-eastern and central China and in adjacent areas of the former USSR, Korea, Japan, and Taiwan. Hymowitz and Singh (1987) infer more than one domestication event in the Shang dynasty (ca. 1700-1100 BC) with the resulting emergence of soybean as a domesticated crop in central and northern China during the period from 1100-700BC. Movement of the crop into India, Nepal, Burma, Thailand, Indochina, Korea, Japan, Malaysia, Indonesia, and the Philippines occurred by the first millenium AD (Smartt and

D. Werner and W. E. Newton (eds.), Nitrogen Fixation in Agriculture, Forestry, Ecology, and the Environment, 25-42.

Hymowitz, 1985) and was accompanied by change in several host genetic traits that affected symbiosis with rhizobia (Pulver *et al.*, 1982; Devine, 1984).

Europe first learned of soybean in 1712 through the writings of the German scientist, Englebert Kaempfer, with seeds sent by missionaries stationed in China and subsequently planted in the Jardin des Plantes, Paris, in 1740 (Probst and Judd, 1973). *Glycine max* was introduced into the USA in 1804, probably as a result of seed interchange between the USA and France (Hymowitz and Newell, 1980), and initially attracted more interest for its potential as a forage species (Probst and Judd, 1973). Significant production of soybean in the United States and Europe dates only from the beginning of the twentieth century (Piper and Morse, 1923; Gray, 1936; Morse, 1950; Hymowitz, 1970), with major germplasm collection and evaluation only initiated between 1927 and 1931 (Probst and Judd, 1973). Even today, the *Glycine* germplasm maintained in the USDA Germplasm Resource Information Network (http://www.ars.grin.gov) includes only 18,765 accessions of *G. max,* 1,117 of *G. soja*, and approximately 1,000 accessions of related *Glycine* spp. Also of concern is the very narrow genetic base used until recently in the majority of soybean-breeding programs. Thus, Delannay *et al.* (1983) noted that 10 introductions provided 80% of the northern US gene pool, whereas only 7 introductions contributed the same percentage to the southern US gene pool.

Soybean today is one of the most important and extensively grown crops in the world. It accounts for 29.7% of the world's processed vegetable oil and is a rich source of dietary protein both for the human diet and for the chicken and pork industries (Graham and Vance, 2003). Isoflavones from soybean may reduce the risks of cancer and lower serum cholesterol (Molteni *et al.*, 1995; Kennedy, 1995). Soybean is also used as a milk substitute in weaning foods and baking and in ink and biodiesel fuels (Anonymous, 2000; 2001; Graham and Vance, 2003).

Although soybean production continues to grow worldwide (Figure 1), with an estimated production in 2002/2003 of 183.28 Tg, a major part of this production comes from only four countries: the USA, Brazil, Argentina, and China (Figure 2). Mercosur, the common market agreement signed by Brazil, Argentina, Paraguay, and Uruguay in 1991, is today responsible for about 45% of world soybean production (Uruguay accounts for 0.03% of the production). Production has increased significantly over the last few years (Figure 3), even though that from the USA has declined from 47% of world production in 1999/2000 to 39% in 2001/2002 (CONAB, 2002a). Bolivia, a partner of Mercosur in the Andean region, now plants 680,000 ha annually with an estimated production of 1.3 Tg in 2002/2003 (USDA, 2002a). Export of soybean from South America continues to grow and to capture most of the increases in world soybean demand with the most dramatic increase being observed in Argentina (USDA, 2002b).

Soybean was probably introduced into Brazil in 1882, in the State of Bahia, as *Soja hispida* (*Glycine soja*) and *Soja ochroleuca* (D´utra, 1882; 1899). A few studies were subsequently undertaken at the experimental-station level, but it was not until the 1940s that soybean was grown commercially in the southern region of Brazil. A significant expansion of soybean production in this region occurred in the 1960s but, even as late as 1975, Brazil produced only 11.24 Tg of soybean annually. This. situation started to change in the 1970s, with the expansion of crop production

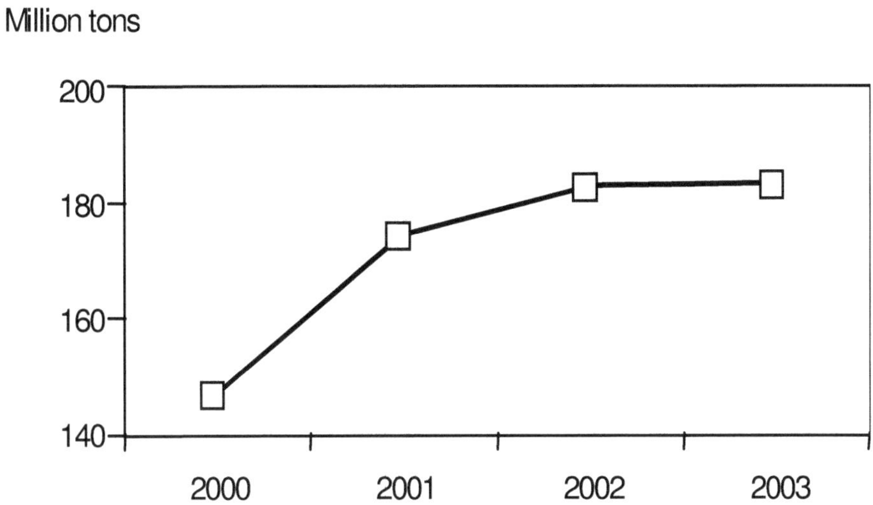

Figure 1. World soybean production in the last three years, with projection for 2002/2003 (CONAB, 2002a).

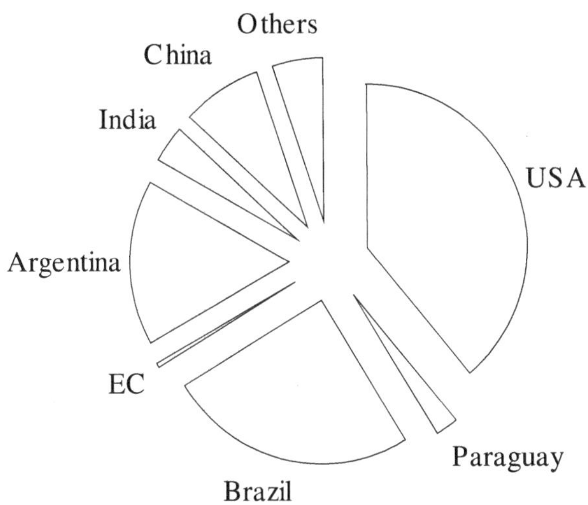

Figure 2. Contribution of different countries to world soybean production (estimated as 183.28 Tg) in 2002/2003 (CONAB, 2002a).

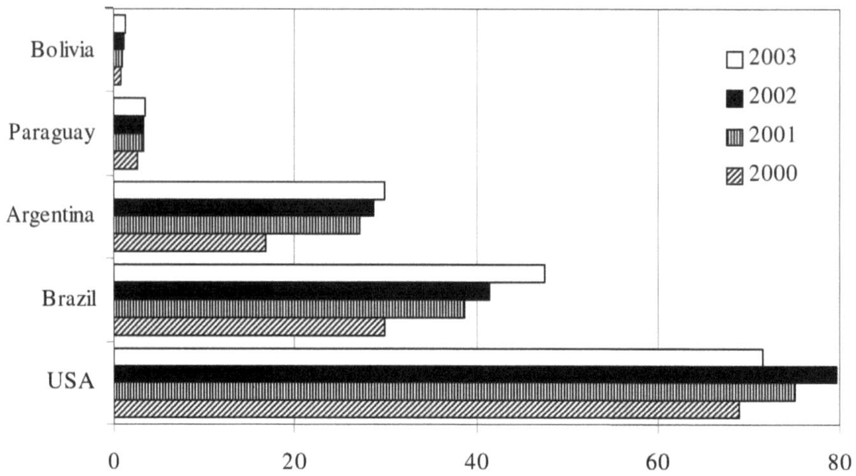

Figure 3. Annual soybean production (Tg) in the United States, and in the four main producers in South America: Brazil, Argentina, Paraguay and Bolivia (CONAB, 2002a; 2002c; 2003; USDA, 2002a).

into the "Cerrados", an edaphic savanna occupying 207 million ha and 25% of the land area of Brazil. The Cerrados are distinct in their soil chemical properties and environment (Goedert, 1985), necessitating the use of varieties that had longer juvenile periods, were aluminum tolerant, and were calcium-use efficient (Spehar, 1995). Initial yields were low but, as appropriate technologies were developed for the area, steadily improved. In 2002/2003, the three main states of this ecosystem, which are Mato Grosso, Mato Grosso do Sul, and Goiás (including the Federal District) cultivated 7.80 million ha and produced 23,329 Tg of soybean, with a yield average (2,991 kg ha^{-1}) greater than the national mean (2,765 kg ha^{-1}). A 10.6% increase in production area is expected in 2002/2003 (CONAB, 2003).

Soybean was known in Argentina in 1880, but the first field experiments there date from around 1908, and only 1,315 ha were planted in 1941/42. As in Brazil, significant expansion in the area cropped to soybean in Argentina dates to the mid 1970s with 169,400 ha cropped in 1972/73, 9.5 million ha in 1988/89, and 12.3 million ha in 2002 (Saumell, 1977; Miró, 1989; Anonymous, 2001). Soybean was introduced into Paraguay in the 1920s with seeds from the United States, Argentina, and Japan, but commercial expansion did not occur until the 1970s (Alvarez, 1989). Importation of Brazilian cultivars adapted to the tropics facilitated crop improvement and continues today (Oliveri *et al.*, 1981).

Breeding for improved productivity and for adaptation to the shorter day length, high temperature, and edaphic constraints common in the Cerrados have been a major part of soybean improvement in Brazil and have resulted in numerous varietal releases. Traditional breeding programs, which include both hybridization of selected parents in single, three/way or multiple combinations and selection by pedigree, mass selection or single/seed descent, have been used with both increased crop yield and yield stability emphasized. Selection for a long juvenile period (Toledo *et al.*, 1994; Spehar, 1995) now allows soybean to be produced even in the state of Roraima, localized between 1°S and 2°N latitudes.

Another decision, made by the National Soybean Commission in the 1960s, was that biological nitrogen fixation (BNF) was an important trait that needed to be considered in breeding activities. Most Brazilian cultivars of this period were derived from North American genotypes, and differences between them in relation to their symbiotic performance had already been identified (Döbereiner and Arruda, 1967; Vargas *et al.*, 1982). Selection of parental lines active in symbiotic N_2 fixation with *Bradyrhizobium* was emphasized in the early breeding activities, but attention to this trait has declined in recent years. A consequence has been a decline in the symbiotic capacity of recently released cultivars (Bohrer and Hungria, 1998; Hungria and Bohrer, 2000). Nicolás *et al.* (2002) studied the genetics of nodulation and nitrogen fixation in Brazilian cultivars with contrasting symbiotic efficiency, and reported narrow-sense heritability (h_n^2) estimates of from 39% to 77% for nodulation and plant growth under low fixed-N soil conditions. These values are high compared to others reported in the literature. Simple sequence repeat (SSR) markers related to symbiotic performance in soybean have also been identified (Nicolás, 2001) and can now be used in the search for cultivars with higher biological N_2 fixation capacity.

3. BIOLOGICAL NITROGEN FIXATION

The area cropped with soybean in Brazil increased from 702,000 ha in 1940/41 to 17.95 million ha in 2002/2003. Soybean now accounts for 42% of national agricultural production and 16% of gross internal product (CONAB, 2002a; 2002b; 2003). The increase in area has been paralleled by a more than four-fold increase in national mean yield over the period from 1968 to the present (Figure 4). Argentina, Paraguay, and Bolivia report similar changes.

As a consequence of the higher yields, plant fixed-N demands have also more than doubled in the last 40 years. A program to select more efficient and competitive strains has been in operation in Brazil since the 1960s and continues to ensure that soybeans in Brazil can satisfy their fixed-N requirements through N_2 fixation (Vargas and Hungria, 1997; Hungria *et al.*, this volume). Bacterial strains are selected for use in commercial inoculant production only after extensive field testing in the main soybean-production areas. Four strains are currently recommended for use; *Bradyrhizobium elkanii* SEMIA 587 and SEMIA 5019 (=29w), since 1979, and *B. japonicum* SEMIA 5079 (=CPAC 15) and SEMIA 5080 (=CPAC 7), 1992 (Vargas and Hungria, 1997; Hungria and Vargas, 2000; Hungria *et al.*, this volume). Each of these strains can fulfill the crops need for fixed-N at yields greater than 4,000 kg ha^{-1}. They are provided free to inoculant producers, and all inoculants must carry two of the four strains. Their genetic relatedness with other well-studied *Bradyrhizobium* strains is shown on Figure 5.

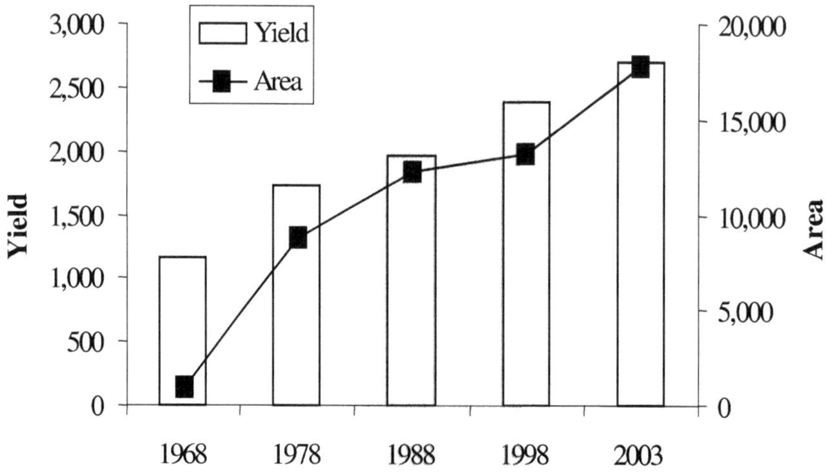

Figure 4. Change in mean yield (kg ha^{-1}) and area (x 1,000 ha) cropped with soybean in Brazil over the last 45 years (Anonymous, 2000; CONAB, 2002c; 2003).

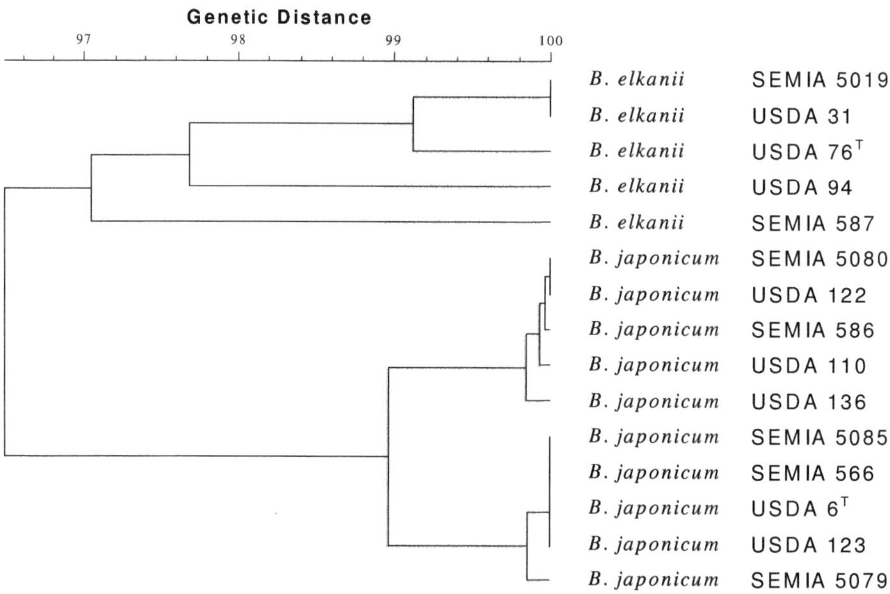

Figure 5. Genetic relatedness among soybean Bradyrhizobium strains recommended for use in commercial inoculants in Brazil (SEMIA 587, SEMIA 5019, SEMIA 5079, SEMIA 5080) and Argentina (SEMIA 5085). After Chueire et al. (2003).

Although more than 90% of the area cropped with soybean in Brazil today has been inoculated before, and generally contains naturalized rhizobia, yield increases averaging 4.5% have been obtained when soils containing 10^3 or more cells g^{-1} soil were reinoculated (Hungria *et al.*, this volume). Table 1 shows some of these results. This finding contrasts with those of studies in the USA (Thies *et al.*, 1991) and the difference in response warrants further detailed studies in both regions. Currently, re-inoculation is practiced by about 60% of farmers in Brazil (ANPI, Associação Nacional dos Produtores de Inoculante, personal communication) and by 50% of those in Argentina (Dr. Enrique R. Moretti, Laboratórios BIAGRO, Argentina, personal communication). As a result, the number of inoculant doses sold each year continues to increase in both countries (Figure 6). This situation is, again, in marked contrast to the USA, where the use of inoculants is a common practice in just 15% of the soybean-production area but, in Canada, about 60% of the farmers have adopted inoculation (Dr. R. Stewart Smith, Nitragin, USA, personal communication). In Argentina, the strain recommended by INTA (Instituto Nacional de Tecnologia Agropecuaria) is *B. japonicum* E109 (=SEMIA 5085, =USDA 138), and its relatedness with other strains is shown in Figure 5.

Table 1. Effects of reinoculation and of addition of N-fertilizer on soybean yield (kg ha^{-1}) in soils with established populations of soybean bradyrhizobia (> 10^3 cells g^{-1} soil). After Hungria et al. (2001).

Treatment	South Region		Cerrados	
	Londrina	Ponta Grossa	Goiânia	Planaltina
Non-inoculated	3,836 a[1]	2,697 b	2,341 a	2,483 a
Non-inoculated + N-fertilizer[2]	3,434 b	2,872 ab	2,432 a	2,660 a
Inoculated with efficient strains	4,025 a	2,912 a	2,462 a	3,119 a
CV (%)	9.9	8.1	8.8	7.0

[1]Values followed by the same letter, in the column, do not show statistical difference (Duncan, $p \leq 0.05$).
[2]200 kg of N ha^{-1} as urea, split at sowing and at early flowering.

In contrast to their Mercosur neighbours, Paraguay has had no strain-selection program and just 15-20% of the soybean crop is inoculated. There is also no quality-control program in the country and locally produced and foreign inoculants used over the past two decades have often been of poor quality. Inoculants now sold in Paraguay come from Argentina, Brazil, and Uruguay. Mercosur legislation permits inoculants produced in one country to be sold in another, provided they carry strains recommended by local rhizobiologists. Despite the less-than-ideal situation, nodulation occurs without inoculation in most areas of Alto Paraná and Itapúa, where 80% of soybean production occurs (Figueredo, 1998). Symbiotic effectiveness varies from site to site but, in a recent survey, some isolates collected in the field were outstanding symbiotic performers, a first step for the identification of strains for use in commercial local inoculants (Chen *et al.*, 2002).

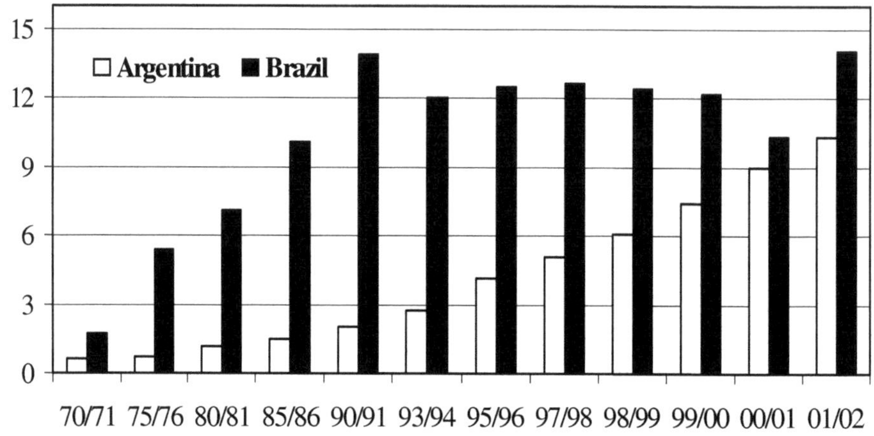

Figure 6. Units of soybean inoculant ($x10^6$) sold in Argentina and Brazil. (Dr. Enrique R. Moretti, Laboratórios BIAGRO, Dr. Solon Araújo, ANPI, and Dr. Laura Machado Ramos, Ministério da Agricultura, Pecuária e Abastecimento, personal communication).

4. ECONOMIC IMPORTANCE OF BIOLOGICAL NITROGEN FIXATION (BNF) IN SOUTH AMERICA

Economic returns from soybeans dependent on BNF in South America are outstanding. Brazil is used here as an example. For each 1,000 kg of soybean produced, approximately 80 kg of N (65 kg N allocated to seeds, which have *ca.* 40% protein, and 15 kg N left in roots, stems, and leaves) are required. When the national mean yield in Brazil in 2002/2003 of 2,765 kg ha^{-1} is considered, the fixed-N requirement would be slightly more than 220 kg ha^{-1}. Where this is supplied through BNF, N losses are minimal. In contrast, the N-fertilizer-use efficiency of plants in the tropics is rarely as much as 60%, meaning that such plants would require *ca.* 360 kg N or 800 kg of urea (at 45% N) to achieve equivalent yields. At US$0.20 kg^{-1} of urea, the cost of applying N-fertilizer would be *ca.* US$160 ha^{-1}, or in excess of US$2.87 billion nationally (for 17.95 million ha). Because Brazilian transportation costs are very high, but total production costs are 40% lower than in the United States (CONAB, 2002a; 2002b), it is evident that BNF plays a key role in lowering Brazilian production costs. Both Unkovich and Pate (2000) and Giller (2001) suggest a potential for N_2 fixation by nodulated soybeans of 360-450 kg of N ha^{-1}. Thus, there is still significant room in Brazil for increased yield without N fertilization. The situation in the other countries in South America is similar.

The data available on the quantification of BNF under field conditions in South America is limited but the results from studies undertaken in Brazil are shown in Table 2. In these studies, the contribution of BNF to N accumulation among cultivars released in recent years is appreciably higher than was achieved in 33 commercial soybean fields in Australia. For these crops, with a mean % of N derived from fixation estimated at 53%, the contribution of BNF to N accumulation was 178 kg N ha^{-1} (Unkovich and Pate, 2000).

Table 2. Contribution of biological N_2 fixation to soybean production in some field experiments performed in Brazil.

Site	*Condition*[1]	*Yield (kg ha^{-1})*	*kg of N_2 fixed ha^{-1}*	*%N from BNF*	*Method*	*Reference*
Rio de Janeiro[2]			250	84.6 (^{15}N) 88.8 (N-difference)	^{15}N isotope dilution, N-difference	Boddey *et al.* (1984)
Londrina, (Paraná)	CT	5,410	*ca.* 300	74.1	Ureides	Zotarelli (2000)
	NT	5,890	*ca.* 300	81.0		
Londrina	CT	4,380	251 (ureide) 242 ($\delta^{15}N$)	72.5 (ureide) 69.3 ($\delta^{15}N$)	Ureides and $\delta^{15}N$	Zotarelli (2000)
	NT	4,510	282 (ureide) 294 ($\delta^{15}N$)	80.4 (ureide) 84.3 ($\delta^{15}N$)		
Londrina	CT	4,128	292	87	Ureides	Hungria *et al.* (2003)
	NT	4,383	245	72		
Brasilia (Cerrados)	CT	1,641	109	79.8	^{15}N isotope dilution	Boddey *et al.* (1990)
Uberlândia (Cerrados)	CT	3,482	234	88	Ureides	Reis *et al.* (2002)
	NT	3,293	208	94		

[1]*CT, conventional tillage; NT, no-tillage system.*
[2]*Plants grown in the field in concrete cylinders and evaluated 92 days after planting.*

There is increasing pressure for farmers that grow high-profit crops to purchase N-fertilizer. However, as discussed by Hungria *et al.* (this volume), starter-N doses as low as 20-40 kg of N ha^{-1} may decrease both nodulation and BNF under Brazilian conditions, with no benefits to yield (Crispino *et al.*, 2001; Mendes *et al.*, 2000; Hungria *et al.*, 2001). Indeed, in more than 50 experiments, where inoculation and fertilization with 200 kg of N ha^{-1} have been compared (split application of N at sowing and flowering), no increase in yield due to N-fertilizer use has been reported. The results of one of these experiments are shown in Table 3. Similarly, there were no benefits when N-fertilizer was applied at a rate of 400 kg N ha^{-1}, split across ten application times (Nishi and Hungria, 1996; Hungria *et al.*, 1997, 2001; Crispino *et al.*, 2001; Loureiro *et al.*, 2001). It is important to emphasize that the application of only 30 kg of N ha^{-1} would imply a cost of US$13.3 ha^{-1} to the farmer and of US$239 million to the country per growing season, which would certainly decrease the competitiveness of the national product.

Table 3. Effects of inoculation and application of N fertilizer on nodule number (NN) and dry weight (NDW), root dry weight (RDW), yield and total seed N (TNG) of soybean, cultivar UFV-18. Experiment performed in Jaciara, Mato Grosso, in a soil with 10^5 soybean bradyrhizobia g^{-1}. Modified from Loureiro et al. (2001).

Treatment[1]	NN (n° pl^{-1})	NDW (mg pl^{-1})	RDW (g pl^{-1})	Yield (kg ha^{-1})	TNG (kg Nha^{-1})
Non-inoculated	38 c	1,22 bc	1,40 b	2,498 b	156 b
Non-inoculated + 200 kg N	40 bc	0,97 c	1,67 ab	2,668 ab	166 ab
SEMIA587	54 a	1,97 a	1,49 ab	2,640 ab	167 ab
SEMIA5080	49 ab	1,54 ab	1,54 ab	2,612 ab	169 ab
SEMIA587+5080	45 abc	1,55 ab	1,49 ab	2,781 a	179 a
SEMIA587+5080 + 30 kg N ha^{-1} at sowing	46 abc	1,24 bc	1,72 a	2,535 b	163 ab
SEMIA587+5080 + 50 kg N ha^{-1} at pre-flowering	50 ab	1,22 bc	1,50 ab	2,580 ab	168 ab
SEMIA587+5080 + 50 kg N ha^{-1} at pod filling	47 abc	1,60 ab	1,66 ab	2,527 b	169 ab
CV (%)	16.0	25.6	14.1	7.2	9.2

[1]Peat inoculants containing 10^8 cells g^{-1} were applied at a rate of 500 g 50 kg^{-1} seeds, as a slurry with 300 ml of a 10% sucrose solution; N fertilizer applied as urea.

5. CROP MANAGEMENT IN SOUTH AMERICA

Soybean is grown in South America as a main summer cash crop. In Argentina, Paraguay, Uruguay and the southern region of Brazil, a rotation with a winter crop, usually wheat (*Triticum aestivum*), is also adopted by many farmers. Benefits to the succeeding crop are well documented, with yields after soybean considerably higher than after the other main summer crop, maize (*Zea mays*). Results from one such experiment in the state of Rio Grande do Sul, Brazil, are shown in Table 4. Most of the benefits are certainly due to BNF, which results in residues with a higher level of N for the following crop. As shown in Table 5, wheat yields in plots previously used for inoculated soybean were higher than those in plots with uninoculated soybean, with or without N-fertilizer (Nishi and Hungria, 1996; Hungria *et al.*, 1997, 2001). This increased yield and consequent profit should be considered in calculating economic benefits from the inoculation of soybean. Where rainfall is not a limiting factor in winter, other crops may be used in crop rotation with soybean, bringing benefits in terms of disease control and improvement of the soil's physical and chemical properties, among others. Unfortunately, many of these crops are not highly profitable in the short term, with farmers often reluctant to use them. Plant species successful in crop rotation with soybean include radish (*Raphanus sativus*), blue lupin (*Lupinus angustifolium*), black oat (*Avena strigosa*), white oat (*Avena sativa*), crotalaria (*Crotalaria juncea*), millet (*Penisetum glaucum*), clovers (*Trifolium* spp.), common vetch (*Vicia sativa*), canola (*Brassica napus*), and linho (*Linum usitatissimum*) (Torres *et al*., 1996, 2001; Gaudencio, 1999; Calegari, 2000; Gaudencio and Costa, 2001). Crop-rotation experiments have shown the benefits to soybean yield (Table 6; ranging from 12-64%), with the lowest yields usually from continuous double crop wheat-soybean or fallow-soybean. In the Brazilian Cerrados, where irregular rainfall limits activities to one crop per year, alternative crops, including millet and forage peanut (*Arachis pintoi*) have been evaluated, shown to grow well, and adopted by some farmers.

Table 4. Mean yield of wheat as a function of the previous crop. After Wiethölter (2000).

Year	Previous crop		Difference
	Soybean	Maize	
	kg ha^{-1}		kg ha^{-1}
1993	3,039	2,744	295
1994	2,632	2,320	312
1995	3,418	2,995	423
1996	4,591	4,172	419
1997[1]	1,998	1,062	936
1998	3,031	2,463	568
Mean[1]	3,342	2,939	403

[1]Excess rainfall in 1997 decreased yield.

Table 5. Residual effects of soybean inoculation in the summer on wheat yield (kg ha^{-1}) in the winter. After Hungria et al. (2001).

Treatment	1994	1998
Non-inoculated	1,827 ab[1]	2,028 a
Non-inoculated + 200 kg N ha^{-1}, split at sowing and at early flowering stage	1,484 b	2,219 ab
Inoculated with efficient strains	2,000 a	2,449 b
CV (%)	12.7	9.3

[1]Values followed by the same letter in a column do not show statistical difference (Duncan, p≤0.05).

Table 6. Compilation of data of experiments with crop-rotation systems in southern Brazil.

N° of systems tested	Location	Best system[a]	Worst system[a]	Year	Yield Increase (%)	References[b]
12	Campo Mourão	Ca/Mz/Rd/Sb Wt/Sb/Wt/Sb	Wt/Sb	2000/01	12	1
8	Londrina	Rd/Mz/Oa/Sb Wt/Sb/Wt/Sb	Wt/Sb	1999/00	17	1
4	Londrina	Oa/Sb	Fallow/Sb	1985/86	24	2
4	Londrina	Oa/Sb	Fallow/Sb	1987/88	24	2
4	Londrina	Sb/Oa/Mz-Mi Sb/Wt/Sb/Wt	Sb/Wt	1997/98	24	3
8	Campo Mourão	Lp/Mz/Oa/Sb Wt/Sb/Wt/Sb	Wt/Sb	1997/98	32	3
9	Passo Fundo	Oa+Cl/Sb/Oa+ Cv/Mz/Wt/Sb	Oa/Sb/Wt/Sb Oa/Sb	1992	64	4
7	Guarapuava	Cv/Mz/Wt/Sb	Li/Sb/Cv/Mz Wt/Sb	1984/89	15	5

[a]*Ca, canola; Mz, maize; Rd, radish; Sb, soybean; Wt, wheat; Oa, black oat; Mi, millet; Lp, lupin; Cl, clover; Cv, common vetch; Li, linho.*
[b]*1, Gaudencio and Costa (2001); 2, Torres* et al. *(1996); 3, Gaudencio (1999); 4, Fontaneli* et al. *(2000); 5, Santos* et al. *(1998).*

No-till (NT) soil management systems are particularly important in many areas of South America, where soils are commonly of low organic matter content, structurally fragile, and of low fertility. In NT, seeds are sown directly through the residue of the previous crop, so protecting the soil against erosion by water, maintaining soil structure, stability and moisture content, helping in the regulation of soil temperature and, with time, increasing soil organic matter content (Derpsch *et al.*, 1991). In 1999, the world area under NT was *ca.* 45.5 million ha (Table 7), but is increasing substantially. By 2002, the NT area in the USA was *ca.* 26.3 million ha, with 13 million ha of soybean (Anonymous, 2002), whereas the NT area in Brazil was 17.356 million ha (with 4.9 million of that in the Cerrados) in 2000/2001 (FEBRAPDP, 2003) and is *ca.* 20 million ha in 2002/2003, which represents half of the cropped area in Brazil. For BNF in soybean, NT results in higher number of bradyrhizobia in the soil; better nodulation, with nodulation extending lower in the soil profile; and higher BNF rates and yields in relation to conventional tillage (CT). Bradyrhizobial strain diversity in the soil is also enhanced (Ferreira *et al.*, 2000; Hungria, 2000; Hungria and Vargas, 2000). Without placing an economic value on biodiversity, Calegari (2000) estimated that a farmer adopting NT for a 50-ha area of soybean in southern Brazil would gain US$8,902 as the result of higher yield and lower costs for maintenance, fuel, labor work, and fertilizers. This result is somewhat at variance with studies in the northern USA, where cooler early-season soil temperatures and high soil-moisture content under conservation tillage can delay nodulation, result in enhanced root rot, and contribute to lowered flavonoid exudation from the root, limiting nodulation (Zhang *et al.*, 1996; 2002; Estevez de Jensen *et al.*, 2002).

Table 7. Estimates of world area under no-till management systems in 1998/99. After Derpsch (2000).

Country	Area (1,000 ha)
United States	19,347
Brazil	11,200
Argentina	7,270
Canada	4,080
Paraguay	790
Bolivia	200
Uruguay	50
Others	2,596
Total	45,533

In an 18-year field trial, Torres *et al.* (2001) found that CT (one disc plough and two light-disc harrow) soybean yield was higher for the first four years, but thereafter soybean yield was always higher under NT (Figure 7). Water and soil temperature conditions are always better under NT. In the first years under NT, some physical (*e.g.*, soil density), chemical, and biological (*e.g.*, lower nutrient

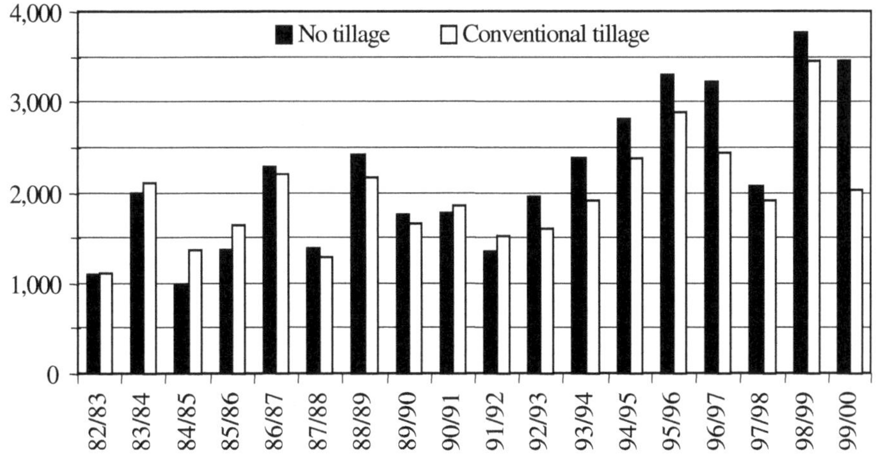

Figure 7. Soybean yield (kg ha^{-1} on left axis) under no-tillage and conventional systems in different agricultural years. Field trial performed in an oxisol of Londrina, State of Paraná. After Torres et al. (2001).

content due to microbial immobilization) changes may result in lower yields (Derpsch *et al.*, 1991; Torres *et al.*, 1996, 2001). Later yields are higher because of both improvements in soil physical properties, which include soil aggregation, and increases in total soil C and labile soil C in the microbial biomass, so regulating mineral cycling (Hungria, 2000; Franchini *et al.*, 2002; Brandão-Junior *et al.*, 2002). Higher soybean yields after the first years under NT were also found in another 11-year study in southern Brazil, where soybean yield was sometimes twice that under non-conservation soil management systems (Figure 8).

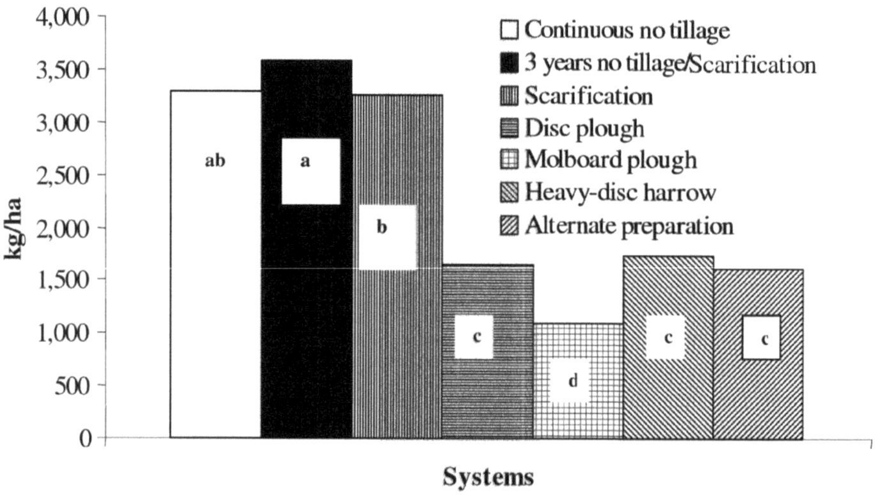

Figure 8. Soybean yields after 11 years under seven different systems of soil preparation. Experiment performed in Londrina, State of Paraná.
Columns with the same letter are not statistically different (Duncan, $p \leq 0.05$).
After Torres et al. (2001).

6. FINAL CONSIDERATIONS

Soybean production in South America has expanded enormously in recent years, both in area and in crop yield. In Brazil, improvements in crop yield and crop economics have been affected very markedly by a research and extension effort focused on BNF. Lines that are active in symbiosis with *Bradyrhizobium* have traditionally been favored in breeding. A strong and consistent effort is expected to select the most efficient inoculant strains and to market high quality inoculants. The result is that more than 50% of Brazilian farmers practice re-inoculation and quantification experiments in Brazil have shown that the crop derives 69-94% of its fixed-N need from symbiosis. In contrast, van Kessel and Hartley (2000) suggest that, for trials undertaken mainly in the USA and Australia, the percentage of fixed-N derived from BNF has declined since 1984 and is now only 54%; however, only about 15% of farmers in the USA inoculate soybean. Some contrast between the South American and US systems of production is inevitable but any comparison should highlight problems to be avoided in South America. There is pressure to increase production in the South America by using fertilizer N. The consequences of such a change, however, need to be considered very carefully. For example, a recent commentary by Randall (2001) suggests that the more intensive corn and soybean production practiced in the mid-western USA is not sustainable. Randall points out several concerns: (i) a dependence of this production system on federal assistance; (ii) significant problems of soil erosion in the region; and (iii) declines in soil and water stewardship. Soybean growers in South America need both to maintain their focus on biological nitrogen fixation as an economic fixed-N source for this crop and to ensure that land quality is maintained (or improved) by continued study of soil quality and the factors affecting it.

ACKNOWLEDGEMENT

The research group in Brazil is supported by CNPq (PRONEX and 520396/96-0). The authors dedicate this chapter to the memory of Raul Martínez Lalis (Nitragin Argentina) and Bernardo Leicach (Laboratórios BIAGRO) who played an important role in improving the quality of the inoculants in Mercosur.

REFERENCES

Alvarez, E. R. (1989). La soja en el Paraguay, retrospectiva y perspectiva. In *II Curso Internacional sobre Producción de Soja* (pp. 551-563). Encarnación, Brazil: Ministerio de Agricultura y Ganaderia.

Anonymous (2000). *Anuário Brasileiro da Soja 2000.* Santa Cruz do Sul, Brazil: Gazeta.

Anonymous (2001). *Anuário Brasileiro da Soja 2001.* Santa Cruz do Sul, Brazil: Gazeta.

Anonymous (2002). Agropecuária-Tele News at http://orbita.starmedia.com/~telenews/agricultura.html. Retrieved December 10, 2002.

Boddey, R. M., Chalk, P. M., Victoria, R. L., and Matsui, E. (1984). Nitrogen fixation by nodulated soybean under tropical field conditions estimated by the ^{15}N isotope dilution technique. *Soil Biol. Biochem., 16*, 583-588.

Boddey, R. M., Urquiaga, S., and Neves, M. C. P. (1990). Quantification of the contribution of N_2 fixation to field-grown grain legumes - a strategy for the practical application of ^{15}N isotope dilution technique. *Soil Biol. Biochem., 22*, 649-655.

Bohrer, T. R. J., and Hungria M. (1998). Avaliação de cultivares de soja quanto à fixação biológica do nitrogênio, *Pesq Agropec Bras., 33*, 937-953.

Brandão Junior, O., Souza, R. A., Crispino, C. C., Franchini, J. C., Torres, E., and Hungria, M. (2002). Variação da biomassa microbiana em sistemas de manejo e rotação de culturas envolvendo a soja. In O.F. Saraiva and C.B. Hoffman-Campo (Eds.), *Resumos do II Congresso Brasileiro de Soja e Mercosoja* (p. 250). Londrina, Brazil: Embrapa Soja.

Brown, A. H. D., Grant, J. E., Bhurdon, J. J., Grace, J. P., and Pullen, R. (1985). Collection and utilization of wild perennial *Glycine*. In R. Shibles (Ed.), *World Soybean Research Conference III* (pp. 345-352). Boulder, CO: Westview Press.

Calegari, A. (2000). Plantas de cobertura e rotação de culturas no sistema plantio direto. In M. Veiga (Coord.), *Memorias de la V Reunión Bienal de la Red Latinoamericana de Agricultura Conservacionista* (CD ROM). Florianópolis, Brazil: EPAGRI/FAO.

Chen, L. S., Figueredo, A., Villani, H., Michajluk, J., and Hungria, M. (2002). Diversity and symbiotic effectiveness of rhizobia isolated from field-grown soybean in Paraguay. *Biol. Fert. Soils, 35*, 448-457.

Chueire, L. M. O., Bangel, E., Mostasso, F. L., Campo, R. J., Pedrosa, F. O., and Hungria, M. (2003). Classificação taxonômica das estirpes de rizóbio recomendadas para as culturas da soja e do feijoeiro baseada no seqüenciamento do gene 16s RNA. *Rev. Bras. Ci. Solo*, *27*, 833-840.

CONAB (Companhia Nacional de Abstecimento) (2002a). Retrieved December 10, 2002, from http//www.conab.gov.br/politica_agricola/ConjunturaSemanal/Semana19a2308/Soja-19a2308.doc.

CONAB (Companhia Nacional de Abstecimento) (2002b). Retrieved December 10, 2002, from http//www.conab.gov.br/politica_agricola/ConjunturaSemanal/Semana2810a0111/ConjunturaSemanaSoja13.doc.

CONAB (Companhia Nacional de Abstecimento) (2002c). Retrieved December 10, 2002, from www.conab.gov.br/politica_agricola/ Safra/Quadro9.xls.

CONAB (Companhia Nacional de Abastecimento) (2003). Retrieved March 20, 2003, from www.conab.gov.br.

Cregan, P., Zhu, Y., Song, Q., and Nelson, R. (2002). Sequence variation, haplotype analysis and linkage disequilibrum in cultivated and wild soybean populations. In *First International Conference on Legume Genomics and Genetics: Translation to Crop Improvement,* June 2-6 (p. 11). Minneapolis-St Paul, MN.

Crispino, C. C., Franchini, J. C., Moraes, J. Z., Sibaldelle, R. N. R., Loureiro, M. F., Santos, E. N., *et al.* (2001). *Adubação Nitrogenada na Cultura da Soja* (Comunicado Técnico, 75). Londrina, Brazil: Embrapa Soja.

Delannay, X., Rodgers, D. M., and Palmer, R. G. (1983). Relative genetic contributions among ancestral lines to North American soybean cultivars. *Crop Sci., 23*, 944-949.

Derpsch, R. (2000). Expansión mundial de la siembra directa y avances tecnológicos. In M. Veiga (Coord.), *Memorias de la V Reunión Bienal de la Red Latinoamericana de Agricultura Conservacionista* (CD ROM). Florianópolis, Brazil: EPAGRI/FAO.

Derpsch, R., Roth, C. H., Sidiras, N., and Kopke, U. (1991). *Controle da erosão no Paraná, Brasil: Sistemas de cobertura do solo, plantio direto e preparo conservacionista do solo* (Sonderpublikation der GTZ, 245). Eschborn, Germany: GTZ.

Devine, T. E. (1984). Genetics and breeding of nitrogen fixation. In M. Alexander (Ed.), *Biological Nitrogen Fixation: Ecology, Technology and Physiology* (pp. 127-154). New York, NY: Plenum Press.

Döbereiner, J., and Arruda, N. B. (1967). Interrelações entre variedades e nutrição na nodulação e simbiose da soja. *Pesq. Agropec. Bras., 2*, 475-487.

D'utra, G. (1882). Soja. *J Agricultor* VII, 185-188.

D'utra, G. (1899). Nova cultura experimental da soja. *Bol. Inst. Agron. Campinas, 10*, 582-587.

Estevez de Jensen, C., Percich, J. A., and Graham, P. H. (2002). Integrated management strategies of bean root rot with *Bacillus subtilis* and *Rhizobium* in Minnesota. *Field Crops Res.* 74, 107-115.

FEBRAPDP (Federação Brasileira de Plantio Direto na Palha) (2003). Retrieved February 24, 2003, from http://www.febrapdp.org.br/pd_area_estados.htm.

Ferreira, M. C., Andrade, D. S., Chueire, L. M. de O., Takemura, S. M., and Hungria, M. (2000). Effects of tillage method and crop rotation on the population sizes and diversity of bradyrhizobia nodulating soybean. *Soil Biol. Biochem., 32*, 627-637.

Figueredo, A. (1998). *Fijación Biológica de Nitrógeno.* Asunción, Paraguay: Universidad Nacional de Asuncion. Dirección de Investigación, Postgrado y Relaciones Internacionales.

Fontaneli, R. S., Santos, H. P., Voss, M. and Ambrosi, I. (2000). Rendimento e nodulação de soja em diferentes rotações de espécies anuais de inverno sob plantio direto. *Pesq. Agropec. Bras., 35*, 349-355.

Franchini, J. C., Crispino, C. C., Souza, R. A., Torres, E., and Hungria, M. (2002). Biomassa microbiana e emissão de CO_2 em sistemas de manejo do solo e rotação de culturas. In O. F. Saraiva and C. B. Hoffman-Campo (Eds.), *Resumos do II Congresso Brasileiro de Soja e Mercosoja* (pp. 133). Londrina, Brazil: Embrapa Soja.

Gaudencio, C. A. (1999). Sistema de rotação de espécies perenes e anuais para recuperação biológica de Lassolos Roxos eutróficos e integração agropecuária, na Região Meridional. In *Resultados de Pesquisa da Embrapa Soja – 1998* (pp. 181-210; Documentos, 125). Londrina, Brazil: Embrapa Soja.

Gaudencio, C. A., and Costa, J. M. (2001). Rotação de culturas anuais no Planalto Meridional Paranaense. In *Resultados de Pesquisa da Embrapa Soja -2000, Manejo do Solo e Plantas Daninhas* (pp. 19-25; Documentos, 161). Londrina, Brazil: Embrapa Soja.

Giller, K. E. (2001). *Nitrogen fixation in tropical cropping systems.* Wallingford, UK: CABI Publishing.

Goedert, W. J. (1985). *Solos dos Cerrados, Tecnologias e Estratégias de Manejo.* Planaltina, Brazil: Embrapa Cerrados.

Graham, P. H., and Vance, C. P. (2003). Legumes: Importance and constraints to greater utilization. *Plant Physiol.*, *131*, 872-877.

Gray, G. E. (1936). *All About the Soya Bean.* London, UK: John Bale.

Hungria, M. (2000). Características biológicas em solos manejados sob plantio direto. In M. Veiga (Coord.), *Memorias de la V Reunión Bienal de la Red Latinoamericana de Agricultura Conservacionista* (CD ROM). Florianópolis, Brazil: EPAGRI/FAO.

Hungria, M., and Bohrer, T. R. J. (2000). Variability of nodulation and dinitrogen fixation capacity among soybean cultivars. *Biol. Fert. Soils, 31*, 45-52.

Hungria, M., and Vargas, M. A. T. (2000) Environmental factors affecting N_2 fixation in grain legumes in the tropics, with an emphasis on Brazil. *Field Crops Res.*, *65*, 151-164.

Hungria, M., Campo, R. J., Franchini, J. C., Chueire, L. M. O., Mendes, I. C., Andrade, D. S., *et al.* (2003). Microbial quantitative and qualitative changes in soils under different crops and tillage management systems in Brazil. In *Annals of Technical Workshop on Biological Management of Soil Ecosystems for Sustainable Agriculture* (CD ROM). Londrina, Brazil: Embrapa Soja/FAO.

Hungria, M., Campo, R. J., and Mendes, I. C. (2001). *Fixação Biológica do Nitrogênio na Cultura da Soja* (Circular Técnica, 13). Londrina, Brazil: Embrapa Soja/Embrapa Cerrados.

Hungria, M., Loureiro, M. F., Mendes, I. C., Campo, R. J., and Graham, P. H. (2005). Inoculant preparation, production and application. This volume.

Hungria, M., Vargas, M. A. T., Campo, R. J., and Galerani, P. R. (1997). *Adubação Nitrogenada na Soja?* (Comunicado Técnico, 57). Londrina, Brazil: EMBRAPA-CNPSo.

Hymowitz, T. (1970). On the domestication of soybean. *Econ. Bot.*, *24*, 408-421.

Hymowitz, T., and Newell, C. A. (1980). Taxonomy, speciation, domestication, dissemination, germplasm resources and variation in the genus *Glycine.* In R.J. Summerfield and A.H. Bunting (Eds.), *Advances in Legume Science* (pp. 251-264). Kew, UK: Royal Botanic Gardens.

Hymowitz, T., and Singh, R. J. (1987). Taxonomy and speciation. In J.R. Wilcox (Ed), *Soybeans, Improvement, Production and Uses.* 2nd Edit. (pp. 23-48). Madison, WI: Amer. Soc. of Agronomy.

Kennedy, A. R. (1995). The evidence for soybean products as cancer preventative agents, *J. Nutr., 125*, S733-S743.

Loureiro, M. F., Santos, E. N., Hungria, M., and Campo, R. J. (2001). *Efeito da Reinoculação e da Adubação Nitrogenada no Rendimento da Soja em Mato Grosso* (Comunicado Técnico, 74). Londrina, Brazil: Embrapa Soja.

Mendes, I. C., Hungria, M., and Vargas, M. A. T. (2000). *Resposta da Soja à Adubação Nitrogenada na Semeadura, em Sistemas de Plantio Direto e Convencional na Região do Cerrado* (Boletim de Pesquisa, 12). Planaltina, Brazil: Embrapa Cerrados.

Miró, D. A. (1989). La expansión del cultivo de soja y su impacto en el comercio exterior argentino. In *IV Conferência Mundial de Investigación en Soja* (pp. 18-27). Buenos Aires, Argentina: Associación Argentina de la Soja.

Molteni, A., Brizio-Molteni, L., and Persky, V. (1995). *In vitro* hormonal effects of soybean isoflavones, *J. Nutr., 125*, S751-S756.

Morse W. J. (1950). History of soybean production. In K.L. Markley (Ed), *Soybeans and Soybean Products*, vol. 1 (pp. 3-59). New York, NY: Interscience Publ. Inc.

Nicolás, M. F. (2001). *Fixação Biológica do Nitrogênio e Nodulação em Cultivares de Soja Brasileiras: Controle Genético e Mapeamento dos QTLs que Controlam esses Caracteres.* Curitiba, Brazil: Universidade Federal do Paraná (Ph.D. Thesis).

Nicolás, M. F., Arias, C. A. A., and Hungria, M. (2002). Genetics of nodulation and nitrogen fixation in Brazilian soybean cultivars. *Biol. Fert. Soils, 36*, 109-117.

Nishi, C. Y. M., and Hungria, M. (1996). Efeito na reinoculação na soja [*Glycine max* (L.) Merrill] em um solo com população estabelecida de *Bradyrhizobium* com as estirpes SEMIA 566, 586, 587, 5019, 5079 e 5080, *Pesq. Agrop. Bras., 31*, 359-368.

Oliveri, N. J., Perucca, C. E., and Morel, F. (1981). *Evolución de Cultivares de Soja (Glycine max L. Merr.) de Origen Brasileño.* Instituto Nacional de Tecnología Agropecuaria - Estación Experimental Agropecuaria INTA, Misiones, (Informe Técnico, 35).

Piper, C. V., and Morse, W. J. (1923). *The Soybean.* New York, NY: McGraw-Hill.

Probst, A. H., and Judd, R. W. (1973). Origin, history and develop, and world distribution. In B.E.Caldwell (Ed.) *Soybean, Improvement, Production and Uses* (pp. 1-15). Madison, WI: American Society of Agronomy.
Pulver, E. L., Brockman, F., and Wein, H. C. (1982). Nodulation of soyabean cultivars with *Rhizobium* spp. and their response to inoculation with *R. japonicum. Crop Sci., 22*, 1065-1070.
Randall, G. W. (2001). Intensive corn soybean production is not sustainable. Retrieved December 10, 2002, from www.extension.umn.edu/extensionnews/2001/IntensiveCornSoybeanAgriculture.html.
Reis, E. H. S., Lara Cabezas, W. A. R., Alves, B. J. R., and Caballero, S. U. (2002). Suplementação de nitrogênio mineral na cultura da soja (*Glycine max*) estabelecida em sistema plantio direto e convencional. In O.F. Saraiva and C.B. Hoffman-Campo (Eds.), *Resumos do II Congresso Brasileiro de Soja e Mercosoja* (pp. 215). Londrina, Brazil: Embrapa Soja.
Santos, H. P., Lhamby, J. C. B., and Wobeto, C. (1998). Efeito de culturas de inverno em plantio direto sobre a soja cultivada em rotação de culturas. *Pesq. Agropec. Bras., 33*, 289-295.
Saumell, H. (1977). *Soja, Información Técnica para su Mejor Conocimiento y Cultivo.* Buenos Aires, Argentina: Editorial Hemisferio Sur.
Smartt, J., and Hymowitz, T. (1985). Domestication and evolution of grain legumes. In R.J. Summerfield and E.H. Roberts (Eds.), *Grain Legume Crops* (pp. 37-72). London, UK: Collins Professional and Technical Books.
Spehar, C. R. (1995). Impact of strategic genes in soybean on agricultural development in the Brazilian tropical savannahs. *Field Crops Res., 41*, 141-146.
Thies, J. E., Singleton, P. W., and Bohlool, B. B. (1991). Influence of the size of indigenous rhizobial populations on establishment and symbiotic performance of introduced rhizobia on field-grown legumes. *Appl. Environ. Microbiol., 57*, 19-28.
Toledo, J. F. F., Almeida, L. A., Kiihl, R. A. S., Panizzi, M. C. C., Kaster, M., Miranda, L. C., and Menosso, O. G. (1994). Genetics and breeding. In *Tropical Soybean Improvement and Production* (pp. 19-36). Rome, Italy: FAO.
Torres, E., Norman, N., and Garcia, A. (1996). Sucessão soja/aveia preta. In *Resultados de Pesquisa de Soja – 1990/91*, vol. 2, pp. 337-341 (Documentos 99). Londrina, Brazil: EMBRAPA-CNPSo.
Torres, E., Saraiva, O. F., Gazziero, D. L. P., Pires, M. S. and Loni, D. A. (2001). Avaliação de sistemas de preparo do solo e manejo do plantio direto envolvendo sucessão e rotaçao de culturas. In *Resultados de Pesquisa da Embrapa Soja - 2000, Manejo do Solo e Plantas Daninhas* (pp. 9-14). Londrina, Brazil: Embrapa Soja.
USDA (United States Department of Agriculture) (2002a). Retrieved December 10, 2002, from http://ffas.usda.gov/wap/circular/2002/02-12/Wap%2012-02.pdf.
USDA (United States Department of Agriculture) (2002b). Retrieved December 10, 2002, from http://ffas.usda.gov/oilseeds/circular/2001/01-12/full.pdf.
Unkovich, M. J., and Pate, J. S. (2000). An appraisal of recent field measurements of symbiotic N_2 fixation by annual legumes. *Field Crops Res., 65*, 211-228.
Van Kessel, C., and Hartley, C. (2000). Agricultural management of grain legumes; has it led to an increase in nitrogen fixation? *Field Crops Res., 65*, 165-181.
Vargas, M. A. T., and Hungria, M. (1997). Fixação biológica do N_2 na cultura da soja. In M. A. T. Vargas and M. Hungria (Eds.), *Biologia dos Solos de Cerrados* (pp. 297-360), Planaltina, Brazil: Embrapa-CPAC.
Vargas, M. A. T., Peres, J. R. R., and Suhet, A. R. (1982). Fixação de nitrogênio atmosférico pela soja em solos de cerrado. *Inf. Agrop., 8*, 20-23.
Wiethölter, S. (2000). Características químicas de solo manejado no sistema plantio direto. In M. Veiga (Coord.), *Memorias de la V Reunión Bienal de la Red Latinoamericana de Agricultura Conservacionista* (CD ROM). Florianópolis, Brazil: EPAGRI/FAO.
Zhang, F., Charles, T. C., Pan, B. and Smith, D. L. (1996). Inhibition of the expression of *Bradyrhizobium japonicum nod* genes at low temperatures. *Soil. Biol. Biochem., 28*, 1579-1583.
Zhang, H., Prithiviraj, B. Souliemanov, A., D'Aoust, F., Charles, T. C., Driscoll, B. T., and Smith, D. L. (2002). The effect of temperature and genistein concentration on lipo-chitooligosaccharide (LCO) production by wild type and mutant strains of *Bradyrhizobium japonicum. Soil Biol. Biochem., 34*, 1175-1180.
Zotarelli, L. (2000). *Balanço de Nitrogênio na Rotação de Culturas em Sistemas de Plantio Direto e Convencional na Região de Londrina, PR.* Seropédica, Brazil: Universidade Federal Rural do Rio de Janeiro (M.Sc. Thesis).

CHAPTER 4

PRODUCTION, REGIONAL DISTRIBUTION OF CULTIVARS, AND AGRICULTURAL ASPECTS OF SOYBEAN IN INDIA

S. K. MAHNA

Department of Botany, Maharshi Dayanand Saraswati University of Ajmer, India

1. INTRODUCTION AND HISTORICAL BACKGROUND

Soybean (*Glycine max* (L.) Merrill) has been one of the man's principle food plants for ages. Soybean has been cultivated since 2800 B.C. in China (more than 5000 years ago), which is also considered to be the center of origin for soybean. Soybean ranks first among the major oilseed crops of the world and has now found a prominent place in India. India is the fifth largest producer of soybean after the United States, Brazil, China, and Argentina (Table 1). Soybean cultivation in India started long ago but its successful cultivation has increased over last two decades. This increased cultivation has revolutionized the rural economy and improved the socio-economic status of farmers. Soybean farming made an unprecedented expansion in India between 1969 and 1996 when an annual growth rate of 15-20% was achieved. Presently, the area covered by soybean cultivation is around 5.7 x 10^6 hectares (ha) as recorded for the winter 2002-03 (SOPA Report, 2002-03). Soybean is grown mainly in Madhya Pradesh, Maharashtra, Rajasthan, and in small pockets in other states, like Uttar Pradesh, Andhra Pradesh, Punjab, Tamil Nadu, Uttaranchal, Gujarat, Karnataka, and Chhattisgarh. Despite being an exotic crop to India, soybean occupies a vital place in its agriculture, edible-oil economy, and foreign exchange. Tiwari *et al.* (1994) identified suitable soybean varieties for the non-traditional regions of India and demonstrated soybean to be a successful crop in northern, eastern, and southern regions of the country.

Soybean cultivation in the Indian subcontinent dates back to 1000 A.D. The crop was introduced from China through the 'silk route' in the Himalayan mountain ranges running across the Tibetan plateau and through the North-East regions (Assam). Around the same time, the crop was introduced to Central India from Japan, South

D. Werner and W. E. Newton (eds.), Nitrogen Fixation in Agriculture, Forestry, Ecology, and the Environment, 43-66.

Table 1. World Production of Soybeans.
Area in million hectare; Yield in tonnes/hectare; Production (Prod.) in million tonnes.
After Oilseed World Market and Trade (USDA, 1999) and Agriculture Statistics at a Glance, Director of Economic and Statistical Ministry of Agriculture, GOI (as given by Paroda, 1999).

Country	1996-1997			1997-1998			1998-1999		
	Area	Yield	Prod.	Area	Yield	Prod.	Area	Yield	Prod.
United States	25.55	2.53	64.84	27.97	2.62	74.22	28.66	2.62	75.03
Brazil	11.80	2.27	26.80	13.00	2.33	30.00	12.90	2.40	31.00
China	7.47	1.77	13.22	8.35	1.67	13.80	8.00	1.73	13.00
Argentina	6.20	1.81	11.20	7.10	2.35	16.00	7.40	2.53	18.70
India	5.23	0.99	5.20	5.86	1.15	6.72	6.30	0.90	5.70/5.90
European Union	0.34	3.44	1.15	0.46	3.37	1.44	0.54	3.26	1.74
Paraguay	1.20	2.25	2.70	1.20	2.23	2.90	1.25	2.64	3.30
Others	5.29	1.22	6.47	6.41	1.12	7.17	5.61	1.51	8.40
Total/(Average)	63.19	(2.08)	131.58	69.34	(2.19)	152.26	70.65	(2.23)	157.70

China and South-East Asia (Hymowitz and Kaizuma, 1981). Black-seeded soybean, under the names Bhatt, Bhatmash, or Kalitur, has since been cultivated for many years in the hilly areas of Assam, Bengal, Manipur, in the hills of Khasi and Naga, and at 6000 ft. elevation in the Kumaun regions as well as the Garhwal hills (Anonymous, 1956). Hooker (1879) clearly described soybean in his book "Flora of British India, Vol. II" and Williams (1932) mentioned the cultivation of soybean in his book "Flora India, Vol. II".

Between the years 1885-1904, attempts were made to cultivate soybean at Nagpur (Maharashtra), Madras (Tamil Nadu), Lahore (Former Punjab of undivided India), Bombay, Pune (Maharashtra), and Surat (Gujarat), but these attempts were not encouraging. However, the work carried out between 1910-1935 in Madhya Pradesh, Assam, Orissa, Bihar, Gujarat, and Uttar Pradesh paved the way for successful cultivation in India (Kaltenbach and Legros, 1936; Kale, 1936). A varietal trial, which was comprised of 33 varieties of Manchurian and Chinese origin, was conducted in 1933 and resulted in the establishment of soybean cultivation in Central India in 1936. Two varieties, "Otootan" (black seeded) and "Easy Cook" were promoted between 1935 and 1952 in addition to the country type (desi) Punjab white. Other varieties from the USA, *viz.*, "Harbinsoy", "Chiquito", "George Washington", "Mammoth Yellow", and "Biloxi" were also evaluated (Tiwari *et al.*, 1999).

In India, large-scale cultivation of soybean started in 1964, using yellow-seeded high-yielding soybean exotic varieties ("Bragg", "Clark-63", and "Lee") received from USA. They were tested at Jawaharlal Nehru Krishi Vishwa Vidhyalaya (JNKVV) Jabalpur, Madhya Pradesh, and almost concurrently at Govind Ballabh Pant (GBP) University of Agriculture and Technology, Pantnagar, Uttaranchal, under a major collaboration with the USA (Paroda, 1999). Subsequently, in the mid-seventies, cultivation practices that were suitable for Indian conditions (Saxena *et al.*, 1971) and several new soybean varieties were introduced (Saxena and Pandey, 1971; Singh and Saxena, 1975). Since then, both the area cultivated and the production of soybean have increased until what was a marginal crop is now a major cash crop and is recognised as the miracle "golden bean" of the 20th century (Singh *et al.*, 2001).

2. ALL-INDIA AREA COVERAGE, PRODUCTIVITY, AND PRODUCTION OF SOYBEAN BETWEEN 1970-2003

Soybean has seen phenomenal growth in both area and production in India (Paroda, 1999). In 1970-71, soybean was grown on 32,000 ha with a production of 14,000 tonnes and a productivity of 426 kg/ha. The area of soybean cultivation, production, and productivity has gradually increased over the years. In 1977-78, a six-fold increase from 32,000 to 195,000 ha in cultivation area occurred with a thirteen-fold increase from 14,000 tonnes to 183,000 tonnes in production and a two-fold increase from 426 to 940 kg/ha in productivity. In 1987-88, the area of soybean cultivation increased almost eight times, to 1,543,000 ha with a significant increase in production, to 898,000 tonnes, however, productivity decreased significantly to 582 kg/ha. This rapid growth in area of soybean cultivation continued in the next decade. In 1997-98, the area under soybean cultivation increased to 5.99×10^6 ha and the yield increased to 1,079 kg/ha, resulting in a record production of 6.463 x 10^6 tonnes (Agricultural Statistics at a Glance, 2001).

During the winter season of 1999-2000, the area of soybean cultivation in India was only 5.645 x 10^6 ha, which was about 11.15 %, lower than in 1998-99 (SOPA Report, 2000 -01). In the subsequent two seasons, an increasing trend in the area of soybean cultivation occurred. During 2001-02, the area increased to 6.0021 x 10^6 ha, which is about 6.32% higher than in 1999-2000 and 3.27% higher than in 2000-01. In the next winter season (2002-03), a decrease of 5.44% in cultivation area was recorded. Production and productivity data for the last five seasons (1998-2003) reveal that a maximum productivity and production of 928 kg/ha and 5.90 x 10^6 tonnes, respectively, were recorded in the 1998-99 season. However, in the four subsequent growing seasons, a variable reduction in both productivity and production was recorded (Table 2).

3. ALL-INDIA STATE-WISE AREA COVERAGE, PRODUCTIVITY, AND PRODUCTION OF SOYBEAN

All-India state-wise area coverage of soybean during the winters of 1998 through 2003 are summarized in Table 2, which also indicates variability in area, production, and productivity in the different states of India. Area coverage, productivity, and production trends over the last five years are critically analysed below for three main soybean-producing states of the country.

3.1. Madhya Pradesh

In India, Madhya Pradesh is recognised as the "Soya state" because of its significant contribution in soybean production (70-80%). In Madhya Pradesh, area coverage, yield estimate, and production of soybean have shown a relatively decreasing trend during the last four seasons (1999-2003). The maxima in yield (945 kg/ha), production (4.18 x 10^6 tonnes) as well as of covered area (44.3 million ha) were all observed in 1998-99, whereas the minima in yield (855 kg/ha), production (3.28 x 10^6 tonnes), and covered area (38.3 million ha) all occurred in 2002-03 (SOPA Reports, 2002-03).

The sowing status of soybean for the year 2000-01 also revealed that, in the Mandsaur and Neemach districts, cotton and maize sowing was the alternative for soybean. Soybean cultivation at present occurs in the 45 districts of seven divisions of the state: Jabalpur, Sagar, Rewa, Indore, Ujjain, Gwalior, and Bhopal (Table 3).

3.2. Maharashtra

In the state of Maharashtra, area covered under soybean cultivation and production was rather variable during last five years (1998-2003). Data for the season 2002-03 indicated that area coverage and production were 1.22 x 10^6 ha and 1.215 x 10^6 tonnes, respectively, which were about 24% and 27.7%, respectively, higher in the previous season. Not much variability was observed in the yield of soybean over the last five years.

Table 2. All India State Wise Area Coverage and Yield Estimates of Soybeans during 1998-2003. Area (A) in 10^5 hectares; Productivity (Pdty) in kg/hectare; Production (Prod.) in 10^5 tonnes. After SOPA Reports (2000-2001; 2001-2002; and 2002-2003).

State (S No.)	1998-1999			1999-2000			2000-2001			2001-2002			2002-2003		
	A	Pdty	Prod	A	Pdty	Prod.	A	Pdty	Prod.	A	Pdty	Prod.	A	Pdty	Prod.
Madhya Pradesh (1)	44.3	945	41.8	38.8	916	35.53	40.3	926	37.30	42.4	920	38.93	38.4	855	32.79
Chhatisgarh (2)	-	-	-	-	-	-	-	-	-	0.54	560	0.30	-	-	-
Maharashtra (3)	10.9	958	10.4	11.3	940	10.62	10.2	980	10.02	9.82	970	9.52	12.2	998	12.15
Rajasthan (4)	6.17	870	5.37	4.84	738	3.57	6.24	800	4.950	6.54	740	4.84	4.17	712	2.967
Uttar Pradesh (5)	0.50	434	0.22	0.28	500	0.74	0.15	600	0.092	-	-	-	-	-	-
Karnataka (6)	0.75	555	0.42	0.40	500	0.20	0.65	535	0.350	-	-	-	-	-	-
Gujrat (7)	0.15	600	0.09	0.31	550	0.17	0.28	500	0.140	-	-	-	-	-	-
Andhra Pradesh (8)	0.22	1495	0.33	0.36	525	0.19	0.28	600	0.168	-	-	-	-	-	-
Other States (9)	0.63	540	0.34	0.15	530	0.08	0.02	525	0.010	0.75	550	0.41	2.06	-	1.798
Total/ Average*	63.5	928*	59.0	56.5	895*	50.50	58.1	912*	53.04	60.0	900*	54.00	56.8	876*	49.71

The sowing status in the state of Maharashtra, as provided in the SOPA Report (2000-01), shows that soybean cultivation is mainly in 29 districts of 7 divisions: Nagpur, Kolhapur, Amarawati, Nasik, Pune, Aurangabad and Latoor (Table 3).

Table 3. Major Soybean-Producing Divisions and Districts of Three States of India.
After the SOPA Report (2002-03).

Madhya Pradesh		*Maharashtra*		*Rajasthan*
Division	*Districts*	*Division*	*Districts*	*Districts*
Bhopal	Bhopal	Nagpur	Nagpur	Kota
	Sehore		Wardha	Bundi
	Raisen		Chandrapur	Baran
	Vidisha		Bhandara	Chittorgarh
	Betul		Gadchiroli	Udaipur
	Raigarh		Gondia	Bhilwara
	Hosangabad/Harda			Banswara
				Sawai
				Madhopur
				Jhalawar
Jabalpur	Jabalpur	Kolhapur	Kolhapur	
	Katni		Sangli	
	Balaghat		Satara	
	Chhindwara			
	Mandla			
	Dindori			
	Seoni			
	Narsinghpur			
Sagar	Sagar	Amrawati	Amrawati	
	Damoh		Yevatmal	
	Panna		Akola	
	Tikamgarh		Washim	
	Chattarpur		Buldhana	
Rewa	Rewa	Nasik	Nasik	
	Sidhi		Jalgaon	
	Satna		Dhulia	
	Shadol		Nandurwar	
	Umaria			
Indore	Indore	Pune	Pune	
	Dhar		Ahmednagar	
	Jhabua		Solapur	
	Khargone			
	Badwani			
	Khandwa			

Ujjain	Ujjain Mandsour Neemuch Ratlam Dewas Shajapur	Aurangabad	Aurangabad Jalna Bid
Gwalior	Gwalior Shivpuri Guna Datia Murena Sheopur kalan Bhind	Latoor	Latoor Osmanabad Parbhani Hingoli Nanded

3.3. Rajasthan

No soybean cultivation in Rajasthan was reported prior to 1952. The cultivation in Rajasthan was initiated in 1981-82 when 10^4 ha were planted (Agriculture in Rajasthan: Some facts, 2001). Both increases and decreases in the soybean cultivation area have occurred over the last five years. The area under cultivation in the years 1998-2003 varied as follows: 6.17 x 10^5 ha > 4.84 x 10^5 ha < 6.24 x 10^5 ha < 6.54 x 10^5 ha > 4.17 x 10^5 ha. Maximum yield (870 kg/ha) and production (5.4 x 10^5 tonnes) were recorded in 1998-99 and, in the subsequent years, a reduction in both yield and production was observed. The maximum decreases of 18.2% and 44.9% in the yield and production, respectively, were observed in 2002-03 as compared to 1998-99.

Soybean cultivation spreads over nine districts of Rajasthan: Kota, Bundi, Baran, Jhalawar, Chittor, Udaipur, Bhilwara, Banswara, and Sawai Madhopur. Of these, Kota and Bundi are the main soybean-producing districts of Rajasthan (Table 3; SOPA Report, 2000-01).

4. REGIONAL DISTRIBUTION OF SOYBEAN CULTIVARS

Under the soybean varietal development programme, intensive research has been conducted by various Universities and Research Institutes in India. These include: JNKVV, Jabalpur (Madhya Pradesh); National Research Center for Soybean (NRCS), Indore (Madhya Pradesh); GBP University of Agriculture and Technology, Pantnagar (Uttaranchal); Indian Council of Agricultural Research (ICAR), New Delhi; Indian Agricultural Research Institute (IARI), New Delhi; Punjab Agriculture University, Ludhiana (Punjab); Birsa Agriculture University, Ranchi (Bihar); Kalyani University, West Bengal; Gujarat Agriculture University, Gujarat, Punjab; Rao Agriculture University, Akola (Maharashtra); Banglore Agriculture University, Banglore (Karnataka); and Tamil Nadu Agriculture University, Coimbatore (Tamil Nadu). The yield, resistance towards pests and diseases, germinability, pod maturity time, improvement of germination, and lodging and shattering resistance were the major targets. A good germplasm collection is maintained and evaluated at NRCS (Indore) to ensure their proper utilization in different zones of the country. From time to time over the last decade, improved varieties of soybean have been released for the different agro-climatic zones of the country. The released varieties consist of those of exotic as well as indigenous origin.

From the point of view of agroclimatic conditions and varietal suitability, different regions of India are grouped into five soybean zones. Bhatnagar and Tiwari (1990), Ram (1996) and Bhatnagar (2002) have tabulated the varieties suitable for these five different soybean zones (Tables 4, 5 and 6).

Table 4. Soybean Varieties Suitable for Different Zones of India.
After Bhatnagar and Tiwari (1990).

S. No.	Zone	Area covered	Suitable Varieties
1	Northern Hill zone	Himachal Pradesh and hills of Uttar Pradesh	Bragg, NRC-2, PK-262, PK-308, PK-327, PK-416, Pusa-16, Pusa-20, Pusa-24, Shilajeet, Shivalik, VL Soya-1, VL Soya-2, VL Soya-21, VL Soya-47 and Hara Soya.
2	Northern Plain zone	Punjab, Haryana, Delhi, North eastern plains of Uttar Pradesh and Western Bihar	Alankar, Ankur, Bragg, , PK-262, PK-308. PK-327, PK-416, PK-564, PK-1024, Pusa-16, Pusa-24, Shilajeet, SL-4 and SL-96.
3	Central zone	Madhya Pradesh, Bhundelkhand region of Uttar Pradesh, Rajasthan, Gujarat and Northern Orissa	Bragg, Durga, Gaurav, Gujarat Soya-1, Gujarat Soya-2, JS 80-21, JS 75-46, JS 71-05, JS-335, MACS-13, MACS-57, MACS-58, Monetta, NRC-2, NRC-12, NRC-7, NRC-37, PK-472, PK 71-21, UPSM-19 and Alankar.
4	Southern zone	Karnataka, Tamil Nadu, Andhra Pradesh, Kerala and Maharashtra	Co-1, Hardee, KHSb-2, MACS-124, Monetta, PK-471, PK-1029, PK-472, Pusa-37, Pusa-40, MACS-450 and LSB-1.
5	Northern Eastern zone	Assam, West Bengal, Bihar, Meghalaya, Sikkim, Arunachal Pradesh, Nagaland and Tripura	JS 80-21, Birsa Soya-1, Bragg, PK-472, Pusa-16, Pusa-22, Pusa-24 and MACS-124.

Description of 1979 germplasms in a published catalogue.
Recommendation of RSC 2 and RSC 3 for cultivation due to their resistance towards stem borer pests.
Enlistment of general characteristic features like germinability, resistance towards leaf shattering and various diseases of the four cultivars Ahilya-1 (NRC-2), Ahilya-2 (NRC-12), Ahilya-3 (NRC-7), Pooja (MAUS-2) and their suitability for cultivation in Himachal Pradesh, Madhya Pradesh, Uttar Pradesh, Rajasthan, Maharashtra, Gujarat.
Recommendation of four rust resistant varieties namely PK-1024, PK-1029, JS-80-21 and Ankur for cultivation in disease affected area of Central India and Maharashtra, Karnataka, Tamil Nadu and Kerala.
Data on an increase of 35% in the national yield in 1997-98 after adopting improved technology by farmers.

Table 5. Soybean Varieties Suitable for Different Zones from 1996. After Ram (1996).

S. No	Zone	Area covered	Varieties recommended
1	Northern Hill zone	Himachal Pradesh and hills of Uttar Pradesh	Bragg, Lee, PK-262, PK-308, PK-327, PK-416, Pant Soybean-564, Pusa-16, Pusa-20, Shilajeet, Shivalik, VL Soybean-1 and VL Soybean-2
2	Northern Plain zone	Punjab, Haryana, Delhi, North Eastern plains of Uttar Pradesh and Western Bihar	Alankar, Ankur, Bragg, Clark-63, PK-262, PK-308. PK-327, PK-416, Pant Soybean-564, Pusa-16, Pusa-22, Pusa-24, Shilajeet, SL-4, and SL-96
3	Central zone	M.P, Bundelkhand region of Uttar Pradesh, Orissa, Rajasthan, Gujarat, Northern and western parts of Maharashtra.	Bragg, Clark-63, Durga, Gaurav, Gujarat Soybean-1, Gujarat Soybean-2, JS-2, JS-80-21, JS-75-46, JS-76-205, MACS-13, MACS-58, Monetta, PK-471, Pusa-16, Pusa-22 and T-49.
4	Southern zone	Karnataka, Tamil Nadu, Andhra Pradesh, Kerala and Southern parts of Maharashtra	Co-1, Davis, Hardee, Improved Palican, PHSb-2, Monetta, PK-471, Pusa-37 and Pusa-40
5	North-Eastern zone	Assam, West Bengal, Bihar and Meghalaya	Birsa Soybean-1, Pusa-16, Pusa-22, Pusa-24, Alankar and Bragg

In the year 2000, the Rajasthan Seed Corporation Ltd., Jaipur, determined the maturity period, yield, and other important characteristic features of three cultivars, namely JS-335, Ahilya (NRC-12), and Ahilya-1 (NRC-2), which were recommended for cultivation in Rajasthan. Culitvars, NRC-7, JS-335, JS 71-05, L129 and NRC 25, were susceptible to insects. NRC 33, which exhibited field resistance, was recommended as a source for insect resistance (Annual Report, DARE/ICAR, 2000-01). Shrivastava and Shrivastava (2001) have listed soybean varieties that are recommended for ten agro-climatic regions of Madhya Pradesh (Table 7) and fifteen other varieties have been described in terms of maturity period, yield, and other important characteristics, suitable for cultivation in Madhya Pradesh (Table 8). Similarly, Khare *et al.* (2000) tabulated soybean varieties, which are important for various agroclimatic zones of Madhya Pradesh (Table 9). Seven soybean varieties, along with their important agronomical characters for various areas of adaptation (Madhya Pradesh, Rajasthan, Uttar Pradesh and Andra Pradesh), were described in 2001-02 (see Table 10).

Khatri *et al.* (2002) have listed ten important pests along with the names of varieties showing resistance against specific pests that cause damage to soybeans (Table 11). Bhatnagar (2002) has tabulated 26 varieties of soybeans with their special characteristic features and duration of maturation (Table 12).

Table 6. Soybean Varieties Suitable for Different Agro-climatic Zones of India from 2002. After Bhatnagar (2002).

S.No.	Zone	Area covered	Suitable varieties
1	Northern hill zone	Himachal Pradesh and hills of Uttar Pradesh	Bragg, PK-262, PK-308, PK-327, PK-416, Pusa-16, Pusa-20, Pusa-24, Shilajeet, Shivalik, VL Soya-2 and NRC-2.
2	Northern plain zone	Punjab, Haryana, Delhi, Eastern plain zones of Uttar Pradesh and Western Bihar	Bragg, PK-262, PK-308, PK-327, PK-416, PK-564, Pusa-16, Pusa-24, Shilajeet, SL-2, SL-4, SL-96, PK-1024 and PK-1042.
3	Central zone	Madhya Pradesh, Bundelkhand region of Uttar Pradesh, Rajasthan and North Eastern Maharashtra	Bragg, JS 80-21, JS 75-46, JS 71-05, JS-335, Monetta, PK-472, Pusa-16, Pusa-24, Punjab-1, PK-416, NRC-2, NRC-12, NRC-7 and NRC-37.
4	Southern zone	Karnataka, Tamil Nadu, Andhra Pradesh, Kerala, and Southern region of Maharashtra	Co-1, Hardee, KHSB-2, KM-1, Monetta, PK-471, ADT-1, MACS-450, PK-472, MACS-58, PK-1029, MAUS-1, MAUS-2 and MACS-450.
5	North Eastern zone	Assam, West Bengal, Bihar, Meghalaya, Chhattisgarh and Orissa	Birsa Soybean-1 and Bragg in Bihar hills, Alankar, Bragg, PK-262, PK-472, Pusa-16, Pusa-24, and JS 80-21 in North-East region.

Table 7. Soybean Varieties Recommended for Different Agroclimatic Zones of Madhya Pradesh. After Shrivastava and Shrivastava (2001).

S. No.	Agroclimatic Zone	Varieties
1.	Plains of *Chhattisgarh	JS 72-280 (Durga), JS 80-21 and JS 335.
2.	Kamour plateau and satpura hills	JS 72-280, JS 72-44 (Gaurav), JS 75-46, JS 80-21, JS 335, JS 90-41, MACS 58 and PK 472.
3.	Plateau of Vindhya	Punjab-1, JS 72-44, JS 76-205, JS 80-21, PK-472.
4.	Central Narmada valley	JS 72-44, JS 72-280, JS 75-46, JS 76-205, JS 80-21, JS 90-41 and PK-472.
5.	Gird region	JS 80-21, JS-335.
6.	Bundelkhand region	JS 72-44, JS 72-280, JS 75-46, JS 76-205, JS 80-21, JS-335, JS 90-41 and PK-472.
7.	Satpura hills	Punjab-1, JS 75-46, JS 80-21, JS-335, JS 90-41, PK-472 and MACS-58.
8.	Malwa	JS 72-44, Punjab-1, JS 71-05, JS 76-205, JS 80-21, JS-335, JS 90-4, PK-472, Ahilya-1, Ahilya-2, Ahilya-3.
9.	Nimar valley	JS -335, JS 90-41.
10.	Jhabua hills	JS-335, JS 90-41.

* Chhattisgarh is separated from Madhya Pradesh as a new state.

Table 8. Developed Varieties for Soybean Cultivation in Madhya Pradesh.
After Shrivastava and Shrivastava (2001).

S. No.	Name of Variety	Maturation Period and Yield	Specific Properties
		(A). Early maturing	
1	Punjab–1	90-98 days, 1.8-2.0 t/ha	High seed germinability
2	JS-71-05	90-97 days, 2.0-2.4 t/ha	Resistant to pod dehiscence
3	JS-335	95-100 days, 2.5-3.0 t/ha	High seed germinability (80%) and high yield
4	JS 90-41	87-98 days, 2.5-3.0 t/ha	Four seeded pods
5	NRC-7 (Ahilya-3)	90-99 days, 2.5-3.5 t/ha	Resistant to various bacterial and viral diseases
6	NRC-12 (Ahilya-2)	96-99 days, 2.5-3.0 t/ha	Resistant to leaf blight and tolerant to YMV.
		(B). Medium maturing	
7	Gaurav (JS 72-44)	106 days, 2.0-2.5 t/ha	Resistant to bacterial pustule and pod blight.
8	Durga (JS 72-280)	102 days, 2.0-2.6 t/ha	High germinability (80%)
9	MACS-13	106-108 days, 2.0-2.6 t/ha	Resistant to YMV and high germinability (80%)
10	JS 75-46	106 days, 2.0-3.0 t/ha	Resistant to pod shattering.
11	JS 76-205	104 days, 2.0-2.5 t/ha	Highest germinability.
12	PK 472	104-106 days, 2.0-3.0 t/ha	Resistant to bacterial pustule, YMV and Rhizoctonia.
13	MACS-58	106-108 days, 2.2-2.6 t/ha	Resistant to YMV and leaf spot.
14	JS 80-81	106 days, 2.2-3.0 t/ha	High germinability and resistant to leaf eating insect.
15	NRC-2 (Ahilya-1)	103-106 days, 25-30 Qui/ha	Resistant to various fungal and bacterial diseases.

Table 9. Soybean Notified Varieties Recommended for Different Agroclimatic Zones of Madhya Pradesh.
After Khare et al. (2000).

S.No.	Agroclimatic Zone	Varieties
I.	Plains of Chhatisgarh	JS 72-280 (Durga), JS 80-21 and JS-335.
II.	Plateau of Baster	JS 80-21, JS–335
III.	Northern hills of Chhatisgarh	JS 80-21, JS–335
IV.	Kamour plateau and Satpura hills	JS 72-280, JS 72-44 (Gaurav), JS 75-46, JS 80-21, JS-335, JS 90-41, MACS 58 and PK-472.
V.	Plateau of Vindhya	Punjab-1, JS 75-46, JS 76-205, JS 80-21, PK-472.
VI.	Central Narmada valley	JS 72-280, JS 75-46, JS 76-205, JS 80-21, JS-335, JS 90-41 and PK- 472.
VII.	Gird region	JS 80-21, JS-335.
VIII.	Bundelkhand region	JS 72-280, JS 75-46, JS 76-205, JS 80-21, JS-335, JS 90-41 and PK-472.
IX.	Satpura hills	Punjab-1, JS 75-46, JS 80-21, JS-335, JS 90-41, PK-472 and MACS-58.
X.	Malwa	Punjab-1, JS71-05, JS 76-205, JS 80-21, JS-335, JS 90-4, PK- 472, Ahilya-1, Ahilya-2, Ahilya-3.
XI.	Nimar valley	JS-335, JS 90-41.
XII.	Jhabua hills	JS-335, JS 90-41.

Table 10. Soybean Varieties Notified in 2000-01.
After Annual Report, DARE/ICAR (2000-01).

Variety	Area of Adaptation	Yield Potential (t/ha)	Remarks
A. Central Releases			
NRCS-37 (Ahilya-4)	Madhya Pradesh, Rajasthan, and Uttar Pradesh	1.8-2.0	Resistant to pod shattering and good germin-ation, early maturity

MAUS-47 (Parbhani Sona)	Maharashtra, Madhya Pradesh, and Rajasthan	1.5-1.8	Culinary purpose
Himso-1563 (Hara Soya)	Maharashtra, Madhya Pradesh, and Rajasthan	1.5-2.0	Culinary purpose

B. State Releases

Indira Soya-9	Madhya Pradesh	1.5-1.8	Moderately resistant to rust, medium in maturity
MAUC-32	Maharashtra	1.8-2.0	Medium duration
Pant Soy 1092	Uttar Pradesh	_	Resistant to YMV, medium duration
LBS-1	Andhra Pradesh	1.2-1.5	Medium duration

Table 11. Insect-resistant Varieties of Soybean.
After Khatri et al. (2002).

S. No.	Common and Scientific Name of Insect	Resistant Varieties
1	Stem fly (*Melanagromyza sojae*)	TAS 1-3-5-2, MAQS-14, SL-459, MACS-617, MACS-716, DSB-I, MACC-124, MACS-534, JS (SH) 90-91, JS 90-35, MAUM-47, JS (SH) 91-33, MAUS 92-3, CO-1, DS 93-39-2, RSC-2, RSC-3, TAS-41, SACS-428, JS-335, JS (SH) 88-86, MACS-380 and MACS-396
2	Girdle beetle (*Obereopsis brevis*)	PK-1162, JS-335
3	Green Semi looper worm (*Chrysodeixis acuta*)	NRC-34, JS (SH) 89-48, MAUS-33, DSB-1, RAC-1, RAC-2, NRC-3, JS 90-9, UGM-52, NRC-12
4	Brown Semi looper worm (*Gisonia gema*)	DSB-1, RCS- 1 and NRC-12, NRC-7, RAC-3, NRC-34
5	Leaf borer insect (*Billonmbata sabsesivella*)	NRC-37, JS (HS) 89-48, MACS-124, MACS 92-47, SL-414, NRC-39, JS71-05, DS 93-35-2, UGM-52, JS-335, MAUS-610, MACS-716, TAS-41, MAUS-33, MAUS-31, JS 92-2, MACS-44

6	Leaf folding caterpillar (*Hedylepta indica)*	HRMSO-1564
7	Tobacco caterpillar (*Spodoptera litura.*)	MRMSO-1564, JS90-41, UGM-52, SL-459, DSK 93-204-101
8	Bean stink bug (*Challiopsis* sp.)	NRC-36, NRC-39, BLS-2
9	Mahu (*Aphis* sp.)	JS-335, JS-7105, JS-88-66, SL-79, CO-2, PK-1092
10	White fly (*Bemisia tabaci*)	PK-1241, JS-91-4, PK-1189, PK-1180, PK-1162, PK-1158, BRAG SL-317, SL-452, SL-443, SL-450, PK-416, SL-427, PKB-25, SL-457, PK-1042, PK-1135.

Table 12. Improved Varieties of Soybean in India.
After Bhatnagar (2002)

S. No.	Variety	Duration Day	Specific Features
1	Ahilya-1 (NRC-2)	103-106	Early maturity, White flower, Yellow seed, Good germination, Limited growth. Best for Malwa plateau and North hills region.
2	Ahilya-2 (NRC-12)	96-99	Purple flower, Yellow seed, Brown hilum, Resistant to Bacterial pustule, Bacterial blight, YMV, GMV, *Rhizoctonia* aerial blight. Best for Bhundelkhund region.
3	Ahilya-3 (NRC-7)	90-99	Limited growth, Purple flower, Yellow seed, Gray hair, Large seed, Early maturity, Resistant to pests and diseases. Best for Madhya Pradesh.
4	Ahilya-4 (NRC-37)	96-101	White flower, Brown hairs, Yellow seed, Light dark brown hilum, Erect growth, Collar rot, Bacterial pustule resistant. Resistant to blight disease of pods and buds, Resistant to stem fly and Leaf minor.

5	CO-2	75-80	Limited growth, Purple flower, deep leaves, yellow seed, Tolerant to YMV and Leafminor. Best for Tamil Nadu.
6	Durga (JS 72-208)	102-105	White flower, Yellow seed, Early maturity and improve variety, Good germination, Resistant to bacterial pustule. Best for Kharif in M.P. and for Rabi in Chhattisgarh.
7	JS 71-05	94-96	Early maturity, Purple flower, Brown hair, Yellow seed, Black hilum, Dwarf determinate 35-40 cms height, Large seeds. Best for Malwa plateau of Madhya Pradesh.
8	JS-335	98-102	Purple flower, Yellow seed, Good germination, Semi-determinate, Tolerance to pod dehiscence, Resistant to bacterial pustules, Blight or *Alternaria* blight. Tolerance to GMV, Stem fly, High yielding variety. Best for Bhundelkhund.
9	JS-80-21	95-109	Purple flower, Yellow seed, good germination, Tolerant to defoliators, Bacterial pustules. Best for central region, North East region and Chhattisgarh.
10	Sneha (KB-79)	85-93	Limited growth, Purple flower, gray hairs, Yellow seeds, Brown hilum, Tolerance to many diseases. Best for Karnataka.
11	MACKS - 13	90-102	Purple flower, Yellow seeds, Good germination, tolerance to bacterial pustules, Viral disease and defoliators. Best for central region of India.
12	MACKS-58	95-105	Purple flower, Yellow seed, Tall plant, Tolerant to Bacterial pustule and leaf spot, YMV. Best for central region.

13	MACKS-124	90-105	Purple flower, Semi-Determinate. Best for southern region and Chhattisgarh.
14	MACKS-450	90-95	Purple flower, Medium height, Semi-Determinate, Yellow seed, Black hilum, Resistant to YMV and different diseases, Moderate resistant to stem fly and Insect pest. Best for southern region.
15	Pooja (MAUS-2)	102-110	Limited growth, Gray hair, Yellow seed, Brown hilum, Resistant to dehiscence of pods. Best for Maharashtra and Southern region.
16	PK-262	120-125	White flower, gray hair, Yellow seed, Brown hilum, Tolerant to YMV and Bacterial pustules. Best for Northern and North Plain region.
17	PK-416	100-109	White flower, Yellow seed, Resistant to Bacterial pustules and YMV, Tolerant to *Rhizoctonia*. Best for Northern region and Madhya Pradesh.
18	PK-472	100-105	White flower, Yellow seed, Deep green plant, Resistant to YMV, Leaf spot and Bacterial diseases. Best for Madhya Pradesh.
19	PK-564	105-115	Whit flower, Yellow seed, Deep brown flower hilum, Limited growth, Strong plant, Resistant to YMV and bacterial pustules, Tolerance to *Rhizoctonia*. Best for Northern region.
20	PK-1024	115-118	White flower, Small leaves, Limited growth, Yellow seed, Brown hilum, Resistant to pod dehiscence, tolerant to rust disease. Best for Northern region.
21	PK-1029	90-95	White flower, Limited growth, Broad green leaves, Yellow seeds, Black hilum and soybean rust disease. Best for Southern region.
22	PK-1042	110-119	Limited growth, White flower, Yellow seed, and Brown hilum, Resistant to pods dehiscence. Best for Northern plain region.
23	PUSA-16	100-105 (Plains), 110-120 (Hills)	Purple flower, Yellow seed, Good germination, Resistant to Rhizoctonia, Bacterial pustules, YMV and Insect pests.

24	Shivalik	120-125	White flower, Light milky hair, Yellow seed, Deep brown hilum, Resistant to middle height (90-100 cms) Yellow mosaic. Best for Himachal Pradesh.
25	VL Soya -2	104-116	Purple flower, Light white hair, Black hilum, Light colour pods. Best for Northern hills.
26	VL Soya -21	120-122	Limited growth, White flower, Brown hair, Yellow seed, Brown hilum, Good germination, Resistant to Bacterial pustules and Cercospora leaf spot.

5. REGIONAL AGRICULTURAL ASPECTS OF SOBEAN CULTIVATION

Soybean, the main winter season crop of India, has wide adaptability to various agro-climatic conditions and can be fitted to all climatic systems (both rain-fed and irrigated). Production and productivity of soybeans sown in various agricultural regions of India depend basically upon rainfall, temperature, sowing, quality of soil, selection of suitable varieties for a specific agro-climatic region, control of diseases, fertilizer management, crop rotation, and intercropping.

The following sections summarize and critically analyse the important agricultural aspects of soybean cultivation in India.

5.1. Environmental Factors

During the soybean cultivation period (June-September), the situation with regards to the monsoon varies in different soybean-growing states of the country. Good rains occur up to the middle of July in most parts of the soybean belt. Rains may appear in the first week of August and may continue until the third week of August. Scattered rains also occur in September. Rainfall generally seizes after the second week of September, which is a very important period for pod filling and development. Most of the soybean cultivation in India depends upon rainfall. In the rain-fed areas, maximum cultivation is completed in two phases, *i.e.*, early sown phase (between the second and third week of June) and timely sown phase (third week of June to the second week of July). As an exception, sometimes early sowing has been reported in the second week of May, as was done in Maharashtra in 2001-02. Timely sowing of soybean, which is accomplished by the arrival of the monsoon at the right time, provides optimum soil moisture for seed germination. Early sowing of the crop invariably gives a good yield, as reported in Maharashtra in winter 2000-01 (SOPA Report, 2001-02).

The late arrival of the monsoon means both less rainfall and late sowing, which leads to the formation of undersized grains and less yield. Scattered rains in September can improve crop prospects (SOPA Report, 2000-01). Assessment of rainfall amounts for the recent years, 2000-2002, for the three main soybean-growing states of India, Madhya Pradesh, Maharashtra, and Rajasthan, indicates the rainfall range at Division/District headquarters as well as the average state rainfall (Table 13).

While discussing the rainfall status and requirement for the cultivation of soybean in different regions of Madhya Pradesh, Mehta and Shrivastava (2002) have pointed out that the cultivation is done in North-west Madhya Pradesh with the average rainfall of *ca.* 800 mm and in South-east Madhya Pradesh with the average rainfall of 600 mm. It is worthwhile to note that the amount of rainfall varies in each monsoon season and, as a consequence, all of normal, excess, deficient and scanty rainfall have been recorded in different years for different regions (SOPA Report, 2001-02).

Table 13. Rainfall Status in Madhya Pradesh and Maharashtrain 2002, and Rajasthan in 2000. After SOPA Report (2002-03) and Agriculture in Rajasthan: Some Facts (2001).

Madyapradesh		
Divisions	Rainfall Range (mm)	Average (mm)
Jabalpur	723-1456	942
Sagar	662-868	700
Rewa	435-1060	690
Indore	505-998	670
Ujjain	291-645	659
Gwalior	168-609	417
Bhopal	566-1318	844
State Avg.	479-993	703
Maharashtra		
Nagpur	782-1223	917
Kolhapur	250-1458	862
Amaravati	600-1293	894
Nasik	504-898	694
Pune	366-601	445
Aurangabad	572-758	639
Latoor	443-1106	744
State Avg.	502-1048	742

Rajasthan - *Rainfall from June 2000 to September2000*		
Districts	Av. Rain fall (Normal; mm)	Av. Rain fall (Actual; mm)
Kota	698	585
Bundi	726	529
Baran	822	814
Jhalawar	801	638
Chittor	792	486
Udaipur	596	391
Bhilwara	642	440
Banswara	903	545
Swai Madhopur	828	475
State Avg.	531	365

The optimum air temperature for seed germination is 30°C (20°C soil temperature) and temperatures above 40°C stop germination (Shrivastava *et al.*, 2002). A temperature range of 25°C-28°C is found to be suitable for soybean growth. Even an increase of 2°C from the threshold temperature (29°C) causes a reduction in yield (Burke and Evett, 1999).

5.2. Quality of the Cropping Field Soil

Cultivation of soybean is done in the well-drained, black loamy soil with a pH between 7.0-7.5 (Bhatnagar, 2002). Very light, sandy, acidic and alkaline soils are not suitable for this crop. In Madhya Pradesh, the soya state of the country, soybean is being cultivated in different soil types, such as shallow-black to dark-black soil (Malwa), medium-black (Vindhyan plateau), dark-black (Central Narmada valley), light-black loamy and silty loam (Satpura plateau), sandy loam to heavy-black (Kamour plateau), red-black mixed (Bhundelkhund), mixed red-black and alluvial (Gird region), and shallow-black (Nimar valley) (Shrivastava and Shrivastava, 2001; Mehta and Shrivastava, 2002). Seeds are sown in the soil at a depth of 2.5-3.0 cm, at a distance of 10-15 cm, and in rows maintaining 45 cm distance between two adjacent rows (Shrivastava *et al.*, 2002).

5.3. Insect Pests, Diseases, and their Control Measures

One of the reasons for the reduced soybean productivity is the increase in diseases caused by various organisms. Various researchers have provided an account of important diseases of soybean and their control measures.

Singh *et al.* (1988) detailed the insect pests of soybeans in India, and then Singh and Singh (1989) classified and described the soybean pests and their control measures. The names and symptoms of the important diseases caused by fungi (seed rot, leaf spot, blight, root and stem diseases), bacteria (soybean rust, bacterial spot), and viruses (YMV), along with their control measures, have been reported by Khare *et al.* (2000). Pests that can spoil the cultivation of soybean due to their attack on different plant parts at different stages of growth have been described (Khatri *et al.*, 2002; Shrivastava *et al.*, 2002; see Table 14). In addition to insect-pest diseases, Khatri *et al.* (2002) also pointed out six main diseases of soybean caused by viruses (Yellow Mosaic Virus), fungi (*Macrofomina* sp., *Cercospora* sp., *Colletotrichum slycence, Rhizoctonia solani, Focopsora pachyrhiza)* and bacteria *(Xanthomonas phaseoli* var. *sojans*). Shrivastava *et al.* (2002), while listing the breeding objectives for soybeans, have pointed out that, with the increase of soybean-cultivation area, diseases are also spreading. Furthermore, out of forty diseases reported to occur in different soybean-growing regions of the country, seven fungal, two viral, eighteen pest, and one bacterial disease have been highlighted by them as being of economic importance.

Biological and chemical treatments are recommended for the control of insects, fungal, bacterial, and viral diseases. The biopesticide (*Bacillus bassiana*) was found effective in controlling major diseases caused by insects (Annual Report,

DARE/ICAR, 1999-00). Khatri *et al.* (2002) have reported that the use of Nuclear Polyhedrosis Virus (NPV) reduces the number of tobacco caterpillars when sprayed with water.

Table 14. Major Insect Pests of Soybean in India.
After Khatri et al. (2002) and Shrivastava et al. (2002).

Common Name	Scientific Name	Damaged plant Parts & Symptoms
A. Seedling insects		
Blue beetle	*Cneorane* sp.	Cotyledonary leaves
Field cricket	*Gryllus* sp.	Seed and seedling
Seed maggot	*Delia platura*	Seedling root
Cut worms	*Agrotis ipsilon*	Cotyledonary leaves
		Seed and seedling
B. Stem borer pests		
Stem fly	*Melanagromyza sojae*	Yellow stem, stunted growth of leaves
Girdle beetle	*Obereopsis brevis*	Hollow stem, stem loss on maturation
C. Foliage feeder insects		
Green semi looper	*Chrysodeixis acuta*	Chlorophyll loss, low yield
Brown striped semi	*Mocis undata*	Cut leaf margins
looper	*Aproaerema modicella*	Reduced growth
Leaf minor caterpillar	*Spodoptera litura*	Cut leaf margins
Tobacco caterpillar	*Spodoptera exigua*	Stuck leaves, cause holes
Linseed caterpillar	*Spilosoma obliqua*	Chlorophyll loss, low yield
Bihar hairy caterpillar	*Amsacta moorei*	Reduced growth, leaf loss
Red hairy caterpillar	*Helicoverpa armigera*	Leaf deformation
Gram pod borer	*Scopula remotataguence*	60-80% leaf loss, cause holes
Looper larva	*Hedylepta indica*	Cut leaf margins
Leaf folding caterpillar		Leaf loss
D. Sap suckers		
White fly	*Bemisia tabaci*	Cause Yellow leaves
Green jassids	*Empoasca terminalis*	Shrunk pods
Thrips	*Caliothrips indicus*	Spread yellow mosaic viral disease
Green stink bug	*Nezara viridula*	Yellow leaves
Brown bug	*Riptortus pedestris*	Shrunk pods

From time to time, various chemicals have been recommended for the control of soybean diseases. For pest diseases, these include Forate, Quinolphos, Aldosulfon, Methylparathion, Thymathaxam 70 (WS), Chloropyrifos 20 EC, or Ethian 50 EC; for fungal diseases, they include Indofill M 45, Bloytox 50, Hexaconezol 5 EC, or Propeconezol 5 EC; and for viral diseases, Mancojeb or Thiram + Carbendazim have been recommended (Annual Report, DARE/ICAR, 1997-98; Khare *et al.*, 2000; Shrivastava and Shrivastava, 2001; Bhatnagar, 2002; Khatri *et al.*, 2002).

5.4. Soybean Weeds and their Control Measures

Weeds and their management during soybean cultivation is a major problem and a cause of productivity loss. Weed species infesting soybean vary according to the agro-climatic region. The main weed species associated with soybeans, as reported by various researchers, are *Cyperus rotundus, Echinochloa colonum, Digitaria sanguinalis, Phyllanthus maderaspatensis, Acalypha indica, Anotis montholoni, Dactyloctenium aegyptium, Commelina benghalensis, Saccharum spontaneum, and Phyllanthus niruri (*Lokras *et al.,* 1987*;* Sharma *et al.*, 1991; Singh *et al.*, 1991; Prabhakar *et al.*, 1992; Upadhyaya *et al.*, 1995; Jain *et al.*, 1998). Mehta and Shrivastava (2002) have found that weeds not only reduce production but also cause the spread of various insect pests.

Hand weeding has been adopted by farmers as the most effective control measure in the soybean-cultivated fields. Weed control is performed manually after a period of 20-25 days (Kurchania and Bhalla, 1999; Shrivastava and Shrivastava, 2001; Mehta and Shrivastava, 2002). Vyas *et al.* (1999) reported that the application of Alachlor as a pre-emergence measure resulted in better weed control and higher grain yield. Further, both Imazethapyr and Propaquizafop have shown promise for controlling weeds after their emergence (Annual Report, DARE/ICAR, 2000). Khare *et al.* (2000) and Shrivastava and Shrivastava (2001) have listed for soybean the herbicides to be applied before sowing (Fluchoralin, Metachlore), after sowing but before germination (Alachlore, Pendymethylin, Acetachlore, Metachlore), and on a standing crop (Imazethapyr).

5.5. Fertilizers

Cow dung and compost fertilizer are used for maintenance of agricultural land for soybean cultivation. Khare *et al.* (2000) pointed out that soil characteristics are maintained, if 10-15 tonnes/ha compost is applied every third year. In general, nitrogen (20 kg/ha), phosphate (60-80 kg/ha), and potash (20 kg/ha) is applied at the time of sowing (Shrivastava and Shrivastava, 2001; Mehta and Shrivatava, 2002; Bhatnagar, 2002). Shrivastava and Shrivastava (2001) suggested that leguminous crops require higher sulfur concentrations compared to cereal crops. Therefore, sulfur-containing fertilizers, like ammonium sulfate, super phosphate, and gypsum, are used for soybean cultivation. Mehta and Shrivastava (2002) have noted that the addition of zinc sulfate to zinc-deficient soil and gypsum to sulfur-deficient soil is necessary to maintain soybean productivity.

5.6. Intercropping and Crop Rotation

By adopting intercropping in the farming of soybeans, higher production can be achieved as compared to other crops. Patra and Chatterjee (1986), Prasad and Shrivatava (1991), and Jagtap *et al.* (1993) reported higher yields and return under soybean-based intercropping systems than with soybean alone. Crop rotation also helps in weed control (Kurchania and Bhalla, 1999). Soybean intercropping provides safe production under unfavorable climatic conditions (Mehta and Shrivastava, 2002). Important intercropping combinations that are beneficial for different soybean-growing regions of Madhya Pradesh have been recommended by Mehta and Shrivastava (2002), with the Soybean-plus-Arhar (*Cajanus cajan*) intercropping system found to be most beneficial in terms of production.

Either soybean-plus-wheat or soybean-plus-chickpea are the major crop rotations, although other crops are also planted after soybeans (Shrivastava *et al.,* 2002; see Table 15). In different agro-climatic regions, sowing of wheat or chickpea or pea or sunflower after soybean cultivation can result in much higher production, but the benefit depends on irrigation facilities (Mehta and Shrivastava, 2002). Although Chui and Shibles (1884) noted reduced yields of soybeans after intercropping with maize, Shrivastava *et al.* (2002) indicated that intercropping of soybean with maize (*Zea mays*), as well as with pigeonpea (*Cajanus cajan*), jowar (*Sorghum vulgare*), fingermillet (*Elusine corocana*), or sugarcane (*Saccharum officinarum*), is beneficial. Goswami (2000) also reported that the soybean-plus-maize intercropping system is productive.

Table 15. Major Intercropping Systems involving Soybean in India.

Region	Intercropping System
Malwa plateau	Soybean-plus-Maize (*Zea mays*) (4:2) or Soybean-plus-Arhar (*Cajanus cajan*) (4:2)
Kamour plateau	Soybean-plus-Arhar (*Cajanus cajan*) (4:2)
Vindhyan Plateau	Soybean-plus-Jowar (*Sorghum vulgare*)(4:2)

ACKNOWLEDGEMENTS

The author thanks the European Union for support of the INCO DEV project "Soybean BNF Mycorrhization for Production in South Asia", ICA4-CT-2001-10057.

REFERENCES

Agricultural statistics at a glance (2001). New Delhi: Department of Agricultural Statistics, Government of India.

Agriculture in Rajasthan: Some facts (2001). Pant Krishi Bhawan, Jaipur, India: Statistical Cell Directorate of Agriculture.

Annual Report (1997 – 1998). (pp. 57-59). New Delhi, India: DARE Ministry of Agriculture, Government of India, ICAR.

Annual Report (1999 – 2000). (pp. 27-28). New Delhi, India: DARE Ministry of Agriculture, Government of India, ICAR.

Annual Report (2000 – 2001). (pp. 30-31). New Delhi, India: DARE Ministry of Agriculture, Government of India, ICAR.

Anonymous (1956). *The wealth of India*, Vol, IV. F-G. *Raw materials* (pp. 142-150). New Delhi, India: Council for Scientific and Industrial Research.

Bhatnagar, P. S. (2002). How to earn more profit in soybean cultivation? Few easy methods and improved varieties of soybean Prasar. The Soybean Processors Association of India (SOPA), Indore, India. *Bulletin 1*, 1–22.
Bhatnagar, P. S., and Tiwari, S. P. (1990). Soybean varieties of India. *NRCS Technical Bulletin 2*, 1-11.
Burke, J. J., and Evett, S. R. (1999). Identification of the temperature threshold for soybean irrigation. In H. E. Kauffman (Ed.), *Proceedings of the World Soybean Research Conference VI*, (pp. 603). Chicago, IL: University of Illinois.
Chui, J., and Shibles, R. (1984). Influence of spatial arrangement of maize on performance of an associated soybean intercrop. *Field Crop Res., 8*, 187–198.
Goswami, D. (2000). *Effect of crop geometry and nutrient management on soybean + maize intercropping system*. Jawaharlal Nehru Krishi Vishna Vidhyalaya, Jabalpur, India. M.Sc. (Ag) Thesis.
Hooker, J. D. (1879). *Flora of British India*, Vol. II. London, UK: L. Reeve and Co.
Hymowitz, T., and Kaizuma, N. (1981). Soybean seed protein electrophoresis profiles from 15 Asian countries or regions: Hypothesis on path of dissemination of soybean from China. *Econ. Bot., 35*, 10–23.
Jagtap, J. G., Gupta, R. K., and Holkar, S. (1993). Evaluation of soybean and pigeonpea varieties for advantage in intercropping on the basis of stability analysis. *Indian J. Agric. Sci., 63*, 327-332.
Jain, K. K. Parorkar, N. R., and Agrawal, K. K. (1998). Major weed flora of soybean field at Powarkheda (MP). *Soybean Abstracts, 21*, 166.
Kale, F. S. (1936). *Soybean, its value in dietetics, cultivation and use*. Baroda, India: Baroda State Press, and New Delhi, India: International Book and Periodicals Supply Service.
Kaltenbach, D., and Legros, J. (1936). Soya selection, classification of varieties, varieties cultivated in various countries. Institutes of Agriculture (Rome, Italy). *Monthly Bulletin of Science and Practical Agriculture 27*, 284.
Khare, D., Rawat, N. D., Bhaley, M. S., and Shrivatava, A. N. (2000). *Soyabean seed production technology* (pp. 1-22). Jawaharlal Nehru Krishi Vishna Vidhyalaya, Jabalpur, India: Seed Technology Unit Transmmision Center.
*Khatri, A. K., Shukla, B. N., and Sharma, A. (2002). *Soybean Phasal mein samanavit keet avum rog prabhandhan* (Integrated pest and disease managements in soybean crops), (pp. 1-22). Jawaharlal Nehru Krishi Vishna Vidhyalaya, Jabalpur, India: Seed Technology Unit Transmmision Center.
*Kurchania, S. P., and Bhalla, C. S. (1999). *Kharif phasal mein neenda niyantran* (Weed control in Kharif crops). Jawaharlal Nehru Krishi Vishna Vidhyalaya, Jabalpur, India: Department of Agronomy, Technical Bulletin 1.
Lokras, V. G., Singh, V. K., and Tiwari, J. P. (1987). Chemical weed control in soybean (*Glycine max* (L.) Merrill). *Indian J. Weed Sci., 17*, 45-48.
*Mehta, S. K., and Shrivastava, V. C. (2002). *Soybean ki unnat kasht* (Improved cropping of soybean). (pp. 1-16). Jawaharlal Nehru Krishi Vishna Vidhyalaya, Jabalpur, India: Department of Plant Breeding and Genetics.
Paroda, R. S. (1999). Status of soybean research and development in India. In H. E. Kauffman (Ed.), *Proceedings of the World Soybean Research Conference VI*, (pp. 13-23). Chicago, IL: University of Illinois.
Patra, A. P., and Chatterjee, B. N. (1986). Intercropping of soybean with rice, maize and pigeonpea in the plains of West Bengal. *Indian J. Agric. Sci., 56*, 413-417.
Prabhakar, Tiwari, S. P., and Bhatnagar, P. S. (1992). Production technology for augmenting soybean productivity in India. *J. Oilseeds Res., 9*, 1-13.
Prasad, K., and Shrivastava, V. C. (1991). Pigeonpea (*Cajanus cajan*) and soybean (*Glycine max*) under intercropping system under rainfed situation. *Indian J. Agric. Sci., 61*, 243-246.
Ram, H. H. (1996) *Soybean breeding*. In J. Nizam, S. A. Farook, and I. A. Khan (Eds.), *Genetic improvement of oilseed crops*, (pp. 27-42). Hyderabad, India: Ukaaz Publications.
Saxena, M. C., Pandey, R. K., and Hymowitz, T. (1971). Agronomic requirement of soybean (*Glycine max* (L.)Merr). *Indian J. Agric. Sci., 41*, 339-344.
Saxena, M. C., and Pandey, R. K. (1971). Characterstics and performance of some promising varieties of soybean at Pantnagar. *Indian J. Agric. Sci., 41*, 355-360
Sharma, R. K., Bangar, K. S., Kanere, G., Singh, O. P., Thakur, G. L., and Sharma, S. R. (1991). Effect of weed control on yield of soybean (*Glycine max* (L.) Merrill). *Indian J. Agron., 37*, 372-373.

*Shrivastava, A. N., and Shrivastava, M. K. (2001). Soybean ki unnat kheti (Improved farming of soybean). Transmission Center, Jawaharlal Nehru Krishi Vishna Vidhyalaya, Jabalpur, India. *Krishi Vishva Kharif Special, 29*, 52-63.

Shrivastava, A. N., Singh, C. B., and Shrivastava, M. K. (2002). Soybean. In C. B. Singh and D. Khare (Eds.), *Genetic improvement of field crops*, (pp. 136-157). Jodhpur, India: Scientific Publishers.

*Singh, A., Pandey, S., and Sharma, Y. K. (2001). *Swasthayavardhak soya aahar* (Health improving soya food). Jawaharlal Nehru Krishi Vishna Vidhyalaya, Jabalpur, India: Department of Food Science and Technology Transmission Center.

Singh, B. B., and Saxena, M. C. (1975). Soybean varieties for different agro-climatic zones in India. *Seed Technology News, 5*, 5-7.

Singh, O. P., and Singh, K. G. (1989). Insect pest control in soybean. In O. P. Singh and S. K. Shirivastava (Eds.), *Soybean*, (pp.113-132). Bikaner, India: Agro Botanical Publishers India.

Singh, O. P., Verma, S. N., and Nema, K. K. (1988). *Insect pests of soybean*. Dehradun, India: India International Book Distributors and Publishers.

Singh, V. K., Bajpai, R. P., Mishra, R. K., and Purohit, K. K. (1991). Chemical weed control in rainfed soybean (*Glycine max* (L.) Merrill). *Indian J. Agron., 36*, 292-294.

SOPA Report (2000-01). *A report on soybean crop estimates*, (pp. 1-8). Indore, India: The Soybean Processor Association of India.

SOPA Report (2001-02). *A report on soybean crop estimates*, (pp. 1-6). Indore, India: The Soybean Processor Association of India.

SOPA Report (2002-03). *A report on soybean crop estimates*, (pp. 1-4). Indore, India: The Soybean Processor Association of India.

Tiwari, S. P., Bhatnagar, P. S., and Prabhakar (1994). Identification of suitable soybean (*Glycine max*) varieties for non-traditional regions of India. *Indian J. Agric. Sci., 64*, 872-874.

Tiwari, S. P., Joshi, O. P., and Sharma, A. N. (1999). *The advent and renaissance of soybean in India - The saga of success*, (pp. 1-58). Indore, India: National Research Council for Soybean (ICAR).

Upadhayay, V. B., Tiwari, J. P., and Kosta, L. D. (1995). Influence of weed control methods on soybean (*Glycine max* (L.) Merrill) yield. *JNKVV Research Journal,* 27, 1-4.

Vyas, M. D., Singh, S., and Singh, P. P. (1999). Efficacy of the herbicide alachlor in soybean. *Soybean Abstracts, 22*, 89.

Williams, R. (1932). Flora of India or Description of Indian plants, 2nd Ed., Vol. II, (pp. 134-315). Calcutta, India.

* Original in Hindi

Chapter 5

SOYBEAN CULTIVATION AND BNF IN CHINA

J. E. RUIZ SAINZ[1], J. C. ZHOU[2], D.-N. RODRIGUEZ-NAVARRO[3], J. M. VINARDELL[1] and J. E. THOMAS-OATES[4]

[1]*Department of Microbiology, Faculty of Biology, University of Sevilla, Apdo-1095, 41080-Sevilla, Spain.*

[2]*Department of Microbiology, Huazhong Agricultural University, Shi Zi Shan Street, P.O. Box 430070, Wuhan, People's Republic of China.*

[3]*Centro de Investigación y Formación Agraria "Las Torres y Tomejil", Apartado Oficial 41200-Alcalá del Río, Sevilla, Spain.*

[4]*Department of Chemistry, University of York, Heslington, York, YO10 5DD, U.K.*

1. SUMMARY

China is the geographical origin of cultured soybeans and is where they have been cultivated for more than 5,000 years. Soybean is grown in nearly all provinces of China, the exceptions being Qinghai and the Tibetan Plateau. Based on climatic differences, China can be divided into five main geographical regions as regards soybean cultivation. As a complementary classification, Chinese soybean varieties have also been classified according to the sowing date; they were divided into winter-, spring-, summer-, and autumn-sowing varieties. In spite of the soybean acreage in China (9.4 million ha in 2001), which ranks fourth in the world, China has also become the largest soybean importer in the world. Although China has greatly improved soybean productivity in the last 20 years, its total production (or average seed yield) is still clearly lower than that reached in western countries. The symbiotic interaction, which soybean forms with different rhizobia, is a key factor in increasing soybean productivity in China within the context of agricultural sustainability. Research efforts in China have mainly concentrated on the identification and characterisation of biological material, both soybean accessions and rhizobial strains that effectively nodulate soybeans. As a result of this activity, a large collection of soybean germplasm is available and the existence of a broad spectrum of bacteria that nodulate soybeans (which we will collectively call "soybean-rhizobia") has been demonstrated. However, although soybean

D. Werner and W. E. Newton (eds.), Nitrogen Fixation in Agriculture, Forestry, Ecology, and the Environment, 67-87.

inoculation is increasing as a soybean-cropping practice in China, the available scientific information on the results obtained in field experiments is negligible. Consequently, this review will be mainly focused on a description of: (i) the Chinese geographical areas devoted to soybean cultivation; (ii) the diversity of soybean-rhizobia and their populations in Chinese soils; and (iii) the collection of soybean germplasm available in China. Information about crop rotation (soybean with non-legumes) and the undesirable effects that frequently accompany continuous soybean cropping is also provided.

2. SOYBEAN CULTIVATION IN CHINA: HISTORICAL ASPECTS AND CURRENT SITUATION

2.1. Historical records and world-wide dissemination of soybean cultivation

China is the geographical origin of cultured soybeans and the use of this crop in agricultural practices is more than 5,000 years old. The first written record is found in the book, "Pen Ts'ao Kong Mu", in which the plants of China are described (2838 B.C.). Further, in similar ancient writings, Zou Zhuan said that "one of Zhou Zi's brothers was so foolish that he could not tell Shu (an ancient name for soybean) from wheat". Soybean was also one of the five sacred grains (or "Wu Ku"), that include rice, soybean, wheat, barley, and millet. In Hubei province, soybean ashes have been found in a Han tomb (second century B.C.). Although it is clear that the soybeans used in agriculture originated from wild soybeans, it is not clear in which location soybean was cultivated for the first time. It may be that the Yellow River valley was the first place ever at which soybean was cultivated.

Although soybean is now grown in many countries, its expansion outside Asia only took place during the last three centuries. 2500 years ago, soybean cultivation was exported to what is now Korea and, 500 years later, to the islands we now call Japan. By the 7th century A.D., soybeans were grown in all southern countries neighbouring China. In 1712, the German botanist Kaempfer brought soybeans to Europe. Soybean was exhibited in the Botanical Garden of Paris in 1739 and in the British Royal Botanic Gardens in 1790. In 1804, the United States of America started to grow soybeans as a forage crop rather than for harvesting for seed consumption. Soybean cropping was initiated in other countries in Europe (1875-1877) as a result of its successful exhibition at an International Exposition in Vienna in 1873. Although soybean was not cropped in Brazil until the 1960s, this country is at present the second largest soybean producer after the United States of America. Soybean cropping was spread through Africa during the twentieth century.

2.2. Current soybean production in China

At the present time, China ranks third internationally in soybean acreage and production. In spite of this position, however, the increasing domestic demand for soybean forced China to start soybean importation in 1995. Since then, Chinese imports of soybean have increased dramatically, with China at present the largest soybean importer in the world. China represents the largest market for U.S.A. soybean exports, which were valued at more than $1 billion in 2001. The other main sources of soybean imports into China are Argentina and Brazil.

Soybean is grown in nearly all provinces of China with the exception of Qinghai and the Tibetan Plateau (see Figure 1 and Table 1). The number of soybean varieties that are available in China is very large; the Chinese Soybean Germplasm collection contains more than 23,000 soybean accessions. This large soybean diversity is the result of cultivation over such a large territory, with wide variations in topography and climate, and a long history of soybean cultivation. Chinese soybean varieties have been divided according to the season of the year in which they are sown. Most of the soybean varieties fall into one of the main three categories: spring-, summer- or autumn-sowing soybeans. There are some winter varieties that can be used in the south of China (in Hainan, Yunnan, Guangxi, and Guangdong provinces).

Figure 1. The Provinces of China.

Table 1. Weather Conditions, Soybean Acreage, and Soybean Production in the Different Regions of China.

Province	Annual rainfall (mm)	Total annual sunshine (h)	Annual average temperature (ºC)	Soybean acreage (ha x 10^3)	Total soybean production (tons x 10^4)	Average yield (kg/ha)
Shanghai	1000	1531	15-16	6.2	1.9	3065
Xinjiang	150	2550-3550	North: -4/9 South: 7-14	62.8	17.5	2787
Jiangsu	780-1170	1980-2640	-----	249.2	67.0	2689
Shandong	550-850	2335-2768	12-14	458.2	104.6	2283
Jilin	400-950	1000-2900	-----	539.0	120.3	2232
Sichuan	400-1300	1100-2700	-1/20	169.6	37.4	2205
Zhejiang	900-1400	1800-2300	15.2-18.2	129.0	28.4	2201
Beijing	550-660	2084-2873	4-12	22.1	4.7	2127
Hunan	1200-1700	1300-1800	16-18	205.8	42.8	2080
Henan	600-1200	2000-2400	12-16	564.7	115.8	2051
Hubei	750-1500	1600-2200	13-18	224.8	45.8	2037
Fujian	1000-2000	1670-2405	14.5-21.3	105.4	20.5	1945
Guang-dong	1350-2600	1450-2300	18-24	96.9	18.7	1930
Jiangxi	1200-1900	1473-2077	16-20	152.5	25.9	1689
Liaoning	485-1130	2285-2870	4.8-10.4	301.9	48.1	1593
Heilong-jiang	400-600	2400-2800	0-4	2868.3	450.1	1569
Gansu	35-810	1700-3300	0-15	88.2	13.4	1519
Hebei	400-800	2500-3100	4-13	423.7	62.9	1485
Yunnan	600-2300	910-2800	4.9-23.7	52.0	7.7	1481
Anhui	770-1700	1730-2500	14-16	682.2	91.5	1341

Shanxi	350-700	-----	3-14	272.5	36.0	1321
Guangxi	1100-2800	1500-1800	16-23	281.4	36.4	1294
Guizhou	1100-1300	1200-1600	14-18	141.0	18.1	1284
Tianjin	900	2204	12	34.6	3.9	1127
Inner Mongolia	50-450	-----	-1/10	793.9	85.8	1081
Shaanxi	400-900	1288-2885	7-16	246.9	22.2	899
Ningxia	178-680	2195-3082	5-9	43.6	2.9	665
Tibet	North: <200 South: 2000	1600-3400	3-12	0.5	0.3	600

Data on weather conditions are those published by The Encyclopaedia of Chinese Agriculture, 1998, China Agricultural Press, Beijing, China (in Chinese).
Data on soybean acreage and production are those corresponding to the year 2000 and being published by China Agriculture Year Book (2001), China Agricultural Press, Beijing, China (in Chinese).

The main soybean-cropping region, in terms of acreage, is located in the Northeast of China and includes Heilongjiang, Jilin, and Liaoning provinces and the Inner Mongolia Autonomous Region (see Figure 1 and Table 1). The total soybean acreage of this area in the year 2000 was 4.5 million ha, which is about 50% of the total soybean acreage in China. The second most important soybean-cropping region is the valleys of the Yellow and Huaihe rivers, including parts of Henan, Hebei, Shandong, and the north of Jiangsu provinces. Soybean cropping in China has been divided into five main geographical regions according to the climatic conditions that govern the sowing date. Table 2 summarises the main characteristics of the Chinese soybean-cultivating regions.

Table 2. Chinese Soybean-cultivating Regions.

Soybean regions[1]	Sub-regions and/or Provinces[1]	Characteristics[1]
I. Spring soybean in North China	North East spring soybean subregion: Heilongjiang, Jilin and Liaoning provinces and East Inner-Mongolia	Non-frost period: 100-170 days. Activity accumulated temperature (>10 °C): 1900-4000 °C. Annual rainfall: 350-1200 mm. Soybean-cultivation period: April-September.

	Loess plateau spring soybean subregion: North Yangtze River, North Shanxi and Shaanxi provinces, Inner-Mongolia Plateau, irrigating area of the Great Bend of Huanghe River and Ningxia province	Non-frost period: 260-280 days. Activity accumulated temperature (>10 °C): 3000-3500 °C. Annual rainfall: 200-250 mm. Soybean-cultivation period: April-September.
	Northwest spring soybean sub-region: Xinjiang and scattered zones in Gansu corridor	Non-frost period: 110-200 days. Activity accumulated temperature (>10 °C): 1500-3000 °C. Annual rainfall: 300-500 mm. Soybean-cultivation period: April-September.
II. Huanghe-Huaihe-Haihe River basin	Middle Hebei and Shanxi subregion: parts of Hebei province south to the Great Wall but north to Shijiazhuang and Tianjin cities, and middle and Southeast parts of Shanxi province.	Non-frost period: 175-220 days. Activity accumulated temperature (>10 °C): 3800-4300 °C. Annual rainfall: 400-800 mm. Soybean-cultivation period: June-September.
	Huanghe-Huaihe-Haihe River basin subregion: the part of Hebei province south to Shijiazhuang and Tianjin City, Shandong province, most parts of Henan province, Hongze district of Jiangsu province, those parts of Anhui province north to Huaihe River, the south western parts and Guangzhong district of Shanxi province and Tianshui district of Gansu province.	Non-frost period: 180-220 days. Activity accumulated temperature (>10 °C): 4000-4800. Annual rainfall: 500-1000 mm. Soybean-cultivation period: June-September.
III. Spring and summer regions of Yangtze River basin	Yangtze basin spring and summer sub-region: Yangtze River basin in Jiangsu, Anhui and Sichuan provinces, Hubei province, southern parts of Henan and Shaanxi province, northern parts of Zhejiang province and Sichuan basin.	Non-frost period: 210-310 days. Activity accumulated temperature (>10 °C): 4500-5500 °C. Annual rainfall: 1000-1500 mm. Soybean-cultivation period: April-July or May-September.

	Yunnan plateau spring soybean subregion: parts of Hunan, Sichuan and Guangxi provinces	Non-frost period: All year. Activity accumulated temperature (>10 ºC): 5000-6000 ºC. Annual rainfall: 750-1000 mm. Soybean-cultivation period: April-July or May-September.
IV. Southeast spring and autumn soybean region	Southern parts of Zhejiang province and most parts of Fujian, Taiwan, Hunan, Guangdong, and Guangxi provinces	Non-frost period: 270-320 days. Activity accumulated temperature (>10 ºC): 5500-7500 ºC. Annual rainfall: 1000-2000 mm. Soybean cultivation period: April-July or July-November
V. South China four-season soybean region	Southern edge of Guangdong, Yunnan and Guangxi provinces, some parts of Fujian, and Hainan province.	Non-frost period: All year. Activity accumulated temperature (>10 ºC): 7500-9000 ºC. Annual rainfall: 1300-2000 mm. Spring soybean: February-June. Summer soybean: June-August. Autumn soybean: July-September. Winter soybean: December-April.

[1]*The information in this Table is gathered from the book "Breeding and Cultivation of Soybean in China" (1987, edited by the Academy of Agricultural Sciences of Jilin province, China) and also from The Atlas of Chinese Climate Resource (1994, China Meteorological Administration, China Map Press, Beijing China).*

Soybean acreage and productivity in China has been increased in the last 20 years. In 1981, soybean acreage was 8 million ha, giving a total soybean production of 9.3 million tons. These figures demonstrate that Chinese soybean productivity was low because the national average yield was only 1163 kg/ha. Results presented in Table 3 clearly show that China has made great efforts in increasing the productivity of soybean cultivation. In comparison to 1981, in 2000, soybean acreage was 9.3 million ha that gave a total production of 15 million tons. This increase in total soybean production is mainly attributable to higher yields (1656 kg/ha) rather than to an increase in soybean acreage. In spite of the improvements in farming practices in combination with improved soybean cultivars, soybean productivity in China is still clearly lower that reached in Western countries (information on world-wide soybean production they can be obtained from different web address, such as http://www.fas.usda.gov/psd/complete tables/OIL-table2-24.htm). In 2001, soybean acreage in the U.S.A. was *ca*. 29.5 million ha, which gave a total production of 78.6 million tons (an average seed yield of 2660 kg/ha).

In the same year, the soybean acreage in China was about 9.4 million ha, which rendered a total production of 15.4 million tons (an average seed yield of 1630 k/ha). In that same year, the average seed yield in the other two main soybean-producing countries (Argentina and Brazil) was 2120 and 2660 kg/ha, respectively. These results clearly show that the main goal for increasing soybean production in China is to increase seed yield.

Table 3. Soybean Production in China in the last Two Decades. This table was constructed by using the information published in China Agriculture Yearbook, 2001, China Agricultural Press, Beijing, China (in Chinese).

Years	Soybean acreage (ha x 10^3)	Total soybean production (tons x 10^4)	Average yield (kg/ha)
1981-1985	7.801	966	1.238
1986-1990	8.090 (3%)	1.134 (17%)	1.402 (13%)
1991-1995	8.212 (5%)	1.302 (34%)	1.585 (28%)
1996-2000	8.317 (6%)	1.455 (50%)	1.749 (41%)

Numbers in brackets refer to increases over the period 1981-1985.

Table 1 also shows the average seed yield of each Chinese province. Shanghai and Xinjiang provinces show the highest yields, but the soybean acreage in these two provinces is very small, so that it does not represent the general situation in China. Jiangsu province is the only one that, having a large soybean acreage (249,000 ha), produces a seed yield that is in line of those obtained by the U.S.A. and Brazil. Most of the other provinces showing seed yields over 2000 kg/ha are located either on the east coast of China (Jilin, Shandong, Jiangsu and Zhejiang) or on a North-South axis in Central China (Henan, Hubei and Hunan), with Sichuan province (Central-West China) being the exception. Neighbouring some of the most productive provinces, there are others (such as Liaoning or Anhui) that have a large soybean acreage but their productivity is clearly much lower. The province with the largest soybean acreage, Heilongjiang, is situated in the North of China and although its seed yield (1569 kg/ha) is clearly lower than that in the above mentioned provinces, it is comparable with that obtained in other cold regions, such as Canada (1520 kg/ha in 2001).

It is clear that the main constrains for increasing soybean productivity in China are of a different nature (such as weather conditions, soil fertility, water availability, farming technology, *etc.*) in the different soybean-cropping areas. Increasing soybean productivity through symbiotic nitrogen fixation is a clear option. However to our knowledge, although soybean inoculation is gaining popularity among Chinese farmers, very little scientific information about the symbiotic performance of soybean inoculants under field conditions is available. Due to this lack of information, we focus our attention on the biological resources (soybean-rhizobia and the soybean germplasm collection) that exist in China and on a description of the soybean-rhizobia populations that are indigenous to soybean-cropping areas.

3. NITROGEN-FIXING BACTERIA THAT NODULATE SOYBEAN

3.1 Diversity of soybean-nodulating rhizobia in China

Because China is the geographical origin of soybeans and they have been cropped in China for at least the last five millenia, it is reasonable to expect that China is the appropriate place for carrying out screenings aimed at the isolation of new soybean-nodulating rhizobia strains. This idea is based on the hypothesis that co-evolution between the two symbionts (the plant and the bacteria) is likely to have taken place and, consequently, that it would be possible to isolate new soybean-nodulating rhizobia species and also bacterial strains showing improved symbiotic performance with soybeans. Taxonomic studies carried out in the last 20 years have shown that bacteria belonging to different genera and species are able to nodulate soybeans. Until 1982, soybeans were believed to be nodulated only by slow-growing bacteria belonging to the species, *Bradyrhizobium japonicum*. Since then, however, new groups of soybean rhizobia have been isolated and classified. Soybean-nodulating bacteria are distributed among different species belonging to three different genera: (i) *Bradyrhizobium japonicum, B. elkanii* and *B. liaoningense*, which are slow-growing bacteria with generation times longer than six hours (van Berkum *et al.*, 1998); (ii) *Mesorhizobium tianshanense*, with a variable generation time (Tan *et al.*, 1997); and (iii) *Sinorhizobium fredii* and *S. xinjiangense*, which are fast-growers with generation times of 1.5 and 4 hours, respectively (Chen *et al.*, 1988).

The slow-growing strains belonging to the genus *Bradyrhizobium* produce alkali in a medium containing mannitol. Those *Bradyrhizobium* strains able to nodulate soybeans have been classified into 17 different serogroups and also into three different DNA homology groups (for a review *Rhizobiaceae* taxonomy, see van Berkum and Eardly, 1998 and volume 3 of this iseries). DNA-homology groups I and Ia appear to be closely related and separated from DNA-homology group II. These differences, as well as differences in other bacterial traits, such as EPS composition, fatty acids, or intrinsic antibiotic-resistance patterns, led to the segregation of those strains belonging to DNA-homology group II into the new species called *B. elkanii*. In these studies, slow-growers isolated from Asia were underrepresented.

Strains belonging to *Bradyrhizobium liaoningense* were isolated from several different Chinese provinces, including Liaoning, Heilonjinag, or Hubei (Xu et al. 1995). Their doubling time is so long (ranging from 19 to 39 hours) that they are commonly referred as extra-slowly growing (ESG) soybean rhizobia. Some strains are highly effective with the Chinese soybean cultivar Heinong 26. *B. liaoningense* poorly nodulated *Phaseolus aureus* and failed to nodulate both *Lotus* sp. and *Astragalus sinicus*. The level of DNA homology between EGS strains and either *B. japonicum* or *B. elkanii* is lower than 40% (Xu *et al.*, 1995).

Mesorhizobium tianshanense strains (originally called *Rhizobium tianshanense*) were isolated from an arid saline desert soil of Xinjiang, the most westerly province of China (Chen *et al.*, 1995). The strains comprising the species *M. tianshanense* were isolated from nodules of different legumes, such as *Glycyrrhiza padilliflora, G. uralensis, Sophora alopecuroides*, or *Glycine max*. *M. tianshanense* strains produced acid in a medium containing mannitol and their generation times varied from 5-15 hours. Those isolates from soybean nodules, strains 009B and 91X01, are slow growers and show a generation time of 10 and 13.5 hours, respectively. Partial 16S rRNA gene sequencing of the type strain CCBAU3306 revealed that this strain is closely related to *M. loti*, but not to *B. japonicum* (Chen *et al.,* 1995).

Fast-growing soybean bacteria were first isolated 20 years ago (Keyser *et al.*, 1982) and were finally clustered into the species, *Sinorhizobium fredii* (Chen *et al.,* 1988). These first soybean isolates are able to form nitrogen-fixing nodules on Asiatic soybean cultivars (such as cultivar Peking) but either fail to nodulate or are very poorly effective with the modern cultivars from North America (Keyser *et al.,* 1982; Buendía-Clavería and Ruiz-Sainz, 1985; Buendía-Clavería *et al.,* 1989). Later, new *S. fredii* strains, which are able to form nitrogen-fixing nodules with both Asiatic and American soybean cultivars, were isolated (Dowdle and Bohlool, 1985; Yang *et al.,* 2001; Camacho *et al.,* 2002). In addition to their marked soybean cultivar specificity, *S. fredii* strains show a very broad host range, being able to nodulate at least 79 different genera of legumes (Pueppke and Broughton, 1999). Most of the well-studied strains of *S. fredii* are from China (Keyser *et al.,* 1982; Dowdle and Bohlool, 1985; Yang *et al.,* 2001; Camacho *et al.,* 2002; Rodríguez-Navarro *et al.*, 2002), although fast-growing soybean microsymbionts have also been isolated from Malasia (Young *et al.,* 1988), Vietnam (Cleyet-Marel, 1987) and Panama (Henández and Focht, 1984). Taxonomic studies have shown that *S. fredii* is closely related to the alfalfa microsymbiont, *Sinorhizobium meliloti* (de Lajudie *et al.,* 1994), although both their host ranges and nodulation-factor structures (Nod factors or LCOs) are different (Dénarié *et al.,* 1996; Gil-Serrano *et al.,* 1997). They are also related to *Sinorhizobium teranga* and *S. saheli* (de Lajudie *et al.,* 1994), two species isolated from nodules of West African trees.

The species *Sinorhizobium xinjiangense* also clusters with fast-growing bacteria that form nitrogen-fixing nodules on soybeans and other legumes, such as *Vigna unguiculata* and *Cajanus cajan* (Chen *et al.,* 1988). It was isolated from soil and soybean nodules collected in Xinjiang province. *S. xinjiangense* strains can be differentiated from those belonging to *S. fredii* by a number of physiological characteristics, such as carbon and nitrogen utilisation, intrinsic antibiotic resistance patterns, or the ability to grow at acid or alkaline pH.

3.2. Studies of the natural soybean-rhizobia populations in Chinese soybean cropping areas

Considering the extensive soybean-cropping areas in China, China's different types of soils and its climatic diversity, there are surprisingly few reports on the indigenous soybean-rhizobia populations, their symbiotic nitrogen-fixation capacity, and their competitive ability to nodulate soybeans (Dowdle and Bohlool, 1985; 1987; Chen *et al.,* 1988; Buendía-Clavería *et al.,* 1994; Xu *et al.*, 1995; Li *et al.*, 1996; Li Fu-Di, 1996; Jiang *et al.,* 1996). All these facts are of relevance to the ultimate goal of producing rhizobial inoculants to increase soybean productivity under field conditions. To reduce this gap in information was one of the main objectives of a project, which was funded by the European Union (1997-2000) and involved research institutions from China, Great Britain, the Netherlands, and Spain. In this project, indigenous fast- and slow-growing soybean rhizobia populations (most probably *Bradyrhizobium japonicum* and *Sinorhizobium fredii*, respectively) in soil samples from Shandong, Henan, Hubei, and Xinjiang autonomous regions were quantitatively analysed using the Most Probable Number (MPN) technique (Yang *et al.*, 2001). Different Asiatic and American soybean cultivars were used as the trapping host. Table 4 shows that all soil samples contained large indigenous soybean-rhizobia populations, which were a mixture of fast- and slow-growing soybean rhizobia.

Table 4. Estimation by the MPN Technique of the Soybean-Rhizobial Populations in Soil Samples from Shandong, Henan, Hubei and Xinjiang Autonomous regions. After Yang et al., 2001

Province of origin of the soil sample [1]	Soybean cultivar[2]	Estimated number of soybean-rhizobia [3]	Ratio fast- / slow-growers (%) [4]
Henan (7.5)	Heinong 33	$4 \times 10^4 - 5 \times 10^5$	73/27 a
	Linzhen	-----	71/29 a
	Williams	$2 \times 10^4 - 3 \times 10^5$	65/35 a
	Bragg	-----	68/32 a
Xinjiang (7.3)	Jing Dou 19	$4 \times 10^4 - 5 \times 10^5$	100/0
	Heinong 33	$6 \times 10^2 - 8 \times 10^3$	99/1
	Williams	-----	8/92
	Kobe		7/93
Hubei (8.0)	Jing Dou 19	$10^4 – 10^5$	81/19
	Williams	$10^4 – 10^5$	90/10
Shandong (7.5)	Heinong 33	$9 \times 10^4 - 10^6$	27/73 a
	Lizhen	-----	32/68 a
	Williams	$2 \times 10^3 - 3 \times 10^4$	29/71 a
	Bragg	-----	36/64 a

[1]The pH of each soil sample is indicated in brackets. [2]Heinong 33, Linzhen, and Jing Dou 19 are Chinese soybean cultivars, whereas Williams, Bragg, and Kobe are American soybean cultivars. [3]The number of soybean-rhizobia in each soil sample was estimated in soybean plants that were grown at the pH of the soil sample. [4]Statistical analyses are among cultivars inoculated with the same soil sample. For each soil sample, numbers on the same column followed by the same letter are not significantly different at the 5% level, using the Fisher test for comparing proportions. Statistical analyses were not carried out for the Xinjiang sample because the numbers are very different.

Apparently, fast-growers are more abundant than slow-growers in the Henan and Hubei soil samples, whereas in Shandong soil, slow-growers appear to be more abundant. Xinjiang soil is composed of soybean-rhizobia populations showing marked soybean-cultivar specificity because the ratio of fast-growers to slow-growers isolated from soybean nodules varies with the soybean cultivar used as the trapping host. With Asiatic cultivars, most of the nodule isolates are fast-growers, whereas with American soybean cultivars, most of the isolates are slow-growers. Hence, the size of the indigenous rhizobial populations can be seriously underestimated if only one cultivar is used to determine the bacterial population. This cultivar specificity exhibited by soybean-rhizobia strains not only discriminates between Asiatic and American soybean cultivars but also among Asiatic cultivars (Jiang *et al.,* 1996; Yang *et al.,* 2001), which it is not surprising when the large diversity of soybean cultivars available in China is considered.

Several thousand isolated soybean rhizobial strains are stored at the Department of Microbiology of the University of Seville in Spain and at the Institute of Biology of the University of Leiden in The Netherlands. Although only a fraction of the collection has received attention so far (Yang *et al.*, 2001; Camacho *et al.*, 2002; Thomas-Oates *et al.,* 2003), the physiological, genetic and symbiotic characteristics a set of 198 fast-growing soybean-nodulating rhizobial strains from the four Chinese regions mentioned above have been studied (Thomas-Oates *et al.,* 2003) and the results compiled in the first extensive catalogue of fast-growing soybean rhizobia, which is available at http://www.soybeanrhizobia.net. Table 5 summarizes the results from the study of some of the bacterial traits.

In a new EU-funded project, an enlarged consortium is evaluating the indigenous population of soybean-rhizobia in soil samples from another 11 provinces that cover an axis from the North to the South of China. Soil samples are from both traditional soybean-cropping regions and also from areas where soybean might be introduced for the first time if the plans of the Chinese Government to increase soybean acreage are implemented. The emerging picture from the MPN experiments indicates that most of the soils contain high levels of soybean-rhizobia populations (in the range of 10^4 bacteria/g of soil), although some locations (for instance, some counties in Chongqin and Inner Mongolia), in which soybean is not cropped, appear to have lower levels (about 10^2 rhizobia/g of soil) of indigenous soybean-rhizobia (our unpublished results). In Heilongjiang, the Chinese province with the largest soybean acreage, the level of soybean-rhizobia populations is extremely high (10^7 bacteria/g of soil). These results are in accord with those previously reported by Jiang *et al.* (1996). They describe Hunan, Shanxi and Liaoning soybean-cropping areas as containing high levels of soybean-rhizobia (*ca.* 10^4/g of soil), whereas a soil sample from a Shandong field, which was not being used for soybean cultivation, had a lower (*ca.* 10^2/g of soil) soybean-rhizobia content.

3.3. Soybean inoculation

Production of rhizobial inoculants and their application as an agricultural practice started in China in the 1950s. The first inoculants produced were for peanut and soybean. In the 1960s and 1970s, inoculants (*Mesorhizobium huakii*) for *Astragalus sinicus* were widely applied to increase rice production in a rotation system. *A. sinicus* is a traditional Chinese green manure, which is grown in winter and dug into the soil the next spring. Field experiments in Hubei, Henan, Anhui and Jiangxi provinces, where soils contained low levels of indigenous *M. huakii* populations, showed significant crop increases.

Table 5. Physiological, Genetic and Symbiotic Properties of Fast-growing Soybean Rhizobia isolated from Soil Samples from Soybean-cropping Areas of Xinjiang, Henan, Shan Dong and Hubei Provinces

Bacterial trait	Remarks
Number of plasmids	All strains contain from 1-6 plasmids. A few strains contain very small plasmids whose size is in line with those previously reported.[1]
Acidification of medium containing mannitol	Only 5% of the isolates did not show a significant acidification of the medium
Growth at different pH	Some strains grew in the pH range 5.5-9.0
Growth at 37°C	Very rare; only 2% of the strains analysed.
Melanin production	Very common; in 57% of the analysed strains.
Generation time	Most (75%) had generation times of 1.5-3.5 hours.
Lipopolysaccharide (LPS) profiles on polyacrylamide gels	A majority of the strains showed an LPS profile which is in line with those already reported for other *S. fredii* strains[2].
PCR finger-printing analyses[3]	Some differences in the RFLP of 16S rDNA or 16S-23S rDNA IGS PCR-amplified fragments. RAPD analyses also showed differences among strains
Symbiotic characteristics	Most are able to form nitrogen-fixing nodules with both American and Asiatic soybean cultivars. Marked bacterial-strain/soybean-cultivar specificity was found among Asiatic soybean cultivars. Some strains are highly effective with Asiatic soybean cultivars, but very few fix N_2 with the American soybean cultivar Williams at a level comparable to that *B. japonicum* USDA110.

[1]In a previous study carried out with a different set of soybean-rhizobia strains, very small indigenous plasmids were found. Strains CH12 and CH154 have indigenous plasmids of about 7.5 and 22 kb, respectively (Camacho et al. 2002). [2]As described by Reuhs et al. 1999: A region (LPS-II) consisting of fast-migrating electrophoretic bands (Rough-LPS) that usually are not well resolved and another region (LPS-I) containing slow-migrating few slow-migrating bands (Smooth-LPS). [3]RFLP, Restriction Fragment Length Polymorphism; IGS, Intergenic Spacer; RAPD, Random Amplified Polymorphic DNA.

In the 1980s, soybean-rhizobia inoculants were used on a large scale in China (on *ca.* 650,000 ha). In field conditions, the use of soybean inoculants has to contend with the problem that most of the soils investigated contain high populations of indigenous soybean rhizobia (Jiang *et al.*, 1996; Yang *et al.*, 2001). In these circumstances, the percentage of nodules occupied by the inoculant strains can be too low to achieve a significant contribution to the total fixed-nitrogen obtained through symbiotic nitrogen fixation (for a review of the competition problem, see Robleto *et al.*, 2000). This problem is particularly serious if the natural rhizobial populations are poor nitrogen fixers, such as appears to be the case in Heilongjiang province, where as many as 22% of the growing soybean plants contain ineffective, or poorly effective, nodules (Li, 1996). Indigenous soybean-rhizobia strains, which also show mediocre nitrogen-fixation capacity, occur in the soils of other provinces (such as Henan, Hubei or Shanxi) (Jiang *et al.,* 1996). Furthermore, soybean inoculation, as a method to increase soybean yield, failed in several experiments conducted in different parts of China, such as the Yellow River valley (Jiang *et al.,* 1996).

Positive responses to inoculation have, however, been reported. For instance, a high percentage of nodule occupancy by the *B. japonicum* strains used as inoculants was observed in a field experiment carried out at the farm of Huazhong Agricultural University (Wuhan, Hubei province), if the inoculant was incorporated into the soil rather than on the surface of the soybean seed (Li, 1996). Because this soil contained indigenous soybean-rhizobia populations (*ca.* 4 x 10^4 bacteria/g of soil), this strategy has the potential to increase the number of nodules formed by the inoculant. Increases of 11% in soybean seed yield have also been reported in field experiments in Heilongjiang, Sichuan and Guangxi, using a genetically engineered *B. japonicum* HN32 inoculant that contains a DNA fragment from *S. fredii* (Zhang *et al.,* 1996). Table 6 shows the increases in soybean seed yield obtained by inoculation with strain HN32 in yearly field experiments conducted during the period 1990-1996. Although the increases are not dramatically high, they are consistent over all the field assays and over a very large acreage. Other soybean-inoculation experiments carried out in Heilongjiang, Jilin, and Liaoning provinces showed increases in soybean yield that varied from 6-33%.

Table 6. Increases in Soybean Seed Yield in Field Experiments using B. japonicum *HN32 as Inoculant.*

Year	Province	Number of locations used	Acreage used in the field experiments	Yield increase (%) over the uninoculated control[1]
1990	Heilongjiang	7	5.797	9.8
1991	Heilongjiang	5	25.665	10.1
1992	Heilongjiang	7	23.997	8.5
1993	Heilongjiang	9	18.678	11.8
1994	Heilongjiang	10	29.329	11.5
1995	Heilongjiang	1	2.666	11.7
1995	Sichuan	1	2.000	13.0
1996	Guangxi	1	11.200	11.7
Total in 1990-1996		41	119.332	11.0

[1]*Numbers refer to mean values for the different locations investigated. Seed yields ranged from 1375 to 2900, most being over 2000 kg/ha.*

Regardless of the competition problem posed by the indigenous soybean-rhizobia populations, soybean inoculation is actively promoted in China. For instance, in 2000, a company operating in Harbin city (Heilongjiang Lufeng Biological Organic Fertilizer LTD) and belonging to the Heilongjiang Academy of Science inoculated 400 ha. Because the company claims that positive responses to inoculation were scored, the production of soybean inoculants has been scaled up by using 10,000 litre fermentors. To our knowledge, however, information on soybean-nodule occupancy by the applied inoculant is not available.

As far as we know, experiments in China, using ^{15}N isotopically labelled nitrogen fertiliser, have only been carried out at the greenhouse level. From these pot experiments using ^{15}N, it was concluded that the percentage of fixed-nitrogen derived from air (Ndfa%) varied from 40% to 60% among the Chinese cultivars analysed. The symbiotic performance of *B. japonicum* strains with 20 different Asiatic soybean cultivars was better than that shown by *S. fredii* strains (Yuangsheng *et al.,* 1995).

4. THE SOYBEAN GERMPLASM COLLECTION IN CHINA

Because soybean is native to China, the soybean germplasm collection is very large. Reports providing information about this collection are already available on the Internet; the Chinese Academy of Agricultural Sciences (CAAS) provides information at its web address http//www.caas.com.cn. An article, entitled *Evaluation of Soybean Germplasm in China* (Zhanyou *et al.*, CAAS, Beijing, China), can be found at the web address http://www.gsf99.uiuc.edu/invited/2-3-05.pdf. The information presented here is a summaryof the information presented on the web pages mentioned above as well as other documents available *via* the internet, such as that presented in http://www.gsf99.uiuc.edu/invited/2-3-07.pdf.

The National Gene Bank at the Institute of Crop Germplasm Resources in CAAS, Beijing, contains 23,587 soybean accessions, including 3000 wild soybean accessions. Twenty-eight provinces (all Chinese provinces except Qinghai and Tianjing) have provided local soybean varieties to the germplasm collection. As mentioned before, Chinese soybean varieties are divided according to the sowing date, into winter, spring, summer and autumn soybean varieties. The collection has 11,206 spring-sowing accessions (47% of the total collection), 11,648 summer-sowing accessions (49%), 733 autumn-sowing accessions (3%), and some winter-sowing accessions. A large proportion of the stored accessions (22,637 out 23,587) have been characterised using six different aspects:

Morphological characteristics: these include growth habit, stem termination, and flower and pubescence colours. Many accessions produce stems that are erect (67.3% of the whole collection) and show a determinate termination (51.7%). The total number of accessions producing either white or purple flowers is similar (10,467 and 12,138, respectively).

Seed characteristics: these include colour, shape and size of seeds. Most of the soybean accessions have yellow cotyledons (96.8%) and yellow seed coats (60.5%). The collection also has accessions showing other seed coat (green, black, brown, and bicolor) or cotyledon (green) colours. Although accessions having ellipse-shaped seeds are the most abundant (61.9%), representatives of other seed shapes (flat ellipse, flat round, long ellipse, and kidney) are also present. In general, seeds of Chinese soybeans are mostly small (32.6%) or medium (42.9%) in size. Accessions having larger size seeds are also present in the collection (including some with extremely large seeds that weigh >30g/100 seeds).

Seed nutrient quality: the protein content of soybean accessions in the collection varies from 29.3% to 52.9%. Accessions containing 40-48% protein are the most abundant (81.9%). The variety, Yan Tian Qinq Pi Dou, has the highest protein content (52.9%). The fat content of most soybean accessions is below 20%.

Tolerance to environmental stress: although the whole collection has not been evaluated for either drought or salt tolerance, 463 accessions proved to be drought tolerant and only nine accessions were salt tolerant at the seedling stage. Most of the accessions were not tolerant to cold conditions, with only nine accessions able to germinate at low temperatures.

Resistance to diseases: especially to soybean cyst nematode (SCN), which is a major disease affecting this crop. Some of the soybean accessions showed high resistance, or even immunity, to some SCN races, whereas all accessions were susceptible to a range of other SCN races. None of the accessions investigated showed high tolerance, or immunity, to soybean rust and only 69 accessions were moderately resistant. Soybean frogeye leaf-spot (SFLS) is a very serious disease that leads to poor seed production quality. In a screening of 961 accessions, 14 of them showed high resistance to SFLS in leaves, stems and pods.

Genetic diversity: DNA finger-printing techniques were applied to 90 soybean accessions. RAPD markers were useful tools to identify soybean germplasm resources. The largest diversity of soybean accessions in the collection was found among those collected from Shanxi province.

5. SOYBEAN IN CROP ROTATION AND IN CONTINUOUS CULTIVATION

Legumes can contribute fixed-nitrogen to cropping systems in several ways. An input of fixed-nitrogen into the soil will occur if the total fixed-nitrogen in the plant residues left after harvesting is greater than the total amount of fixed-nitrogen absorbed from the soil. Among all legumes cultivated in China, soybean is that one most commonly cropped. Higher economic benefits and environmental conditions account for this preference. Crop rotation between soybean and other relevant non-legume crops is also used in China to enrich the soil N status for the following non-legume crop. In Jilin province, for instance, corn yield was 18% higher in fields where soybean had been previously cropped than in those where rice was previously grown. Similarly, rice and wheat yields increased (16% and 31%, respectively) if soybean was previously cultivated. The crop-rotation strategies involving soybean vary according to the Chinese geographical area. Table 7 indicates some of the soybean/non-legume combinations most commonly used.

In addition of the well-known benefits to the non-legume crop, rotation of legume and other crops avoids the decrease in soybean yield that is observed if soybean is continuously cropped in the same area. Unfortunately, in some areas, in

Table 7. The Use of Soybean in Crop-Rotation Practices in Different Regions of China. From Zhang Mengcheng, 1999. Problems and Solutions on Cultivation of Soybean High Yield. China Braille Press, Beijing, China. (In Chinese).

Region	Crop-rotation strategy	Remarks
Northeast China	Soybean – corn – broomcorn – corn Soybean – broomcorn (or rice) – corn Spring soybean – spring wheat	Only one crop per year.
Valley of the Yellow and Huaihe River	Winter wheat – summer soybean – winter wheat - summer soybean – corn Winter wheat – summer soybean – no crop in winter – cotton	One crop per year or three crops in two years.
Valley of the Yangze River	Winter wheat – summer soybean – winter wheat – summer corn Winter wheat – summer soybean – winter corn – cotton Winter wheat – summer soybean – winter wheat – rice Winter wheat – summer soybean and corn – winter wheat – sweet potato	Two or three crops per year.
Southeast China	Early rice – autumn soybean – winter wheat or corn Early rice – autumn soybean – green manure or fallow in Winter Spring soybean – late rice – winter wheat or corn	Water supply is low in autumn. Soybean is frequently sown after the rice harvest.

Corn, *Zea mays;* broomcorn, *Sorghum bicolour;* sweet potato, *Ipomoea batatas;* wheat, *Triticum aestivum;* rice, *Oryza sativa.*

which the soil is poorly drained, normal strategies for crop rotation cannot be applied, so that soybean appears as the only possibility to local farmers. Research projects, supported by the Chinese Academy of Sciences (in 1986), The Department of Science and Technology of Heilongjiang (in 1988), and The Ministry of Science and Technology of China (in 1995), have been carried out in order to understand the main reasons underlying soybean and soil deterioration through continuous soybean-cropping practices (for a review, see Liu and Herbert, 2002). Table 8 compiles some of the negative effects that have been observed when soybean is continuously cultivated.

Different crop-management practices have been proposed as possible solutions to alleviate the decline of soybean productivity when rotation is not possible (Liu and Herbert, 2002). These include: the adoption of resistant or tolerant soybean cultivars; a better control of soybean diseases and pest insects (for instance, by seed coating with fungicides and/or insecticides); the correction of the imbalance of macro- and micronutrients in the soil; or the application of suitable tillage practices.

Table 8. Negative Effects on Soybean Plants, Soil Microbial Communities, and Soil Characteristics as a Result of Continuous Soybean Cropping.
After Liu and Herbert, 2002 (The publications listed in the table are in Chinese)

Type of effect	Description	References
Plant growth and development	Shorter stem; yellowing of leaves; reduction of root development and leaf area; poor canopy development; reduction of the number of nodules; reduction of root dry matter; increase of flower and pod shedding.	Liu *et al.*, 1990 Wang *et al.*, 1995a Wang *et al.*, 1995b Xu and Wang, 1995 Xu *et al.*, 1999
Seed quality and yield	Decrease of pod and seed number; decrease of seed size; increased number of diseased seeds by infection with fungi and virus; increase in seed protein content; decrease in seed oil	Liu and Yu, 2000 Xu *et al.*, 1997 Xu *et al.*, 1999 Yang, 1997 Zheng, 1999
Physiological and biochemical plant characteristics	Increase in praline and phenolic compounds; decrease of superoxide dismutase and catalase activities; decrease in chlorophyll *a* and *b*; decrease in root sugar content; decrease in P, K, Zn, Mo, and B content; increase in Ca absorption; imbalance in K/Ca ratio; decrease in lipid membrane plasticity.	Han and Xu, 1996 Han and Xu, 1997 Jia and Yu, 1995 Liu *et al.*, 1997 Zhao *et al.*, 1998 Zheng *et al.*, 1995
Plant diseases	Increase in phytopathogenic fungi (such as *Penicillium, Botrytis, Fusarium* and *Gliocladium*), soybean cyst nematodes, and pest insects	Xu and Wang, 1995 Wang *et al.*, 1989 Hu *et al.*, 1996 Han and Xu, 1997
Microbial communities of the rhizosphere.	Decrease in bacterial populations; increase in fungi populations; decrease in total microbial populations.	Yu *et al.*, 1988 Liu *et al.*, 1990 Xu and Wang, 1995 Jia and Yu, 1995
Soil charact-eristics, enzymatic activities, and soil compounds	Acidification of soil; decrease in phosphatase, urease, invertase and dehydrogenase activities; increase in soil phenol content; decrease in polysaccharides and soil aggregates.	Fu and Yang, 1999 Jia and Yu, 1995

6. CONCLUSIONS

Soybean production in China is key to improving living standards for many farmers. Although Chinese soybean acreage ranks fourth in the world, its productivity is clearly lower than that of western soybean-producing countries. Research efforts have been mainly directed at collecting and characterising soybean varieties and soybean-rhizobia strains that form nitrogen-fixing nodules with this legume. Most of the soybean-cropping areas of China contain high levels of indigenous soybean-rhizobia, which represent a constraint for the successful use of soybean inoculants. Information from field experiments in China using soybean inoculants is not abundant. In contrast to reports on the failure of the inoculants to increase soybean productivity, there are also some reports that indicate that soybean inoculants might be successfully used for such a purpose. Due to the strong cultivar specificity exhibited by soybean-rhizobia and the fact that large collections of bacterial strains and soybean varieties are available, it might be possible to identify soybean cultivars that, while highly effective with a particular set of soybean-rhizobial strains, are also poorly nodulated by the indigenous soybean-rhizobia populations of a particular soil. The quality of the inoculants used and the way they are applied are also important questions that need to be thoroughly evaluated if this sustainable option is to become preferred over chemical fertilisation.

ACKNOWLEDGEMENTS

Part of the work presented here was founded by EU grants ERBIC18CT970191 and ICA4-CT-2001-10056.

REFERENCES

Buendía-Clavería, A. M., and Ruiz-Sainz, J. E. (1985). Isolation of mutants of fast-growing soybean strains that are effective on commercial soybean cultivars. *Physiol. Plant., 64*, 507-512.

Buendía-Clavería, A. M., Chamber, M. M., and Ruiz-Sainz, J. E. (1989). A comparative study of the physiological characteristics, plasmid content and symbiotic properties of different *Rhizobium fredii* strains. *Syst. Appl. Microb., 12*, 203-209.

Buendía-Clavería, A. M., Rodríguez-Navarro, D. N., Santamaría-Linaza, C., Ruiz-Sainz, J. E., and Temprano Vera, F. (1994). Evaluation of the symbiotic properties of *Rhizobium fredii* in European soils. *Syst. Appl. Microbiol., 17*, 155-160.

Camacho, M., Santamaría, C., Temprano, F., Rodríguez-Navarro, D. N., Daza, A., Espuny, R., Bellogín, R., Ollero, F. J., Lyra, M. C. C. P., Buendía-Clavería, A., Zhou, J., Li, F. D., Mateos, C., Velázquez, E., Vinardell, J. M., and Ruiz-Sainz, J. E. (2002). Soils of the Chinese Hubei province show a very high diversity of *Sinorhizobium fredii* strains. *Syst. Appl. Microbiol., 12*, 592-602.

Chen, W. X., Yang, G. H., and Li, J. L. (1988). Numerical taxonomy study of fast-growing soybean rhizobia and a proposal that *Rhizobium fredii* be assigned to *Sinorhizobium* gen. nov. *Int. J. Syst. Bacteriol., 38*, 392-397.

Chen, W., Wang, E., Wang, S., Li, Y., Chen, X., and Li, Y. (1995). Characteristics of *Rhizobium tianshanense* sp. nov., a moderately and slow growing root nodule bacterium isolated from an arid saline environment in Xinjiang, People's Republic of China. *Int. J. Syst. Bacteriol., 45*, 153-159.

Cleyet-Marel, J. C. (1987). *Dynamique des populations de Rhizobium et de Bradyrhizobium dans le sol et la rhizosphere*. These d'Etat. University Claude Bernard, Lyon, France.

Dénarié, J., Debellé, F., and Promé, J. C. (1996). *Rhizobium* lipo-quitooligosaccharide nodulation factors: Signalling molecules mediating recognition and morphogenesis. *Annu. Rev. Biochem., 65*, 503-535.

De Lajudie, P., Willems, A., Pot, B., Dewettnick, D., Maestrojuan, G., Neyra, M., Collins, M. D. Dreyfus, B., Kersters, K., and Gillis, M. (1994). Polyphasic taxonomy of rhizobia: Emendation of the genus *Sinorhizobium* and description of *Sinorhizobium meliloti* comb. Nov., *Sinorhizobium saheli* sp. nov., and *Sinorhizobium teranga* sp. nov. *Int. J. Syst. Bacteriol., 44*, 715-733.

Dowdle, S. F., and Bohlool, B. B. (1985). Predominance of fast-growing *Rhizobium japonicum* in a soybean field in the People's Republic of China. *Appl. Environ. Microbiol., 33*, 990-995.

Dowdle, S. F., and Bohlool, B. B. (1987). Intra- and inter-specific competition in *Rhizobium fredii* and *Bradyrhizobium japonicum* as indigenous and introduced organisms. *Can. J. Microbiol., 33,* 990-995.

Fu, H. L., and Yang, M. Z. (1999). Effect of continuous soybean on soil cellulase activities. *Soybean Science 18,* 81-84.

Gil-Serrano, A. M., Franco-Rodríguez, G., Tejero-Mateo, P., Thomas-Oates, J., Spaink, H. P., Ruiz-Sainz, J. E., Megías, M., and Lamrabet, Y. (1997). Structural determination of the lipochitin oligosaccharide nodulation signals produced by *Rhizobium fredii* HH103. *Carbohydrate Research 303,* 435-443.

Han, Z. Z., and Xu, Y. L. (1996). Reasons and control of nutrition imbalance in continuous soybean. *Research in Agricultural Modernization 16,* 118-122.

Han, Z. Z. and Xu, Y. L. (1997). N, P, K accumulation characteristics in the plants of continuous soybean. *Research in Agricultural Modernization 17,* 215-219.

Hernández, B. S., and Focht, D. D. (1984). Invalidity of the concept of slow growth and alkali production in cowpea rhizobia. *Appl. Environ. Microbiol., 48,* 206-210.

Hu, J. C., Gao, Z. Q., and Zhang, S. X. (1996). Toxicity of *Penicilium purpurogenum* in continuous soybean. *Chin. J. Appl. Ecol., 7,* 422-427

Jia, X. M., and Yu, Q. L. (1995). Study on soil polyphenol oxidase in continuous soybean. *J. Heilongjiang Land Reclamation Univ., 8,* 40-43.

Jiang, M. L. Song, Y. P., Zhang, X. J., and Hu, X. J. (1996). Dominant soybean rhizobia population and their nitrogen fixation effectiveness in China. In F. D. Li, T. A. Lie, W. X. Chen and J. C. Zhou (Eds.), *Diversity and Taxonomy of Rhizobia* (pp. 263-269). Beijing, China: China Agricultural Scientech Press.

Keyser, H. H., Hu, T., Bohlool, B. B., and Weber, D. F. (1982). Fast-growing rhizobia isolated from root nodules of soybean. *Science, 215,* 1631-1632.

Li, F. D. (1996). Diversity and distribution of rhizobia among soil in relation to the competitive nodulation. In F. D. Li, T. A. Lie, W. X. Chen and J. C. Zhou (Eds.), *Diversity and Taxonomy of Rhizobia* (pp. 231-238). Beijing, China: China Agricultural Scientech Press.

Li, J., Xu, L. M., Fan, H., Cui, Z., and Ge, C. (1996). Diversity of rhizobia isolated from the root nodules of soybean. In F. D. Li, T. A. Lie, W. X. Chen and J. C. Zhou (Eds.), *Diversity and Taxonomy of Rhizobia* (pp. 141-146). Beijing, China: China Agricultural Scientech Press.

Liu, X., and Herbert, S. J. (2002). Fifteen years of research examining cultivation of continuous soybean in northeast China: a review. *Field Crops Res., 79,* 1-7.

Liu, Z. T., and Yu, L. S. (2000). Influence of alternate and continuous soybean on yield and quality. *Soybean Science, 19,* 229-237.

Liu, X. B., Yu, G. W., and Xu, Y. L. (1990). System analysis on the responses of continuous soybean. *Systematics Sciences of Comprehensive Studies in Agriculture, 3,* 40-44.

Liu Y. Y., Luo, S. G., and Liu, S. J. (1997). Nutrient uptake in continuous soybean. *J. Northeast Agricultural Univ., 28,* 209-215.

Pueppke, S. G., and Broughton, J. (1999). *Rhizobium* sp. strain NGR234 and *R. fredii* USDA257 share exceptionally broad, nested host ranges. *Mol. Plant-Microbe Interact., 12,* 293-318.

Reuhs, B. L., Stephens, S. B., Geller, D. P., Kim, J. S., Glenn, J., Przytycki, J., and Ojanen-Rehus, T. (1999). Epitope identification for a panel of anti-*Sinorhizobium meliloti* monoclonal antibodies and application to the analysis of K antigens and lipopolysachharides from bacteroids. *Appl. Environ. Microbiol., 65,* 5186-5191.

Robleto, E. A., Scupham, A. J., and Triplett, E. W. (2000). Solving the competition problem: Genetics and field approaches to enhance the effectiveness of the *Rhizobium*-legume symbiosis. In E. Triplett (Ed.), *Prokaryotic Nitrogen Fixation. A model System for the Analysis of a Biological Process.* Norfolk, England: Horizon Scientific Press.

Rodríguez-Navarro, D.N., Bellogín, R., Camacho, M., Daza, A., Medina, C., Ollero, F. J., Santamaría, C., Ruiz-Sainz, J. E., Vinardell, J. M., and Temprano, F. J. (2002). Field assessment and genetic stability of *Sinorhizobium fredii* strain SMH12 for commercial soybean inoculants. *Eur. J. Agron., 19,* 301-311.

Tan, Z. Y., Xu, X. D., Wang, E. T., Gao, J. L., Martínez-Romero, E., and Chen, W. X. (1997). Phylogenetic and genetic relationships of *Mesorhizobium tianshanense* and related rhizobia. *Int. J. Syst. Bacteriol., 47,* 874-879.

Thomas-Oates, J., Bereszczak. Edwards, E., Gill, A, Noreen, S., Zhou, J. C., Chen, M. Z., Miao, L. H., Xie, F. L., Yang, J. K., Zhou, Q., Yang, S. S., Li, X. H., Wang, L., Spaink, H. P., Schlaman H. R. M., Harteveld, M., Díaz, C. L., van Brussel, A. A. N., Camacho, M., Rodríguez-Navarro, D. N., Snatamaría, C., Temprano, F., Acebes, J. M., Bellogín, R. A., Buendía-Clavería, A. M., Cubo, M. T., Espuny, M. R., Gil, A. M., Gutiérrez, R., Hidalgo, A., López-Baena, F. J., Madinabeitia, N., Medina, C., Ollero, F. J., Vinardell, J. M., and Ruiz-Sainz, J. E. (2003). A catalogue of molecular, physiological and symbiotic properties of soybean-nodulating rhizobial strains from different soybean cropping areas of China. *Syst. Appl. Microbiol., (in revision).*

Van Berkum, P., and Eardly, D. (1998). Molecular evolutionary systematics of the *Rhizobiaceae*. In H.P. Spaink, A. Kondorosi, and P.J.J. Hooykaas (pp. 1-24). *The Rhizobiaceae, Molecular Biology of Model Plant-associated Bacteria*. Dordrecht, The Netherlands: Kluwer Academic Press.

Wang, Z. Y., Wang, Y. X., and Chen, T. L. (1989). Preliminary studies of continuous soybean on growth and development. *Soybean Science, 10*, 31-36.

Wang, G. H., Xu, Y. L., and Liu, X. B. (1995a). Root exudates in relation to continuous soybean barriers. *Soybean Science, 14*, 158-161.

Wang, G. H., Xu, Y. L., and Jiang, X. Y. (1995b). Plant residuals in relation to continuous soybean barriers. *Systematics Sciences of Comprehensive Studies in Agriculture, 4*, 8-13.

Xu, L. M., Ge, C., Cui, Z., Li, J., and Fan, H. (1995). *Bradyrhizobium liaoningense* sp. nov., isolated from the root nodules of soybeans. *Int. J. Syst. Bacteriol., 45*, 706-711.

Xu, Y. L., and Wang, G. H. (1995). Yield comparison of sterilized CS soil with control. *Heilongjiang Agricultural Science, 6*, 8-12.

Xu, Y. H., He, Z. H., and Liu, Z. T. (1997). Effect of continuous soybean on seed protein and oil content. *Soybean Science, 16*, 319-327.

Xu, Y. L., Liu, X. B., and Han, X. Z. (1999). Effect of continuous soybean on growth and development and yield. *Sci. Agric. Sinica, 32* (Suppl.), 64-68.

Yang, S.S., Bellogín, R.A., Buendía, A., Camacho, M., Chen, M., Cubo, T., Daza, A., Díaz, C.L., Espuny, M.R., Gutiérrez, R., Harteveld, M., Li, X.H., Lyra, M.C.C.P., Madinabeitia, N., Medina, C., Miao, L., Ollero, F.J., Olsthoorn, M.M.A., Rodríguez, D.N., Santamaría, C., Schlaman, H.R.M., Spaink, H.P:, Temprano, F., Thomas-Oates, J.E., van Brussel, A.A.N., Vinardell, J.M., Xie, F., Yang, J., Zhang, H.Y., Zhen, J., Zhou, J., and Ruiz-Sainz, J.E. (2001). Effect of pH and soybean cultivars on the quantitative analyses of soybean rhizobia populations. *J. Biotechnol., 91*, 243-255.

Yang, X. J. (1997). Screening of tolerant soybean cultivars (lines) to continuous soybean on biomass of soil microorganism. *Soybean Science, 15*, 357-361.

Young, C. C., Chang, J. Y., and Chao, C. C. (1988). Physiological and symbiotic characteristics of *Rhizobium fredii* isolated from subtropical-tropical soils. *Biol. Fertil. Soils, 5*, 350-354.

Yu., G. R., Liu, J. X., and Su, G. Y. (1988). Number of fungi and bacteria in topsoils of continuous soybean and sunflower system. *Chin. J. Appl. Ecol., 7*, 1-8.

Yuansheng, M., Xuanjun, F., and Min, L. (1995). *Biochemistry and Genetics of Nitrogen Fixation*. Beijing, China: Chinese Science and Technology Press (in Chinese).

Zhang, X. X., Zhou, J. C., Ma, L. X. Zhang, Z. M., and Chen H. K. (1996). Cloning and sequencing of a 3.7 kb enhancing fragment from *Rhizobium fredii* B52. In F. D. Li, T. A. Lie, W. X. Chen and J. C. Zhou (Eds.). *Diversity and Taxonomy of Rhizobia* (pp. 255-262). Beijing, China: China Agricultural Scientech Press.

Chapter 6

SOIL STRESS FACTORS INFLUENCING SYMBIOTIC NITROGEN FIXATION

M. J. SADOWSKY
Department of Soil, Water, and Climate, and The BioTechnology Institute, University of Minnesota, St. Paul, MN 55108, U.S.A.

1. INTRODUCTION

The soil environment is under a constant state of change and, as such, can be relatively stressful for both macro- and microorganisms. Fluctuations in pH, nutrient availability, temperature, and water status, among other factors, greatly influence the growth, survival, and metabolic activity of soil microorganisms and plants, and their ability to enter into symbiotic interactions. Despite this situation, soils represent one of Earth's most productive ecospheres, accounting for a majority of primary and successional productivity. Consequently, microbes, plants, and other soil inhabitants have evolved to adapt to the ever changing and often inhospitable soil environment. In this Chapter, I will discuss stress factors in soils that influence symbiotic nitrogen fixation and I will do so from the perspective of both the host plant and the microsymbiont. However, the reader should be aware that, whereas some stress factors simultaneously affect both symbiotic partners, *e.g.*, water stress, others may differentially influence each partner to a seemingly different degree by different mechanisms. Moreover, both plants and microbes have often adopted different strategies for dealing with these stress factors.

2. IMPORTANCE OF SYMBIOTIC NITROGEN FIXATION

Nitrogen (N) is one of the major limiting nutrients for most crop and other plant species (Newbould, 1989). Moreover, fixed-N acquisition and assimilation is second in importance only to photosynthesis for plant growth (Vance, 1998; Graham and Vance, 2000). From several indications and by many estimations, the

D. Werner and W. E. Newton (eds.), Nitrogen Fixation in Agriculture, Forestry, Ecology, and the Environment, 89-112.

world will no doubt face severe food shortages in the not-too-distant future, in part due to excessive population growth and the negative environmental impact associated with this increased growth. Although Waggoner (1994) suggested that the Earth's population will reach nearly twice its current level of over 10 billion people by the year 2035, there is still some debate on when this will actually occur. However, there is little debate that this will indeed occur sometime in the not-too-distant future. In addition, populations in developing (and less developed) countries, which reside in the tropical and subtropical regions of Asia, Africa, and Latin America, may account for *ca.* 90% of the projected world population. Today, in tropical countries, plant materials provide *ca.* 80% of the caloric and dietary protein needs of individuals and this situation is not expected to change in the near future. In the fairly recent past, humans used *ca.* 10% of the total fixed carbon that is produced by plants through photosynthetic activity (Golley *et al.*,, 1992); today, humans use *ca.* 40% of that carbon. Moreover, it is estimated that, by 2030, humans will require *ca.* 80% of all photosynthetically-fixed carbon to meet their dietary requirements. Taking it as a given that enhanced agricultural production will require the utilization of large areas of land that are now considered to be either marginally productive or even non-arable, several alternate strategies are needed to meet these considerable and increasing human dietary needs in the future.

Many diverse biological associations contribute to N_2 fixation (BNF) in both soil and aquatic systems (Sprent, 1984). However, in most agricultural systems, the primary source of biologically-fixed N (*ca.* 80%) occurs *via* the symbiotic interactions of legumes and soil bacteria of the genera *Rhizobium*, *Bradyrhizobium*, *Sinorhizobium*, *Allorhizobium*, *Mesorhizobium*, and *Azorhizobium* (Sadowsky and Graham, 1998; Vance, 1998). The other 20% is contributed mainly by the actinorrhizal (*e.g.*, by *Frankia*) and *Anabena-Azolla* types of symbiotic interactions. It has been estimated that legumes provide approximately 35% of the worldwide protein intake and that *ca.* 250 million ha of legumes are currently grown world wide. These symbiotic partners together fix an astounding 90 Tg N per year (Kinzig and Socolow, 1994) which points to the obvious conclusion that enhancing the both the use and management of biologically-fixed N will result in huge environmental and economic benefits. To put this situation in some perspective, it has been estimated that *ca.* 288 Tg of fuel (at a cost of $30 billion U.S. annually) would be required to replace the N fixed by legumes with anhydrous ammonia produced by the Haber-Bosch process. For the U.S. alone, either decreasing or eliminating the use of synthetic N-fertilizers could save an estimated $1.0-4.5 billion annually (Tauer, 1989). This reduction in the amount of fuel consumed will also reduce undesirable impacts of the increased use of industrially-derived fix N.

3. SYMBIOTIC INTERACTION OF LEGUMES WITH RHIZOBIA

In order to properly discuss how environmental stress factors influence symbiotic nitrogen fixation, it is important to understand how the micro- and macro-symbionts interact at the cellular and molecular levels. In some instances, environmental perturbation independently influences the nodulation and nitrogen-fixation processes. Although, in this chapter, I only discuss this topic in broad terms, several recent review articles cover this area in much greater detail (Hirsch *et al.*, 2001; Roy *et al.*, 2002; Vitousek *et al.*, 2002).

3.1. The Nitrogen-fixation Process

Environmental symbiotic N_2 fixation requires the coordinate interaction of two major classes of genes present in rhizobia, the *nif* genes and *fix* genes. Only the *nif* genes, which encode the molybdenum-based enzyme system, are found in rhizobia and they have structural and functional-relatedness to the N_2-fixation genes found in *Klebsiella pneumoniae*. The structural *nif* genes from taxonomically diverse microbes are nearly identical and function in a similar manner to encode nitrogenase (Ruvkin and Ausubel, 1980). A majority of the *nif* genes are plasmid borne in most rhizobia, but are located on the chromosome in the bradyrhizobia. Nitrogen fixation in symbionts and free-living microbes is catalyzed by nitrogenase, an enzyme complex encoded by the *nifDK* and *nifH* genes. Nitrogenase itself consists of a molybdenum-iron protein (MoFe), sometimes called component 1, and an iron-containing protein (Fe), component 2. The MoFe protein subunits are encoded by *nifK* and *nifD* and an FeMo-cofactor (called FeMoco) is required for activation of the MoFe protein. This cofactor is assembled through the activity of the *nifB,V,N,H* and *E* genes. The Fe-protein subunit is encoded by the *nifH* gene. The organization and complexity of *nif* genes in microorganisms varies tremendously (Downie 1998). For example, in the free-living *K. pneumoniae,* at least 20 *nif* genes are organized in about 8 operons (Dean and Jacobson 1992). In most systems, however, the regulation of all *nif* genes is controlled by NifA (a positive activator of transcription) and NifL (the negative regulator).

Environmentally, *nif*-gene expression is regulated by both O_2 and fixed-nitrogen levels (Merrick and Edwards, 1995). For example, elevated soil ammonia (NH_3 or NH_4^+) concentrations allow NifL to act as a negative controller of gene expression by preventing NifA acting as an activator. In addition, elevated O_2 concentrations inhibit FixL from activating the transcriptional activator FixJ, which in turn prevents increases in NifA. Because NifA is the transcriptional activator of all other *nif* genes, elevated O_2 results in a net decrease in the synthesis of nitrogenase and a decrease in, or abolition of, symbiotic N_2 fixation (Monson *et al.,* 1995).

In addition to the *nif* genes, many other microbial genes are involved in symbiotic nitrogen fixation, these are collectively referred to as *fix* genes. Moreover, several other genes in the microsymbiont, including those for exopolysaccharide (Leigh and Walker, 1994; Glazebrook and Walker, 1989), hydrogen uptake (Baginsky *et al.*, 2002), glutamine synthase (Carlson *et al.*, 1987), dicarboxylate transport (Finan *et al.*, 1983; Jiang *et al.*, 1989), nodulation efficiency (Sanjuan and Olivares, 1989), β-1,2-glucans (Breedveld and Miller, 1994), and lipopolysaccharides (Carlson *et al.*, 1987), either directly or indirectly influence symbiotic N_2 fixation. Excellent in-depth reviews of the regulation of nitrogen fixation in free-living and symbiotic bacteria can be found in Merrick and Edwards (1995), Dean and Jacobson (1992), and Patriarca *et al.* (2002).

3.2. The Nodulation Process

The infection and nodulation process involves an intimate interaction of micro- and macro-symbiont, and is mediated by a bidirectional molecular communication between both symbiotic partners. The rhizobia induce two types of nodules on legumes, either determinant or indeterminant nodules (Franssen *et al.*, 1992). The indeterminant nodules are most commonly formed on temperate legumes (pea, clover, and alfalfa) inoculated with the fast-growing nodule bacteria, whereas determinate nodules are normally induced by bradyrhizobia on tropical legumes, such as soybean and common bean. Rhizobia infect host plants, and induce root- or stem-nodules, using three fundamentally different mechanisms: (i) *via* root hairs (Kijne, 1992); (ii) entry through wounds, cracks, or lesions (Boogerd and van Rossum, 1997); or (iii) *via* cavities located around root primorida of adventitious roots (Boivin *et al.* 1997).

In the root-hair mode of infection, rhizobia attach, often perpendicular, to susceptible root hairs within minutes of inoculation. Subsequent penetration of root-hair cell walls by rhizobia leads to root-hair curling, usually within 6-18 hours of inoculation. Rhizobia are enclosed within a plant-derived infection thread within the root-hair cell and move down the root hair towards the root cortex. Cell division within the root cortex, ahead of the approaching infection thread, eventually leads to the production of nodule primordia (Kijne, 1992). The infection thread spreads among cells of the nodule primordium and rhizobia are released into the host cortex by endocytosis. Within the host cytoplasm, the rhizobia are surrounded by a host-derived peribacteroid membrane, compartmentalizing the rhizobia into a symbiosome. Over time, the nodules expand and are usually visible 6-18 days after inoculation. Although nodulation initially is often heaviest in the crown of the root, secondary nodules frequently appear on lateral roots as the early crown nodules begin to senesce. The number and size of nodules on each legume host is controlled by the genotype of both the host and rhizobial partner, by the efficiency of the symbiotic interaction, by the presence of existing nodules, and by environmental factors, such as soil-nitrogen level and soil-moisture status (Caetano-Annoles, 1997; Sagan and Gresshoff, 1996; Singleton and Stockinger, 1983).

4. NODULATION AND NITROGEN-FIXATION GENETICS IN THE RHIZOBIA AND BRADYRHIZOBIA

The majority of genetic studies on rhizobia and bradyrhiziobia have concentrated on the genetics and molecular biology of nodulation and N_2 fixation. Over the last 17 years, advances in molecular biology and genetics have helped elucidated a large number of genes having symbiotic functions. In the fast growing species, symbiosis-related genes tend to be clustered on one or several relatively large symbiotic plasmids (Broughton *et al.*, 1984; Hombrecher *et al.*, 1981; Kondorosi *et al.*, 1989), whereas in the bradyrhizobia, these genes are chromosomally located. The review articles listed below should be consulted for more detailed information on the genetics of nodulation and nitrogen fixation (Bladergroen and Spaink, 1998; Boivin *et al.*, 1997; Debelle *et al.*, 2001; Gualtieri and Bisseling, 2000; Pueppke,

1996; Schultze and Kondorosi, 1998; Spaink, 1995; Spaink, 2000; van der Drift *et al.*, 1998).

Many of the bacterial genes involved in the formation of nodules on legumes have been identified and a complete description of the function of many of these genes can be found in several recent articles (Bladergroen and Spaink, 1998; Debelle *et al.,* 2001; Niner and Hirsch, 1998; Pueppke, 1996). Taken together, more than 70 nodulation genes have been identified in rhizobia, although only a subset of these may be found in any single strain. Interestingly, despite this relatively large number, only a relatively few genes are required for nodulation of legumes (Göttfert, 1993; Long *et al.,* 1985; Long, 1989; van Rhijn and Vanderleyden, 1995). Nodulation genes can be divided three broad groups based on their relationship to host specificity; they are common nodulation genes, host-specific nodulation genes (hsn), and genotype specific nodulation (gsn) genes (Bachem *et al.,* 1986; Bassam *et al.,* 1986; Broughton *et al.,* 1984; Davis *et al.*, 1988; Djordjevic *et al.*, 1985; Heron *et al.*, 1989; Horvath *et al.*, 1986; Lewin *et al.,* 1987; Lie, 1978; Lewis-Henderson *et al.*,1991; Meinhardt *et al.*, 1993, Nieuwkoop *et al.*, 1987; Sadowsky *et al.,* 1991; Sadowsky *et al.*, 1995; Wijffelman *et al.,* 1985).

4.1. Environmental Influences on Nodulation Genes

As might be expected from the number of genes involved, the induction and repression of bacterial nodulation genes is under tight regulatory control and is a major factor influencing host specificity (Spaink *et al.*, 1987) and response to environmental variables. The *nodD* gene can be viewed as a global regulatory gene which, together with plant flavonoid-signal molecules, activates transcription of other inducible nodulation genes (Banfalvi *et al.*, 1988; Boundy-Mills *et al.*, 1994; Djordjevic *et al.,* 1987; Fellay *et al.,* 1995; Göttfert *et al.*, 1988; Innes *et al.*, 1985; Kosslak *et al.*, 1987; Long 1989; Long, 2001; Martinez and Palacios, 1990; Mulligan and Long, 1985; Olson *et al.*, 1985; Peters *et al.*, 1986; Price *et al.*, 1992; Sadowsky *et al.*, 1988; van Brussel *et al.*, 1990; Zaat *et al.*, 1987). The flavonoid *nod*-gene inducers are specific for a particular legume-*Rhizobium* interaction (Schlaman *et al.*, 1998) and their production is influenced by environmental variables, like plant fertility, pH (Hubac *et al.*, 1994), and Nod factors (Schmidt *et al.*, 1994). Repressor proteins also play a role in *nod*-gene regulation (Kondorosi *et al.*, 1988; Kondorosi *et al.*, 1989; Kondorosi *et al.*, 1991; Stacey *et al.*, 2002).

The *nod* genes of the microsymbiont are involved in the production of extracellular nodulation factors, called lipochitinoligosaccharides (LCOs) (Carlson *et al.*, 1993; 1994; Downie, 1998; Lerouge *et al.*, 1990; Pueppke, 1996). LCOs stimulate the plant to produce more *nod*-gene inducers, to deform root hairs on their respective host plant, and to initiate cell division in the root cortex (Banfalvi *et al.*, 1989; Faucher *et al.,* 1989; Lerouge *et al.,* 1990; Price *et al.*, 1992; Relic *et al.,* 1993; Schultze *et al.*, 1992; Spaink *et al.,* 1991; van Brussel *et al.*, 1990). Purified Nod factors have been shown to induce nodules on the specific host plant (Downie, 1998; Mergaert *et al.,* 1993; Relic *et al.,* 1993, Schultze *et al.,* 1992; Truchet *et al.,* 1991). The functions of *nod* genes and the basic structure of Nod factors for *B.*

japonicum and several species of the genus *Rhizobium* have been reviewed (Carlson *et al.*, 1993; Debelle *et al.*, 2001; D'Haeze and Holster, 2002; Downie, 1998; Sanjuan *et al.*, 1992). Recently, Endre and coworkers (2002) reported that a receptor kinase, NORK, in *Medicago sativa* is essential for Nod-factor perception in alfalfa and that the NORK system initiates a signal cascade, which leads to nodulation. The production of Nod factors by rhizobia is influenced by pH, temperature, and both phosphorus and nitrogen concentration (McKay and Djordjevic, 1993). Production and excretion of *nod* metabolites by *Rhizobium leguminosarum* bv. *trifolii* are disrupted by the same environmental factors that reduce nodulation in the field.

Although much has been learned about legume nodulation over the last 7 years using genetic and molecular methods, the genomic revolution will no doubt substantially increase our understanding of this process at a very rapid rate. To date, the complete genomic sequence of the symbiotic plasmid from *Rhizobium* sp. strain NGR-234 (see http://genome.imb-jena.de/other/cfreiber/ pNGR234a2.html), *S. meliloti* (see http://cmgm.stanford.edu/~mbarnett/genome.htm and http://sequence.toulouse.inra.fr/meliloti.html), and the genomes of *M. loti* (see http://www.kazusa.or.jp/en/database.html) and *B. japonicum* (see http://www.genome.clemson.edu /~twood/projects/brady.html and http://www.kazusa.or.jp/en/database.html, and http://www.kazusa.or.jp/rhizobase/Bradyrhizobium/index.html) have been determined. For further information, please also see volume 3 of this series entitled *Genomes and Genomics of Nitrogen-fixing Organisms*.

5. RHIZOBIA IN THE SOIL ENVIRONMENT

Rhizobia can exist in two fundamentally different modes. They can live in soils either as free-living saprophytic heterotrophs or as legume-host-specific nitrogen-fixing symbionts. This dual mode of existence gives rhizobia several distinct advantages with respect to survival and persistence over most other soil bacteria. The bulk soil that surrounds legumes contains relatively large numbers of rhizobia, often approaching 10^6 cells g^{-1} in soils of the American Midwest (Ellis *et al.*, 1984) and sometimes up to 10^8 cells g^{-1} (Bottomley, 1992). The growth of rhizobia in the rhizosphere may be stimulated by plant root exudates (Van Egeraat, 1975), although more research is needed in this area. Interestingly, Phillips *et al.* (1999) reported that rhizobia can also stimulate growth and respiration of leguminous plants.

Generally speaking, rhizobia in soils are associated with aggregates (Mendes and Bottomley, 1998; Postma *et al.*, 1990), which gives them some degree of protection from perturbations by environmental and biotic factors. However, the nodule environment affords the rhizobia a unique niche in which to multiply while being protected. Nodules can contain more than 10^{10} rhizobia g-1 (McDermott *et al.*, 1987). Nodule senescence at the end of the growing season leads to the release of a large number of rhizobia into soils. Numerous studies have shown that a legume host is not needed for persistence (saprophytic competence) of rhizobia in soils (Bottomley, 1992; Brunel *et al.*, 1988; Chatel *et al.*, 1968; Kucey and Hynes, 1989). Although nodule bacteria and bacteroids, after release into the environment, are often susceptible to osmotic and other soil stress factors (Sutton, 1983), many of the released rhizobia survive and persist in the soil as free-living, heterotrophic saprophytes for long periods until they again come into contact with susceptible legume host (Diatloff, 1977; Brunel *et al.*, 1988; Lindstrom *et al.*, 1990).

6. STRESS FACTORS IN THE SOIL ENVIRONMENT THAT INFLUENCE N_2 FIXATION

Virtually any environmental factor that negatively influences either the growth of rhizobia or the host plant itself has a dramatic impact on symbiotic N_2 fixation. These factors can independently negatively influence the nodulation process itself, and thereby indirectly affect nitrogen fixation, or directly influence plant growth and vigor during post-nodulation events and so affect the efficient functioning of the nitrogenase enzyme complex. To facilitate discussion, I have separated these fundamentally different sets of factors below. However, the reader should be aware that, in some instances, some factors simultaneously affect both rhizobia and the host plant.

6.1. Soil Water Content and Stress

Soil water influences the growth of soil microorganisms through processes of diffusion, mass flow, and nutrient concentration (Paul and Clark, 1988). Soil water is related to soil pore space, and soils containing larger pores and pore spaces retain less water. Thus, soil aggregates having smaller internal pore spaces are more favorable environments for the growth of rhizobia and most soil microbes (Papendick and Campbell, 1981; Turco and Sadowsky, 1995). Soil-water content also directly influences the growth of rhizosphere microorganisms, like rhizobia, by decreasing water activity below critical tolerance limits and indirectly by altering plant growth, root architecture, and exudations. Poor nodulation of legumes in arid soils is likely due to decreases in population levels of rhizobia during the dry season. However, the influence of soil-water activity on plant growth and vigor, and hence nodulation, should not be ignored.

Water activity (A_w) and water potential (ψ) values are parameters often used for describing water relations with respect to microorganisms (Harris, 1981). Rhizobia vary widely in their tolerance to water stress, and there is little apparent correlation to taxonomic criteria and phylogenetic relationships. Although bradyrhizobia have been suggested to be more resistant to water stress than the rhizobia (Bushby and Marshall, 1977), the reverse has also been reported (Mahler and Wollum, 1981). In solute-controlled water stress conditions, such as those controlled by salts, bacteria are able to grow and survive when Aw values are in the range of 0.76 - 0.99. This range corresponds to water potential (ψ) values of -15 to -350 bars. In rhizobia (and all bacteria), both cell membranes and walls play a pivotal role in tolerance to water-potential stress. Under conditions of elevated salt or other osmolytes, water stress is dominated by the movement of water in response to the ψ gradient (Harris, 1981). The total ψ of a microbial cell is due to the sum of the ψ from components of intermediary metabolism, and the intracellular accumulation of stress and compatible solutes. Many microbes, including rhizobia, accumulate compatible solutes, such as amino acids, salts, and betaines, as a means to equilibrate internal and external osmotic concentrations (Csonka, 1991; Csonka and Epstein, 1996).

Osmoregulation is a complex problem for rhizobia because they must be able to adapt to unfavorable and changing environmental conditions, as well as to osmotic-stress conditions associated with the infection process itself and life in the nodule. The detrimental effects of salt stress on inoculum viability, nodulation and nitrogen fixation have been reported for many *Rhizobium* spp. strains (Israel, 1988). Thus, elevated osmotic conditions can limit symbiosis by affecting survival and proliferation of rhizobia in the soil, inhibiting the infection process, or by directly affecting root-nodule function.

Rhizobia have evolved a variety of mechanisms for adapting to osmotic stress, mostly by the intracellular accumulation of inorganic and/or organic solutes. For example, *R. meliloti* overcomes osmotic stress-induced growth inhibition by accumulating compatible solutes, such as K^+, glutamate, proline, glycine betaine, proline betaine, trehalose, and the dipeptide, N-acetylglutaminylglutamine amide (Bernard e*t al.*, 1986; Botsford, 1984; Botsford, 1990; LeRudulier and Bouillard, 1983; Smith *et al.*, 1988; Boscari *et al.*, 2002). Some compatible solutes can be used as either nitrogen or carbon sources for growth, suggesting that their catabolism may be regulated to prevent degradation during osmotic stress (Smith and Smith, 1989).

6.2. Desiccation Tolerance

Some species of *Rhizobium* are also susceptible to non solute-mediated desiccation, referred to here as matric-mediated drought conditions. Species of *Rhizobium* differ in their susceptibility to the detrimental effects of desiccation in natural soils. Slow-growing rhizobia are generally thought to survive desiccation in a sandy soil better than fast-growing types (Bushby and Marshall, 1977; Bushby and Marshall, 1977a). Mahler and Wollum (1980; 1981) reported that moisture level was the dominant factor influencing both short- and long-term survival of *B. japonicum* strains inoculated into a loamy sand. Both soil type and temperature are important factors influencing survival of rhizobia in desiccated soils (Boumahdi *et al.*, 2001; Mahler and Wollum, 1980; Trotman and Weaver, 1995). There is no information available concerning the genetics and physiology of desiccation resistance in rhizobia, however, Boumahdi *et al.* (2001) reported that a reduction in medium Aw led to a decrease in unsaturation of cellular fatty acids in rhizobia (Boumahdi, *et al.*, 2001).

On addition to soils, rapid moisture loss is responsible for rhizobial death on the seed surface (Vincent *et al.*, 1962). In addition, and generally speaking, rhizobia are more resistant to soil-water deficit (drought) than the plant itself. Nevertheless, rhizobial strains that are superior under drought conditions have been reported (Athar and Johnson, 1996; Athar and Johnson, 1997; Hunt *et al*, 1981).

The impact of drought conditions on N_2 fixation is also due to direct influences on the plant partner (see reviews by Serraj *et al.,* 1999 and Hungria and Vargas, 2000). Soil-water deficit influences several aspects of the host legume, including but not limited to, nodule establishment, carbon and nitrogen metabolism, nitrogenase functioning, and photosynthetically-derived energy supply. Drought stress should be thought of as influencing the host in a global manner, rather than a collection of individual processes. Results from studies over the last 30 or so years have shown that the N_2-fixation process itself is more sensitive to drought conditions than gas exchange from leaves (Durrand *et al.*, 1987; Sinclair *et al.*, 1986), the accumulation of plant dry matter and carbon assimilation (Sinclair *et al.*, 1987; Wery *et al.*, 1994), photosynthesis, and nitrate assimilation (Purcell and King, 1996; Wery *et al.*, 1988). However, as discussed above, soil moisture also indirectly influences total plant N_2 fixation by decreasing nodule mass and number (Sangakkara *et al.*, 1996; Sinclair *et al.*, 1988). The decline in nodule number under drought conditions is most likely due to impacts on the infection process itself. Root-hair infections and infection-thread formation are negatively influenced by drought conditions (Sprent, 1971; Graham, 1992).

6.3. Nutrient Stress

As might be expected, soil nutrient status has a tremendous influence on the symbiosis, as well as on the independent growth and survival of both partners. It should be noted, however, that in some cases, nutrient stresses are indirectly caused by changes in soil matric potential or acidity, which limit nutrient bioavailability, rather than to the lack of the presence of nutrients *per se*. When considering nutrient limitations to symbiotic nitrogen fixation, one must clearly separate factors affecting growth of the host from those influencing the microbe or the symbiotic interaction. For example, acid and water stress causes alterations in root growth, which can indirectly affect both nodulation and nitrogen fixation. This effect is thought to be mediated by abscisic acid (Zhang *et al.*, 2001).

Stress conditions apparently increase requirements for essential elements, such as Ca^{2+}, P, and N, in both plants and microsymbionts (Beck and Munns, 1984; O'Hara *et al.*, 1989; Zahran, 1999). Ca^{2+} might, in some instances, offset the deleterious influence of low pH on root growth and ion uptake (Torimitsu *et al.*, 1985) and increase *nod*-gene induction and expression (Richardson *et al.*, 1988). Calcium deficiency, with or without the confounding influence of low pH, also affects attachment of rhizobia to root hairs (Caetano-Anolles *et al.*, 1989; Smit *et al.*, 1992), and nodulation and nodule development (Alva *et al.*, 1990). Lastly, a calcium-spiking phenomenon is initiated in root-hair cells of legumes by nodulation factors and rhizobia (Wais *et al.*, 2002), suggesting that Ca^{2+} plays a pivotal role in symbiotic interactions at the molecular level.

Phosphorous supply and availability remains a severe limitation to nitrogen fixation and symbiotic interactions. About 33% of the ariable land in the world is limited by P availability (Graham and Vance, 2000; Pereira and Bliss, 1989; Sanchez and Euhara, 1983). This situation is especially true in soils impacted by low pH. There are marked differences in rhizobial and plant requirements for P (Beck and Munns, 1985; Pereira and Bliss, 1989) with the slow-growers being more tolerant to low P than the fast-growing rhizobia (Beck and Munns, 1985). Nitrogen-fixing plants have an increased requirement for P over those receiving direct nitrogen fertilization, owing perhaps to nodule development and signal transduction

(Graham and Vance, 2000), and to P-lipids in the large number of bacterioids. Moreover, in white lupin, P deficiency leads to enhanced P acquisition by the formation of proteoid roots (Johnson *et al.*, 1996), changes in carbon metabolism, enhanced secretion of citrate and malate from roots, and release of a novel acid phosphatase into the rhizosphere proteoid roots (Gilbert *et al.*, 1999). Nodules themselves are sinks for P (Hart, 1989) and nodulation and N_2 fixation are influenced strongly by P availability (Leung and Bottomley, 1987; Singleton *et al.*, 1985; Saxena and Rewari, 1991). The rhizobia respond to P stress in a manner analogous to the host plant; that is, low P induces expression of genes involved in P acquisition (phosphatases, phosphate transporters) and acidification of the root zone (Al-Niemi *et al.*, 1997; Smart *et al.*, 1984; Torriani-Gorini *et al.*, 1987).

In addition to macro nutrients, the growth and persistence of rhizobia in soils is also influenced by several other nutritional factors (reviewed by Brockwell *et al.*, 1995 and Bottomley, 1992). The rhizobia are metabolically diverse and have been shown to use a variety of both plant- and soil-derived compounds for growth. Interestingly, some of the same compounds that support growth have also been shown to be chemotactic and induce *nod* genes (Parke and Ornston, 1986; Sadowsky and Graham, 1998). In addition, supplementation of soil and inoculants with gutamate, glycerol, and organic matter has been shown to enhance the survival and numbers of rhizobia in soils and increase both early nodulation and N_2 fixation (Rynne *et al.*, 1994). This result indicates that, although rhizobia can surely persist in soils, their efficacy can be enhanced by carbon addition, which suggests that they are C limited in the natural state.

6.4. Soil pH Stress

The influence of soil pH on the nodulation process has been extensively examined, in part due to the World's large number of acid soils (see reviews by Graham, 1992 and Hungria and Vargas, 2000). Worldwide, more than 1.5 Gha of acid soils limit agriculture production (Edwards *et al.*, 1991; Graham and Vance, 2000) and as much as 25% of the earth's croplands are impacted by problems associated with soil acidity (Munns, 1986). Brockwell *et al.* (1991) reported a nearly 10^{-3} decrease in the number of *S. meliloti* in soils with a pH < 6 compared to those with a pH > 7.0. As found for soil moisture, there is a range of effects of soil pH on rhizobia, but relatively few grow and survive well below pH values of 4.5-5.0 (Graham *et al.*, 1994; Vargas and Graham, 1988).

Competitive interactions have been shown to be influenced by soil pH. Generally speaking, the bradyrhizobia are more acid tolerant than the rhizobia (Brockwell *et al.*, 1991; Date and Halliday, 1979; Keyser and Munns, 1979; Sadowsky and Graham, 1998), although some strains of *R. tropici* are very acid tolerant (Graham *et al.*, 1992). The influence of soil pH on the behavior of rhizobia in soils can be dramatic. Generally, *S. fredii* is more competitive than *B. japonicum* in soybean nodulation in neutral soils in Spain, but *B. japonicum* USDA 110 out-competed *S. fredii* in soil at pH 4.9 (Triplett and Sadowsky, 1992). Furthermore, despite the fact that *R. etli* is more competitive than *R. tropici* in nodule formation with beans (Chaverra and Graham, 1992; Martinez Romero and Rosenblueth,

1990), acidification of soils led to the replacement of *R. etli* by introduced, acid-tolerant *R. tropici* (Anyango *et al.*, 1995; Hungria *et al.*, 1997). However, the relationship between soil acidity, competitiveness, and the ability to survive in acid soils is not always straightforward and due to resistance to acidity. For example, Richardson and Simpson (1989) reported that many rhizobia from acid soils are sensitive to acidity. They suggested that soil microniches protect these rhizobia from extremes of soil pH. So, merely isolating a rhizobial strain from root nodules of plants grown in acid conditions does not guarantee that the isolated strain will be acid resistant (Graham, 1992). Furthermore, metals, such as Al, Cu, and Mn that become more soluble at lower pH, may also secondarily contribute to the inhibition of the growth and persistence of rhizobia in acid soils (Cooper *et al.*, 1983; Coventry and Evans, 1989; Hungria and Vargas, 2000; Lal, 1993; Reeve *et al.*, 2002). Moreover, soil Ca and P levels are also influenced by soil pH (Bell *et al.*, 1989; Munns, 1970) and may secondarily influence the growth and survival of rhizobia. Nevertheless, results from several studies have indicated that tolerance to acid conditions in rhizobia is often correlated to the strains ability to maintain an internal pH approaching neutrality (pH 7.2-7.5) (Graham *et al.*, 1994; Kashket, 1985; O'Hara *et al.*, 1989). This ability has been suggested to be due to proton exclusion (Graham, 1991), enhanced cytoplasmic-buffering capacity (Krulwich *et al.*, 1985), the presence of acid-shock responses (Bhagwat and Apte, 1989), the presence of glutathione (Riccillo *et al.*, 2000), the maintenance of elevated cellular potassium and glutamate concentrations (Aarons and Graham, 1991; Graham *et al.*, 1992), membrane permeability (Chen *et al.*, 1993), and calcium metabolism (Howieson *et al.*, 1992).

Although the microsymbiont appears more pH sensitive than the host partner (Hungria and Vargas, 2000), acidity also influences both the growth of the legume plant and the infection process (Munns, 1986). This effect is, in part, most likely due to both a disruption of signal exchange between macro- and micro-symbionts (Hungria and Stacey, 1997) and repression of nodulation genes and excretion of Nod factor in the rhizobia (Richardson *et al.*, 1988). Interestingly, nodulated legumes appear more sensitive to metal toxicity by Mn and Al than to their N-fed control counterparts (Hungria and Vargas, 2000).

Recently, using molecular techniques and proteomics, Glenn and colleagues have shown that rhizobial genes, such as *actA, actP, exoR, lpiA, actR, actS,* and *phrR*, are essential for growth at low pH (Dilworth *et al.*, 2001; Glenn *et al.*, 1999; Reeve *et al.*, 2002). Vinuesa and coworkers (2003), using a *Tn*5-mutagenesis approach, isolated and characterized pH-responsive genes, *lpiA* and *atvA*, from *Rhizobium* tropici CIAT899. Complementation analyses indicated that *atvA*, an ortholog of the *A. tumefaciens acvB* gene, is required for acid tolerance.

6.5. Soil Temperature

Temperature has a marked influence on survival and persistence of rhizobial strains in soils. For example, cowpea rhizobia from the hot dry Sahel-savannah of West Africa grow at 37°C, and more than 90% of the strains isolated from this region grew well to 40°C (Eaglesham *et al.*, 1991). The influence of temperature on rhizobia appears to be both strain and soil dependent. For example, *Bradyrhizobium* sp. (lupins) was less susceptible than *R. leguminosarum* bv. *trifolii* to high soil temperatures, but addition of montmorillonite and illite remediated this problem in sandy soils (Marshall, 1964). Soil temperature also greatly influences competition for nodulation (Kluson *et al.*, 1986; Triplett and Sadowsky, 1996). This effect may, in part, be due to a temperature-induced delay in nodulation or the restriction of nodules to the sub-surface region (Munns *et al.*, 1977).

Both temperature extremes need to be examined when considering the influence of soil temperature on the growth and survival of rhizobia, and the latter two parameters need to be separated. For example, whereas rhizobia isolated from temperate regions often survive at 4°C, little growth occurs at this temperature (Trinick, 1982). Nevertheless, there have been reports of growth of rhizobia from the Canadian high arctic at 5°C (Prévost *et al.*, 1987) and 10°C (Caudry-Reznick *et al.*, 1986). However, the symbiosis itself is sensitive to low temperatures; cooler root-zone temperatures limit nodulation and nitrogen fixation in the soybean-*B. japonicum* symbiosis (Lynch and Smith, 1994; Zhang *et al.*, 1995).

There are a few generalities that can be made. Although high soil temperatures may lead to death of many rhizobia isolated from temperate climates, strains from tropical region generally survive better at high soil temperatures. Nevertheless, Somasegaran *et al.* (1984) reported that incubation of inoculant strains at 37°C led to a gradual decline in population levels over an 8-week period. Some temperature-tolerant *Bradyrhizobium* sp. strains that nodulate cowpea in Nigeria have also been noted (Hartel and Alexander, 1984). In addition, temperature-tolerant strains of rhizobia can be either artificially (Karanja and Wood, 1988; Hartel and Alexander, 1984) or naturally selected for (Zahran *et al.*, 1994). However, excessive temperature shock has been shown to cure plasmids in fast-growing strains and some strains, which were isolated from high-temperature environments, have a Fix^- phenotype (Hungria and Franco, 1993; Moawad and Beck, 1991). Effective high-temperature (40°C) tolerant rhizobia that are capable of nodulating and fixing nitrogen with *Phaseolus vulgaris* (Hungria *et al.*, 1993; Michiels *et al.*, 1994), *Acacia* (Zerhari *et al.*, 2000), and *Prosopis* (Kulkarni and Nautiyal, 1999) have also been reported. Although there have been several attempts to adapt rhizobia to higher temperatures for inoculation of legumes in tropical regions, in most cases, incubation of strains at elevated temperatures results in the loss of either infectivity or effectivity (Segovia *et al.*, 1991; Wilkins, 1967).

Relatively high-root temperature has also been shown to influence infection, N_2-fixation ability, and legume growth (Arayankoon *et al.*, 1990; Hungria and Franco, 1993; Kishinevsky *et al.*, 1992; Michiels *et al.*, 1994; Munevar and Wollum, 1982) and it has a strong influence on specific strain and cultivar interactions (Arayankoon *et al.*, 1990; Munevar and Wollum, 1982). It appears that every legume/*Rhizobium* combination has an optimum temperature relationship, which is around 30°C for clover and pea, between 35-40°C for soybean, peanut and cowpea, and between 25-30°C for common bean (Michiels *et al.*, 1994; Piha and Munns, 1987). Exposure of both symbiotic partners to temperature extremes much above or below these critical temperatures impairs infection, nodulation, nodule development, and general nodule functioning (Gibson, 1971; Roughley, 1970) as well as both plant growth and productivity. High soil tempratures also restrict nodulation to sub-surface regions where cooler temperatures prevail (Graham, 1991).

The physiological basis feor the temperature sensitivity of the symbiosis is most likely complex because many cellular functions in both host and microbe are affected by elevated and low temperature. Nevertheless, elevated temperature directly influences the production or release of *nod*-gene inducers from soybean and bean (Hungria and Stacey, 1997), it alters nodule functioning due to leghemoglobin synthesis, nitrogenase activity, and hydrogen evolution, and, in addition, hastens nodule senescence (Hungri and Vargas 2000). Although heat-shock proteins have been found in rhizobia (Aarons and Graham, 1991; Labidi *et al.*, 2000; Michiels *et al.*, 1994; Munchbach *et al.*, 1999) and heat stress alters the mobility of LPS (Zahran *et al.*, 1994), their direct role in either heat tolerance or sensitivity has not been demonstrated.

7. CONCLUDING REMARKS

Despite many decades of progress and the acquisition of a large amount of useful information, the physiological and molecular bases for the tolerance of legume-microbe symbiotic systems to environmental stress remains largely unknown and empirical in nature. Although understanding these processes was originally thought to be straightforward and tractable, we have learned that we now have more questions than answers. This situation is perhaps due to the fact that abiotic stresses independently and differentially influence the host legume, the rhizobia, and the symbiotic couple. So where do we go from here? Clearly, more work needs to be done on the underlying molecular bases for tolerance to stress factors in both legume and microbe.

Recent advances in the genomics and proetomics of macro- and micro-symbionts will accelerate progress in this area by providing a wealth of information on how both host and microbe respond to environmental perturbations. For example, proteome analysis has been used to investigate oxidative stress in the *Rhizobium etili-Phaseolus vulgaris* symbiosis (del Carmen Vargas *et al.*, 2003), to define bacterial genes involved in growth at low pH (Dilworth *et al.*, 2001; Glenn *et al.*, 1999; Reeve *et al.*, 2002), and to examine cultivar-specific interactions between *Rhizobium leguminosarum* bv. *trifolii* and subterranean clover (Morris and Djordjevic, 2001). Similarly, Saalbach and coworkers (2002) and Wienkoop and Saalbach (2003) have used proteome analysis to investigate the proteins in the pea and lotus peribacteroid membrane, respectively, and Mathesius *et al.* (2001) have established a root proteome reference map of *Medicago truncatula* that can be used with expressed sequence tag databases (Fedorova *et al.*, 2002; Lamblin *et al.*, 2003) to investigate molecular mechanisms of root symbioses in legumes. This type of global organismal information at the genomic and proteomic levels, however, now needs to be coupled to traditional plant breeding and microbial selection efforts in order to rapidly define and utilize microbial and host genetic loci that are involved in tolerance to a large number of environmental stresses.

REFERENCES

Aarons, S. R., and Graham, P. H. (1991). Response of *Rhizobium leguminosarum bv. phaseoli* to acidity. *Plant Soil, 134*, 145-151.

Al-Niemi, T. S., Kahn, M. L., and McDermott, T. R. (1997). P metabolism in the bean-*Rhizobium tropici* symbiosis. *Plant Physiol., 113*, 1233-1242.

Alva, A. K., Assher, C. J., and Edwards, D. G. (1990). Effect of solution pH, external calcium concentration, and aluminum activity on nodulation and early growth of cowpea. *Aust J. Agric. Res., 41*, 359-365.

Anyango, B., Wilson, J. K., Beynon, J. L., and Giller, K. E. (1995). Diversity of rhizobia nodulating *Phaseolus vulgaris* in two Kenyan soils with contrasting pHs. *Appl. Environ. Microbiol., 61*, 416-421.

Arayankoon, T., Schomberg, H. H., and Weaver, R. W. (1990). Nodulation and N_2 fixation of guar at high root temperature. *Plant Soil, 126*, 209-213.

Athar, M., and Johnson, D. A. (1996). Nodulation, biomass production, and nitrogen fixation in alfalfa under drought. *J. Plant Nutr., 19*, 185-199.

Athar, M., and Johnson, D. A. (1997). Effect of drought on the growth and survival of *Rhizobium meliloti* strains from Pakistan and Nepal. *J. Arid Environ., 35*, 335-340.

Bachem, C. W., Banfalvi, Z., Kondorosi, E., Schell, J., and Kondorosi, A. (1986). Identification of host range determinants in the *Rhizobium* species MPIK3030. *Mol. Gen. Genet., 203*, 42-48.

Baginsky, C., Brito, B., Imperial, J., Palacios, J.-M., and Ruiz-Argüeso, T. (2002). Diversity and evolution of hydrogenase systems in rhizobia. *Appl. Environ. Microbiol., 68*, 4915-4924.

Banfalvi, Z., and Kondorosi, A. (1989). Production of root hair deformation factors by *Rhizobium meliloti* nodulation genes in *Escherichia coli,* HsnD (*nodH*) is involved in plant host-specific modification of the nodABC factor. *J. Mol. Biol., 13*, 1-12.

Banfalvi, Z., Nieuwkoop, A., Schell, M., Best, L., and Stacey, G. (1988). Regulation of *nod* gene expression in *Bradyrhizobium japonicum. Mol. Gen. Genet., 214*, 420-424.

Bassam, B. J., Rolfe, B. G., and Djordjevic, M. A. (1986). *Macroptilium atropurpureum* (siratro) host specificity genes are linked to a *nodD*- like gene in the broad host range *Rhizobium* strain NGR234. *Mol. Gen. Genet., 203*, 49-57.

Beck, D. P., and Munns, D. N. (1985). Effect of calcium on the phosphorus nutrition of *Rhizobium meliloti. Soil Sci. Soc. Am. J., 49*, 334-337.

Beck, D. P, and Munns, D. N. (1984). Phosphate nutrition of *Rhizobium* sp., *Appl. Environ. Microbiol., 47*, 278-282.

Bell, W., Edwards, D. G., and Asher, C. J. (1989). External calcium requirements for growth and nodulation of six tropical food legumes grown in flowing solution culture. *Aust. J. Agric. Res., 40*, 85-96.

Bernard, T., Pocard, J. A., Perroud, B. and LeRudulier, D. (1986). Variations in the response of salt-stressed *Rhizobium* strains to betaines. *Arch. Microbiol., 143*, 359-364.

Bhagwat, A. A., and Apte, S. K. (1989). Comparative analysis of proteins induced by heat shock salinity and osmotic stress in the nitrogen-fixing *Cyanobacterium anabaena*-sp strain. *J. Bacteriol., 171*, 5187-5189.

Bladergroen, M. R., and Spaink, H. P. (1998). Genes and signal molecules involved in the rhizobia-*Leguminoseae* symbiosis. *Curr. Opin. Plant Biol., 1*, 353-359.

Boivin C., Ndoye, I., Molouba, F., Delajudie, P., Dupuy, N., and Dreyfus, B. (1997). Stem nodulation in legumes - diversity, mechanisms, and unusual characteristics. *Critical Rev. Plant Sci., 16*, 1-30.

Boogerd, F. C., and van Rossum, D. (1997). Nodulation of groundnut by *Bradyrhizobium* - a simple infection process by crack entry. *FEMS Microbiol. Rev., 21*, 5-27.

Boscari, A., Mandon, K., Dupont, L., Poggi, M.-C., and Le Rudulier, D. (2002). BetS is a major glycine betaine/proline betaine transporter required for early osmotic adjustment in *Sinorhizobium meliloti. J. Bacteriol., 184*, 2654-2663.

Botsford, J. L. (1984). Osmoregulation in *R.meliloti*: Inhibition of growth by salts. *Arch. Microbiol., 137*, 124-127.

Botsford, J. L. (1990). Osmoregulation in *R.meliloti*: Production of glutamic acid in response to osmotic stress. *Appl. Environ. Microbiol., 56*, 488-494.

Bottomley, P. (1992). Ecology of *Rhizobium* and *Bradyrhizobium*. In G. Stacey, R. H. Burris, and H. J. Evans (Eds.), *Biological Nitrogen Fixation* (pp. 292-347). New York: Chapman & Hall.

Boumahdi, M, Mary, P and Hornez, J. P. (2001). Changes in fatty acid composition and degree of unsaturation of (brady)rhizobia as a response to phases of growth, reduced water activities and mild desiccation. *Antonie Van Leeuwenhoek, 79*, 73-79.

Boundy-Mills, K. L., Kosslak, R. M., Tully, R. E., Pueppke, S. G., Lohrke, S. and Sadowsky, M. J. (1994). Induction of the *Rhizobium fredii* nod box-independent nodulation gene *nolJ* requires a functional *nodD1* gene. *Mol. Plant-Microbe Interact., 7*, 305-308.

Breedveld, M. W. and Miller, K. J. (1994). Cyclic beta-glucans of members of the family *Rhizobiaceae. Microbiol. Rev., 58*, 145-161.

Brockwell, J., Bottomley, P. J. and Thies, J. E. (1995). Manipulation of rhizobia microflora for improving legume productivity and soil fertility, a critical assessment. *Plant Soil, 174*, 143-180.

Brockwell, J., Pilka, A. and Holliday, R. A. (1991). Soil pH is the major determinant of the numbers of naturally-occurring *Rhizobium meliloti* in non-cultivated soils in New South Wales. *Aust. J. Exp. Agric., 31*, 211-219.

Broughton, W. J., Heycke, N., Meyer, Z. A. and Pankhurst, C. E. (1984). Plasmid linked *nif* and *nod* genes in fast-growing rhizobia that nodulate *Glycine max, Psophocarpus tetragonolobus* and *Vigna unguiculata. Proc. Natl. Acad. Sci. USA, 81*, 3093-3097.

Brunel, B., Cleyet-Marel, J. C., Normand, P., and Bardin, R. (1988). Stability of *Bradyrhizobium japonicum* inoculants after introduction into soil. *Appl. Environ. Microbiol., 54*, 2636-2642.

Bushby, H. V. A., and Marshall, K. C. (1977). Water status of rhizobia in relation to their susceptibility to desiccation and to their protection by montmorillonite, *J. Gen. Microbiol., 99*, 19-27.

Bushby, H. V. A., and Marshall, K. C. (1977a). Some factors affecting the survival of root-nodule bacteria on desiccation. *Soil Biol. Biochem., 9*, 143-147.

Caetano-Anolles, G., Lagares, A., and Favelukes, G. (1989). Adsorption of *Rhizobium meliloti* to alfalfa roots, dependence on divalent cations and pH. *Plant Soil,* 117, 67-74.

Caetano-Anolles, G. (1997). Molecular dissection and improvement of the nodule symbiosis in legumes. *Field Crops Res., 53*, 47-68.

Carlson, R. W., Sanjuan, J., Bhat, U. R., Glushka, J., Spaink, H. P., Wijfjes, A. H. M., *et al.* (1993). The structures and biological activities of the lipo-oligosaccharide nodulation signals produced by type I and II strains of *Bradyrhizobium japonicum. J. Biol. Chem., 268*, 18372-18381.

Carlson, R. W., Kalembasa, S., Turowski, D., Pachori, P., and Noel, K. D. (1987). Characterization of the lipopolysaccharide from a *Rhizobium phaseoli* mutant that is defective in infection thread development. *J. Bacteriol., 169*, 4923–4928.

Carlson, R. W., Price, N. P. J., and Stacey, G. (1994). The biosynthesis of rhizobial lipo-oligosaccharide nodulation signal molecules. *Mol. Plant Microbe Interact.,* 7, 684-695.

Caudry-Reznick, S., Prevost, D., and Schulman, H. M. (1986). Some properties of arctic rhizobia. *Arch. Microbiol., 146*, 12-18.

Chatel, D. L., Greenwood, R. M., and Parker, C. A. (1968). Saprophytic competence as an important characteristic in the selection of *Rhizobium* for inoculation. *Proc. IXth Intern. Cong. Soil Sci. Adelaide, 2*, 65-73.

Chaverra, M. H., and Graham, P. H. (1992). Cultivar variation in traits affecting early nodulation of common bean. *Crop Sci., 32*, 1432-1436.

Chen, H., Richardson, A. E., and Rolfe, B. G. (1993). Studies on the physiological and genetic basis of acid tolerance in *Rhizobium leguminosarum* bv. *Trifolii. Appl. Environ. Microbiol., 59*, 1798-1804.

Cooper, J. E., Wood, M., and Holding, A. J. (1983). The influence of soil acidity factors on rhizobia. In D. G. Jones and D. R. Davies (Eds.), *Temperate Legumes. Physiology, Genetics and Nodulation* (pp. 319-335). London: Pittman.

Coventry, D. R., and Evans, J. (1989). Symbiotic nitrogen fixation and soil acidity. In A. D. Robson (Ed.), *Soil Acidity and Plant Growth* (pp. 103-137). Sydney: Academic Press.

Csonka, L. N. (1991). Prokaryotic Osmoregulation: Genetics and Physiology. *Ann. Rev. Microbiol., 45*, 569-606.

Csonka, L. N., and Epstein, W. (1996). Osmoregulation. In R. Curtiss III, *et al.* (Eds.), *Escherichia coli and Salmonella: Cellular and Molecular Biology, 2nd Ed.* (pp. 1210-1223). Washington D.C.:American Society for Microbiology.

D'Haeze, W., and Holsters, M. (2002). Nod factor structures, responses, and perception during initiation of nodule development. *Glycobiol., 12*, 79R-105R.

Date, R. A., and Halliday, J. (1979). Selecting *Rhizobium* for acid, infertile soils of the tropics. *Nature (Lond.), 277*, 62-64.

Davis E. O., Evans I. J., and Johnston, A. W. B. (1988). Identification of *nod*X, a gene that allows *Rhizobium leguminosarum* biovar *viceae* strain TOM to nodulate Afghanistan peas. *Mol. Gen. Genet., 212*, 531-535.

Dean, D. R., and Jacobson, M. R. (1992). Biochemical Genetics of Nitrogenase. In G. Stacey, R. H. Burris, and H. J. Evans (Eds.), *Biological Nitrogen Fixation* (pp. 763-834). New York: Chapman and Hall.

Debelle, F., Moulin, L., Mangin, B., Denarie, J., and Boivin, C. (2001). Nod genes and Nod signals and the evolution of the *Rhizobium* legume symbiosis. *Acta Biochim. Pol., 48*, 359-65.

Del Carmen Vargas, M., Encarnacion, S., Davalos, A., Reyes-Perez, A., Mora, Y., Garcia-De Los Santos, A., Brom, S., and Mora, J. (2003). Only one catalase, *katG*, is detectable in *Rhizobium etli*, and is encoded along with the regulator OxyR on a plasmid replicon. *Microbiol., 149*, 1165-1176.

Diatloff, A. (1977). Ecological studies of root nodule bacteria introduced into field environments. 6. Antigenic and symbiotic stability in *Lotononis* rhizobia over a 12-year period. *Soil Biol. Biochem., 9*, 85-88.

Dilworth, M. J., Howieson, J. G., Reeve, W. G., Tiwari, R. P., and Glenn, A. R. (2001). Acid tolerance in legume root nodule bacteria and selecting for it, *Aust. J. Experimen. Agric., 41*, 435-446.

Djordjevic, M. A., Schofield, P. R., and Rolfe, B. G. (1985). *Tn5* mutagenesis of *Rhizobium trifolii* host-specific nodulation genes results in mutants with altered host-range ability, *Mol. Gen. Genet., 200*, 463-471.

Djordjevic, M. A., Redmond, J. W., Batley, M., and Rolfe, B. G. (1987). Clovers secrete specific phenolic compounds which either stimulate or repress *nod* gene expression in *Rhizobium trifolii. EMBO J., 6*, 1173-1179.

Downie, J. A. (1998). The *Rhizobiaceae.* In H. P. Spaink, A. Kondorosi, and P. J. J. Hooykaas (Eds.), *Functions of Rhizobial Nodulation Genes* (pp. 387-402). Dordrecht: Kluwer Academic Publishers.

Durrand, J. L., Sheehy, J. E., and Minchin, F. R. (1987). Nitrogenase activity, photosynthesis, and nodule water potential in soybean plants experiencing water deprivation. *J. Exper. Bot., 38*, 311-321.

Eaglesham, A., Seaman, B., Ahmad, H., Hassouna, S., Ayanaba A. and Mulongoy, K. (1991). High temperature tolerant "cowpea" rhizobia. In A. H. Gibson and W. E. Newton (Eds.), *Current Perspectives in Nitrogen Fixation* (p. 356). Canberra: Austral. Acad. Sci.

Edwards, D. G., Sharifuddin, H. A. H., Yusoff, M. N. M., Grundon, N. J., Shamshuddin, J., and Norhayati, M. (1991). The management of soil acidity for sustainable crop production. In R. J. Wright *et al.* (Eds.), *Plant-Soil Interaction at Low* pH (pp. 383-396). Dordrecht: Kluwer Academic Publishers.

Ellis, W. R., Ham, G. E., and Schmidt, E. L. (1984). Persistence and recovery of *Rhizobium japonicum* inoculum in a field soil. *Agron. J., 76*, 573-576.

Endre, G., Kereszt, A., Kevei, Z., Mihacea, S., Kalo, P., and Kiss, G. B. (2002). A receptor kinase gene regulating symbiotic nodule development. *Nature, 417*, 962-966.

Faucher, C., Camut, S., Denarie, J., and Truchet, G. (1989). The *nodH* and *nodQ* host range genes of *Rhizobium meliloti* behave as avirulence genes in *R. legumiosarum bv viceae* and determine changes in the production of plant-specific extracellular signals. *Mol. Plant-Microbe Interact., 2*, 291-300.

Fedorova, M., van de Mortel, J., Matsumoto, P. A., Cho, J., Town, C. D., VandenBosch, K. A., *et al.* (2002). Genome-wide identification of nodule-specific transcripts in the model legume *Medicago truncatula. Plant Physiol., 130*, 519-537.

Fellay, R., Perret, X., Viprey, V., and Broughton, W. J. (1995). Organization of host-inducible transcripts on the symbiotic plasmid of *Rhizobium* sp. NGR234. *Molec. Microbiol., 16*, 657–667.

Finan, T. M., Wood, J. M., and Jordon, D. C. (1983). Symbiotic properties of C4-dicarboxylic acid transport mutants of *Rhizobium leguminosarum. J. Bacteriol., 154*, 1403–1413.

Franssen, H. J, Nap, J.-P., and Bisseling, T. (1992). Nodulins in root nodule development. In G. Stacey, R. H. Burris, and H. J. Evans (Eds.), *Biological Nitrogen Fixation* (pp. 598-624). New York: Chapman and Hall.

Gibson, A. H. (1971). Factors in the physical and biological environment affecting nodulation and nitrogen fixation by legumes. *Plant Soil (Spec. Vol.)*, 139–152.

Gilbert, G. A., Knight, J. D., Vance, C. P., and Allan, D. L. (1999). Acid phosphatase activity in phosphorus-deficient white lupin roots. *Plant Cell Environ., 22*, 801-810.

Glazebrook, J., and Walker, G. C. (1989). A novel exopolysaccharide can function in place of the calcofluor-binding exopolysaccharide in nodulation of alfalfa by *Rhizobium meliloti. Cell, 56*, 661–672.

Glenn, A. R., Reeve, W. G., Tiwari, R. P., and Dilworth, M. J. (1999). Acid tolerance in root nodule bacteria. *Novartis Found Symp., 221*, 112-126.

Golley, F., Baudry, J., Berry, R., Bornkamm, R., Dahlberg, K., Jansson, *et al.* (1992). What is the road to sustainability. *INTECOL Bull., 20*, 15-20.

Göttfert, M. (1993). Regulation and function of rhizobial nodulation genes. *FEMS Microbiol. Lett., 104*, 39-64.

Göttfert, M., Webber, J., and Hennecke, H. (1988). Induction of a *nodA-lacZ* fusion in *Bradyrhizobium japonicum* by an isoflavone. *J. Plant Physiol., 132*, 394-397.

Graham, P. H. (1992). Stress tolerance in *Rhizobium* and *Bradyrhizobium* and nodulation under adverse soil conditions. *Can. J. Microbiol., 38*, 475-484.

Graham, P. H., and Vance, C. P. (2000). Nitrogen fixation in perspective, an overview of research and extension needs. *Field Crops Res., 65*, 93-106.

Graham, P. H., Draeger, K. J., Ferrey, M. L., Conroy, M. J., Hammer, B. E., Martinez, E., *et al.* (1994). Acid pH tolerance in strains of *Rhizobium* and *Bradyrhizobium*, and initial studies on the basis for pH tolerance of *Rhizobium tropici* UMR1899. *Can. J. Microbiol., 40*, 198-207.

Gualtieri, G., and Bisseling, T. (2000). The evolution of nodulation. *Plant Mol Biol., 42*, 181-194.

Harris, R. F. (1981). Effect of water potential on microbial growth and activity. In *Water potential Relations in Soil Microbiology*. SSSA special publication No. 9, (p. 23). Madison, WI: Soil Science Society of America.

Hart, A. L. (1989). Nodule phosphorus and nodule ctivity in white clover. *N. Z. J. Agric. Res., 32*, 145-149.

Hartel, P. G., and Alexander, M. (1984). Temperature and desiccation tolerance of cowpea rhizobia. *Can. J. Microbiol., 30*, 820-823.

Heron D. S., Ersek, T., Krishan, H. B., and Pueppke, S. G. (1989). Nodulation mutants of *Rhizobium fredii* USDA 257. *Molec. Plant-Microbe Interact., 2*, 4-10.

Hirsch, A. M., Lum, M. R., and Downie, J. A. (2001). What makes the rhizobia-legume symbiosis so special? *Plant Physiol., 127*, 1484-1492.

Hombrecher, G., Brewin, N. J., and Johnston, A. W. B. (1981). Linkage of genes for nitrogenase and nodulation ability on plasmids in *Rhizobium leguminosarum* and *R. phaseoli. Mol. Gen. Genet., 182*, 133-136.

Horvath, B., Kondorosi, E., John, M., Schmidt, J., Torok, I, Gyorgypal, *et al.* (1986). Organization, structure, and symbiotic function of *Rhizobium meliloti* nodulation genes determining host specificity for alfalfa. *Cell, 46*, 335-343.

Howieson, J. G., Robson, A. D., and Abbott, L. K. (1992). Calcium modifies pH effects on the growth of acid-tolerant and acid-sensitive *Rhizobium meliloti. Aust. J. Agric. Res., 43*, 765-772.

Hubac, C., Ferran, J., Tremolieres, A., and Kondorosi, A. (1994). Luteolin uptake by *Rhizobium meliloti,* Evidence for several steps, including an active extrusion process. *Microbiol., 140*, 2769-2774.

Hungria, M., and Franco, A. A. (1993). Effects of high temperature on nodulation and nitrogen fixation by *Phaseolus vulgaris* (L.). *Plant Soil, 149*, 95-102.

Hungria, M., Franco, A. A., and Sprent, J. I. (1993). New sources of high-temperature tolerant rhizobia for *Phaseolus vulgaris* (L.). *Plant Soil, 149*, 103-109.

Hungria, M. and Stacey, G. (1997). Molecular signals exchanged between host plants and rhizobia, basic aspects and potential application in agriculture. *Soil Biol. Biochem., 29*, 519–830.

Hungria, M., Vargas, M. A. T., and Araujo, R. S. (1997). Fixação biológica do nitrogênio em feijoeiro. In M. A. T. Vargas and M. Hungria (Eds.), *Biologia dos Solos dos Cerrados* (pp. 189-295). Planaltina, Brazil: EMBRAPA-CPAC.

Hungria, M., and Vargas, M. A. T. (2000). Environmental factors affecting N_2 fixation in grain legumes in the tropics, with emphasis on Brazil. *Field Crops Res., 65*, 151-164.

Hunt, P. J., Wollum, A. G., and Matheny, T. A. (1981). Effects of soil water on *Rhizobium japonicum* infection nitrogen accumulation and yield in Bragg soybean. *Agric. J., 73*, 501-505.

Innes, R. W., Kuempel, P. L., Plazinski, J., Canter-Cremers, H., Rolfe, B. G., and Djordjevic, M. A. (1985). Plant factors induce expression of nodulation and host-range genes in *R. trifolii.Mol. Gen. Genet., 201*, 426-432.

Israel D. W., *et al.* (1988). Relative performance of *Rhizobium* and *Bradyrhizobium* strains under different environmental conditions. *ISI Atlas of Science; Animal and Plant Sciences*, pp. 95-99. Philadelphia, PA: Institute for Scientific Information.

Jiang, J., Gu, B. H., Albright, L. M., and Nixon, B. T. (1989). Conservation between coding and regulatory elements of *Rhizobium meliloti* and *Rhizobium leguminosarum dct* genes. *J. Bacteriol., 171*, 5244-5253.

Johnson, J. F., Vance, C. P., and Allan, D. L. (1996). Phosphorus deficiency in *Lupinus albus*, altered lateral root development and enhanced expression of phosphoenolpyruvate carboxylase. *Plant Physiol., 112*, 31-41.

Karanja, N. K., and Wood, M. (1988). Selecting Rhizobium phaseoli strains for use with beans (*Phaseolus vulgaris* L.) in Kenya. Tolerance of high temperature and antibiotic resistance. *Plant Soil, 112*, 15-22.

Kashket, E. (1985). The proton motive force in bacteria. A critical assessment of methods. *Ann. Rev. Microbiol., 39*, 219–242.

Keyser, H. H., and Munns, D. N. (1979). Effects of calcium, manganese and aluminum on growth of rhizobia in acid media. *Soil Sci. Soc. Amer. J., 43*, 500-503.

Kijne, J. W. (1992). The *Rhizobium* infection process. In G. Stacey, R. H. Burris, and H. J. Evans (Eds.), *Biological Nitrogen Fixation* (pp. 349-398). New York: Chapman & Hall.

Kinzig, A. P., and Socolow, R. H. (1994). Is nitrogen fertilizer use nearing a balance - reply. *Phys. Today, 47*, 24-35.

Kishinevsky, B. D., Sen, D., and Weaver, R. W. (1992). Effect of high root temperature on the *Bradyrhizobium*-peanut symbiosis. *Plant Soil, 143*, 275-282.

Kluson, R. A., Kenworthy, W. J., and Weber, D. F. (1986). Soil temperature effects on competitiveness and growth of *Rhizobium japonicum* and on *Rhizobium*-induced chlorosis of soybean. *Plant Soil, 95*, 202-207.

Kondorosi, E., Gyuris, J., Schmidt, J., John, M., Duda, E., Schell, J., and Kondorosi, A. (1988). Positive and negative control of nodulation genes in *Rhizobium meliloti* strain 41. In D. P. S. Verma and R. Palacios (Eds.), *Molecular Microbe-Plant Interactions* (p. 73). St. Paul, MN: APS Press.

Kondorosi, E., Gyuris, J., Schmidt, J., John, M., Duda, E., Hoffman, B., Schell, J., and Kondorosi, A. (1989). Positive and negative control of nod gene expression in *Rhizobium meliloti* is required for optimal nodulation. *EMBO. J., 8*, 1331-1340.

Kondorosi, E., Pierre, M., Cren, M., Haumann, U., Buire, M., Hoffman, B., Schell, J., and Kondorosi, A. (1991). Identification of *nolR*, a negatively transacting factor controlling the *nod* regulon in *Rhizobium meliloti. J. Mol. Biol., 222*, 885-896.

Kosslak, R. M., Bookland, R., Barkei, J., Paaren, H. E., and Applebaum, E. R. (1987). Induction of *Bradyrhizobium japonicum* common *nod* genes by isoflavones isolated from *Glycine max. Proc. Natl. Acad. Sci. USA, 84*, 7428-7432.

Krulwich, T. A., Agus, R., Schneir, M., and Guffanti, A. A. (1985). Buffering capacity of bacilli that grow at different pH ranges. *J. Bacteriol., 162*, 768-772.

Kucey, R. M. N., and Hynes, M. F. (1989). Populations of *Rhizobium leguminosarum* biovars *phaseoli* and *viciae* in fields after bean or pea in rotation with nonlegumes. *Can. J. Microbiol., 35*, 661-667.

Kulkarni, S., and Nautiyal, C. S. (1999). Characterization of high temperature-tolerant rhizobia isolated from *Prosopis juliflora* grown in alkaline soil. *J. Gen. Appl. Microbiol., 45*, 213-220.

Lal, R. (1993). The role of no-till farming in sustainable agriculture in the tropics, *Anais do I Encontro Latinoamericano sobre Plantio Direto na Pequena Propriedade, IAPAR* (pp. 29–62). 22–26 Novembro, Ponta Grossa, Brazil.

Labidi, M., Laberge, S., Vezina, L. P., and Antoun, H. (2000). The *dnaJ* (hsp40) locus in *Rhizobium leguminosarum bv. phaseoli* is required for the establishment of an effective symbiosis with *Phaseolus vulgaris. Mol. Plant-Microbe Interact., 13*, 1271-1274.

Lamblin, A. F., Crow, J. A., Johnson, J. E., Silverstein, K. A., Kunau, T. M., Kilian, A., *et al.* (2003). MtDB: A database for personalized data mining of the model legume *Medicago truncatula* transcriptome. *Nucl. Acids Res., 31*, 196-201.

Lerouge, P., Roche, P., Faucher, C., Maillet, F., Truchet, G., Prome, J. C., and Denarie, J. (1990). Symbiotic host specificity of *Rhizobium meliloti* is determined by a sulphated and acylated glucosamine oligosaccharide signals. *Nature, 344*, 781-784.

Leung, K., and Bottomley, P. J. (1987). Influence of phosphate on the growth and nodulation characteristics of *Rhizobium trifolii. Appl. Environ. Microbiol., 53*, 2098-2105.

Lewin, A., *et al.* (1987). Multiple host-specificity loci of the broad host-range *Rhizobium* sp. NGR234 selected using the widely compatible legume *Vigna unguiculata. Plant Mol. Biol., 8*, 447-459.

Lewis-Henderson, W. R., and Djordjevic, M. A. (1991). A cultivar-specific interaction between *Rhizobium leguminosarum* biovar *trifolii* and subterranean clover is controlled by *nodM*, other bacterial cultivar specificity genes, and a single recessive host gene. *J. Bacteriol., 173*, 2791-2799.

Lie, T.A. (1978). Symbiotic specialization in pea plants. The requirement of specific *Rhizobium* strains for peas from Afghanistan. *Ann. Appl. Biol., 88*, 462-465.

Leigh, J. A., and Walker, G. C. (1994). Exopolysaccharides of *Rhizobium:* Synthesis, regulation and symbiotic function. *Trends Genet., 10*, 63-67.

LeRudulier, D., and Bouillard, L. (1983). Glycine betaine, an osmotic effector in *K. pneumonniae* and other members of *Enterobacteriaceae*. *Appl. Environ. Microbiol., 46*, 152-159.

Lindstrom, K., Lipsanen, P., and Kaijalainen, S. (1990). Stability of markers used for identification of two *Rhizobium galegae* inoculant strains after five years in the field. *Appl. Environ. Microbiol., 56*, 444-450.

Long, S. R., Egelhoff, T., Fisher, R. F., Jacobs, T. W., and Mulligan, J. T. (1985). Fine structure studies of *R. meliloti nodDABC* genes. In H. J. Evans, P. J. Bottomley, and W. E. Newton (Eds.), *Nitrogen Fixation Research Progress* (p. 87-94). Boston: Martinus Nijhoff Publishers.

Long, S. R. (1989). *Rhizobium*-legume nodulation, life together in the underground. *Cell, 56*, 203-214.

Long, S. R. (2001). Genes and signals in the *Rhizobium*-legume symbiosis. *Plant Physiol., 125*, 69-72.

Lynch, D. H., and Smith, D. L. (1994). The effects of low root-zone temperature stress on 2 soybean (*Glycine max*) genotypes when combined with *Bradyrhizobium* strains of varying geographic origin. *Physiol. Plant., 90*, 105-113.

Mahler, R. L., and Wollum, A. G. (1980). Influence of water potential on the survival of rhizobia in Goldsboro loamy sand. *Soil Sci. Soc. Am. J., 44*, 988-992.

Mahler, R. L., and Wollum, A. G. (1981). The influence of soil water potential and soil texture on the survival of *Rhizobium japonicum* and *Rhizobium leguminosarum* isolates in the soil. *Soil Sci. Soc. Am. J., 45*, 761-766.

Marshall, K. C. (1964). Survival of root nodule bacteria in dry soils exposed to high temperatures. *Aust. J. Agric. Res., 15*, 273-281.

Martinez-Romero, E., and Rosenblueth, M. (1990). Increased bean *Phaseolus vulgaris* L. nodulation competitiveness of genetically modified strains. *Appl. Environ. Microbiol., 56*, 2384-2388.

Martinez, E., Romero, D., and Palacios, R. (1990). The *Rhizobium* genome, *Crit. Rev. Plant Sci., 9*, 59-93.

Mary, P., Dupuy, N., Dolhembiremon, C., Delfives, C., and Tailliez, R. (1994). Differences among *Rhizobium meliloti* and *Bradyrhizobium japonicum* strains in tolerance to desiccation and storage at different relative humidities. *Soil Biol. Biochem., 26*, 1125-1132.

Mathesius, U., Keijzers, G., Natera, S. H., Weinman, J. J., Djordjevic, M. A., and Rolfe, B. G. (2001). Establishment of a root proteome reference map for the model legume *Medicago truncatula* using the expressed sequence tag database for peptide mass fingerprinting. *Proteomics, 1*, 1424-1440.

McDermott, T. R., Graham P. H., and Brandwein, D. M. (1987). Viability of *Bradyrhizobium japonicum* bacteroids. *Arch. Mikrobiol., 148*, 100-106.

McKay, I. A., and Djordjevic, M. A. (1993). Production and excretion of *nod* metabolites by *Rhizobium leguminosarum bv. trifolii* are disrupted by the same environmental factors that reduce nodulation in the field. *Appl. Envir. Microbiol., 59*, 3385-3392.

Meinhardt, L. W., Krishnan, H. B., Balatti, P. A., and Pueppke, S. G. (1993). Molecular cloning and characterization of a Sym-plasmid locus that regulates cultivar-specific nodulation of soybean by *Rhizobium fredii* USDA 257. *Mol Microbiol., 9*, 17-27.

Mendes I. C., and Bottomley, P. J. (1998). Distribution of a population of *Rhizobium leguminosarum* bv *trifolii* among different size classes of soil aggregates. *Appl. Environ. Microbiol., 64*, 970-975

Mergaert, P., van Montagu, M., Prome, J.-C., and Holsters, M. (1993). Three unusual modifications, a D-arabinosyl, an N-methyl, and a carbamoyl group, are present on the Nod factors of *Azorhizobium caulinodans* strain ORS571. *Proc. Natl. Acad. Sci. USA, 90*, 1551-1555.

Merrick, M. J., and Edwards, R. A. (1995). Nitrogen control in bacteria. *Microbiol. Rev., 59*, 604-622.

Michiels, J., Verreth, C., and Vanderleyden, J. (1994). Effects of temperature on bean-nodulating *Rhizobium* strains. *Appl. Environ. Microbiol., 60*, 1206-1212.

Moawad, H., and Beck, D. (1991). Some characteristics of *Rhizobium leguminosarum* isolates from uninoculated field-grown lentil. *Soil Biol. Biochem., 23*, 917-925.

Monson, E., Ditta, G., and Helinski, D. (1995). The oxygen sensor protein, FixL, of *Rhizobium meliloti*. Role of histidine residues in heme binding, phosphorylation, and signal transduction. *J. Biol. Chem., 270*, 5243-5250.

Morris, A. C., and Djordjevic, M. A. (2001). Proteome analysis of cultivar-specific interactions between *Rhizobium leguminosarum* biovar *trifolii* and subterranean clover cultivar Woogenellup. *Electrophoresis, 22*, 586-98.

Mulligan, J. T., and Long, S. R. (1985). Induction of *Rhizobium meliloti nodC* expression by plant exudate requires *nodD*. *Proc. Natl. Acad. Sci. USA, 82*, 6609-6613.

Munchbach, M., Nocker, A., and Narberhaus, F. (1999). Multiple small heat shock proteins in rhizobia. *J. Bacteriol., 181*, 83-90.

Munevar, F., and Wollum, A. G. (1982). Response of soybean plants to high root temperature as affected by plant cultivar and *Rhizobium* strain. *Agron. J., 74*, 138-142.

Munns, D. N. (1970). Nodulation of *Medicago sativa* in solution culture. V. Calcium and pH requirements during infection. *Plant Soil, 32*, 90-102.

Munns, D. N. (1986). Acid soils tolerance in legumes and rhizobia. *Adv. Plant Nutr., 2*, 63-91.

Munns, D. N., Fogle, V. W., and Hallock, B. G. (1977). Alfalfa root nodule distribution and inhibition of nitrogen fixation by heat. *Agron. J., 69*, 377-380.

Newbould, P. (1989). The use of nitrogen fertilizer in agriculture. Where do we go practically and ecologically? *Plant Soil, 115*, 297-311.

Nieuwkoop, A. J., Banfalvi, Z., Deshmane, N., Gerhold, D., Schell, M., Sirotkin, K., and Stacey, G. (1987). A locus encoding host range is linked to the common nodulation genes of *Bradyrhizobium japonicum*. *J. Bacteriol., 169*, 2631-2638.

Niner, B. M., and Hirsch, A. M. (1998). How many *Rhizobium* genes, in addition to *nod, nif/fix*, and *exo*, are needed for nodule development and function. *Symbiosis, 24*, 51-102.

O'Hara, G. W., Goss, T. J., Dilworth, M. J., and Glenn, A. R. (1989). Maintenance of intracellular pH and acid tolerance in *Rhizobium meliloti*. *Appl. Environ. Microbiol., 55*, 1870-1876.

Olson, E. R., Sadowsky, M. J., and Verma, D. P. S. (1985). Identification of genes involved in the *Rhizobium*-legume symbiosis by Mu-dI(*kan, lac*)-generated transcription fusions. *Bio/Tech., 3*, 143-149.

Papendick, R. I., and Campbell, G. S. (1981). Theory and measurement of water potential in soil, organic materials, plants, seeds, and microorganisms. In J. F. Parr, W. R. Gardner, and L. F. Elliott (Eds.), *Water potential relations in soil microbiology: Proceedings of a symposium sponsored by Divisions S-1 and S-3 of the Soil Science Society of America* (pp. 1-22). Madison, WI: Soil Science Society of America.

Parke, D., and Ornston, L. N. (1986). Enzymes of the β-ketoadipate pathway are inducible in *Rhizobium* and *Agrobacterium*, and constitutive in *Bradyrhizobium* spp. *J. Bacteriol., 165*, 288-292.

Patriarca, E. J., Tate, R., and Iaccarino, M. (2002). Key role of bacterial NH_4^+ metabolism in *Rhizobium*-plant symbiosis. *Microbiol. Mol. Biol. Rev., 66*, 203-222.

Paul, E. A., and Clark, F. E. (1988). *Soil Microbiology and Biochemistry.* San Diego, CA: Academic Press.

Pereira, P. A. A., and Bliss. F. A. (1989). Selection of common bean (*Phaseolus vulgaris* L.) for N_2 fixation at different levels of available phosphorus under field and environmentally-controlled conditions. *Plant Soil, 115*, 75-82.

Peters, N. K., Frost, J. W., and Long, S. R. (1986). A plant flavone, luteolin, induces expression of *Rhizobium meliloti* nodulation genes. *Science, 233*, 977-979.

Phillips, D. A., Joseph, C. M., Yang, G. P., Martinez-Romero, E., Sanborn, J. R., and Volpin, H. (1999). Identification of lumichrome as a *Sinorhizobium* enhancer of alfalfa root respiration and shoot growth. *Proc. Natl. Acad. Sci. USA, 96*, 12275-12280.

Piha, M. I., and Munns, D. N. (1987). Sensitivity of the common bean (*Phaseolus vulgaris* L.) symbiosis to high soil temperature, *Plant Soil, 98*, 183-194.

Postma, J., Hok-a-Hin, C. H., and van Veen, J. A. (1990). Role of microniches in protecting introduced *Rhizobium leguminosarum* biovar *trifolii* against competition and predation in soil, *Appl. Environ. Microbiol., 56*, 495-502.

Prévost, D., Bordeleau, L. M., Caudry-Reznick, S., Schulman, H. M., and Antoun, H. (1987). Characteristics of rhizobia isolated from three legumes indigenous to the high arctic, *Astragalus alpinus*, *Oxytropis maydelliana*, and *Oxytropis arctobia*. *Plant Soil, 98*, 313-324.

Price, N. J. P., Relic, B., Talmont, F., Lewin, A., Prome, D., Pueppke, S. G., *et al.* (1992). Broad-host-range *Rhizobium species* strain NGR234 secretes a family of carbamoylated and fucosylated nodulation signals that are *O*-acylated or sulphated. *Mol. Microbiol., 6*, 3575-3584.

Pueppke, S. G. (1996). The genetic and biochemical basis for nodulation of legumes by rhizobia. *Crit. Rev. Biotechnol., 16*, 1-51.

Purcell, L. C., and King, C. A. (1996). Drought and nitrogen source effects on nitrogen nutrition, seed growth, and yield in soybean. *J. Plant Nutrit., 19*, 969-993.

Reeve, W. G., Tiwari, R. P., Kale, N. B., Dilworth, M. J., and Glenn, A. R. (2002). ActP controls copper homeostasis in *Rhizobium leguminosarum* bv. *viciae* and *Sinorhizobium meliloti* preventing low pH-induced copper toxicity. *Mol. Microbiol., 43*, 981-991.

Relic, B., Talmont, F., Kopcinska, J., Golinowski, J. W., Prome, J.-C., and Broughton W. J. (1993). Biological activity of *Rhizobium* sp. NGR234 Nod factors on *Macroptilium atropurpureum. Mol. Plant-Microbe Interact., 6*, 764-774.

Riccillo, P. M., Muglia, C. I., de Bruijn, F. J., Roe, A. J., Booth, I. R., and Aguilar, O. M. (2000). Glutathione is involved in environmental stress responses in *Rhizobium tropici*, including acid tolerance. *J. Bacteriol., 182*, 1748-53.

Richardson, A. E., Simpson, R. J., Djordjevic, M. A., and Rolfe, B. G. (1988). Expression of nodulation genes in *Rhizobium leguminosarum* bv. *trifolii* is affected by low pH and by Ca^{2+} and Al ions. *Appl. Environ. Microbiol., 54*, 2541-2548.

Richardson, A. E., and Simpson, R. J. (1989). Acid-tolerance and symbiotic effectiveness of *Rhizobium trifolii* associated with a *Trifolium subterraneaum* L. based pasture growing in an acid soil. *Soil Biol. Biochem., 21*, 87-95.

Roughley, R. J. (1970). The influence of root temperature, *Rhizobium* strain and host selection on the structure and nitrogen-fixing efficiency of the root nodules of *Trifolium subterraneum. Ann. Bot., 34*, 631-646.

Roy, R. N., Misra, R. V., and Montanez, A. (2002). Decreasing reliance on mineral nitrogen--yet more food. *Ambio., 31*, 177-83.

Ruvkin, G. B., and Ausubel, F. M. (1980). Interspecies homology of nitrogenase genes. *Proc. Natl. Acad. Sci. USA, 77*, 191–195.

Rynne, F. G., Glenn, A. R., and Dilworth, M. J. (1994). Effect of mutations in aromatic catabolism on the persistence and competitiveness of *Rhizobium leguminosarum* bv. *Trifolii. Soil Biol. Biochem., 26*, 703-710.

Saalbach, G., Erik, P., and Wienkoop, S. (2002). Characterisation by proteomics of peribacteroid space and peribacteroid membrane preparations from pea (*Pisum sativum*) symbiosomes. *Proteomics, 2*, 325-337.

Sadowsky, M. J., Olson, E. R., Foster, V. E., Kosslak, R. M., and Verma, D. P. S. (1988). Two host-inducible genes of *Rhizobium fredii* and the characterization of the inducing compound. *J. Bacteriol., 170*, 171-178.

Sadowsky, M. J., Cregan, P. B., Gottfert, M., Sharma, A., Gerhold, D., Rodriguez-Quniones, F., *et al.* (1991). The *Bradyrhizobium japonicum nolA* gene and its involvement in the genotype-specific nodulation of soybeans. *Proc. Natl. Acad. Sci. USA, 88*, 637-641.

Sadowsky, M. J., and Graham, P. H. (1998). Soil Biology of the *Rhizobiaceae.* In H. P. Spaink, A. Kondorosi, and P. J. J. Hooykaas (Eds.), *The Rhizobiaceae* (pp. 155-172). Dordrecht: Kluwer Academic Publishers.

Sadowsky, M. J., Kosslak, R. M., Golinska, B., Madrzak, C. J., and Cregan, P. B. (1995). Restriction of nodulation by *B. japonicum* is mediated by factors present in the roots of *Glycine max. Appl. Environ. Microbiol., 61*, 832-836.

Sagan, M., and Gresshoff, P. M. (1996). Developmental mapping of nodulation events in pea (*Pisum sativum* L.) using supernodulating plant genotypes and bacterial variability reveals both plant and *Rhizobium* control of nodulation regulation. *Plant Sci., 117*, 167-179.

Sanchez, P. A., and Euhara, G. (1983). Management considerations for acid soils with high phosphorus fixation capacity. In F. E. Kharawuch *et al.* (Eds.). Madison, WI: American Society for Agronomy.

Sangakkara, U. R., Hartwig, U. A. and Nosberger, J. (1996). Soil moisture and potassium affect the performance of symbiotic nitrogen fixation in faba bean and common bean. *Plant Soil, 184*, 123-130.

Sanjuan, J., and Olivares, J. (1989). Implication of *nifA* in regulation of genes located on a *Rhizobium meliloti* cryptic plasmid that effect nodulation efficiency, *J. Bacteriol., 171*, 4154–4161.

Saxena, A. K., and Rewari. R. B. (1991). The influence of phosphate and zinc on growth, nodulation and mineral composition of chickpea (*Cicer arietinum* L.) under salt stress. *World J. Microbiol. Biotechnol., 7*, 202-205.
Schlaman, H. R. M., Phillips, D. A., and Kondorosi, E. (1998). Genetic organization and transcriptional regulation of rhizobial nodulation genes. In H. P. Spaink, A. Kondorosi, and P. J. J. Hooykaas (Eds.), *The Rhizobiacea* (pp. 351-386). Dordrecht: Kluwer Academic Publishers.
Schmidt, P. E., Broughton, W. J., and Werner, D. (1994). Nod factors of *Bradyrhizobium japonicum* and *Rhizobium* sp. NGR 234 induce flavonoid accumulation in soybean root exudates. *Molec. Plant-Microbe Interact., 7*, 384-390.
Schultze, M., and Kondorosi, A. (1998). Regulation of symbiotic root nodule development. *Annu. Rev. Genet., 32*, 33-57.
Schultze, M., Quiclet-Sire, B., Kondorosi, E., Virelizier, H., Glushka, N., Endre, G., *et al.* (1992). *Rhizobium meliloti* produces a family of sulphated lipooligosaccharides exhibiting different degrees of plant host specificity. *Proc. Natl. Acad. Sci. USA, 89*, 192-196.
Segovia, L., Pinero, D., Palacios, R., and Martinez-Romero, E. (1991). Genetic structure of a soil population of nonsymbiotic *Rhizobium leguminosarum. Appl. Environ. Microbiol., 57*, 426-43.
Serraj, R. and Sinclair, T. R. (1997). Variation among soybean cultivars in dinitrogen fixation response to drought. *Agron. J., 89*, 963-969.
Serraj, R., Sinclair, T., and Purcell, L. (1999). Symbiotic N_2 fixation response to drought. *J. Exp. Bot., 50*, 143-155.
Sinclair, T. R. (1986). Water and nitrogen limitation in soybean grain production. I. Model development. *Field Crops Res., 15*, 125-141.
Sinclair, T. R., Muchow, R. C., Bennett, J. M., and Hammond, L. C. (1987). Relative sensitivity of nitrogen and biomass accumulation to drought in field-grown soybean. *Agron. J., 79*, 986-991.
Sinclair, T. R., Zimet, A. R., and Muchow, R. C. (1988). Changes in soybean nodule number and dry weight in response to drought. *Field Crops Res., 18*, 197-202.
Singleton, P. W., and Stockinger, K. R. (1983). Compensation against ineffective nodulation in soybean (*Glycine max*). *Crop Sci., 23*, 69-72.
Singleton, P. W., Abel Magid, H. M., and Tavares, J. W. (1985). Effect of phosphorus on the effectiveness of strains of *Rhizobium japonicum. Soil Sci. Soc. Am. J., 49*, 613-616.
Smart, J. B., Dilworth, M. J., and Robson, A. D. (1984). A continuous culture study of the phosphorus nutrition of *Rhizobium trifolii* WU95, *Rhizobium* NGR234 and *Bradyrhizobium* CB756. *Arch. Microbiol., 140*, 276-280.
Smit, G., Swart, S., Lugtenberg, B. J. J., and Kijne, J. W. (1992). Molecular mechanisms of attachment of *Rhizobium* bacteria to plant roots. *Mol. Microbiol., 6*, 2897-2903.
Smith, L., and Smith, G. M. (1989). An osmoregulated dipeptitide in stressed *R. meliloti. J. Bacteriol., 171*, 4714-4717.
Smith, L. T., Pocard, J. A., Bernard, T., and LeRudulier, D. (1988). Osmotic control of glycine betaine biosynthesis and degradation in *R. melilot. J. Bacteriol., 170*, 3142-3149.
Somasegaran, P., Reyes, V. G., and Hoben, H. J. (1984). The influence of high temperatures on the growth and survival of *Rhizobium* spp. in peat inoculants during preparation, storage and distribution. *Can. J. Microbiol., 30*, 23-30.
Spaink, H. P., Sheeley, D. M., van Brussel, A. A. N., Glushka, J., York, W. S., Tak, T., *et al.* (1991). A novel highly unsaturated fatty acid moiety of lipooligosaccharide signals determines host specificity of *Rhizobium. Nature, 354*, 124-130.
Spaink, H. P. (1995). The molecular basis of infection and nodulation by rhizobia - the ins and outs of sympathogenesis. *Ann. Rev. Phytopath., 33*, 345–368.
Spaink, H. P. (2000). Root nodulation and infection factors produced by rhizobial bacteria. *Annu. Rev. Microbiol., 54*, 257-288.
Spaink, H. P., Wijffelman, C. A, Pees, E., Okker, R. J. H., and Lugtenberg, B. J. J. (1987). *Rhizobium* nodulation gene *nodD* as a determinant of host specificity. *Nature, 328*, 337-340.
Sprent, J. I. (1971). Effects of water stress on nitrogen fixation in root nodules. *Plant Soil (Spec. Vol.)*, 225–228.
Sprent, J. I. (1984). Nitrogen fixation. In M. B. Wilkins (Ed.), *Advances in Plant Physiology* (pp. 249-276). London: Pitman.

Stacey, G., *et al.* (2002). Signal exchange during the early events of soybean nodulation. In T. M. Finan, M. R. O'Brian, D. B. Layzell, J. K. Vessey, and W. E. Newton (Eds.), *Nitrogen Fixation Global Perspectives* (pp. 118-122). Wallingford, UK: CABI Publishing.
Sutton, W. D. (1983). Nodule development and senescence. In W. J. Broughton (Ed.), *Nitrogen Fixation, Vol. 3 Legumes* (pp. 144-212). Oxford, UK: Clarendon Press.
Tauer, L. (1989). Economic impact of future biological nitrogen fixation technologies on United States Agriculture. *Plant Soil, 119*, 261-270.
Torimitsu, K., Hayashi, M., Ohta, E., and Sakata, M. (1985). Effect of K^+ and H^+ stress and role of Ca^{2+} in the regulation of intracellular K^+ concentration in mung bean roots. *Physiol. Plant., 63*, 247-252.
Torriani-Gorini, A., Rothman, F. G., Silver, S., *et al.* (1987). *Phosphate metabolism and cellular regulation in microorganisms*. Washington, D.C.: American Society for Microbiology.
Trinick, M. J. (1982). Biology. In W. J. Broughton (Ed.), *Nitrogen Fixation, Vol. 2* Rhizobium (pp. 76-146). Oxford, UK: Clarendon Press.
Triplett, E. W., and Sadowsky, M. J. (1992). Genetics of competition for nodulation. *Ann. Rev. Microbiol., 46*, 399-428.
Trotman, A. P., and Weaver, R. W. (1995). Tolerance of clover rhizobia to heat and desiccation stresses in soil. *Soil Sci. Soc. Am. J., 59*, 466-470.
Truchet, G., Roche, P., Lerouge, P., Vasse, J., and Camut, S. (1991). Sulphated lipo-oligosaccharide signals of the symbiotic procaryote *Rhizobium meliloti* elicit root nodule organogenesis on the host plant *Medicago sativa. Nature, 351*, 670-673.
Turco, R. F., and Sadowsky, M. J. (1995). Understanding the microflora of bioremediation. In H. D. Skipper and R. F. Turco (Eds.), Bioremediation: Science and Applications. *Soil Science (Special Publication) 43* (pp. 87-103). Madison, WI: Soil Science Society of America.
van Brussel, A., Recourt, K., Pees, E., Spaink, H. P., Tak, T., Wijffelman, C., Kijne, J. W., and Lugtenberg, B. J. J. (1990). A biovar specific signal of *Rhizobium leguminosarum* bv. *viceae* induces increased nodulation gene-inducing activity in root exudate of *Vicia sativa* subsp. *nigra. J. Bacteriol., 172*, 5394-5401.
van der Drift, K. M. G. M, Olsthoorn, M. M. A. L., Brull, P., Blok-Tip, L., and Thomas-Oates, J. E. (1998). Mass spectrometric analysis of lipo-chitin oligosaccharide signal molecules mediating the host-specific legume-*Rhizobium* symbiosis. *Mass Spec. Rev., 17*, 75–95.
Van Egeraat, A. W. S. M. (1975). The possible role of homoserine in the development of *Rhizobium leguminosarum* in the rhizosphere of pea seedlings. *Plant Soil, 42*, 381-386.
van Rhijn, P., and Vanderleyden, J. (1995). The *Rhizobium*-plant symbiosis, *Microbiol. Rev., 59*, 124-142.
Vance, C. P. (1998). Legume symbiotic nitrogen fixation, agronomic aspects. In H. P. Spaink, A. Kondorosi, and P. J. J. Hooykaas (Eds.), *The Rhizobiaceae* (pp. 509-530). Dordrecht: Kluwer Academic Publishers.
Vargas, A. A. T., and Graham, P. H. (1988). *Phaseolus vulgaris* cultivar and *Rhizobium* strain variation in acid-pH tolerance and nodulation under acid conditions. *Field Crops Res., 19*, 91-101.
Vincent, J. M, Thompson, J. A., and Donovan, K. O. (1962). Death of root nodule bacteria on drying. *Australian J. Agric. Res., 13*, 258-270.
Vinuesa, P., Neumann-Silkow, F., Pacios-Bras, Ch., Spaink, H. P., Martinez-Romero, E., and Werner, D. (2003). Genetic analysis of a pH regulated operon from *Rhizobium tropici* CIAT 899 involved in acid tolerance and nodulation competitiveness, *Mol. Plant-Microbe Interact., 16*, 159-168.
Vitousek, P. M., Hattenschwiler, S., Olander, L. and Allison, S. (2002). Nitrogen and nature. *Ambio., 31*, 97-101.
Waggoner, P. E. (1994). How much land can ten million people spare for nature? *Task Force Report 121* (p. 64). Ames, IA: Council on Agricultural Science and Technology.
Wais, R. J., Keating, D. H., and Long, S. R. (2002). Structure-function analysis of *nod* factor-induced root hair calcium spiking in *Rhizobium*-legume symbiosis, *Plant Physiol., 129*, 211-224.
Wery, J., Deshamps, M., and Leger-Cresson, N. (1988). Influence of some agro-climatic factors and agronomic practices on nitrogen nutrition of chickpea. In D. P. Beck and L. A. Materon (Eds.), *Nitrogen fixation by legumes in Mediterranean agriculture* (pp. 287-301). Dordrecht: ICARDA-Martinus Nijhoff.
Wery, J., Silim, S. N., Knight, E. J., Malhorta, R. S., and Cousin, R. (1994). Screening techniques and sources of tolerance to extremes of moisture and air temperature in cool season food legumes. *Euphytica, 70*, 487-495.

Wienkoop, S., and Saalbach, G. (2003). Proteome analysis. Novel proteins identified at the peribacteroid membrane from *Lotus japonicus* root nodules. *Plant Physiol.*, *131*, 1080-1090.

Wijffelman, C. A., Pees, E., van Brussel, A. A., Priem, M., Okker, R., and Lugtenberg, B. J. J. (1985). Analysis of the nodulation region of the *Rhizobium leguminosarum* Sym plasmid pRL1JI. In H. J. Evans, P. J. Bottomly, and W. E. Newton (Eds.), *Nitrogen Fixation Research Progress* (pp. 127). Boston, MA: Martinus Nijhoff Publishers.

Wilkins, J. (1967). The effects of high temperature on certain root-nodule bacteria. *Aust. J. Agric. Res.*, *18*, 299-304.

Zaat, S. A. J., Wijffelman, C. A., Spaink, H. P., van Brussel, A. A. N., Okker, R. J. H., and Lugtenberg, B. J. J. (1987). Induction of the *nodA* promoter of *Rhizobium leguminosarum* Sym plasmid pRL1JI by plant flavanones and flavones. *J. Bacteriol.*, *169*, 198–204.

Zahran, H. H. (1999). *Rhizobium*-legume symbiosis and nitrogen fixation under severe conditions and in arid climate, *Microbiol. Molec. Biol. Rev.*, 63, 968-989.

Zahran, H. H., Rasanen, L. A., Karsisto, M., and Lindstrom, K. (1994). Alteration of lipopolysaccharide and protein profiles in SDS-PAGE of rhizobia by osmotic and heat stress. *World J. Microbiol. Biotechnol.*, *10*, 100-105.

Zerhari, K., Aurag, J., Khbaya, B., Kharchaf, D., and Filali-Mattouf, A. (2000). Phenotypic characteristics of rhizobia isolates nodulating *Acacia* species in the arid and Saharan regions of Morocco. *Lett. Appl. Microbiol.*, *30*, 351-357.

Zhang, F., Lynch, D. H., and Smith, D. L. (1995). Impact of low root temperature in soybean [*Glycine max* (L. Merr.)] on nodulation and nitrogen fixation. *Environ. Exper. Bot.*, *35*, 279-285.

Zhang, S. Q., Outlaw, W. H., Jr., and Aghoram, K. (2001). Relationship between changes in the guard cell abscisic-acid content and other stress-related physiological parameters in intact plants, *J. Exp. Bot.*, *52*, 301-308.

Chapter 7

NODULATED LEGUME TREES

J. I. SPRENT
32 Birkhill Avenue, Wormit, Fife DD6 8PW
Scotland, UK

1. INTRODUCTION

The Leguminosae is the third largest family of flowering plants. One of its most widely recognised features is the ability of many species to form a symbiotic relationship with soil bacteria, leading to the formation of nitrogen-fixing root or (rarely) stem nodules. This chapter considers only those legume genera that contain tree species. As many legume trees lack the ability to nodulate and many others have not been checked for this ability, the emphasis is on the known nodulating genera. Because the information on these genera is not readily available in a concise form, a major part of the chapter will be on generic descriptions. Following this section, both nodule characteristics and nodulating bacteria will be discussed. Because mycorrhizal formation is an integral part of the nutrient-uptake systems and is important in nature and for management, these will also be briefly considered. A consideration of the problems of measuring nitrogen fixation by trees will precede a discussion of the role of nitrogen-fixing legume trees in natural ecosystems and how they may be exploited in a sustainable way.

2. LEGUMINOSAE

Although the way in which legumes should be classified is generally agreed, the terminology is not. Many people still prefer to use Leguminosae as the family name, but for reasons of historical precedence, Fabaceae is now widely accepted. Within either of these terms, three subfamilies are usually recognised, although some workers endow these with family status. Thus, we may have Caesalpinioideae (Caesalpiniaceae), Mimosoideae (Mimosaceae) and Papilionoideae (Papilionaceae). In this chapter, I shall use Leguminosae and three subfamilies. The tribes and genera within the subfamilies are also generally agreed on, although some workers still retain

D. Werner and W. E. Newton (eds.), Nitrogen Fixation in Agriculture, Forestry, Ecology, and the Environment, 113-141.

the tribe, Swartzieae, within the Caesalpinioideae, rather than the more common position in the Papilionoideae. The spelling of some names has changed over the years. Here, the spelling agreed to and used by the International Legume Data Base and Information Service (ILDIS, web site at http://www.ildis.org/LegumeWeb/) will be used. Earlier spellings can be found in Sprent (2001).

Table 1 summarises the main features of the subfamilies and indicates the number of genera with confirmed reports of nodulation. Throughout this chapter, only those genera and species, where nodulation has been confirmed, will be discussed. Many reports of swellings on roots, which appear superficially to be nodules, have not been confirmed, for example, on the tropical tree genera, *Eperua* and *Mora*. Examples of such structures are illustrated in Sprent (2001). Criteria used in establishing nodulation include appearance, location on roots, internal coloration and structure, nitrogenase activity, isolation of rhizobia and establishment of Koch's postulates, and immunochemical location of key enzymes, particularly nitrogenase. Not all of these criteria have been used in all cases, but a minimum of three is considered by the author to be essential. An additional complication is that many legumes, particularly tropical trees are difficult to identify. Where specimens have been lodged in herbaria, it is possible to check identification at a later stage and this procedure is strongly recommended. It is essential, when giving either generic or specific names, to quote the authority for these names (as here) because so many have been subjected to name changes and a single species may still be known by several names, under more than one genus.

Table 1. Major Features of Legume Subfamilies.

	Caesalpinioideae	Mimosoideae	Papilionoideae
Number of genera - approximate total	157	78	479
Confirmed as nodulating	8	42	297
Distribution	Mainly moist tropics	Tropics/subtropics, often in dry areas	Tropics to arctic, dry to flooded
Habit	Mainly trees	Mainly trees and shrubs	Trees, shrubs, and herbs
Flower/infloresence	Zygomorphic, but very variable	Regular-frequently in globose or spicate heads with prominent stamens	Classically pea-with marked standard and wings
Nodulation	Rare, nodule structure usually primitive	Common, but with important exceptions	Very common, with some exceptions

Many woody legumes, particularly in the tropics, produce lenticels on stems. These have been recorded as nodules, *e.g.*, on *Cassia tora*, now classified as *Senna tora* (Subba Rao and Yatazawa, 1984), and have shown nitrogenase activity. Recent evidence suggests that nitrogen-fixing bacteria may be found in airspaces in such lenticels, but that these are not symbiotic (E. Wang, personal communication) and they will not be included in this chapter. There are no confirmed reports of stem nodules on trees, although many trees produce adventitious roots on even large trunks and, in humid forests, these roots may be abundantly nodulated, *e.g.*, *Pterocarpus* species in Cameroon as illustrated in Sprent (2001).

With very few exceptions, which will be discussed later, nodulation appears to be a generic character. Indeed, genera that have recently been sub-divided on classical taxonomic grounds have proved to be coincident with either the presence or absence of nodules, *e.g.*, *Chamaecrista* (nodulating) from both *Cassia* and *Senna* (non-nodulating), and *Sophora* (nodulating) from *Stypholobium* (non-nodulating). Further details and references on these genera can be found in Sprent (2000; 2001) and Lavin *et al.* (2001). In the following taxonomic sections, details of nodulating species are from Sprent (2001). Numbers in brackets following many entries are for species that are either potentially or actually threatened and are taken from Oldfield *et al.* (1998). However, it should be noted that the latter is not an exhaustive text as many species remain to be evaluated. Other references are given for particular genera. Many genera are composed of both shrubs and small trees, whereas others have a broader range of habits. The distinction between a shrub and a small tree depends not only on the species, but also on the habitat, so they have not been treated separately here when both forms occur within a genus.

For most of the genera, information on uses and tree size has been taken from Allen and Allen (1981). It is very surprising that very few tree legumes (rather than the more shrubby species used in agroforestry), with the major exception of *Acacia* species, have been either planted or managed with a view to maximising their nitrogen-fixing potential (Sprent and Parsons, 2000). For this reason, I have tried to highlight some genera that should be more widely studied. In some cases, attention has been drawn to them by the excellent books produced by the US National Academy of Sciences, for example, Anon (1979). With current concerns over deforestation and the need for low-input agriculture and forestry, perhaps more consideration will be given to nodulated legume trees. In order to paint as complete a picture of them as possible, the first sections will cover all genera known to have nodulated trees as members, arranged alphabetically within subfamilies. In this section, only references not included in Sprent (2001) are cited.

2.1. Caesalpinioideae

Nodulation is rare in this subfamily, as has been known for many years (Allen and Allen, 1981). However, there are still many genera that have not been examined for nodules. Nodules are always of indeterminate growth. Except for some shrubs and all herbaceous species of *Chamaecrista* (Naisbitt *et al.*, 1992) that have been examined, the nodules have bacteroids retained within infection-thread-like structures, now usually termed fixation threads, throughout the period of nitrogen fixation (Faria *et al.*, 1987). The genera are confined to certain tribes within the sub-family, details of which can be found in Sprent (2001).

2.1.1. Campsiandra Benth.
Now considered to have 22 species of both shrub and tree, one of which is known to be nodulated, this genus is mainly found in inundated regions, such as the Amazon and Orinoco basins. It is relatively unusual in having functional nodules under water (Sprent, 1999). Some species have durable timber and some are used medicinally.

2.1.2. Chamaecrista Moench.
This genus currently contains 287 species, making it the largest nodulating genus in the Caesalpinioideae. It is unusual in having a wide range of forms, from trees to annual herbs, and in being the only genus in the subfamily with nodulating species (all herbaceous) in temperate areas. Of the 47 species known to nodulate, a few are trees. These may fix N_2 in their natural habitat (Sprent *et al.*, 1996; Perreijn, 2002), the only caesalpinioid species for which such evidence is available.

2.1.3. Dimorphandra Schott.
This genus has 26 species of tree of various heights. It is found in tropical S America, where it is widespread. Nine of these species are reported as nodulated and several are potentially endangered. This genus is sometimes included in the subfamily, Mimosoideae, *e.g.*, in Lorenzi (1992). Although there is little available information on its uses, some species have beautifully colored wood (Ribeiro *et al.*, 1999).

2.1.4. Erythrophleum Afzel. ex G.Don.
This genus has eight species of tree in Africa and Asia plus one in Australia. Three of the African species are known to nodulate and there is indirect evidence that the Australian species may do so. Apart from some herbaceous pantropical species of *Chamaecrista*, this is the only caesalpinioid genus outside the Americas with confirmed ability to nodulate. The typical caesalpinioid structure of infected cells has recently been shown for three African species (Sprent, 2001; A. Kiam *et al.*, unpublished data). Species are used for timber and pharmaceutical purposes and at least one (*E. fordii* from Asia) is in such demand for timber that it is listed as threatened.

2.1.5. Melanoxylon Schott.
The genus has two species in tropical Brazil, one of which, *M. brauna*, has been reported to nodulate. This species has very fine, long-lasting timber and, although not listed as threatened, it is now becoming very scarce.

2.1.6. Moldenhowera Schrad.
The seven species of this genus are found in tropical Brazil and Venezuela and one has been reported to nodulate. Its timber is used for construction purposes and it is also widely grown as an ornamental (Lorenzi, 1992).

2.1.7. Sclerolobium Vogel.
This genus of forty species of tree is found in northern tropical South America. Ten species have reports of good nodulation. Some are extensively exploited for timber and five species are threatened.

2.1.8. Tachigali Aubl.
From tropical central and South America, there are reports of good nodulation for three of the 24 species, with one species threatened.

2.2. Mimosoideae

Although nodulation is more widespread in this subfamily, it is by no means universal (Table 2). A group of the basal genera appears not to nodulate. Arguably, they never had this character, whereas other less-basal genera may have lost the ability to nodulate (Sprent, 2001). The taxonomy of this subfamily has been the subject of major revisions in recent years. These changes have been incorporated as far as possible by Sprent (2001) and the same classification will be used here. As with the previous section, this section will take the genera in turn, alphabetically. There are many tree genera for which there are no reports of nodulation and these are not included here, but their details are available elsewhere (Sprent, 2001).

Table 2. Some Mimosoid Tree Genera that Appear to Lack the Ability to Nodulate.

Calpocalyx Harms	*Parkia* R.Br
Cylicodiscus Harms	*Tetrapleura* Benth
Dinizia Ducke	*Zapoteca* H.M.Hern
Newtonia Baill.	

2.2.1. Abarema Pittier
This genus of 44 species, mainly trees, is from the New World tropics. It is largely made up of species formerly included in *Pithecellobium* Mart, but also includes some species from other genera. Ten species have been reported as nodulated. Nodules are indeterminate, but not as freely branched as in *Acacia*. The genus is not widely used for timber, but some species have medicinal properties and others are used for tannin. Twenty species are threatened.

2.2.2. Acacia Mill.
This is by far the largest genus of nodulating legume trees. It is also one of the three legume genera known to have non-nodulating species. These sixteen (so far) closely related non-nodulating species, in the sub-genus *Aculeiferum*, have probably lost the nodulation character because some of the relevant genetic material has been retained (Sprent, 2001). So far, no Australian species has been confirmed not to nodulate. This fact is, perhaps, not surprisingly because the Australian species are generally quite distinct from the African and American species that do not nodulate.

It is likely that, in the next few years, the genus will be divided into a number of individual genera, rather than the three subgenera now commonly used (Table 3), but this process is proving to be rather difficult. However, there are several differences in symbiotic properties that relate to geographical origin as well as to subfamily. In particular, many Australian species have both arbuscular and ecto- mycorrhizae, an important property in terms of their ability to be exploited for land reclamation and also become invasive weeds (see later). Nodules on *Acacia* are indeterminate, often considerably branched. They vary greatly in effectiveness, a fact that is probably related to the wide range of rhizobia that nodulate many species (see later).

Table 3. Features of the Subgenera of Acacia.
(Phyllodineae is sometimes known as Heterophyllum).

Subgenus	Number of species	Main distribution	Characters
Acacia	120-130	Pan tropical	Bipinnate leaves; stipular spines; some have swollen spines associated with ants
Aculeiferum	180-190	Pan tropical, mainly Africa and New World	Bipinnnate leaves; with prickles or unarmed
Phyllodineae	>900	Largely Australian	Bipinnate leaves in seedlings; usually phyllodes in adults; no prickles; some have stipular spines

Almost all species are shrubs or trees. Many occur as understory plants in forests, particularly eucalypt forests in Australia; others can develop either into large trees in open moist areas, *e.g.*, *A. xanthophloea* Benth. in Africa or into smaller trees in extremely dry habitats, *e.g.*, some species in the Negev desert in the Middle East. These species are important in many ways. They are used as very high quality timber (*e.g.*, *A. melanoxylon* R.Br. in temperate Australia), as fuel wood (many African species), as forage and browse for wild and domestic animals (many African species), for honey (almost all species from dry areas), as food additives (gum arabic from *A. senegal* (L.) Wild.), for tannins (from bark of various species, particularly *A. mearnsii* De Wild.), and in land reclamation (dual mycorrhizal Australian species, such as *A. auriculiformis* Benth.). Thirty-six species, from various parts of the world, but in particular dry areas of Africa, are listed as potentially threatened.

2.2.3. Albizia Durazz.

This genus is currently undergoing taxonomic changes, but now consists of about 120 species, many of which are trees. They are most numerous in the Old World tropics, especially Africa. Forty-four species are reported to nodulate and their nodules are abundant and often profusely branched. Some species produce valuable timber, others medicinal products, but possibly the best known use is as shade for crops, such as cocoa and coffee. Sixteen species are potentially threatened.

2.2.4. Anadenanthera Speg.
This genus consists of two species of large tree, native to tropical South America, of which one has been reported to nodulate, with much branched nodules (Gross *et al.*, 2002). It produces excellent timber and has a number of other uses. It is not to be confused with the related genus, *Adenanthera* L., which cannot nodulate.

2.2.5. Archidendron F. Muell.
A genus that is sometimes included with *Pithecellobium*. It has 94 species of shrub or tree and is native to Indo-Malaysia and parts of Queensland, Australia. Four species have been reported as nodulated with nodules that are branched and indeterminate. Some species are used for timber and three species are threatened.

2.2.6. Archidendropsis I. Nielsen
This genus is a segregant from *Albizia* and contains 14 species of tree from the South-West Pacific area. One species has been reported to have indeterminate nodules and three are listed as threatened, largely as a result of habitat clearance.

2.2.7. Balizia Barneby and Grimes
Another segregant of *Pithecellobium*, this genus has three species occurring in wet tropical New World areas, one of which has been reported to nodulate.

2.2.8. Calliandra Benth.
Currently, this genus has 132 species of shrub and small tree from tropical or subtropical America. Fourteen species have been reported as nodulating and the shrubby/small tree, *C. calothyrsus* Meisn., is now being widely used in agroforestry (see later). Madagascan and South Asian species have now been transferred to *Viguieranthus* Villiers, but species combinations for all of these are not yet published, so it is not possible to give their nodulation status. The genus is not in the latest version of ILDIS (6.05, July 15 2002), but see Du Puy (2001).

2.2.9. Cedrelinga Ducke
A monotypic genus of nodulated tree from Amazonia and one of the largest to grow there, reaching a height of 50m.

2.2.10. Chloroleucon (Benth.) Britton and Rose
The genus consists of ten species of shrub and tree, which were formerly included in the *Pithecellobium*, and are native to tropical America. Four species are reported to

nodulate with indeterminate branched nodules. Some species are xeromorphic and five are potentially threatened.

2.2.11. Cojoba Britton and Rose

Another segregant of *Pithecellobium*, this genus one has twelve species of shrub or tree in tropical America, one of which has been reported to have indeterminate nodules.

2.2.12. Dichrostachys (A.DC.) Wight and Arn

Currently, this genus contains 14 species of shrub or small tree, which are native to the Old World tropics. It is, however, thought to be polyphyletic, so the number of species may well be reduced in future, if some are transferred to other genera. The species, *D. cinerea*, nodulates and has a number of subspecies. It can be very invasive when introduced into new areas, *e.g.*, in some of the forest reserves in Cuba. On the other hand, two species from Yemen, Somalia and Ethiopia are at risk.

2.2.13. Elephantorrhiza Benth.

Fourteen species of shrub or small tree, which are native to tropical and South Africa, constitute this genus. Four species are known to nodulate and have indeterminate branched nodules.

2.2.14. Entada Adans.

Twenty-five to 30 species of shrub, small tree or liana, which are found in tropical America, Africa and Australia, make up this genus. Seven species are recorded as nodulated. Some have pods that are over 1m long and are said to have been used as clubs by policemen in the West Indies.

2.2.15. Enterolobium Mart.

This genus consists of *ca.* ten species of tree in tropical America, eight of which have been reported to nodulate. Many of the trees are large (40m high) with spreading canopies and broad buttressed trunks. They are used for many purposes, however, no species is currently listed as threatened. It is surprising that this fast-growing tree has not been more developed for commercial purposes, especially as their potential has been recognized (Anon, 1979).

2.2.16. Falcataria (Nielsen) Barneby and Grimes

Three species of tree, one of which nodulates, make up this genus from the Pacific area and Queensland. All species have been included in other genera in the past.

2.2.17. Havardia Small

Five species of tree from warm America, previously classified under *Pithecellobium*, constitute this genus, with one species having indeterminate, branched nodules.

2.2.18. Hesperalbizia Barneby and Grimes
A monotypic nodulated tree from Central America, formerly included in *Albizia*.

2.2.19. Hydrochorea Barneby and Grimes
Four species of shrub or tree, formerly included in *Pithecellobium,* with one able to nodulate.

2.2.20. Inga Mill.
This large and important genus of shrub or tree from tropical and subtropical America has recently been the subject of much taxonomic study. Currently, about 350 species are recognised. Fifty-six species have been reported to nodulate; nodules in some species start out more or less spherical, but later branch. A monograph has been published of its many uses (Pennington and Fernandes, 1998), which include that of a shade tree for coffee, where its ability to fix nitrogen is considered important (van Kessel, 1983).

2.2.21. Faidherbia A. Chev.
This monotypic nodulating genus, also known as *Acacia albida* Del., is native to Africa and has many of the same uses as *Acacia*, plus the additional useful property of shedding its leaves in the wet season and producing them in the dry season.

2.2.22. Leucaena Benth.
About 22 species, which occur in areas ranging from Texas south to Peru, make up this genus. Nodulation has been reported for five species. One species, *L. leucocephala* (Lam) de Wit, has been used extensively in agroforestry because of its many useful properties, which include a degree of drought and salinity tolerance, plus the ability to nodulate freely and fix significant amounts of nitrogen. Unfortunately, these same characters, coupled with its ability to produce large numbers of seed, have led to it becoming an invasive weed in many areas where it has been introduced. *Calliandra calothysus* (see above) is now usually preferred for agroforestry in many areas, especially in Africa (Kanmegne and Degrande, 2002).

2.2.23. Lysiloma Benth.
Currently, this genus is considered to have about nine species, with many others having been reassigned to other genera, such as *Acacia*. Recognised species are shrubs and trees native to Central America and the West Indies. Five have been reported to nodulate and the nodules are indeterminate.

2.2.24. Macrosamanea Britton and Rose ex Britton and Killip
A genus of 11 species, previously classified under *Pithecellobium* or *Inga*, which are shrubs or slender trees found in seasonally flooded soils in tropical South America. They are, therefore, of interest for the possible flood tolerance of nitrogen fixation. Five have been reported to nodulate with branched indeterminate nodules.

2.2.25. *Mimosa* L.
This large genus (*ca.* 480 species) has habits ranging from herbs to large trees. Fifty species can nodulate. The nodules are indeterminate and often branched. *M. caesalpinifolia* Benth. is xerophytic and widely used for timber, fuel, and charcoal production and is one of three tree species listed as threatened. *M. scabrella* Benth. lacks root hairs and rhizobial infection occurs between epidermal cells, but other species are infected classically *via* root hairs (E. K. James, personal communication).

2.2.26. *Mimozyganthus* Burkhart
A monotypic nodulated xerophytic shrub or tree from Argentina.

2.2.27. *Parapiptadenia* Brenan
Six species of shrub or tree from tropical eastern Brazil make up this genus. Two species have been reported as nodulated.

2.2.28. *Paraserianthes* I.C.Nielsen
Currently, this genus has two tree species from Malaysia and tropical Australia, which were previously assigned to other genera. One species has indeterminate nodules.

2.2.29. *Pentaclethra* Benth.
The genus has only two or three species and is one of those known to have variable nodulation. There is good evidence that the African species, *P. macrophylla* Benth., does not nodulate and equally good evidence that the south America *P. macroloba* does. The genus is thought to be fairly basal within the Mimosoideae and close to some of the non-nodulating genera listed in Table 2, but some taxonomists believe that these two species are definitely from the same genus. They produce good timber, have protein-rich edible (after cooking) seeds, and medicinal uses.

2.2.30. *Piptadenia* Benth.
About 21 species of tree, shrub or liana, found from Mexico south through Peru to Argentina, constitute this genus. Six species have been reported to nodulate and one of the Peruvian species is listed as threatened.

2.2.31. *Piptadeniastrum* Brenan
A monotypic large tree from West Africa, which has indeterminate branched nodules, produces excellent, though odoriferous, timber that has some termite resistance.

2.2.32. *Pithecellobium* Mart.
Now a much smaller genus than previously constituted, it has about 18 species of shrub or tree from tropical America. Six species have been reported to nodulate and seven are threatened.

2.2.33. Plathymenia Benth

The two species of shrub or tree that make up this genus are found in tropical South America; both are nodulated. Although small, the trees produce fine wood, which is much exploited and, for this reason, one species is listed as threatened.

2.2.34. Prosopis L.

Forty-four species of shrub and tree from various warm parts of the world, particularly Central America, constitute this genus. Sixteen species are reported to nodulate. Some species are among the most drought- and salt-tolerant woody plants known. They are exploited for agroforestry and land reclamation. They are used widely for wood, honey, medicines, tannin, forage, hedges, and many other purposes.

2.2.35. Pseudosamanea Harms

This is a monotypic genus of one nodulated tree from Central and South America. In Sprent (2001), *P. guachapele* is given as nodulated. This species is now accepted in *Albizia* with the one remaining species of *Pseudosmanea*, *P. cubana* (Britton and Rose) Barneby and Grimes, having no record on nodulation, but listed as threatened.

2.2.36. Pseudopiptadenia Rauschert

This genus has eleven species of shrub or tree, which range from northern South America to southern Brazil. The two nodulating American species of *Newtonia* Baill., an otherwise non-nodulating genus from Africa, have been transferred to a single species, *P. contorta* (DC.) G.P.Lewis and M.P. Lima, and a second species is also known to nodulate.

2.2.37. Samanea Merr.

This genus is thought by some to be part of *Albizia*. It has three species of tree from northern South America. Two of these species are reported to nodulate, including *S. saman* (Jacq.) Merr., the so called 'rain tree', which closes its leaves when it rains. This tree can grow to great heights, over 60m, and its canopy may cover 0.2ha. It is found on many different types of soil and, although widely grown, is not exploited in a major way for commercial purposes.

2.2.38. Schleinitzia Warb.

Of its four species of shrub or tree from the Pacific region, formerly included in other genera, one is reported to nodulate.

2.2.39. Stryphnodendron Mart.

About 30 species of tree from Central and tropical South America, of which four are known to nodulate, constitute this genus.

2.2.40. Wallaceodendron Koord

This monotypic nodulated tree from the Philippines and Sulawesi, produces excellent dark wood.

2.2.41. Xerocladia Harv.

This monotypic nodulated shrub or small tree occurs in Namibia and South Africa.

2.2.42. Zygia P. Browne

This genus contains many species that have been transferred from *Pithecellobium* and other genera. The current number of species is 57, mainly trees and shrubs from tropical America. Ten are known to nodulate and two are listed as threatened.

2.3. Papilionoideae

This is generally considered as the most advanced subfamily. It consists mainly of small shrubs and annual or perennial herbs, however, there is a significant number of trees, some of which cannot nodulate (Table 4). Like those in the Mimosoideae, most of the non-nodulating genera are considered to be in the basal part of the Papilionoideae. This section covers those genera with tree species known to nodulate. As with other subfamilies, there are numerous genera for which there are no reports.

Table 4. Some Papilionoid Tree Genera that Appear to lack the Ability to Nodulate.

Aldina Endl.	*Lecointea* Ducke
Alexa Moq.	*Luetzelburgia* Harms
Amburana Schwacke and Taub.	*Milbraediodendron* Harms
Amphimas Pierre ex Harms	*Myroxylon* L.f
Angylocalyx Taub.	*Pterodon* Vogl
Castanospermum A. Cunn. ex Hook.	*Styphnolobium* Schott ex Engl
Cladastris Raf.	*Taralea* Aubl
Cordyla Lour.	*Vatairea* Aubl.
Dipteryx Schreb.	*Vataireopsis* Ducke
Exostyles Schott	*Xanthocersis* Baill.
Harleyodendron R.S.Cowan	*Zollernia* Wied.-Neuw. and Nees

2.3.1. Acosmium Schott

This genus has been the subject of taxonomic research, together with the related genus, *Sweetia* Sprengl. At present, *Acosmium* is considered to have 17 species of shrub or tree from tropical America. Reports on nodulation are very varied, but the evidence for *A. nitens* (Vogel) Yakovlev is strong. Nodules are indeterminate and may be found under water (Barrios and Herrera, 1994).

2.3.2. Andira Juss.
It includes thirty species of tree, occurring in tropical America and West Africa. Reports on nodulation have been mixed, but there are ten positives that have been confirmed. One of the problems here is that the nodules vary greatly in size and are usually black and woody on the outside, even when small. Further, their internal structure is of the primitive type, typical of the Caesalpinioideae, with bacteroids retained within fixation threads. These factors make nodules difficult to recognise and also make isolation of rhizobia tricky. Trees may be large; many are attractive and a good source, *inter alia*, of honey, timber and drugs. This genus has more potential than has yet been realised. One species is threatened.

2.3.3. Baphia Afzel. ex Lodd.
Forty-six species of shrub, tree or liana, found in tropical Africa, South Africa and Madagascar, constitute this genus. Three species are reported to nodulate. Nodules are indeterminate but, unusually, may have lenticels when young. Some species produce useful wood and other products, such as dyes, and sixteen are threatened.

2.3.4. Bergeronia M. Micheli
This genus consists of a monotypic nodulated shrub/small tree from northern South America.

2.3.5. Bobgunnia J.H. Kirkbr. and Wiersema
This genus of two species was recently segregated from *Swartzia*; one is nodulated.

2.3.6. Bolusanthus Harms
This monotypic nodulated tree from tropical South Africa has insect-resistant wood.

2.3.7. Bowdichia Kunth
Two nodulated species of tree from tropical South America make up this genus. The trees may grow up to 50m in height and produce course, heavy wood with good resistance to decay.

2.3.8. Brongniartia Kunth.
Currently, this genus contains about 56 species of shrub and tree, which are found on drier soils from Texas south through Central America to the Andes. Surprisingly, only one has been reported to nodulate so far. Their natural habitats suggest that they could be useful additions to the known suite of woody plants for dry, alkaline areas.

2.3.9. Brya P. Browne
This genus has twelve species of shrub or small tree from the West-Indian region. Two species are known to nodulate with nodules of the aeschynomenoid type (see below).

2.3.10. Calia Berland
This segregant of *Sophora* consists of six species of shrub or tree from Mexico and the southern USA. One species is reported to nodulate.

2.3.11. Callerya Endl.
The genus has recently been enlarged to include species from other genera, including *Millettia*. It currently consists of about 19 species of shrub, tree and liana from East and South-East Asia and Australia. Two species have been reported to nodulate.

2.3.12. Calpurnia E. Mey
This genus has seven species of shrub or small tree that are found in southern India and in the eastern Cape region of South Africa. Three species have been reported as nodulating.

2.3.13. Centrolobium Mart. ex Benth.
Of these six species of tree from South America, three have been reported to nodulate with nodules of the aeschynomenoid type. The wood is high quality and is used especially for furniture. One species is threatened.

2.3.14. Chadsia Bojer
This genus of about 17 species of shrub or small tree is confined to Madagascar. One species has been reported to have indeterminate branched nodules.

2.3.15. Chamaecytisus Link
A genus of 30 species of shrub or small tree, which is native to Europe and the Canary Islands, was formerly included in *Cytisus* Desf. Six species have been reported to nodulate. There are unusual features both in the infection by rhizobia and in nodule development of the shrubby species *C. proliferus* Link (Vega-Hernàndez *et al.*, 2001).

2.3.16. Clathrotropis (Benth.) Harms
It consists of six species of tree from tropical South America, of which two are reported to nodulate. Although the trees produce hard wood, it has not proven useful commercially.

2.3.17. Coursetia DC.
There are thirty-eight species of shrub and small tree, all are found in the warmer parts of America. Three species have been reported to nodulate, including the tall tree species, *C. ferruginea* (Kunth) Lavin, which is useful for wood. Two species are on the threatened list, including one that may produce compounds effective against the AIDS virus.

2.3.18. Cullen Medik.
This genus has undergone various taxonomic changes recently. It currently contains about 36 species ranging from annual herbs to small trees, which are widely distributed in the Old World tropics. Four species are known to nodulate.

2.3.19. Cyclolobium Benth.
Now considered as a monotypic genus, it is nodulated with bacteroids retained within fixation threads.

2.3.20. Dahlstedtia Malme
This genus consists of two species of shrub or small tree that are found in Argentina and Brazil. One species has been reported to nodulate with bacteroids retained within fixation threads.

2.3.21. Dalbergia L.f.
This is a large genus of tree, shrub or liana from tropical regions. All species examined to date nodulate with aeschynomenoid nodules (see below) that can be formed on quite old roots as long as they retain the potential to form laterals. Foresters, however, have shown very little interest in the nitrogen-fixing activity of this genus (Sprent and Parsons, 2000). Many of the tree species produce very fine timber, which is widely exploited. Seventy-two species are on the threatened list.

2.3.22. Dalbergiella Baker f.
This genus consists of three species of shrub or small tree that are found in tropical Africa, one of which has been reported to nodulate.

2.3.23. Dalea L.
A few of the 160 species of this genus are trees and at least some of these are known to have indeterminate nodules. They are found from southern Canada down to Argentina. Plants are browsed by wild mammals and have some local uses, for example, in basket making.

2.3.24. Dendrolobium (L.) Benth.
These are twelve species of shrub or small tree found in Indo-Malaysia and northern Australia. Three have been reported as nodulating. This is one of the few papilionoid tree genera from the tribe Desmodieae (now being included in the Phaseoleae). Their nodules are of determinate growth.

2.3.25. Derris Lour.
Recently the subject of many taxonomic changes, the genus currently consists of 50-60 species, most of which are either shrubs or lianas, but a few are trees. Of the

species reported to nodulate, only three remain in the genus, one of which is a tree. Best known for the insecticidal properties of the roots of some species, tree species have been used on a local scale for many purposes, such as culinary, turnery, shade and green manure.

2.3.26. Diphysa Jacq.

This genus has fifteen species of shrub or small tree from tropical America, mostly in the uplands, with one species reported as nodulating.

2.3.27. Diplotropis Benth.

Four of the twelve species of tree of this genus from tropical South America are reported to nodulate. The larger tree species produce excellent wood.

2.3.28. Erythrina L.

This is a large genus of about 120 species of shrub or tree with pantropical distribution. It is the only genus, which has large trees, that is found in the tribe, Phaseoleae, which contains important grain legumes, such as soybean and cowpea. Like them, its nodules may be determinate although, in *Erythrina*, they sometimes become lobed. The tree species usually have brilliant flowers, but the seeds are poisonous and their toxins have been used as fish poisons. Wood of the genus is of limited use, but it has been exhaustively studied for its alkaloids and other pharmaceuticals. Some research has been performed on its nitrogen-fixing potential (see references in Sprent and Parsons, 2000), but there is scope for much more. Twelve species are on the threatened list.

2.3.29. Etaballia Benth

This monotypic nodulated tree from tropical South America produces high-quality wood.

2.3.30. Eysenhardtia Kunth.

About 10 species of shrub or small tree from Central America constitute this genus with one species reported as nodulating. Some species produce good timber and several are browsed by animals, such as cattle and horses.

2.3.31. Geoffroea Jacq.

It consists of two species of shrub or tree from tropical South America, the West Indies, and Panama, one of which is reported to nodulate.

2.3.32. Gliricidia Kunth

This is one of the few tree genera, which have been intensively studied for features that include nitrogen fixation. It has four species of tree from tropical America, which are fast growing, produce a variety of marketable products, and are widely used in agroforestry in Africa and elsewhere.

2.3.33. Harpalyce Moçino Sessé

This genus has about 24 species of shrub or tree, which grow in hot dry and often rocky areas of Mexico, Cuba and Brazil. This genus presents another opportunity for species that could be useful for management of dry areas. Two species have recently been reported to have indeterminate nodules.

2.3.34. Hebestigma Urb.

This monotypic nodulated tree grows in Cuba and its wood is used locally.

2.3.35. Hesperolaburnum Maire

This monotypic nodulated tree is found in the mountains of Morocco.

2.3.36. Hymenolobium Benth.

This genus has about 12 species of tree from tropical South America, of which six have been confirmed to nodulate. Nodule bacteroids are retained within fixation threads, as in the Caesalpinioideae. Many species produce strong timber. This is a genus with high potential.

2.3.37. Kotschya Endl.

This genus consists of 30 species of herb or small tree from tropical Africa; five have been reported to nodulate with aeschynomenoid nodules. Some tree species occur up to 2000m altitude, a comparatively rare feature among nodulated legume trees.

2.3.38. Laburnum Fabr.

Both species of nodulated tree from southern Europe have toxic seeds and are widely grown as ornamentals. The fine textured wood is little exploited commercially.

2.3.39. Leptoderris Dunn

From about 20 species of shrub or small tree, which originate from tropical West Africa, only one has been reported as having indeterminate nodules.

2.3.40. Lonchocarpus Kunth.

This large genus, which currently has about 130 species of shrub or tree, originates mainly from tropical and subtropical America, but with one species in Africa. Twenty-one species have been reported as nodulated and in one species, at least, nodulation is unusual in that infection occurs between epidermal cells. The wood is not considered to be a commercially viable product, but the genus is well known for the range of insecticidal products it produces. Thirteen species are threatened.

2.3.41. Maackia Rupr.
With just six species of tree from East Asia and East Siberia, one has been reported to have indeterminate nodules. This genus produces heavy dark wood that is used in construction. One species is threatened.

2.3.42. Machaerium Pers.
This fairly large genus, containing about 120 species of tree and liana, originates mostly from tropical America, but with one species in West Africa. Twenty-four species have been reported to nodulate. Tree species are relatively few in number and not widely exploited for commercial use. Five species are threatened.

2.3.43. Millettia Wight and Arn
Currently, with *ca.* 150 species of tree, shrub and liana, the genus is probably polyphyletic and may be divided in the future. It is found in the Old World tropics and 12 species appear to nodulate. Like its relative, *Lonchocarpus*, *Millettia* species produce insecticides and other toxic agents. Some produce good timber and most have attractive flowers. It can be found in dense rainforests, where it may play an important role. Twenty-nine species are on the threatened list.

2.3.44. Mundulea (DC.) Benth.
Twelve species of shrub or tree from Madagascar and other parts of the Old World tropics make up this genus. One species has been reported to nodulate. Members of this genus produce various toxins, including insecticides.

2.3.45. Olneya A. Gray
This monotypic nodulated genus of shrub or small tree from south-western North America has hard heavy wood that is used locally.

2.3.46. Ormocarpum P. Beauv.
Twenty species of shrub and small tree from tropical and subtropical areas of the Old World, three of which have been reported as nodulated, constitute this genus. It is used as browse, for shade, and other purposes. Three species are threatened.

2.3.47. Ormosia G. Jacks
About 145 species of shrub or tree from tropical America, Asia and Australia, of which twelve species are reported to nodulate, make up this genus. Nodules may be very large, woody, and profusely branched.

2.3.48. Ortholobium C.H. Stirt.
Formed from part of *Psoralea*, the genus has 53 species of herb, shrub, or tree, which are found in East and South Africa and in Soouth America. Six nodulate.

2.3.49. Paramachaerium Ducke
It consists of five species of tree from tropical South America, one of which has been reported to nodulate and another is on the threatened list.

2.3.50. Pericopsis Thwaites
Previously classified as *Afrormosia* Harms, all three species, two of which are from tropical Africa and one from Sri Lanka, are nodulated. Nodules are indeterminate. The trees produce fine wood that has been widely used as a substitute for teak. For this reason, two of the species are listed as threatened. This genus appears to be an excellent target for properly managed plantation use, taking into account its ability to fix nitrogen, which until now has been neglected.

2.3.51. Philenoptera Fenzl. ex A. Rich.
These twelve species of tree, shrub or liana are from tropical Africa and Madagascar and were previously mainly included in *Lonchcarpus*. Two are known to nodulate.

2.3.52. Piscidia L.
Of the about seven species of shrub and, occasionally tree, found in Florida, Central America, and south to northern Peru, only two species are known to nodulate.

2.3.53. Platycyamus Benth.
Of two tree species from tropical South America, one is reported to nodulate.

2.3.54. Platymiscium Vogel
Consisting of eighteen species of tree from tropical Central and South America, seven so far have been reported as nodulated with nodules of the aeschynomenoid type. Some of the trees exceed 30m in height with a girth of 10m. They have showy flowers and produce high quality timber. Two are threatened.

2.3.55. Platypodium Vogel
It consists of only one or two species of tree from Central and South America, one of which is reported to nodulate.

2.3.56. Poecilanthe Benth.
This genus has about ten species of shrub or tree (two threatened) from tropical South America. Five species nodulate with bacteroids retained within fixation threads.

2.3.57. Psoralea L.
Now restricted to 50 species of herb, shrub or tree from South Africa (see also *Ortholobium* above), this genus has 27 species that are known to nodulate.

2.3.58. Pterocarpus Jacq.
This pan-tropical genus of tree has both nodulated and non-nodulated species. The latter are confined to Brazil, even though Brazilian soils have rhizobia that can nodulate African species. Sixteen species are reported to nodulate, with nodules of the aeschynomenoid type. Trees may be very large with buttress roots and even large trees may be profusely nodulated in the Cameroonian rain forests (see below). A characteristic feature of the genus is that, when cut, the xylem exudes a bright red sap. Wood of the African and Asian species is highly prized. Six species are threatened.

2.3.59. Robinia L.
This genus has 4 species of shrub or tree from North America. *R. pseudoacacia* L. is nodulated and has been widely introduced into other areas, such as central Europe. It fixes nitrogen well, but also suckers freely and may become invasive. It produces useful wood and is a good source of honey, but some parts produce substances toxic to animals, such as cattle and sheep.

2.3.60. Sophora L.
This genus now has about 50 species of herb, shrub or tree, after other species, including all the non-nodulating ones, were transferred to *Styphnolobium*. Nineteen species have been reported to nodulate. Nodules are indeterminate. Many woody species are grown for the flowers but they also have many other uses, although some have leaves that are toxic to cattle, sheep and goats. Eight species are threatened.

2.3.61. Swartzia Schreb.
This genus has about 150 species of tree, shrub or liana, most of which are found in tropical South America, but with some in tropical Africa and Madagascar. Twenty-five species can nodulate. Nodules are indeterminate and may grow quite large. The trees are very ornamental and some species produce compounds used to stupefy fish. At present, they do not appear to have much commercial potential, although they may be important in their ecosystems. Seven species are threatened.

2.3.62. Tipuana (Benth.) Benth.
This monotypic genus of nodulated tree from subtropical South America has nodules that are aeschynomenoid. It is a fast-growing species that has been introduced to many countries where it is used for various purposes ranging from ornamental to carpentry, tanning and dying.

2.3.63. Virgilia Poir.
Two species of tree, both reported to nodulate, constitute this genus from the Cape region of South Africa.

2.3.64. Xeroderris Roberty
This monotypic genus of nodulated tree from tropical Africa is used as stock feed, for making canoes, and in tanning.

3. RHIZOBIA THAT NODULATE LEGUME TREES

Other chapters in this book series cover rhizobial taxonomy. Here, I shall only cover the most important general considerations for trees and then give some information specific to particular trees. Most of the nodulated legume trees are currently found in tropical/subtropical regions. The widest range of studies on their rhizobia has been by Moreira and co-workers, using mainly Brazilian species. However, I believe that their finding that there is no close correlation between host and rhizobial genotypes is of general application (Moreira *et al.*, 1998). This research group is the only one so far to consider members of all three subfamilies, but a number of others has looked at members of the Mimosoideae and Papilionoideae. Bala and Giller (2001) isolated rhizobia from soils in seven counties on three continents and looked at their ability to nodulate the woody legumes, *Calliandra calothyrus*, *Leucaena leucocephala*, and *Sesbania sesban*. As with earlier workers, most notably Barnet and Catt (1991) with Australian *Acacia* species, they found that soil properties were just as important as the host in determining which strains were present and potentially nodulating. Further, in contrast to much of the earlier work cited in Allen and Allen (1981), these and other workers have found that many rhizobia, which nodulate tropical trees, are fast- as well as slow-growing.

There are numerous reports that host range and nodule effectiveness vary greatly with both host species and rhizobial strain. One of the first papers in this area was that of Turk and Keyser (1992), but their work has been supplemented by that of others from several continents, including Bala and Giller (2001), Burdon *et al.* (1999), and Odee *et al.* (2002). The overall picture is that many trees may be nodulated by a variety of rhizobial genera and species, but that some have a greater degree of host specificity than others. This range of specificity is found within host genera. Further, within a host species that is nodulated by many rhizobia, there is a wide range of effectiveness; from ineffective, where nodules are essentially parasitic, to highly effective, where nodulated plants may fix more nitrogen than can be assimilated by control plants on combined nitrogen (*e.g.*, Burdon *et al.* (1999) for work with Australian acacias). Nodule mass on plants with ineffective strains may be far greater than on plants with effective strains. For examples, see McInroy (1997) and the illustration on p. xii of Sprent (2001).

A further point to stress is that nodule structure is host determined. This was first shown for peanut and cowpea, both of which can be nodulated by the same strain of slow-growing *Bradyrhizobium*, where in the former case, the nodules are aeschynomenoid and the export product of nitrogen fixation is amino acids, but in the latter, the nodules are desmodioid and export ureides (Sen and Weaver, 1984). This observation has since been extended to trees, for example, caesalpinioid species with fixation threads and mimosoid species without these structures (Moreira *et al.*, 1998).

There are numerous examples of soils from one country having rhizobia that will nodulate exotic species. These include *Pterocarpus* (as mentioned above), where rhizobia in Brazilian soils nodulate African species, even though the Brazilian species cannot nodulate (Faria and Lima, 1998) and the Australian acacias, *A. auriculiformis* and *A. holosericea*, which nodulate with slow-growing rhizobia in Kenya whereas native species nodulate with fast-growing rhizobia (Odee *et al.*, 1998).

Most of the detailed work on rhizobia that nodulate trees has been carried out with species used in agroforestry and land reclamation/soil amelioration, although the large genus *Acacia* has been studied more widely in both Africa and Australia. Many species of *Acacia* may be nodulated by both fast- and slow-growing rhizobia, although one type tends to predominate in a particular soil. *A. senegal* L. Willd. is one of the more promiscuous species known, being nodulated by at least 6 species of rhizobia from the three genera, *Mesorhizobium, Rhizobium*, and *Sinorhizobium* (Nick, 1998). Within the genus, there may be differences, for example, the wide-host-range NGR234, which can nodulate members of all three subfamilies, only formed effective nodules on the largely Australian sub-genus, *Phyllodineae*, but it could form effective nodules on the related African species, *Faidherbia albida* (Pueppke and Broughton, 1999). Even within Australian species, Burdon *et al.* (1999) found wide variation in host/rhizobial specificity and effectiveness. These problems have led to the suggestion that multi-strain inoculants may sometimes be preferable to single strains (Sutherland *et al.*, 2000).

A rather similar situation obtains for *Calliandra*, although most of the available data are only for *C. calothyrsus*. This species is nodulated by a range of both fast- and slow-growing rhizobia, but with the latter predominant and including species similar to those of the genera, *Agrobacterium, Mesorhizobium, Rhizobium* and *Sinorhizobium* (Bala and Giller, 2001). As with *Acacia,* effectiveness varies widely. It is not nodulated by NGR 234 (Pueppke and Broughton, 1999). One of the few temperate species to be studied in detail is *Robinia pseudoacacia*, which is also nodulated with both fast- and slow-growing rhizobia (Batzli *et al.*, 1992; Röhm and Werner, 1992).

4. TYPES OF NODULE FORMED ON TREES

Legumes have been known for many years to have a range of nodule morphological types and these were found by Corby (*e.g.*, 1988) to have taxonomic value. Taking internal structure into consideration as well as morphology, the categories used by Corby were modified by Sprent and co-workers. The categories found in trees, which do not include the lupinoid form, are given in Table 5. The majority of legume nodules from all latitudes have indeterminate growth, but significant groups have determinate forms. One of the reasons that complicated the development of the classification of Corby was the discovery that all genera (but not all species, see above) in the Caesalpinioideae and some in the Papilionoideae do not release bacteroids into symbiosomes in the active nitrogen-fixing stage, but instead retain them in fixation threads. Many of the indeterminate nodules on trees are woody and often have a more extensive vascular system than their herbaceous counterparts (Sprent *et al.,* 1989).

Table 5. External Features of Nodule Types found in Tree Legumes. For further detail, see Sprent (2001).

Type	Features	Examples
Aeschynomenoid	Associated with lateral roots; determinate, oblate.	*Dalbergia, Pterocarpus*
Desmodioid	Determinate, with lenticels.	*Dendrolobium, Erythrina*
Indeterminate	Apical meristem, often branched, sometimes extensively.	Common in all subfamilies

5. MYCORRHIZAS AND OTHER NUTRIENT-ACQUISITION SYSTEMS

The mycorrhizal status of legume trees is almost certainly important for the optimal functioning of nodules (see Barea *et al.*, this volume). Nodulated legume trees may be arbuscular mycorrhizal (AM), ectomycorrhizal, or both. However, there are distinct variations with both taxonomic position and with geographical location.

Within the Caesalpinioideae, all confirmed cases of ectomycorrhizal species lack the ability to nodulate. Arbuscular mycorrhizal species may or may not nodulate. Interactions between AM and nodulation have not been studied in this sub-family. Classical ectomycorrhizas, with a Hartig net, have not been reported in the Mimosoideae and are rare in the Papilionoideae. Some shrubby endemic Australian legumes are both ectomycorrhizal and nodulated. In Africa, the papilionoid species, *Pericopsis angolensis* (Baker) van Meeuwen, in some areas forms ectomycorrhizas and then does not nodulate; in other areas, it forms AM and may then nodulate (Högberg and Piearce, 1986), thus confirming an early observation by Bonnier (1960) that African legumes are rather versatile in their symbioses.

Some Australian species of *Acacia* produce both AM and also a form of ectomycorrhiza, which is less highly structured than the classical type, particularly in having a much looser fungal sheath. Purists may not accept this structure as a true ectomycorrhizal structure but, because it is formed using standard ectomycorrhizal fungal species, many workers accept them as a form of ectomycorrhiza. Dual mycorrhizal acacias are confined to parts of the sub-genus *Phyllodineae* and all can nodulate. Some, *e.g.*, *A. mangium* Willd., are important timber species that have been introduced into Africa, Brazil, and elsewhere. In these countries, they can associate with both AM and ectomycorrhizal fungi from the local soil, although native acacias are only AM. The symbioses of these acacias allow them to grow in very poor soils (see below), but the ectomycorrhizal partner may also confer other benefits, such as reducing the plant parasitic nematode populations (Founoune *et al.*, 2002). There are numerous publications showing that AM and nodulation may be complementary for acquiring nutrients in legume trees. The overall picture is similar to that in crop and pasture legumes, where AM may help plants obtain P from poorly available sources (Sprent, 1999; Giller, 2001; and references therein).

Cluster roots occur in a number of shrubby legumes of Australia and South Africa, some of the former also have both AM and classical ectomycorrhizas. An early suggestion that cluster roots form on acacias has recently been confirmed for *A. mangium* in the Pilbara region of Western Australia (P. Landsman and M. Adams, personal communication). Compared with the cluster roots found on either shrubby legumes or non-legume trees, such as species of *Grevillea* R. Br. ex J. Knight, those on acacias are very small and their significance remains to be established. They may emerge as being more common among acacias and to be important in accessing poorly available soil nutrients.

6. MEASUREMENT OF NITROGEN FIXATION BY TREES

Accurate measurement of nitrogen fixation in large field-grown trees is virtually impossible. Thus, attempts to quantify nitrogen fixation in natural ecosystems and to understand its role in them are still at a very early stage (Vitousek *et al.*, 2002). General methods of measuring nitrogen fixation are considered elsewhere in these volumes (see Dilworth, in Volume 1, *Catalysts for Nitrogen Fixation*); of these, the acetylene-reduction assay is useful for confirming that nodules are nodules and that they are active. Nitrogen-balance methods are useful when soil N is very low and the ^{15}N-dilution method has been shown to be useful for work on potted trees or young trees in plots and agroforestry systems. For example, Ståhl *et al.* (2002) used this method to compare the effects of nodulated trees as fallows in a maize crop-rotation system in Kenya. *Sesbania sesban* (L.) Merr. fixed 280-360 kg N ha^{-1} and *Calliandra calothyrsus* Meisn. 120-170 kg N ha^{-1} in 22 months, which are figures comparable to those for crop legumes.

In natural ecosystems, the measurement of ^{15}N at natural abundance is the only feasible method currently available. Although it has its problems, in some circumstances and when used with care, it can give very useful data (Boddey *et al.*, 2000). Sprent *et al.* (1996) used it to show nitrogen fixation in a number of woody plants from the Cerrado region of Brazil. Perreijn (2002) made the first attempt to use the method in combination with a number of others to look at the magnitude of different reactions in the nitrogen cycle in a lowland tropical rainforest in Guyana. He concluded that there was wide variation in the amount of nitrogen fixed by various nodulated trees from areas with different soils and other conditions. In West Africa (Côte d'Ivoire), Galiana *et al.* (2002) used the method successfully to show variations in nitrogen fixation by *Acacia mangium* in soils of different nutrient status.

7. ROLE OF LEGUME TREES IN NATURAL AND MANAGED SYSTEMS

7.1. Tropical Rain Forests.

There appears to be a longitudinal gradient in the occurrence of legume trees in tropical rain forests, with fewest in Asian forests. In addition, the proportion of those legume trees with the potential to nodulate increases when going from Asia *via* Africa to tropical America. In Africa, most legume trees lack the ability to nodulate but, in Amazonia, a significant number can (Sprent, 2001). However, recent unpublished work by the author and others in Cameroon, supported by the work of Perreijn (2002) in Guyana, suggests that nodulated legumes may play a bigger part than once thought. Contrary to perceived wisdom that mature trees recycle most of their nitrogen and, therefore, need few nodules, extensive networks of nodulated roots were observed on some very large trees and lianas in Cameroon and on lianas, in particular, in Guyana. Thus, we have the interesting possibility that a few trees may fix nitrogen and share it, possibly *via* mycorrhizas, with other trees in the ecosystem. If this is the case, then selective logging could be even more disastrous than now thought.

Many tropical rainforests are close to major rivers. In South America, nodulated legumes trees from all sub-families are found bordering the Orinoco, often with their nodules being under water (Sprent, 1999). Shrubby legumes, such as *Discolobium pulchellum* Benth., in the flooded Pantanal of Brazil, may only nodulate on stems when submerged (Loureiro *et al.*, 1994). The role of submerged nodules in tropical ecosystems is a subject needing further study.

7.2. Legume Trees in Dry and Saline Areas.

The genera, *Acacia* and *Prosopis* L., have many species that can grow and nodulate in dry and/or saline soils. For example, in the Negev desert, almost the only trees to be found are species of *Acacia* and these may enhance soil-nutrient status but, at the same time, may increase soil salinity (Zuzana and Ward, 2002). Some aspects of management of such trees will be considered in the next section.

7.3. Management of Nodulated Legume Trees.

In traditional forestry, nitrogen fixation by legumes has largely been ignored (Sprent and Parsons, 2000), even though many legume genera, *e.g.*, *Dalbergia* spp., produce valuable timber (see above). In agroforestry, land reclamation, and other situations, nitrogen fixation and other attributes of legume trees are considered to be important for management and some examples are given below. For the humid tropics, work has been reviewed recently by Schroth *et al.* (2001).

7.3.1. Shade

Legume trees have been used as shade plants for crops such as coffee (*e.g.*, species of *Inga* in Central and South America) and cocoa (*e.g.*, *Albizzia lebbek* (L.) Benth. in Ghana) for many years. A number of studies has shown that nitrogen fixation by *Inga* can be important for the associated crop (see references in Pennington and Fernandez, 1998). Moreover, recent work in Mexico (Romero-Alvarado *et al.*, 2002) found that mixtures of shade trees, some legume and some not, were of equal benefit to the nutrient balance of the system when compared to mono-cropped legumes but, in addition, also enhanced biodiversity, including number of bird species present.

7.3.2. Soil Properties

Legume trees, *e.g.*, *Acacia senegal* (L.) Willd in Senegal, can lead to improved soil nutrient content, a fact which is important in their use as fallow systems (Deans *et al.*, 1999). In Costa Rica, growth of the timber tree, *Terminalia amazonia* (J.F.Gmel.) Exell (Combretaceae), planted on eroded, degraded pastures was considerably improved when interplanted with legume trees, especially *Inga edulis* Mart. (Nichols *et al.*, 2001).

One of the problems of soil cultivation for crops may be the formation of impermeable layers. As an alternative to mechanical breakage of such layers, Yunusa *et al.* (2002) looked at the effects of planting six native (to Australia) woody species, including three acacias. All, but the acacias in particular, had a marked effect in improving the hydraulic conductivity of the soil, allowing *inter alia* better root growth of subsequent crop species.

On extremely sodic (pH 10.2) soils in northwestern India, a silvopastoral system, involving grasses and the legume trees, *Acacia nilotica* (L.) Del., *Dalbergia sissoo* Roxb. and *Prosopis juliflora* (Sweet) DC., showed the legume trees to lead to marked soil improvement (Kaur *et al.*, 2002).

Land reclamation, particularly following mining activity, is a priority in many areas. Australian acacias with dual mycorrhizas and nodules are proving to be particularly useful for this purpose. For example, when grown on residues from bauxite mines in Brazil, both *A. holosericea* G. Don and *A. mangium* in 22 months produced 4-to-5-times more above-ground biomass than several other species including the endemic *A. angustissima* (Mill.) Kuntze that only forms AM (Franco and Faria, 1997). At the opposite end of the scale, legumes, including acacias, may be used for biogeochemical prospecting. A recent study in a mountainous region of Pakistan found that a number of deep-rooted woody plants, including *A. arabica* Willd. (now usually known as *A. nilotica* subsp. *indica* (Benth.) Brenan), can take up Zn and Cu to varying amounts, which reflect the concentrations in the underlying rock, thus, enabling accurate targeting of mining activities (Naseem and Sheikh, 2002).

In contrast, the ability of some tree legumes, especially dual mycorrhizal acacias, to grow on poor soils has led to severe problems in areas such as the Cape province of South Africa, where they have invaded the fragile Fynbos vegetation which is unique to the area (Stock *et al.*, 1995; Holmes and Cowling, 1997). In other parts of South Africa, where these acacias (and other non-legume trees) have been grown for commercial timber, their fast growth is associated with high rates of water usage. This, in turn, is leading to major reductions in river flow in some catchment areas and so impeding economic growth (Le Maitre *et al.*, 2002). The invasiveness of these acacias is enhanced by their ability to reproduce freely by both seed and vegetative methods. Prolific seeding has proved to be the downfall of *Leucaena leucocephala* in many countries, *e.g.*, Cameroon (Kanmegne and Degrande, 2002), where it was introduced as an agroforestry plant and has turned out to be an invasive weed.

Finally, the fact that nitrogen fixation can impact other reactions in the nitrogen cycle should not be ignored (see the chapters herein by van Spanning and Fiencke and by Bock). Nitrate run-off from agricultural soils following ploughing of pastures is well known, but a recent report from an experiments in Uganda, where nitrogen fixing and non-fixing trees were compared, found that the soils around the former produced up to four-times as much NO and N_2O, both greenhouse gases, following rain (Dick *et al.*, 2001).

REFERENCES

Allen, O. N., and Allen, E. K. (1981). *The Leguminosae: A source book of characteristics, uses and nodulation.* Madison, WI: University of Wisconsin Press, and London: Macmillan Publishers Ltd.

Anon (1979). *Tropical Legumes: Resources for the future.* Washington, D.C.: National Academy of Sciences.

Bala, A., and Giller, K. E. (2001). Symbiotic specificity of tropical tree rhizobia for host legumes. *New Phytol., 149*, 495-507.

Barnet, Y. M., and Catt, P. C. (1991). Distribution and characteristics of root nodule bacteria isolated from Australian *Acacia* spp. *Plant Soil, 135*, 109-120.

Barrios, E., and Herrera, R. (1994). Nitrogen cycling in a Venezuelan tropical seasonally flooded forest: soil nitrogen mineralization and nitrification. *J. Trop. Ecol., 10*, 399-116.

Batzli, J. M., Graves, W. R., and van Berkum, P. (1992). Diversity among rhizobia effective with *Robinia pseudoacacia* L. *Appl. Environ. Microbiol., 58*, 2137-2143.

Boddey, R. M., Peoples, M. B., Palmer, B., and Dart, P. J. (2000). Use of ^{15}N natural abundance technique to quantify biological nitrogen fixation by woody perennials. *Nutrient Cycling in Agroecosystems, 57*, 235-270.

Bonnier, C. (1960). Symbiose *Rhizobium*-légumineuses: Aspects particuliers aux régions tropicales. *Ann. Inst. Pasteur, 98*, 537-556.

Burdon, J. J., Gibson, A. H., Searle, S. D., Woods, M. J., and Brockwell, J. (1999). Variation in the effectiveness of symbiotic associations between native rhizobia and temperate Australian *Acacia*: Within-species interactions. *J. Appl. Ecol., 36*, 398-408.

Corby, H. D. L. (1988). Types of rhizobial nodule and their distribution among the Leguminosae. *Kirkia, 13*, 53-123.

Deans, J. D., Digne, O., Lindley, D. K.. Dione, M., and Parkinson, J. A. (1999). Nutrient and organic-matter aacumulation in *Acacia Senegal* fallows over 18 years. *For. Ecol. Manag., 124*,153-167.

Dick, J., Skiba, U., and Wilson, J. (2001). The effect of rainfall on NO and N_2O emissions from Ugandan agroforest soils. *Phyton-Annales Rei Botanicae, 41*, 73-80.

DuPuy, D. (2001). *Legumes of Madagascar.* Kew, UK: Royal Botanic Gardens.

Faria, S. M. de, and Lima, H. C. de (1998). Additional studies of the nodulation status of legume species in Brazil. *Plant Soil, 200*, 185-192.

Faria, S. M. de, McInroy, S. G., and Sprent, J. I. (1987). The occurrence of infected cells, with persistent infection threads, in legume root nodules. *Can. J. Bot., 65*, 553-558.

Founoune, H., Duponnois, R., and Bâ, R. (2002). Ectomycorrhization of *Acacia mangium*, Willd. and *Acacia holosericea* A. Cunn. ex G. Don in Senegal. Impact on plant growth, populations of indigenous microorganisms and plant parasitic nematodes. *J. Arid Environ., 50*, 325-332.

Franco, A. A., and Faria, S. M. de. (1997). The contribution of N_2-fixing tree legumes to land reclamation and sustainability in the tropics. *Soil Biol. Biochem., 29*, 897-903.

Galiana, A., Balle, P., Kanga, A. N., and Domenach, A. M. (2002). Nitrogen fixation estimated by the N-15 natural abundance method in *Acacia mangium* Willd. inoculated with *Bradyrhizobium* sp and grown in silvicutural conditions. *Soil Biol. Biochem., 34*, 251-262.

Giller, K. E. (2001). *Nitrogen Fixation in Tropical Cropping Systems.* 2nd Edition. Wallingford, UK: CAB International.

Gross, E., Cordeiro, L., and Caetano, F. H. (2002). Nodule ultrastructure and initial growth of *Anadenanthera peregrina* (L.) Speg. var. *falcate* (Bneth.) Altschul plants infected with rhizobia. *Ann. Bot., 90*, 175-183.

Högberg, P., and Piearce, G. D. (1986). Mycorrhizas in Zambian trees in relation to host taxonomy, vegetation type and successional patterns. *J. Ecol., 74*, 775-785.

Holmes, P. M., and Cowling, R. M. (1997). The effects of invasion by *Acacia saligna* on the guild structure and regeneration capabilities of South African fynbos. *J. Appl. Ecol., 34*, 317-332.

Kanmegne, J., and Degrande, A. (2002). From alley cropping to rotational fallow: Farmers involvement in the development of fallow management techniques in the humid forest zone of Cameroon. *Agroforestry Systems, 54*, 115-120.

Kaur, B., Gupta, S. R., and Singh, G. (2002). Carbon storage and nitrogen cycling in silvopastoral ststems on a sodic soil in northwestern India. *Agroforestry Systems, 54*, 21-29.

Lavin, M., Pennington, R. T., Klitgaard, B. B., Sprent, J. I., de Lima, H. C., and Gasson, P. E. (2001). The Dalbergioid legume (Fabaceae): Ddelimitation of a pantropical monophyletic clade. *Amer. J. Bot., 88*, 503-533.

Le Maitre, D. C., Wilgen, B. W. van, Gelderblom, C. M., Bailey, C., Chapman, R. A., and Nel, J. A. (2002). Invasive alien trees and water resources in South Africa: Case studies of the costs and benefits of management. *Forest Ecol. Management, 160*, 143-159.

Lorenzi, H. (1992). *Árvores Brasileiras*. SP Brasil: Editora Plantarum LTDA.

Loureiro, M. F., Faria, S. M. de, James, E. K., Pott. A. A., and Franco, A. A. (1994). Nitrogen-fixing stem nodules of the legume, *Discolobium pulchellum* Benth. *New Phytol., 128*, 283-295.

McInroy, S. G. (1997). *Rhizobia for African species of* Acacia: *Characterisation, conservation and selection of inoculants.* University of Dundee, UK: M.Sc. Thesis.

Moreira, F. M. S, Haukka, K., and Young, J. P. W. (1998). Biodiversity of rhizobia isolated from a wide range of forest legumes in Brazil. *Mol. Ecol., 7*, 889-895.

Naisbitt, T., James, E. K., and Sprent, J. I. (1992). The evolutionary significance of the genus *Chamaecrista* as determined by nodule structure. *New Phytol., 122*, 487-492.

Naseem, S., and Sheikh, S. A. (2002). Biogeochemical prospecting of sulphide minerals in Winder Valley, Balochistan, Pakistan. *Res. Geol., 52*, 59-66.

Nichols, J. D., Rosemeyer, M. E., Carpenter, F. L., and Kettler, J. (2001). Intercropping legume trees with native timber rapidly restores cover to eroded tropical pastures without fertilization. *Forest Ecol. Management, 152*, 195-209.

Nick, G. (1998). *Polyphasic taxonomy of rhizobia isolated from tropical tree legumes.* University of Helsinki, Finland: Ph.D. Thesis.

Odee, D. W., Njoroge, J., Machua, J., and Dart, P. (1998). Selective preference for nodulation and symbiotic nitrogen fixing potential of indigenous rhizobia with African and Australian acacias. In C. Elmerich, A. Kondorosi, and W. E. Newton (Eds.), *Biological Nitrogen Fixation for the 21st Century*, (pp. 673-674). Dordrecht, The Netherlands: Kluwer Acacemic Publishers.

Odee, D. W., Haukka, H., McInroy, S. G., Sprent, J. I., Sutherland, J. M., and Young, J. P. W. (2002). Genetic and symbiotic characterization of rhizobia isolated from tree and herbaceous legumes grown in soild from ecologically diverse sites in Kenya. *Soil Biol. Biochem., 34*, 801-811.

Oldfield, S., Lusty, C., and MacKinven, A. (1998). *The World List of Threatened Trees*. Cambridge, UK: World Conservation Press.

Pennington, T. D., and Fernandez, E. C. M. (Eds.) (1998). *The Genus Inga: Utilization*. Kew, UK: Royal Botanic Gardens.

Peoples, M. B., Faizah, A. W., Rerkasen, B. and Herridge, D. F. (Eds.) (1989). *Methods for Evaluating Nitrogen Fixation by Nodulated Legumes in the Field*. Canberra, Australia: ACIAR.

Perreijn, K. (2002) Symbiotic *Nitrogen Fixation by Leguminous trees in Tropical Rain Forest in Guyana.* Ph.D. Thesis, University of Utrecht, The Netherlands: Tropenbos-Guyana Series 11.

Pueppke, S. G., and Broughton, W. J. (1999). *Rhizobium* sp. strain NGR234 and *R. fredii* USDA257 share exceptionally broad, nested host-ranges. *Mol. Plant-Microbe Interact., 12*, 293-318.

Ribeiro, J. E. L. da S., Hopkins, M. J. G., Vicenti, A., Sother, C. A., Costa, M. A. da S., Britto, J. M. de, *et al.* (1999). *Flora da Reserva Ducke*. Manaus, Brazil: INPA.

Röhm, M., and Werner, D. (1992) *Robinia pseudoacacia-Rhizobium* symbiosis: Isolation and characterization of a fast nodulating and efficiently nitrogen fixing, *Rhizobium* strain. *Nitrogen Fixing Tree Research Reports, 10*, 193-197.

Romero-Alvarado, Y., Soto-Pinto, L., García-Barrios, L., and Barrera-Gaytán, J. F. (2002). Coffee yields and soil nutrients under the shades of *Inga* sp. vs. multiple species in Chiapas, Mexico. *Agroforestry Systems, 54*, 215-224.

Schroth, G., Lehmann, J., Rodrigues, M. R. L., Barros, E., and Macêdo, J. L. V. (2001). Plant-soil interactions in multistrata agroforestry in the humid tropics. *Agroforestry Systems, 53*, 85-102.

Sen, D., and Weaver, R. W. (1984). A basis of different rates of N_2-fixation by some strains of *Rhizobium* in peanut and cowpea root nodules. *Plant Sci. Lett., 34*, 239-246.

Sprent, J. I. (1999). Nitrogen fixation and growth of non-crop species in diverse environments. *Persp. Plant*

Ecol. Evol. System., *2*, 149-162.

Sprent, J. I. (2000). Nodulation as a taxonomic tool. In P. S. Herendeen and A. Bruneau (Eds.), *Advances in Legume Systematics*, Part 9 (pp. 21-44). Kew, UK: Royal Botanic Gardens.

Sprent, J. I. (2001). *Nodulation in Legumes*. Kew, UK: Royal Botanic Gardens.

Sprent, J. I., Geoghegan, I. E., Whitty, P. W., and James, E. K. (1996). Natural abundance of ^{15}N and ^{13}C in nodulated legumes and other plants in the Cerrado and neighbouring regions of Brazil. *Oecologia*, *105*, 440-446.

Sprent, J. I., and Parsons, R. (2000). Nitrogen fixation in legume and non-legume trees. *Field Crops Res.*, *65*, 183-196.

Sprent, J. I., Sutherland, J. M., and Faria, S. M. de (1989). Structure and function of nodules from woody legumes. In C. H. Stirton and J. L. Zarucchi (Eds.) *Monographs in Systematic Botany 29* (pp. 559-578). St Louis, MO: Missouri Botanical Gardens.

Ståhl, L., Nyberg, G., Högberg, P., and Buresh, R. J. (2002). Effects of planted tree fallow on soil nitrogen dymanics, above-ground and root biomass, N_2-fixation and subsequent maize crop productivity in Kenya. *Plant Soil*, *243*, 103-117.

Stock, W. D., Wienand, K. T., and Baker, A. C. (1995). Impacts of invading N_2-fixing *Acacia* species on patterns of nutrient cycling in two Cape ecosystems - Evidence from soil incubation studies and ^{15}N natural abundance values. *Oecologia*, *101*, 375-382.

Subba Rao, N. S., and Yatazawa, M. (1984). Stem Nodules. In N. S. Subba Rao, N.S. (Ed.), *Current Developments in Biological Nitrogen Fixation* (pp. 101-110). New Delhi, India: Edward Arnold (see page 2, Yatazawa; Yatazawa is correct).

Sutherland, J. M., Odee, D. W., Muluvi, G. M., McInroy, S. G., and Patel, A. (2000). Single and multi-strain rhizobial inoculation of African acacias in nursery conditions. *Soil Biol. Biochem.*, *32*, 323-333.

Turk, D., and Keyser, H. K. (1992). Rhizobia that nodulate tree legumes: Specificity of the host for nodulation and effectiveness. *Can. J. Microbiol.*, *38*, 451-460.

Van Kessel, C. J. H. (1983). *Effects of environmental and physiological factors in N_2 fixation by Inga jinicuil and Trifolium species*. University of Utrecht, The Netherlands. Ph.D. Thesis.

Vega-Hernàndez, M., Pérez-Galdona, R., Dazzo, F. B., Jarabo-Lorenzo, A., Alfayate, M. C., and Leòn-Barrios, M. (2001). Novel infection process in the indeterminate root nodule symbiosis between *Chamaecytysis proliferus* (tagasaste) and *Bradyrhizobium* sp. *New Phytol.*, *150*, 707-721.

Vitousek, P. M., Cassman, K., Cleveland, C., Crews, T., Field, C. B., Grimm, *et al.* (2002). Towards an ecological understanding of biological nitrogen fixation. *Biogeochem.*, *57/58*, 1-45.

Yunusa, I. A. M., Mele, P. M., Rab, M. A., Schefe, C. R., and Beverly, C. R. (2002) Priming of soil structural and hydrological properties by native woody species, annual crops, and a permanent pasture. *Austral. J. Soil Res.*, *40*, 207-219.

Zuzana, M., and Ward, D. (2002). *Acacia* trees as keystone species in Negev desert ecosystems. *J. Veget. Sci.*, *13*, 227-236.

Chapter 8

NITROGEN-FIXING TREES WITH ACTINORHIZA IN FORESTRY AND AGROFORESTRY

R. O. RUSSO
EARTH University, Costa Rica

1. INTRODUCTION

Nitrogen-fixing trees with actinorhiza (or actinorhizal trees) form a group that is a key component in many natural ecosystems, agro-ecosystems, and agroforestry systems in the world, and that provides an important source of fixed nitrogen in these ecosystems. The general characteristics of the actinorhizal symbiosis, including aspects of nodule formation, co-evolution of both partners, and nitrogen fixation rates at field level of some selected species, as well as mycorrhizal associations in actinorhizal trees, will be described. In addition, topics related to actinorhizal trees in agroforestry, plus the role these woody perennial components may play in agroforestry systems, will also be explored, with emphasis on two genera, Casuarina and Alnus, widely known and used in this kind of production systems. The experience of the Central America Fuelwood Project, the case of Alnus acuminata in tropical highlands, and other uses of actinorhizal trees are also covered. Just as actinorhizal trees open new horizons for research and applications aimed toward sustainability, this chapter attempts to make a modest complementary contribution to the knowledge generated in the last four decades.

Actinorhiza is the result of a symbiotic relationship between a special soil N_2-fixing actinomycete (filamentous bacteria of the order Actinomycetales of the genus Frankia) and fine plant roots; it is neither the actinomycete nor the root, but rather the structure or root nodule formed from these two partners. Because the result of the symbiotic association is a root nodule, the host plant is known as an actinomycete-nodulated plant or actinorhizal plant (Torrey and Tjepkema, 1979; Huss-Danell, 1997). Actinorhizal associations range from the arctic to the tropics and from the semi-desert to rainforest ecosystems. Actinorhizal plants can be found in forest, swamp, riparian, shrub, prairie, and desert ecosystems. In moving from warmer to colder climates, actinorhizal plants become more prevalent and seem to

D. Werner and W. E. Newton (eds.), Nitrogen Fixation in Agriculture, Forestry, Ecology, and the Environment, 143-171.

fill the niche dominated by woody legumes in the tropics (Dawson, 1986). The actinorhizal trees are much less numerous than the vast group of the nitrogen-fixing legume trees. Even so, they are an important source of fixed nitrogen in their ecosystems and can fix from 2 to 300 kg N ha^{-1} yr^{-1}.

Despite of the size of the group, actinorhizal trees are important because: (a) their nitrogen-fixing ability is not restricted to one family but it expands to eight families; (b) most of the actinorhizal trees are good and sometimes aggressive colonizers that are capable of regenerating poor soils or disturbed sites, which are rapidly increasing in many tropical countries; and (c) tropical actiinorhizal trees, especially Casuarina and Alnus, produce not only timber but also firewood and charcoal, sometimes also providing shelter to the cattle, and are used for the protection and recovery of degraded soils (Diem et al., 1984). Many actinorhizal plants are pioneering species, like Alnus, which grows in moist environments, or Myrica, which grows on landslides, eroded slopes, and mined areas, or Casuarinas, which have been identified as nitrogen-fixing trees for adverse sites (NRC, 1984).

This chapter will assess the realities and opportunities of actinorhizal trees in forestry and agroforestry, with an emphasis on tropical regions where the author has expertise and working experience.

2. GENERAL CHARACTERISTICS OF THE ACTINORHIZAL SYMBIOSIS

Actinorhizal trees are defined by their ability to form N_2-fixing root-nodule symbioses with the genus, Frankia. This association is typical for angiosperms but has not been reported in gymnosperms. Frankia is a Gram-positive, branched, septate, filamentous bacterium of the order Actinomycetales, which shows a complex pattern of differentiation in vitro and in the nodule (Berry, 1986). Frankia has also been defined as a diverse group of soil actinomycetes that have in common the formation of multi-locular sporangia, filamentous growth, and nitrogenase-containing vesicles that are enveloped in multi-laminated lipid envelopes (Benson and Silvester, 1993). Frankia infects a host plant to form a nodule on the root system. In the symbiosis, Frankia provides the plant with a source of fixed nitrogen and, in return, the plant provides the Frankia with a carbon source. Although some 40 attempts had been made earlier (Baker and Torrey, 1979), the first confirmed Frankia strain was isolated in 1978 from nodules of Comptonia, using an enzymatic method (Callaham et al., 1978). Other methods, e.g., sucrose-density-gradient centrifugation (Baker et al., 1979), serial dilution (Lalonde, 1979), selective incubation (Quispel and Burggraaf, 1981), and osmium-tetroxide treatment (Lalonde et al., 1981), have successfully produced Frankia in pure culture.

2.1. Host Specificity Groups.

Early on, compatibility groups of host plants and *Frankia* were thought to exist (Becking, 1974), however, later studies found that some isolates were capable of nodulating hosts from different “compatibility” groups (Lechevalier and Ruan, 1984) and this finding opened new horizons for research concerning host specificity

(Berry, 1986). These newer studies often apply molecular-biology tools to questions related to the genetics, ecology, and evolution of actinorhizal symbiotic systems. Molecular phylogeny groupings of host plants were correlated with morphological and anatomical features of actinorhizal nodules in the past, but host-plant phylogenies that are based on molecular data soon revealed different relationships among host plants than have previously been assumed (Berry, 1994; Swensen and Mullin, 1997a).

Three host-specificity groups, *Alnus*, *Elaeagnus*, and *Casuarina*, have been identified and the majority of *Frankia* strains studied fell into one of these three cluster groups. The agreement between the phenotypic clusters and the genospecies described previously shows that the grouping may reflect the taxonomic structure of the genus, *Frankia*. Dobritsa (1998) validated this concept with a study in which thirty-nine selected *Frankia* strains, which belonged to different genomic species, were clustered on the basis of their *in vitro* susceptibility to 17 antibiotics, pigment production and ability to nodulate plants of the genus *Alnus* and/or the family Elaeagnaceae or the family Casuarinaceae. The author found differentiating phenotypic characters for some clusters that may be useful for species definition.

Simultaneously, Huguet *et al.* (2000) carried out a study to identity *Frankia* strains from nodules of *Myrica gale*, *Alnus incana* subsp. *rugosa*, and *Shepherdia canadensis* growing in close proximity and also from nodules of two additional stands with *M. gale* as the sole actinorhizal component. Using a gene-polymorphism technique, they demonstrated that, at the first site, each host-tree species was nodulated by a different phylogenetic group of *Francia*. At the second site, they found that the *M. gale*-strains belonged to yet another group and were relatively low in diversity for a host genus considered promiscuous with respect to *Frankia* microsymbiont genotype. It should be remembered here that plants in the Myricaceae family are considered to be promiscuous hosts because several species are effectively nodulated by most isolated strains of *Frankia* in the greenhouse. This observation has led to the hypothesis that plants in the Myricaceae family have sufficient diversity to serve as a reservoir host for *Frankia* strains that infect plants from other actinorhizal families (Clawson and Benson, 1999).

These issues have contributed to the development of new hypotheses on the origin and evolution of actinorhizal symbiotic systems and have initiated the notion that nitrogen-fixing endosymbionts of actinorhizal plants can interact with a very broad range of unrelated host-plant genotypes (Kohls *et al.*, 1994). Furthermore, a model for actinorhizal specificity has been proposed that includes different degrees of specificity of host-symbiont interactions, ranging from fully compatible to incompatible, and also that actinorhizal plants undergo feedback regulation of symbiosis, involving at least two different consecutive signals that control root nodulation (Wall, 2000). Further, the use of heterologous probes in combination with nucleotide-sequence analysis have allowed a number of *nif* genes to be mapped on the *Frankia* chromosome and this work will ultimately contribute to our understanding of *nif*-gene regulation in *Frankia* (Mullin and Dobritsa, 1996).

Actinorhizal nodules are ontogenically related to lateral roots and may result from modification of the developmental pathway that leads to lateral-root formation

(Goetting-Minesky and Mullin, 1994). This type of root nodule is characterized by differentially expressed genes, which supports the idea of the distinctiveness of this new entity, called "Actinorhiza", a name that refers to both the filamentous bacteria and to the root location of nitrogen-fixing nodules (Wall, 2000).

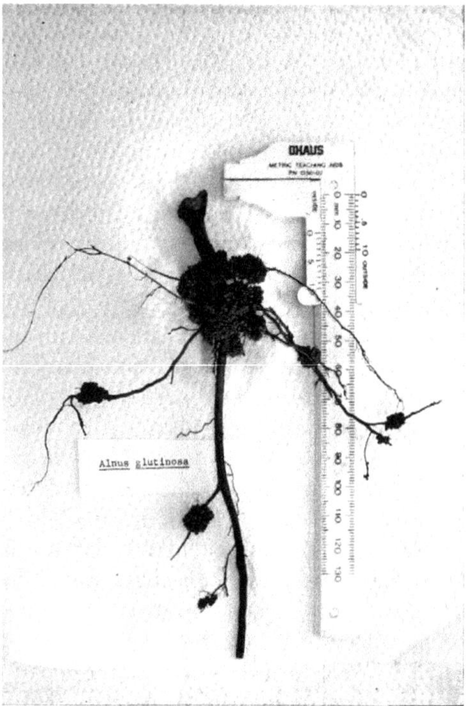

Figure 1. Root Nodules on Alnus glutinosa.
(Photo: R.O. Russo)

2.2. Infection Pathways.

Two pathways for root infection have been described for compatible *Frankia* interactions; they are either root-hair infection or intercellular penetration. The functional physiology of nodule morphogenesis has been reviewed by Berry (1994). As stated by this autor, nodules are initiated in actinorhizal plants either *via* root infection by *Frankia* or by intercellular invasion (Miller and Baker, 1985) and that the mode of infection is host-determined because single strains of *Frankia* have been shown to nodulate different hosts by different infection mechanisms (Miller and Baker, 1986). Premature stages of the developing infection embrace Esther deformations of the root-hair wall or, in the case of intracellular penetration, the release of host extracellular pectic polysaccharides (Liu and Berry, 1991). Host cortical-cell proliferation is followed by expansion and even hypertrophy of groups of cells infected by *Frankia*. The endosymbiont penetrates cortical cells after meristematic growth, probably during cell expansion. Within the host cells, *Frankia* proliferates and vesicles are formed, wherein nitrogen fixation takes place (except with the Casuarinaceae family). *Frankia* strains can be classified in two groups, those that form spores (sp+) within the nodules (spore-positive nodules) and those that do not (sp-) and so produce spore-negative nodules (Schwintzer, 1990). It appears that sp+ strains are less common than sp-.

The morphology of the vesicles is relatively constant in culture but, in symbiotic conditions when interacting with the host plant, it is modified to give a diverse array of vesicles shapes (Benson and Silvester, 1993). Another line of research involves the function of flavonoid compounds on the patterns of nodulation by *Frankia*, *e.g.*, Benoit and Berry (1997) found that flavonoid-like compounds from seeds of red alder (*Alnus rubra*) influenced host nodulation. Independent of the infection mechanism, some authors support the concept that the opening (or closing) of the window of susceptibility for infection and nodule development in the growing root are signal-mediated processes (Wall, 2000), even though rhizosphere nodulation signals have not been as well characterized for actinorhizal plants as they have been for legumes. Recently, the steps in nodule formation for actinorhizal and rhizobial symbioses have been compared and phylogenetic aspects of nodulating both actinorhizal and legume plants described (Gualtieri and Bisseling, 2000).

2.3. External Factors That Impact Nodulation.

Many factors intervene in the process of nodulatin and nodule formation. As for legumes, fixed-nitrogen levels in the substrate inhibit both nodulation and N_2 fixation in actinorhizal plants (Huss-Danell, 1997). This effect had been observed in 1975 by Bond and Mackintosh, who reported that nitrate inhibited both nodule biomass and nitrogenase activity in *Hippophaë rhamnoides* and *Coriaria arborea* (Gentili and Huss-Danell, 2002). The same pattern was also observed by Arnone *et al.* (1994), who used a split-root system that allowed them to distinguish between local and systemic effects of fixed-N compounds. They found that nodulation was inhibited locally by nitrate in *Casuarina cunninghamiana*.

But fixed-N compounds are not the only impacting factor. Phosphorus compounds interact with fixed-N compounds and together affect nodulation. This interaction effect between ammonium nitrate and phosphate on nodulation by the intercellular infection was studied by Gentili and Huss-Danell (2002). They found that phosphate modifies the effects of fixed-nitrogen compounds on nodulation in split-root systems of *Hippophaë rhamnoides*. Further, the inhibition of nodulation by fixed-N compounds was systemic for both nodule number and nodule biomass and the high phosphate level had a systemic stimulation on nodule number and biomass. In this study, phosphate prevented the systemic inhibitoin by fixed-N compounds. The stimulation by phosphate was specific to nodulation and not simply mediated *via* plant growth. This study, however, did not allow the authors to make conclusions on the effect of both fixed-N and P compounds on N_2 fixation.

In addition to root nodulation, aerial nodulation has been observed in *Casuarina* by a number of authors. Prin *et al.* (1991a; 1991b) reported aerial nodules on the trunks of *Casuarina cunnimghamiana.* Then, Valdés and Cruz Cisneros (1996) described stem nodulation of *Casuarina* in Mexico and Bertalot and Mendoza (1998) observed aerial nodulation in *Casuarina equisetifolia* that was shading pastures in the inlands of the State of Sao Paulo in Brazil. The authors described them as stem protuberances, which contained actinorhizal nodules and adventitious roots that were 3-8 cm thick. They covered irregularly up to 20 percent of the trunk surface below breast height and tested positively by the acetylene-reduction assay.

2.4. Co-Evolution of the Symbionts.

Taking into consideration that two genomes, one from the host and one from the actinomycete, are involved in the host-endophyte interactions, this is an intricate symbiotic relationship. Many factors are involved in the actinorhizal symbiosis; these include host-plant physiology, host compatibility, host specificity in terms of both infectivity or affectivity, ability to promote host-plant growth, and interactions of the microsymbiont with other microorganisms inside or outside the host root. All these interactions have led to a coevolution process between the two genomes. Considerable evidence exists to support the hypothesis that coevolution has taken place through mutual interaction of host plants and *Frankia* populations in the soil to produce the most efficient symbiotic associations (Lie *et al.,* 1984). Several research groups interested in coevolution have studied *Frankia*, when associated with host plants of different genera and families, and have found some correspondence between the bacterial and plant phylogenies, which suggests coevolution (Benson *et al.,* 1996). Although an exploratory approach, based on genome interactions, is helpful in looking at this problem, the situation is much more complicated than it appears, mainly because the host-plant genome is more complex than that of the prokaryote. Studies on co-evolution between *Frankia* populations and host plants in the family Casuarinaceae were carried out by Simonet *et al*., (1999), who deduced patterns of global dispersal.

The ecological significance of considering the evolution of both symbiotic partners is that the complete symbiotic unit can explore a wider environmental range. This is the situation for those actinorhizal trees that grow in: (i) poor, disturbed, or degraded soils; (ii) in nitrogen-poor sites, such as sandy soils, disturbed soils, and wet soils; or (iii) as pioneer vegetation at early stages of plant succession following disturbances, such as eruptions, flooding, landslides, and fires.

3. HOST BOTANICAL FAMILIES

Actinorhizal hosts described to date are trees and shrubs that, according to the classification scheme of Cronquist (1981), which is based primarily on the analysis of morphological characters, are distributed among four subclasses, Rosidae, Hamamelidae, Magnoliidae and Dilleniidae, in eight families of angiosperms that include 25 genera and over 200 plant species. This broad phylogenetic distribution of nodulated actinorhizal plants has created the notion that nitrogen-fixing endosymbionts, particularly those of actinorhizal plants, can interact with a very broad range of unrelated host-plant genotypes (Swensen and Mullin, 1997b). Several reviews on the biology and development of these actinorhizal associations have been published in recent years (Benson and Silvester, 1993; Berry, 1994; Schwintzer and Tjepkema, 1990). Not all genera in these families and not all species in the 25 genera fix atmospheric N_2. In certain environments, species of these genera may become more abundant and ecologically significant. According to Benson and Clawson (2000), nodulating families occur sporadically within the families; in the Casuarinaceae, Coriariaceae, Datiscaceae, Elaeagnaceae, and Myricaceae, all or most genera are nodulated, whereas nodulation occurs only occasionally in the Betulaceae, Rhamnaceae, and Rosaceae. Table 1 summarizes the taxonomy of the actinorhizal genera with their distribution.

4. NITROGEN FIXATION IN ACTINORHIZAL TREES

Measurement of N_2 fixation in actinorhizal trees (Table 2) is still a difficult task because no simple and accurate methods to determine annual rates are available. Conventionally, N_2 fixation has been measured by the acetylene-reduction assay, which is based on measuring the production of ethylene from acetylene, catalyzed by nitrogenase, with a gas chromatograph. The enzyme, within the N_2-fixing organisms, can also reduce a number of other compounds besides N_2 and acetylene. However, several other methods have been used, including N accretion, N difference, ^{15}N isotope dilution, and ^{15}N natural abundance. Each of the methods has its own assumptions and inaccuracies (Table 3).

Of these, the ^{15}N stable-isotope methods are currently considered to have the least drawbacks for the quantitative measurement of biological nitrogen fixation (BNF) (Warembourg, 1993). Their advantages are based on: (i) the capacity to measure BNF cumulatively over more than one growing season; (ii) the high degree of precision of stable isotopes; and (iii) the capability of assessing the relative efficiency of N_2 fixation, *i.e.*, the proportion of plant-N derived from fixation (N_{dff}). Stable-isotope methods are not without their limitations, however. Several assumptions, which primarily involve the selection of the reference plant, must be satisfied to ensure an accurate measurement of BNF. Reference plants are required in order to estimate the relative proportion of plant-N derived from fixation *versus* uptake of soil-N. They must have a similar rooting profile, a similar timing of N-uptake, and a similar internal isotopic discrimination to those of the N_2-fixing plants. Finding an appropriate reference plant is always difficult. In the case of actinorhizal plants, the N_{dff} determined by ^{15}N isotope-dilution methods averaged 87% for *Ceanothus velutinus* and 83% for *Purshia tridentata* (Busse, 2000a; 2000b). Other studies have reported values of 68 to 100 percent for *Alnus glutinosa* for N derived from fixation (Beaupied *et al.*, 1990; Domenach *et al.*, 1989) and 48-67% for *Casuarina equisetifolia* (Gauthier *et al.*, 1985, Parrotta *et al.*, 1994).

Table 1. Taxonomy of the Actinorhizal Genera with theirDistribution.

Subclass*	Family	Genus	Present distribution
Hamamelidae	Betulaceae	*Alnus*	Europe, N. America, Highlands of Central and South America, scattered in North Africa.
	Casuarinaceae	*Allocasuarina*	Australia, primarily S and SW
		Casuarina	Australia, tropical Asia, Pacific Islands.
		Ceuthostoma	Sumatra to New Caledonia, Fiji and Cape York Pen., Australia.
		Gymnostona	
	Myricaceae	*Comptonia*	Europe, Asia, North America.
		Myrica	Europe, Asia, North America.
Rosidae	Elaeagnaceae	*Elaeagnus*	North America.
		Hippophae	Widespread tropical, subtropical, temperate.
		Shepperdia	
	Rhamnaceae	*Adolphia*	Southern South America.
		Ceanothus	North America.
		Colletia	Temperate and tropical South America.
		Discaria	South America, New Zealand, Australia.
		Kenthrothamnus	Argentina, Bolivia.
		Retanilla	Southern Soth America.
		Talguena	Chile.
		Trevoa	Andes of South America.
	Rosaceae	*Cercocarpus*	West and S-W USA, Mexico.
		Chamaebatia	Sierra Nevada of West USA.
		Cowania	USA, México
		Dryas	Artic mountains, North Temperate Zone
		Purshia	North America.
		Rubus	North Temperate Zone
Magnolidae	Coriariaceae	*Coriaria*	Mediterranean, Japan, China, New Zealand, Chile, Mexico.
Dillenidae	Datiscaceae	*Datisca*	Mediterranean to Himalayas and Central Asia, S-W USA.

Classification according to Cronquist (1981), Brewbaker et al. (1990), and Benson and Clawson (2000).

Table 2. Rates of Symbiotic N_2 Fixation of Some Actinorhizal Trees and Shrubs at Field Level

Actinorhizal species	kg N fixed ha^{-1} yr^{-1}	References
Alnus acuminata (A. jorullensis)	279	Carlson and Dawson, 1985
Alnus glutinosa	40 – 53	Cote and Camire 1984; Hansen and Dawson 1982
Alnus rubra	85 – 320	Cole *et al.* 1978; Newton *et al.* 1968
Casuarina equisetifolia	12 – 110	Dommergues 1987 ; Gauthier *et al.* 1985 ; Diem and Dommergues 1990 ; Baker 1990a
Coriaria arborea	192	Dommergues, 1990
Ceanothus velutinus	4 -100	McNabb and Cromack 1983; Youngberg and Wollum 1976; Zavitkovski and Newton 1968; Busse 2000a; 2000b ^{15}N
Purshia tridentata	1	Busse 2000a; 2000b ^{15}N

Table 3. Comparison of Methods of Estimating Nitrogen Fixation.
After I. Watanabe (2000) lecture in Cantho University, Vietnam. Retrieved June 6, 2002, from http://www.asahi-net.or.jp/~it6i-wtnb/BNF.html#ch4

Method	Advantages	Disadvantages	Sensitivity
Total N balance	Simplest	Low sensitivity including other inputs.	Lowest
$^{15}N2$ incorporation	Most direct	Expensive, only for short period	High-moderate
Acetylene reduction	Simple, highly sensitive	Indirect, semi-quantitative	High
^{15}N dilution	Throughout growing season	Only N Fixation in plant Varies with reference plants	High-low
Natural abundance	Simple, no disturbance to system	Only slight difference in ^{15}N content	Low
Substrate addition	Difference in ^{15}N content is large	Change of ^{15}N in time and space in soil	Moderate

The rates of nitrogen fixation measurable by the acetylene-reduction assay (ARA) in nodules of actinorhizal genera, such as *Alnus* and *Eleagnus*, range between 10-90 μmol ethylene g^{-1} nodule fresh weight h^{-1} (Berry, 1994). Other values of ARA reported for 120-days-old *Alnus acuminata* seedlings, which were inoculated with a crushed nodule suspension, were between 32.5 and 86.4 μmol ethylene g^{-1} nodule fresh weight h^{-1} (Russo and Berlyn, 1989; see Figure 2) and are consistent with the above mentioned range. However, variability is very high and accuracy has been challenged because acetylene, in some cases, inhibits nitrogenase activity. In other cases, the ARA technique is used inappropriately. Furthermore, conversion from acetylene-reduction to N_2-fixation values often uses the routinely assumed conversion ratio of 3:1-4:1 (due to H_2 production under N_2), however, conversion rates can fluctuate depending on a number of factors and are not constant. Finally, to extrapolate laboratory or greenhouse values to field conditions in terms of kg N ha^{-1} yr^{-1} can be a significant source of error (Giller, 2001).

Figure 2. In Alnus acuminata *seedlings inoculated with* Frankia, *nodules appear two weeks after inoculation and thereafter the rate of growth and interval of appearance of leaves is modified. Nodulated seedlings (inoculated) at 50 days have higher total height ($P<0.05$), higher leaf length ($P<0.05$), higher shoot biomass ($P<0.05$), and lower root/shoot ratio ($P<0.01$) than non-inoculated seedlings (Personal observations). Left side: non-inoculated; right side:* Frankia-*inoculated seedlings. (Photo: R.O. Russo)*

Environmental effects on N_2 fixation have been well described by Huss-Danell (1997). Both nodulation and N_2 fixation are influenced by environmental factors but, unlike legume nodules, actinorhizal nodules are not O_2 limited and Frankia itself appears able to protect nitrogenase from O_2 inactivation. Five major effects, which themselves are affected by other factors, were highlighted (Figure 3).

At the forest-stand level, the rates of symbiotic N_2 fixation reported for actinorhizal trees range between 12 and 320 kg N ha^{-1} yr^{-1} (Table 2). By comparison, asymbiotic N fixation by free-living soil prokaryotes typically contributes only 1 kg N ha^{-1} yr^{-1} or less in forest ecosystems (Jurgensen et al., 1992), whereas associative N_2 fixation, although controversial, has been suggested to fix up to 50 kg ha^{-1} yr^{-1} in the rhizosphere of conifer roots (Bormann et al., 1993).

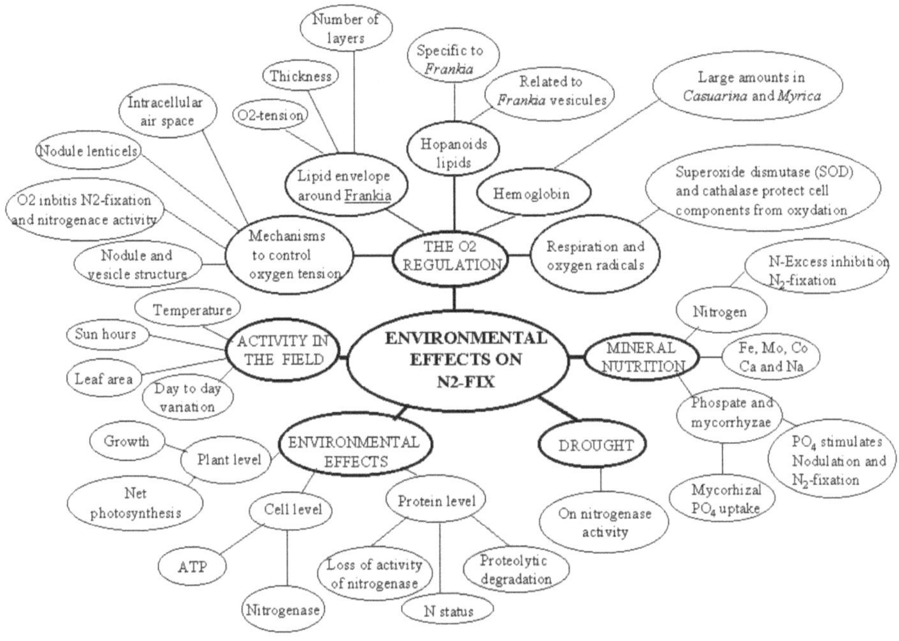

Figure 3. Concept Map of Environmental Effects on N_2 fixation by Actinorhizal Systems. Based on Huss-Danell (1997).

As Dommergues (1997) states, the nitrogen-fixing potential of a number of actinorhizal plants, e.g., Casuarina sp. and Alnus sp., is high but the amount of N actually fixed in the field is often low because the expression of this potential is limited by either unfavourable environmental conditions or improper management practices. In consequence, some strategies can be adopted to increase the input of fixed N into ecosystems by using management practices to optimize actinorhizal N_2 fixation. Actinorhizal plants should receive serious consideration as soil improvers in a number of situations and sites where they are not yet used. Furthermore, under-exploited actinorhizal trees and shrubs should be domesticated to exploit their ability to contribute to the rehabilitation of wasted lands and possibly to the phytoremediation of polluted sites (Baker, 1990a; 1990b).

5. MYCORRHIZAL ASSOCIATIONS WITH ACTINORHIZAL TREES

The existence of tripartite symbiotic associations among actinorhizal plants, *Frankia*, and mycorrhizal fungi has been reported for *Alnus* spp. (Rose and Trappe, 1980), for *Ceanothus velutinus* (Rose and Youngberg, 1981), for *Casuarina equisetifolia* (Gauthier *et al.*, 1983), for *Hippophae rhamnoides* (Gardner *et al.*, 1984), and for *Alnus acuminata* (Russo, 1989; 1992; see Figure 4). In these tripartite associations of tree-*Frankia*-mycorrhiza, the double symbiosis is established on the exchange of carbon, phosphate, and fixed-nitrogen between plant host and both bacterial and fungal microsymbionts. In addition, ectomycorrhizal associations of *Casuarina* have been reported (Warcup, 1980; Dell *et al.*, 1994) and the development and function of *Pisolithus* and *Scleroderma* ectomycorrhizas in *Casuarina* described. For *Alnus acuminata*, ectomycorrhizal associations with the basidiomycetes, *Phylloporus caballeroi* and *Gyrodon monticola*, have been reported (Singer and Gomez, 1984).

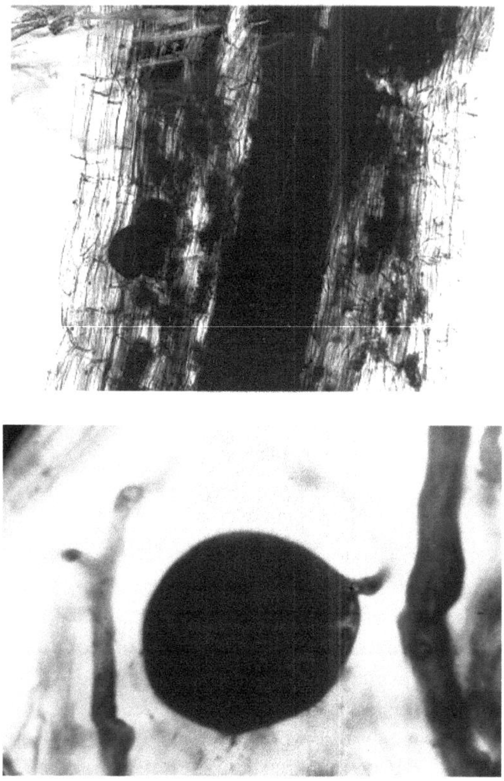

Figure 4. Arbuscular Mycorrhizas (AM) in Alnus acuminata roots and Glomus intraradices internal spore in Alnus acuminata inoculated seedlings. (Photos: R.O. Russo)

Several authors have reviewed the functional aspects of the relationship between the mycorrhizal and actinorhizal symbioses in non-legumes (*e.g.*, Barea and Azcón-Aguilar, 1983; Rodríguez-Barrueco, 1984; 1992; Gardner, 1986; Cervantes and Diem, 1997; Dommergues, 1997). Most of them agree that actinorhizal plants are predisposed to form both ecto- and endo-mycorrhizal symbiotic associations, either tripartite or tetrapartite, that are as essential for plant growth as the associations they form with *Frankia*. As expected, mycorrhizas on the same roots as *Frankia* stimulate both the development and the nitrogen-fixing activity of the actinorhizal symbiosis by improving mineral nutrition of the host plant, particularly, those aspects related to phosphorus uptake (Diem, 1997; Gentili and Huss-Danell, 2002; Yang, 1995). It remains to be proved whether the development of root nodules and endomycorrhizal symbioses involves processes and genes recruited from a common plant origin or whether each has its own unique properties (Gualtieri and Bisseling, 2000).

Early research done by Gauthier *et al.* (1983) with *Casuarina equisetifolia,* which were simultaneously inoculated with both *Frankia* and *Glomus mosseae*, found that total dry weight was more than 80% greater in doubly inoculated seedlings than in those inoculated with *Frankia* alone, whereas no differences between the control and the VAM-inoculated plants were found. All treatments used 10 μg P g^{-1} with one exception in which inoculation was by crushed nodules and 90 μg P g^{-1} was applied to the plants. These results have been validated for other actinorhizal species (Gardner *et al.*, 1984; Gardner, 1986; Chatarpaul *et al.*, 1989; Cervantes and Rodríguez-Barrueco, 1992).

The effect of arbuscular mycorrhizae (AM; *Glomus intraradices*) on both acetylene reduction and the growth response, with three different phosphorus levels (10, 50 and 100 μg g^{-1}), of *Alnus acuminata* seedlings, which were inoculated with *Frankia* strain ArI3 (originally isolated from *A. rubra* at Harvard Forest), was studied. Significant differences in acetylene reduction were observed after 120 days at the lowest P level (10 μg g^{-1}) between *Frankia*-inoculated and *Frankia*+AM-inoculated plants. In addition, significant differences in seedling growth at the middle P level (50 μg g^{-1}) were observed in favor of the doubly inoculated plants (Russo, 1989). The data suggest that the presence of AM favors acetylene reduction at lower P levels where the AM improves P uptake, but also can compete with the nodules as a carbohydrate sink at the middle levels.

Although the mycorrhizal symbiosis should promote plant growth in actinorhizal plants, certain aspects need to be considered. When nitrogen is a limiting factor, photosynthate allocation could be modified. With non-nodulated plants that are supplied with a growth-limiting level of combined-N, a higher proportion of dry mass is allocated to roots than in nodulated plants (Arnone and Gordon, 1990). The situation with *Frankia*-inoculated plants involves a different response, which can be related to the ability of the nodulated plants to get more fixed nitrogen than the amount supplied. Thus, the predictions that can be made from experimental studies in controlled conditions may not be realistic for all situations; they would be mainly applicable to growth conditions characterized by a sub-optimal nitrogen supply.

For *Casuarina*, the AM endophytes did not invade nodular cortical tissues (Khan, 1993), suggesting the presence of an exclusion mechanism that needs further study. Further, the highest endomycorrhizal infection occurred in nodulated specimens and a relationship was found between soil-moisture gradient and AM infection in *C. cunnimghamiana*. In this species, typical vesicles and arbuscules were found in roots from drier soils, however, there was a lack of arbuscules in relatively wet soils where large lipid-filled intracellular vesicles were present. Both mycorrhizal infection and AM spore number seem to be related to redox-potential;

they are lower at sites, such as swamps, water or sediments, with lower Eh values than in terrestrial soils with higher Eh values.

Recently, the genetic and molecular mechanisms of the development of N_2-fixing nodules (both *Rhizobium*- and *Frankia*-based) were compared with those of arbuscular mycorrhiza (Provorov *et al.*, 2002). The new primordium developing from root tissues is common for all known types of N_2-fixing nodules. However, their structure varies greatly with respect to: (i) tissue topology (the location of vascular bundles is peripherical in legumes or central in non-legumes); (ii) the position of the nodule primordium (in the inner or outer cortex in legumes and the pericycle in non-legumes); and (iii) the stability of the apical meristem (persistent in the indeterminate nodules, transient in the determinate ones). The origin of actinorhizal symbioses is suggested to be based on a set of pre-adaptations, *e.g.*, inter- and intra-cellular maintenance of symbionts, their control *via* defense-like reactions, and recognition of chitin-like molecules, many of which had been evolved in angiosperms during coevolution with arbuscular mycorrhizal fungi.

In addition to mycorrhizas, actinorhizal plants, including *Alnus* and *Casuarina,* are also able to form an unique type of root, called either "proteoid roots" or "cluster roots" in response to the detrimental effects of soil nutrient deficiencies. Proteoid roots, which are formed by clusters of closely spaced lateral rootlets of limited growth (Hurd and Schwintzer, 1996; Arahou and Diem, 1997), have been reported in 28 species from the Betulaceae, Casuarinaceae, Eleagnaceae, Leguminosae, Moraceae, and Myricaceae families, four of which are actinorhizal species (Dinkelaker *et al.*, 1995). Skene (1998) reviewed both the functional aspects and the ecological considerations of proteoid-root formation, emphasizing that the species with this kind of root formation can grow in soils with low nutrient availability. Watt and Evans (1999), who reviewed the physiology and development of proteoid roots, cite that auxin probably works alongside other hormones, such as ethylene and cytokinin, during proteoid-root development. In this way, P nutrition has been clearly implicated in the formation of proteoid roots, which decline as P availability to roots increases. By analogy, P may alter nodule formation and function in the *Casuarina-Frankia* symbiosis (Yang, 1995). Once developed, the proteoid roots: (i) mobilize mineral P that is bound to cations, such as Fe, Al, and Ca; (ii) extract P from organic layers in soil; (iii) obtain Fe and Mn from alkaline soils; and (iv) take up organic forms of N (Dinkelaker *et al.*, 1995).

Diem and Arahou (1996) visualize proteoid roots as a primary strategy of the Casuarinaceae to overcome soil nutrient deficiencies. They, thus, explain the ability of the family to grow in marginal soils by their adaptability to soil constraints. These authors also underline the likely role of proteoid roots as an alternative to mycorrhizas. These anatomical root structures improve the absorption of nutrients, other than N, from the soil, especially those needed for nitrogen fixation and growth (Diem *et al.*, 1999). Proteoid roots also occur for *Myrica gale*, an actinorhizal shrub that occurs in the higher latitudes of the Northern hemisphere and that dominates in waterlogged soils (Herdman *et al.*, 1999).

6. ACTINORHIZAL TREES IN AGROFORESTRY

Agroforestry is defined by the International Centre for Research in Agroforestry (ICRAF, 2002) as a dynamic, ecologically-based, natural-resources management system that, through the integration of trees on farms and in the agricultural landscape, diversifies and sustains production for increased social, economic and environmental benefits for land users at all levels. This definition is an up-dated response to current needs of social and environmental accountability. Previous definitions emphasized the fact that agroforestry systems are deliberate combinations of a woody perennial component with agricultural components (crops, pastures and/or livestock) interacting with each other at the same site as an integrated agroecosystem. During the 1980s, agroforestry systems were considered to be a solution for solving land-use problems in the tropics. Today, it is generally accepted that to be called agroforestry, a land-use practice must satisfy some minimum criteria. These are that the combination must be intentional or deliberate, interactions among components must be at least biological, and the components have to be integrated on the same unit of land. The goal and justification of agroforestry systems were to put emphasis on the positive interaction in order to obtain a higher total, a more diversified and/or more sustainable production from available resources than is possible with other forms of land-use under prevailing ecological, technological and socio-economic conditions (Nair, 1990).

Actinorhizal trees may play the role of the woody perennial in agroforestry systems. In fact, they have been used historically in agricultural systems in several ways (Baker, 1990). Probably one of the most known species of actinorhizal trees is *Casuarina,* which is used for firewood production in the tropics or as a component of a multipurpose agroforestry plantation or when large plantations are interplanted (Taungya system) with agricultural crops during the first few years of the rotation (Diem and Dommergues, 1990; Pinyopusarerk *et al.*, 1996). Other actinorhizal tree species have been used in agroforestry; these include: *Alnus*, which is cultivated as a primary crop for timber and pulpwood; *Elaeagnus spp.*, which is interplanted as a nurse plant for other more-valuable species; and *Elaeagnus, Shepherdia*, and *Purshia* spp, which are planted for soil reclamation (Kohls *et al.*, 1994; Paschke, 1997).

Several authors have listed a number of desirable characteristics of trees to be used in agroforestry. Basically, these traits are a function of the goals, production objectives, and needs of the user. A first consideration is compatibility with soil and local climate. Also, the tree should be a fast or moderately-fast-growing species that is either valuable for markets or biodiversity maintenance; it should be a nitrogen-fixing tree; it should have good shade characteristics appropriate to companion crops; and it should produce nuts or fruits that can have an acceptable local market. Additionally, it is desirable that the tree does not produce growth-inhibitory chemicals (allelochemicals), which would prevent some crops from growing near them, and that the trees should have minimal root competition (Beer, 1987; Hodge *et al.*, 1999; Perry, 1989, Werner and Müller, 1989).

Many of these traits are accommodated by *Casuarina* and *Alnus*, the main two genera of actinorhizal trees with uses and applications in agroforestry. Both genera have been reported as shade trees in various agroforestry combinations. However, as in field conditions, variation exists in *Frankia* strains in terms of infectiveness (the ability of a strain to nodulate a given host) and effectiveness (the relative capacity of the actinorhiza, once established, to fix N_2). A matter still to be answered in these agroforestry system is whether the level of nitrogen fixation achieved in actinorhizal nodules reflects the infectiveness and effectiveness of the *Frankia* strain or rather the inhibition (or promotion) by the level of N-fertility in the soil (Russo, 1990). This debate continues, even though it is known, under experimental conditions, that the combined-nitrogen level inhibits N_2 fixation (Arnone *et al.*, 1994; Huss-Danell, 1997). Dommergues (1997) states that the nitrogen-fixing potential of a number of actinorhizal plants, *e.g. Casuarina* sp. and *Alnus* sp., is high, but the amount of N_2 actually fixed in the field is often low because the expression of this potential is limited by either unfavourable environmental conditions or improper management practices. For instance, current N-fertilization in pastures associated with *Alnus acuminata,* as practiced by Costa Rican farmers (92 kg of ammonium nitrate every five weeks per hectare), may be depressing nodulation and nitrogen fixation (Russo, 1990).

7. THE GENUS *CASUARINA*

The genus *Casuarina* is one of the better known and disseminated actinorhizal trees. Today, *Casuarina* plantations are of large social and economic importance throughout the tropical regions of the world, given that there is a global trend towards greater reliance on plantations as a source of industrial wood. The global area planted with different *Casuarina* species reached 1.4 million ha around the world (Vercoe, 1993). Another more modest figure (Krishnapillay, 2000), based on unpublished data from D. Pandey, is 787 x 10^3 hectares in the tropical regions.

The world-wide interest in several species of the Casuarinaceae family has resulted in three international Casuarina Workshops. The First Workshop was held in Canberra, Australia, in 1981; the second in Cairo, Egypt, in 1990, and the third in Da Nang, Vietnam, in 1996. At the third workshop, thirty five papers, which covered genetics and tree breeding, reproductive biology, nitrogen fixation, silviculture, diseases, natural distribution and ecology, socio-economic aspects, and utilization, were presented (Pinyopusarerk *et al.*, 1996). *Casuarina equisetifolia* has been planted extensively in the coastal areas of southern China for wood production and for stabilizing moving sands. There are currently about 300,000 ha of *Casuarina* plantations in the coastal areas of Guangdong, Hainan, Fujian, Zhejiang, and Guangxi Autonomous Region, bordering the South China Sea (Bai Jiayu and Zhong Chongu, 1996).

Bourke (1985) described an agroforestry farming system from the Papua New Guinea highlands (1400-2100m altitude), which was developed by village growers in about 1960 and has expanded rapidly since about 1970, where the tree component is *Casuarina oligodon* associated with numerous species of annual and perennial food crops (especially bananas) and arabica coffee. The system provides food, a cash crop, and timber for construction and fuel.

Another agroforestry experience involving two actinorhizal trees was in Guatemala, in two sites at the upper watershed area of the Achiguate River. Here, research plots were established in 1987 and included *Casuarina equisetifolia*, *Alnus acuminata*, and *Eucalyptus globulus*, all associated and non-associated to annual crops (maize and beans). After four years, it was reported that crop production decreased in all treatments where the crops were associated with trees, but the diminishing rates were less when the crop was associated with *Casuarina equisetifolia* when compared with the other tree species (Leiva and Borel, 1995). It is not clear whether the higher yields of the annual crops can be attributed to the relatively poorer growth of the *Casuarina* trees. No measurements of nitrogen fixation are available, although relatively high fixation rates have generally been reported for both actinorhizal species.

Erosion control, sand dune stabilization, and soil reclamation have been the most frequently mentioned environmental uses of *Casuarina* (NCR, 1984). More recently, in an extensive review on ecological restoration of land, with particular reference to the mining of metals and industrial minerals, Cooke and Johnson (2002) present an example of pioneer work done on the coast of Kenya on ecosystem restoration of exhausted quarries. They report that the land restoration is attributable to the establishment of 26 tree species and grasses, centered around *Casuarina equisetifoli*a. They also mention that this nitrogen-fixing tree is salt-tolerant and acted as the catalyst for ecosystem development, perhaps by favoring the increase of a millipede population (*Epilobus pulchripe*s). This arthropod fed on dry needles of *Casuarin*a, which when digested initiated humus formation and secondary colonization by plants and animals. Over 250 plant species now inhabit the quarry habitat, an achievement that was largely made without fertilizers or soil amendments. In Argentina, *Casuarina cunninghamiana* was planted along the river borders (in the delta of the Parana River) to protect the banks from erosion by waves (personal observation). In Colombia, *Casuarina equisetifolia* trees were planted along the borders of a drainage channel to protect the side slope in the "El Hatico" Farm in the Cauca River Valley. However, the Farm Manager decided to eliminate all the trees, some of them with diameters over 30 cm, after seeing that where *Casuarina* leaf litter was present "no other plants had grown under those trees" (Prof. Raúl Botero, personal communication). This field observation on allelopathy was validated by a study of Jadhav and Gaynar (1995) in which C*asuarina equisetifolia* leaf litter substances showed allelopathic effects on germination and seedling growth of rice and cowpea.

8. THE EXPERIENCE OF THE CENTRAL AMERICA FUELWOOD PROJECT

The genus *Casuarina* in America has experienced a high degree of deforestation and degradation of its natural environment, resulting in a scarcity of forest products, such as timber and fuelwood. As a possible solution, the development of agroforestry systems has become a priority in the region. Facing this challenge, The

Tropical Agricultural Research and Training Center (CATIE, 1991) responded by developing (1981-1984) a project in Central American, which is known as the Fuelwood and Alternative Energy Sources Project. In a second phase (1985-1990), the project was renamed as the Multi-purpose Tree Cultivation or Madeleña (timber and fuelwood) Project. In a third phase (1991-95), the project was committed to dissemination of multi-purpose tree cultivation (Belaunde and Rivas, 1993).

The Project became part of the Forestry Action Plan for Central America (PAFCA), under which a set of complementary projects were implemented in each Central American country. More than 100 forestry species were identified, both exotic and native, with good potential for widespread cultivation in the region. On-farm research methodologies were implemented and training was offered to technical staff of the different forestry institutions of the region. As a result, multi-purpose, fast-growing tree species began to be included in project research and demonstration sites. *Casuarina equisetifolia* was one of the successful species included in these trials due to its characteristics of fast growth, nitrogen-fixation capability and good fuelwood quality (CATIE, 1991). The initial growth results of the species in Costa Rica demonstration plots are summarized in Table 4.

Table 4. Growth of Casuarina equisetifolia *L. ex J.R. Forst. and G. Forst in Central America Country Plots. After CATIE, 1991.*

Country	Site	Age (months)	Diameter		Height	
			Mean (cm)	MAI* (cm/yr)	Mean (m)	MAI (m/yr)
Guatemala	Amatitlán	244	14.60	0.72	15.50	0.76
	San Pedro	42	3.95	1.13	4.45	1.27
	La Conora	95	8.20	1.04	10.90	1.38
Honduras	F. Morazán	41	4.68	0.97	4.50	0.93
	V. Angeles	29	2.84	1.17	3.79	1.65
Costa Rica	Piedades S.	57	6.81	1.43	6.23	1.32
	Piedades N	53	7.81	1.77	6.68	1.51
	Sarchí	68	11.18	1.97	9.96	1.75
Panama	Coclé	39	2.30	0.71	2.60	0.80
	Coclé	35	2.40	0.82	300	1.03
Nicaragua	El Gurú	43	5.47	1.52	6.97	2.08
	Tipitapa	31	3.84	1.48	4.30	1.66
	UCA	58	9.69	2.01	7.80	1.61

* MAI = mean annual increment.

The best results were observed in Nicaragua, where height reached 2.08 meter per year, followed by those in Costa Rica and Honduras (Table 4). When the species was associated with other crops in agroforestry systems, growth in height was higher than reached in monoculture plots independently of tree density. In some cases, growth reached 2.8 m height in the first year. *Casuarina* was associated with beans and maize in Costa Rica and El Salvador. It was also planted both as a windbreak and associated with ornamental plants (*Dracaena* sp.).

9. THE CASE OF *ALNUS ACUMINATA* IN TROPICAL HIGHLANDS

Alnus acuminata is a fast-growing species valued for its wood, watershed protection, and soil improvement. Native from Mexico to Northern Argentina, it is known as: aliso (Mexico, Argentina, Colombia, Ecuador, and Peru); aile, ilite (Mexico); ramrám, lambdn (Guatemala, Costa Rica, and Peru); jaúl (Costa Rica); palo de lama (Guatemala); and cerezo and chaquiro (Colombia). Easily propagated either from seed or by natural regeneration, *A. acuminata* is a popular agroforestry species in its native range (Figures 5 and 6). It has been successfully introduced into southern Chile and southern New Zealand.

Figure 5. Alnus acuminata in a pasture in San Jerónimo, Costa Rica. Photo: R.O. Russo.

It grows in moist soil environments, usually along the banks of streams, rivers, ponds, and swamps, where it typically forms dense pure stands. It also associates with wet flood plains, or moist mountain slopes. It may be adapted to somewhat drier conditions, however, it is usually restricted to zones with extra soil moisture, such as cool tropical highlands and cool high-latitude regions with abundant rainfall where mist and cloud cover can be a source of fog-drip precipitation. In the tropical highlands of Central and South America, clouds and mist are important in supporting *Alnus acuminata* and grass, when associated, through the dry season. *Alnus acuminata* prefers deep, well-drained soils with high organic matter content, however, it is commonly found growing on shallow soils, such as landslides. Rojas *et al.* (1991) report that it will grow in soil with pH as low as 4.5.

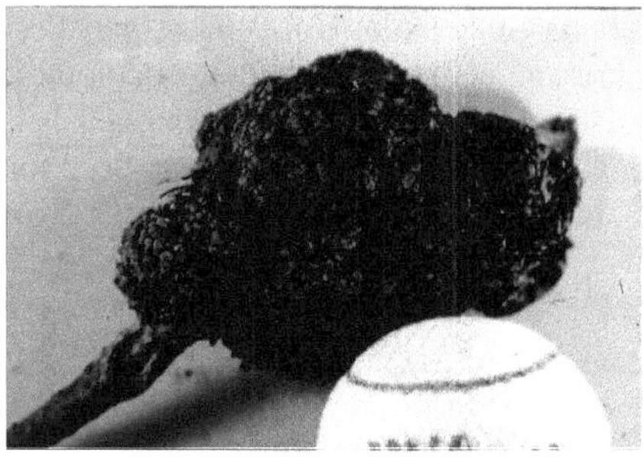

Figure 6. Nodule of Alnus acuminata from San Jerónimo, Costa Rica. Photo: R.O. Russo.

Farmers in Costa Rica have grown *Alnus acuminata* in pastures and as a shade tree for coffee crops for more than 100 years. Trees are either regenerated naturally or planted from nursery stock at a spacing of 8-14 m (about 100 trees/ha). According to the farmers, one benefit of including trees in cattle pastures is greater milk production.

Alnus acuminata is also cultivated in plantations mainly in Colombia and Costa Rica, but in other countries as well. In Colombia, an initial spacing of 2.6 x 2.6 m (1,480 trees/ha) is common (Sicco Smit, 1971). In Costa Rica, an initial spacing of 3 x 3 m is preferred. At least two thinnings are recommended, the first after the third year and the second after 10-15 years, leaving 250-350 trees per hectare. Trees are harvested in rotations of about 20 years. Average annual wood production is 15-20 m^3 per hectare. According to Canet (1985), a stand of 30-year-old trees with a density of 35 trees/ha yielded 70 m^3/ha of timber, 18.3 ton/ha of dry fuelwood, and 3.6 ton/ha of leaves and fine branches. *Alnus acuminata* resprouts vigorously from the stump after cutting.

Another interesting experience in the eastern Himalayas is the *Alnus*-cardamom agroforestry system, where cardamom (*Amomum subulatum*), the most important perennial cash crop in the region, is cultivated predominantly under the shade of N_2-fixing *Alnus nepalensis* (Sharma *et al.,* 2000). The biomass, net primary productivity, energetics, and energy efficiencies of the system at different ages (from 5 to 40 years) were determined and the impact of stand age on the performance of mixtures *Alnus* and cardamom plantations was evaluated. The results showed that net primary productivity was lowest (7 t ha^{-1} $year^{-1}$) in the 40-

year-old stand and was more than three-times higher (22 t ha^{-1} $year^{-1}$) in the 15-year-old stand. The agronomic yield of large cardamom peaked between 15 and 20 years of age, whereas cardamom productivity doubled from the 5- to the 15-year-old stand and then decreased with plantation age to reach a minimum in the 40-year-old stand. The annual net energy fixation was highest (444 x 10^6 kJ ha^{-1} $year^{-1}$) in the 15-year-old stand, which was 1.4-times that of the 5-year-old stand and 2.9-times that of the 40-year-old stand. The results indicate that the younger plantations are more productive and also suggest that the beneficial association can last 20 years or longer if replanting is undertaken.

10. OTHER USES OF ACTINORHIZAL TREES

The environmental benefits provided by the actinorhizal trees also include soil reclamation and erosion control on steep slopes; other uses have been reported in the literature. For instance, *Alnus rubra* (Red alder) is considered the most important commercial hardwood of the Pacific Northwest of the United States (Leney *et al.*, 1978). Red alder also contains salicin, which chemically is closely related to acetylsalicylic acid (commonly known as aspirin), which is probably why Native Americans used various preparations for medicinal purposes. Native Americans also used the wood for various utensils (Arno and Hammerly, 1977).

Casuarina wood is known as the best firewood in the world with a calorific value greater than 20,000 kJ/kg, and also is excellent for charcoal making. The wood also has many other uses, including for poles, posts, roundwood for fencing, beams, roofing shingles, paneling, furniture, marine pilings, tool handles, and cabinets. The wood, however, is subject to cracking and splitting. *Casuarina equisetifolia* is used for paper pulp (CATIE, 1991; NRC, 1984). Additionally, some medicinal uses have been reported for *Casuarina equisetifolia.* According to Duke (1972), the fruit is mixed with powdered nutmeg to treat toothaches; ashes may be used to make soap; and the bark is rich in tannin and it is said to be antidysenteric and emmenagogic, and is also used in gargles for sore throat.

Many of the actinorhizal species of the New World, such as *Alnus rhombifolia* (white alder), *Cercocarpus betuloides* (birchleaf mountain-mahogany), *C. ledifolius* (Curlleaf mountain-mahogany), *Elaeagnus angustifolia* (Russian-olive), *E. umbellata* (autumn olive), *Myrica cerifera* (southern bayberry), *Purshia mexicana* var. *stansburiana* (Stansbury cliffrose), and *Sheperdia argentea* (silver buffaloberry), are important to wildlife such as deer, mule deer, elk, pronghorn, grizzly bear, a variety of birds and mammals, and also livestock (Davis, 1990; Sampson and Jespersen, 1963; USFS, 1937)

The development of an insect repellent made with an oil from *Myrica gale* has created a need for cultivation of the species. In addition to oil production, other compounds, such as pharmacologically active flavonoids, can also be obtained by extraction of the byproducts, so adding more value to the species. Because *M. gale* thrives on well-aerated acid peatland, it could become a valuable crop on lands of low agricultural value. *M. gale* could also be used either with other crops in agroforestry systems or combined with softwood forestry because those crops and trees would benefit from the soil-nitrogen enrichment due to the symbiotic association of *M. gale and Frankia* (Simpson *et al.*, 1996).

The production of antibiotics is another potential use for *Frankia.* Some studies demonstrate that *Frankia* has the potential to inhibit growth of competing soil microbes by producing antibiotic compounds. Lang (1999) showed that *Frankia* strains, which were isolated from different *Casuarina* sp., produced metabolites active against Gram-negative bacteria. More recently, Haansuu *et al.* (2001) and Klika *et al.* (2001) isolated and described the biological activity of frankiamide, an antibiotic isolated from *Frankia.* This compound, also known as demethyl (C-11) cezomycin, showed strong activity against Gram-positive bacteria, particularly pathogenic *Clavibacter michiganensis* subsp. *Sepedonicus*, as well as against several plant pathogenic fungal strains (Haansuu, 2002).

11. CONCLUDING CONSIDERATIONS

Actinorhizal trees, in general, have practical importance because of their characteristic features, such as the environmental benefits they provide. These benefits include nitrogen fixation, erosion control, soil conservation, regeneration of degraded soils, climate-change mitigation through carbon sequestration, microclimate regulation through a windbreak effect, pollination, and biodiversity and wild-life conservation.

With reference to the symbiosis itself, Huss-Danell (1997) remarks that: (i) *Frankia* actinorhiza shows host-determined variations the infection pathway, the morphology and the anatomy of nodules, and in the differentiation process of *Frankia* cells in the nodules; (ii) nitrogenase in *Frankia* is localized in vesicles, except for *Casuarina* and *Allocasuarina* nodules where only filaments (hyphae) are formed; (iii) both nodulation and nitrogen fixation are influenced by environmental factors; and (iv) unlike legume nodules, actinorhizal nodules are not O_2 limited and *Frankia* itself appears able to protect nitrogenase from O_2 inactivation. All these features are remarkable in themselves.

In the case of *Alnus* and *Casuarina* species, there are countless forestry and agroforestry applications. As agroforestry is a land-use system that involves the growing of crops and woody perennials on the same land unit in space or time, both genera may be useful as tree components in these systems. Many scientists agree that the actinorhizal-tree properties, which are based on both their nitrogen-fixation and soil-restoration capabilities, are real and not just conceptual.

Alnus and *Casuarina* are important to rural households throughout the tropics, providing a variety of products and services. Fast-growing *Casuarina* makes excellent fuelwood and charcoal. The deeply penetrating and wide spreading roots of *Casuarina* can reach moisture during dry seasons and also protect stream banks from erosion. Windbreaks of *Casuarina* planted in single or multiple rows on windward field boundaries help prevent soil desiccation and yield secondary tree products. *Casuarina* has been planted for dune stabilization in over 1,000,000 hectares in China. In addition to *Alnus* and *Casuarina*, species of *Eleagnus* and *Shepherdia* also have the potential to add large amounts of fixed N and C to soils and they should be given more consideration as soil-improvement tools. Nitrogen dynamics is favored by tree litter in agroforestry systems where actinorhizal trees (*Alnus* and *Casuarina*) are interplanted with agricultural crops. For instance, alder leaves decompose very rapidly and their nutrients are incorporated into the soil.

Research on nitrogen-fixing, actinorhizal trees has basically focused on the physiology of the endosymbiont. However, taking into consideration that sustainability is a desired property in current agricultural systems, research and development of agroforestry combinations, including actinorhizal trees, should receive more attention, especially as they relate to crop yield, the biomass yield, and the interaction of crops and trees. Additionally, depending on local needs and preferences, a variety of different planting schemes with actinorhizal trees can be utilized to yield a wide variety of products.

A nursery stock of actinorhizal species has been experimentally inoculated with *Frankia* and mycorrhizal fungi prior to planting in order to increase successful nodulation and mycorrhizal infection. However, this experiment has not been attempted on a large scale. Some authors question whether techniques for the mass-culturing of *Frankia* and the inoculation of rangelands might become a possible management tool for stimulating the vigor of actinorhizal shrub stands.

The study of the actinorhizal symbiosis is still in an early phase. Because of the abundance of wild actinorhizal plants and their importance in natural ecosystems, the group is gaining recognition as valuable trees and shrubs for diverse ecological roles in natural ecosystems, including food and cover for wildlife and livestock. Such is the case of many of the actinorhizal shrubs in the Rhamnaceae and Rosaceae families, which are adapted to survive on the harsh rangelands in the western regions of North America.

REFERENCES

Arahou, A., and Diem, H. G. (1997). Iron deficiency induces cluster (proteoid) root formation in *Casuarina glauca*. *Plant Soil*, *196*, 71–79.

Arno, S. F., and Hammerly, R. P. (1977). *Northwest trees* (222 p.). Seattle, WA: The Mountaineers.

Arnone, J. A., III., and Gordon, J. C. (1990). Effect of nodulation, nitrogen fixation and CO_2 enrichment on the physiology, growth and dry mass allocation of seedlings of *Alnus rubra* Bong. *New Phytol.*, *116*, 55-66.

Arnone, J. A., III., Kohls, S. J., and Baker, D. D. (1994). Nitrate effects on nodulation and nitrogenase activity of actinorhizal *Casuarina* studied in split-root systems. *Soil Biol. Biochem.*, *26*, 599–606.

Bai J., and Zhong C. (1996). Management of *Casuarina* plantations in China. In K. Pinyopusarerk, J. W. Turnbull and S. J. Midgley (Eds.), *Current research and development in Casuarinas. Proceedings of the Third International Casuarina Workshop*. (pp 196-199). Canberra, Australia: CSIRO Forestry and Forest Products.

Baker, D. (1990a). Actinorhizal plants: Underexploited trees and shrubs for forestry and agroforesty. *Nitrogen Fixing Tree Research Reports*, *8*, 3-7.

Baker, D. (1990b). Optimizing actinorhizal nitrogen fixation and assessing actual contribution under field conditions. In: Werner, D. and Müller, P. (Eds.), *Fast growing trees and nitrogen fixing trees*. (pp. 291-299). Stuttgart, Germany: G. Fischer Verlag.

Baker, D., and Torrey, J. G. (1979). The isolation and cultivation of actinomicetous root nodule endophytes. In J. C. Gordon, C. T. Wheeler, and D. A. Perrin (Eds.), *Symbiotic nitrogen fixation in the management of temperate forest*. (pp. 38-56). Corvallis, OR: Oregon State University.

Baker, D., Torrey, J. G., and Kidd, G. H. (1979). Isolation by sucrose density fractionation and cultivation *in vitro* of actinomycetes from nitrogen-fixing root nodule. *Nature, 281,* 76-78.

Baker, D., and Miller, N.G. (1980). Ultrastructural evidence for the existence of actinorhizal symbioses in the late Pleistocene. *Can. J. Bot., 58,* 1612-1620.

Baker D. D., and Schwintzer C. R. (1990). Introduction. In D. D. Baker, J. D. Tjepkema, and C. R. Schwintzer (Eds), *The biology of* Frankia *and actinorhizal plants.* San Diego, CA: Academic Press, Inc.

Baker D. D., and Mullin B. C. (1992). Actinorhizal symbioses. In G. Stacey, R. H. Burns and H. J. Evans (Eds.). *Biological nitrogen fixation.* New York, NY: Routledge, Chapman and Hall.

Barea, J. M., and Azcón-Aguilar, C. (1983). Mycorrhizas and their significance in nodulating nitrogen-fixing plants. *Adv. Agron., 36,* 1-54.

Beaupied, H., Moiroud, A. Domenach, A. M. Kurdali, Fawaz, A., and Lensi, R. (1990). Ratio of fixed and assimilated nitrogen in a black alder (*Alnus glutinosa*) stand. *Can. J. Forest Res., 20,* 1116-1119.

Becking, J. H. (1974). Family III. Frankiaceae. In R. E. Buchanan, and N. E. Gibbon (Eds.), *Bergey's manual of determinative bacteriology,* 8th ed. (pp. 701-706). Baltimore, MD: The Williams and Wilkins Co.

Beer, J. (1987). Advantages, disadvantages and desirable characteristics of shade trees for coffee, cacao and tea. *Agroforestry Systems, 5,* 3-13.

Benoit L. F., and Berry A. M. (1997). Flavonoid-like compounds from seeds of red alder (*Alnus rubra*) influence host nodulation by *Frankia* (Actinomycetales). *Physiol. Plant., 99,* 588-593.

Benson, D. R., Stephens, D. W., Clawson, M. L., and Silvester, W. B. (1996). Amplification of 16S rRNA genes from *Frankia* strains in root nodules of *Ceanothus griseus, Coriaria arborea, Coriaria plumose, Discaria toumatou,* and *Purshia tridentata. Appl. Environ. Microbiol., 62,* 2904-2909.

Benson, D. R., and Silvester, W. B. (1993). Biology of *Frankia* strains, actinomycetes symbionts of actinorhizal plants. *Microbiol. Rev., 57,* 293–319.

Benson, D. R., and Clawson, M. L. (2000). Evolution of the actinorhizal plant symbiosis. In E. Triplett (Ed.), *Prokaryotic nitrogen fixation: A model system for the analysis of a biological process.* (pp. 207-224). Wymondham, UK: Horizon Scientific Press.

Berry, A. M. (1987). Cellular aspects of root nodule establishment in *Frankia* symbioses. In T. Kosuge, and E. W. Nester (Eds.), *Plant-microbe interactions: Molecular and genetics perpectives, Vol. 2.* (pp. 194-213). New York, NY: Macmillan Publishing Company.

Berry, A. M. (1994). Recent developments in the actinorhizal symbioses. *Plant Soil, 161,* 135-145.

Berry A. M. (1998). Oxygen relations in *Frankia* and in *Actinorhizal* nodules. In C. Elmerich, A. Kodorosi, and W. E. Newton, (Eds.). *Biological nitrogen fixation for the 21st century.* (pp. 357-358). Dordrecht, The Netherlands: Kluwer Academic Publishers.

Bertalot, M. J. A., and Mendoza, E. (1998). An observation of aerial nodulation of *Casuarina equisetifolia* in an agroforestry system in Brazil. *Forest, Farm, and Community Tree Research Reports, 3,* 61-65.

Bino, B., and Kanua M. B. (1996) Growth performance, litter yield and nutrient turnover of *Casuarina oligodon* in the Highlands of Papua New Guinea. In K. Pinyopusarerk, J. W. Turnbull, and S. J. Midgley (Eds.), *Recent* Casuarina *research and development: Proceedings of the Third International Casuarina Workshop,* Da Nang, Vienam. (pp 167-170). Canberra, Australia: Forestry and Forest Products, CSIRO.

Bourke, R.M. (1985). Food, coffee and casuarina: An agroforestry system from the Papua New Guinea highlands. *Agroforestry Systems,* 2, 273-279.

Bourke, R. M. (1997). *Management of fallow species composition with tree planting in Papua New Guinea.* Canberra, Australia: Resource Management in Asia-Pacific Project, Division of Pacific and Asian History, Research School for Pacific and Asian Studies, The Australian National University.

Brewbaker, J. L., Willers, K. B., and Macklin, W. (1990). Nitrogen fixing trees; Validation and prioritization. *Nitrogen Fixing Tree Research Reports, 8,* 8-16.

Busse M. D. (2000a). Ecological significance of nitrogen fixation by actinorhizal shrubs in interior forests of California and Oregon. In R. F. Powers, D. L. Hauxwell, and G. M. Nakamura (Eds.), *Proceedings of the California Forest Soils Council conference on forest soils biology and forest management,* Sacramento, CA. Gen. Tech. Rep. PSW-GTR-178. (pp.23-41). Albany, CA: Pacific Southwest Research Station, Forest Service, U.S. Department of Agriculture.

Busse M. D. (2000b). Suitability and use of the ^{15}N-isotope dilution method to estimate nitrogen fixation by actinorhizal shrubs. *Forest Ecology and Management, 136,* 85-95.

Callaham, D., Del Tredici, P., and Torrey., J. G. (1978). Isolation and cultivation *in vitro* of the actinomycete causing root nodulation in *Comptonia*. *Science*, *199*, 899-902.

Canet, G. C. (1985). Caracteristicas del sistema silvo-pastoril jaúl (*Alnus acuminata*) con lechería de altura en Costa Rica. In R. Salazar (Ed.), *Técnicas de producción de leña en fincas pequenas. Actas de los simposios.* (pp. 241-249). Turrialba, Costa Rica: CATIE.

Carlson, P. J., and Dawson, J. O. (1985). Soil nitrogen changes, early growth, and response to soil internal drainage of a plantation of *Alnus jorullensis* in the Colombian highlands. *Turrialba*, *35*, 141-150.

Centro Agronómico Tropical de Investigación y Enseñanza – CATIE. (1991). *Casuarina: Casuarina equisetifolia L. Ex J.R. Forst and Forst., árbol de uso múltiple en América Central* (53 p.). Turrialba, Costa Rica: CATIE.

Cervantes, E., and Rodríguez-Barrueco, C. (1992). Relationships between the mycorrhizal and actinorhizal symbioses in non-legumes. *Methods in Microbiology*, *24*, 417-432.

Cervantes, E., and Rodriguez-Barrueco, C. (1994). Relationships between the mycorrhizal and actinorhizal symbioses in non-legumes. In J. R. Norris, D. Read, and A. K. Varma (Eds.), *Techniques for mycorrhizal research methods in microbiology*. (pp. 877-891). San Diego, CA: Academic Press, Inc.

Chatarpaul, L., Chakravarty, P., and Subramaniam, P. (1989). Studies in tetrapartite symbioses. I. *Role of ecto- and endomycorrhizal fungi and* Frankia *on the growth performance of* Alnus incana. *Plant Soil, 118*, 145-150.

Clawson, M. L., and Benson D. R. (1999). Natural diversity of *Frankia* strains in actinorhizal root nodules from promiscuous hosts in the family Myricaceae. *Appl. Environ. Microbiol.*, *65*, 4521–4527.

Cote, B., and Camire, C. (1984). Growth, nitrogen accumulation, and symbiotic dinitrogen fixation in pure and mixed plantings of hybrid poplar and black alder. *Plant Soil, 78*, 209-220.

Cronquist, A. (1981). *An integrated system of classification of flowering plants.* New York, NY: Columbia University Press.

Cruz-Cisneros, R., and Valdés, M. (1991). Actinorhizal root nodules on *Adolphia infesta* (H.B.K.) Meissner (Rhamnaceae). *Nitrogen Fixing Tree Res. Rep.*, *9*, 87-89.

Davis, J. N. (1990). General ecology, wildlife use, and management of the mountain mahoganies in the intermountain West. In: K. L. Johnson (Ed.), *The genus* Cercocarpus, *Proceedings of the 5th Utah Shrub Ecology Workshop* (pp. 1-13). Logan, UT: Utah State University, College of Natural Resources.

Dawson J. O. (1986). Actinorhizal plants: Their use in forestry and agriculture. *Outlook on Agriculture,* 15, 202-208.

Dell, B., Malajczuk, N., Bougher, N. L., and Thomson, G. (1994). Development and function of *Pisolithus* and *Scleroderma* ectomycorrhizas formed in vivo with *Allocasuarina*, *Casuarina* and *Eucalyptus*. *Mycorrhiza*, *5*, 129-138.

Diem, H. G. (1997). Mycorrhizae of antinorhizal plants. *Acta Bot. Gall.*, *143*, 581-592.

Diem, H. G., and Dommergues, Y. R. (1990). Current and potential uses and management of *Casuarinaceae* in the tropics and subtropics. In C. R. Schwintzer, and J. D. Tjepkema (Eds.), *The biology of Frankia and actinorhizal plants*. New York, NY: Academic Press, Inc.

Diem, H. G., and Arahou, M. (1996). A review of cluster root formation: A primary strategy of Casuarinaceae to overcome soil nutrient deficiency. In K. Pinyopusarerk, J. W. Tyrnbull, and S. J. Midgley (Eds.), *Recent Casuarina research and development. Proceedings of the Third International Casuarina Workshop, Da Nang, Vietnam.* Canberra, Australia: CSIRO, Forestry and Forest Products.

Diem, H. G., Duhoux, E., Zaid, H., and Arahou, M. (1999). Cluster roots in *Casuarina*: Role and relationship to soil nutrient factors. *XVI International Botanical Congress, Abstracts,* 4236.

Dinkelaker, B., Hengeler, C., and Marschner, H. (1995). Distribution and function of proteoid roots and other root clusters. *Bot. Acta*, *108*, 183–200.

Dobritsa, S. V. (1998). Grouping of *Frankia* strains on the basis of susceptibility to antibiotics, pigment production and host specificity. *Int. J. Syst. Bacteriol.*, *48*, 1265–1275.

Domenach, A. M., Kurdali, F., and Bardin, R. (1989). Estimation of symbiotic dinitrogen fixation in alder forest by the method based on natural ^{15}N abundance. *Plant Soil, 118*, 51-59.

Dommergues, Y. R. (1987). The role of biological nitrogen fixation in agroforestry. In H. A. Steppler, and P. K. R. Nair (Eds.), *Agroforestry: A decade of development*. Nairobi, Kenya: ICRAF.

Dommergues, Y. R. (1990). *Casuarina equisetifolia*: An old-timer with a new future. *Nitrogen Fixing Tree Association Highlights*,. May 1990, 1.

Dommergues, Y. R. (1997). Contribution of actinorhizal plants to tropical soil productivity and rehabilitation. *Soil Biol. Biochem.*, *29*, 931-941.

Doran, J. C., and Turnbull, J. W. (1997). *Australian trees and shrubs: Species for land rehabilitation and farm planting in the tropics*. Canberra, Australia: Australian Centre for International Agricultural Research, ACIAR.

Duke, J. A. (1972). *Isthmian ethnobotanical dictionary*. Fulton, MD: Published by the author. (Currently online as *Tico Ethnobotanical Dictionary* at http://www.ars-grin.gov/duke/dictionary/tico).

Gardner, I. C. (1986). Mycorrhizae of actinorhizal plants. *Mircen J. Appl. Microbiol. Biotechnol.*, *2*, 147-160.

Gardner, I. C., Clelland, D. M., and Scott, A. (1984). Mycorrhizal improvement in non-leguminous nitrogen fixing association with particular reference to *Hippophae rhamnoides* L. *Plant Soil, 78*, 188-200.

Gauthier, D., Diem, H. G., and Dommergues, Y. (1983). Preliminary results of research on *Frankia* and endomycorrhizae associated with *Casuarina equisetifolia*. In S. J. Midgeley, J. W. Turnbull and R. D. Johnson (Eds.), Casuarina *ecology, management and utilization*. (pp. 211-217). Melbourne, Australia: CSIRO.

Gauthier, D. L., Diem, H. G., and Dommergues Y. R. (1984). Tropical and subtropical actinorhizal plants *Pesquisa Agropecuaria Brasileira, 19*, 119-136.

Gauthier D., Diem, H. G., Dommergues, Y. R., and Gantry, F. (1985). Assessment of N_2 fixation by *Casuarina equisetifolia* inoculated with *Frankia* ORS02001 using ^{15}N methods. *Soil Biol. Biochem.*, *17*, 375-379.

Gauthier, D., Jaffre, T., and Prin, Y. (1999). Occurrence of both *Casuarina*-infective and *Elaeagnus*-infective *Frankia* strains within actinorhizae of *Casuarina collina*, endemic to New Caledonia. *Eur. J. Soil Biol.*, *35*, 9-15.

Gauthier, D., Navarro, E., Rinaudo, G., Jourand, P., Jaffre, T., and Prin, Y. (1999). Isolation, characterisation (PCR-RFLP) and specificity of *Frankia* from eight *Gymnostoma* species endemic to New Caledonia. *Eur. J. Soil Biol.*, *9*, 199-205.

Gentili, F., and Huss-Danell, K. (2002). Phosphorus modifies the effects of nitrogen on nodulation in split-root systems of *Hippophaë rhamnoides*. *New Phytol.*, *153*, 53 –61.

Giller, K. (2001). Assessment of the role of N_2 fixation. In *Nitrogen fixation in tropical cropping systems, 2nd edit.* (pp. 71-92). Wallingford, UK: CABI *Publishing*.

Goetting-Minesky, M. P., and Mullin, B. C. (1994). Differential gene expression in an actinorhizal symbiosis: Evidence for a nodule-specific cysteine proteinase. *Proc. Natl. Acad. Sci. USA, 91*, 9891-9895.

Gualtieri, G., and Bisseling, T. (2000). The evolution of nodulation. *Plant Mol. Biol.*, *421*, 81-94.

Haansuu, J. P. (2002). *Demethyl (C-11) cezomycin - a novel calcimycin antibiotic from the symbiotic, N_2-fixing actinomycete* Frankia. Academic Dissertation in General Microbiology. Helsinki, Finland: Graduate School in Microbiology, University of Helsinki.

Haansuu J. P., Klika, K. D., Söderholm, P. P., Ovcharenko, V. V., Pihlaja, K., Haahtela, K. K., and Vuorela, P. M. (2001). Isolation and biological activity of frankiamide. *J. Ind. Microbiol. Biotechnol.*, *27*, 62-66.

Herdman, L., Skene, K. R., and Raven, J. A. (1999). Structural-functional relations in cluster roots of *Myrica gale* L. *XVI International Botanical Congress, Abstracts,* 4478.

Hodge, S., Garrett, H. E., and Bratton, J. (1999). Alley cropping: An agroforestry practice. *Agroforestry Notes*, *12*, 1-4.

Huguet, V., Batzli, J. M., Zimpfer, J. F., Normand, P., Dawson, J. O., and Fernandez, M. P. (2000). Diversity and specificity of *Frankia* strains in nodules of sympatric *Myrica gale*, *Alnus incana*, and *Shepherdia canadensis* determined by rrs gene polymorphism. *Plant Mol. Biol.*, *42*, 181-194.

Hurd, T. M., and Schwintzer, C. R. (1996). Formation of cluster roots in *Alnus incana* ssp. *rugosa* and other *Alnus* species. *Can. J. Bot.*, *74*, 1684–1686.

Hurd, T. M., and Schwintzer, C. R. (1997). Formation of cluster roots and mycorrhizal status of *Comptonia peregrina* and *Myrica pensylvanica* in Maine, U.S.A. *Physiol. Plant.*, *99*, 680-689.

Huss-Danell, K. (1997). Actinorhizal symbioses and their N_2 fixation. *New Phytol.*, *136*, 375-405.

International Center for Research in *Agroforestry* - ICRAF. (2002), Did you know? Agroforestry facts. ICRAF Website. Retrieved June 12, 2002, from http://www.icraf.cgiar.org/ag_facts/ag_facts.htm.

Jadhav, B. B., and Gaynar, D. G. (1995). Effect of *Casuarina equisetifolia* leaf litter leachates on germination and seedling growth of rice and cowpea. *Allelopathy J.*, *2*, 105-108.

Jurgensen, M. F., Graham, R. T., Larsen, M. J., and Harvey, A. E. (1992). Clear-cutting, woody residue removal, and nonsymbiotic nitrogen fixation in forest soils of the inland Pacific Northwest. *Can. J. Forest Res.*, *22*, 1172-1178.

Khan, A. G. (1993). Occurrence and importance of mycorrhizae in aquatic trees of New South Wales, Australia. *Mycorrhiza, 3*, 31-38.

Klika, K. D., Haansuu, J. P., Ovcharenko, V. V., Haahtela, K. K., Vuorela, P. M., and Pihlaja, K. (2001). Frankiamide, a highly unusual macrocycle containing the imide and orthoamide functionalities from the symbiotic actinomycete *Frankia*. *J. Org. Chem.*, *66*, 4065-4068.

Kohls, S. J., Thimmapuram, J., Buschena, C. A., Paschke, M. W. and Dawson, J. O. (1994). Nodulation patterns of actinorhizal plants in the family Rosaceae. *Plant Soil*, *162*, 229-239.

Krishnapillay, B. (2000). Silviculture and management of teak plantations. *Unasylva*, *51*, 14-21.

Lalonde, M. (1979). A simple and rapid method for the isolation and cultivation *in vitro* and characterization of *Frankia* strains from *Alnus* root nodules. In J. C. Gordon, C. T. Wheeler, and D. A. Perrin (Eds.), *Symbiotic nitrogen fixation in the management of temperate forest.* (p. 1180). Corvallis, OR: Oregon State University.

Lalonde, M., Calvert, H. E., and Pine, S. (1981). Isolation and use of *Frankia* strains in actinorhizae formation. In A. H. Gibson and W. E. Newton (Eds.), *Current perspectives in nitrogen fixation.* (pp. 296-299). Canberra, Australia: Australian Academy of Sciences.

Lang, L. (1999). Inhibition of bacterial wilt growth by *Frankia* isolated from Casuarinaceae. *Forest Res.*, *12*, 47-52.

Leiva, J. M., and Borel, R. (1995). Evaluación de tres especies forestales en plantación pura y sistema taungya: Crecimiento de los árboles y producción de los cultivos. *Nitrogen Fixing Tree Research Reports*, Special Issue, 85-93.

Lechevalier, M. P., and Ruan, J. (1984). Physiology and chemical diversity of *Frankia* spp. isolated from nodules of *Comptonia peregrina* (L) Coul and *Ceanothus americanus* L. In A. D. L. Akkermann, D. D. Baker, K. Huss-Danell, and J. D. Tjepkema (Eds.), Frankia *symbiosis.* (pp. 15-22). The Hague, The Netherlands: Martinus Nijhoff/Dr. W. Junk.

Leney, L., Jackson, A., and Erickson, H. D. (1978). Properties of red alder (*Alnus rubra* Bong.) and its comparison to other hardwoods. In D. G. Briggs, D. S. DeBell and W. A. Atkinson (Edds.), *Utilization and management of alder: Proceedings of a symposium* (pp. 25-33) Ocean Shores, WA: U.S. Department of Agriculture, Forest Service.

Lie, T. A., Akkermans, A. D. L., and van Egeraat, A. W. S. M. (1984). Natural variation in symbiotic nitrogen fixing *Rhizobium* and *Frankia* spp. *Antonie van Leeuwenhoek*, *50*, 489-503.

Liu, Q., and Berry, A. M. (1991). Localization and characterization of pectin polysaccharides in roots and root nodules of *Ceanothus* spp. during intercellular infection by *Frankia*. *Protoplasma*, *163*, 93-101.

McNabb, D. H., and Cromack, K., Jr. (1983). Dinitrogen fixation by a mature *Ceanothus velutinus* (Dougl.) stand in the western Oregon Cascades. *Can. J. Microbiol.*, *29*, 1014-1021.

Miller I. M., and Baker D. D. (1985). The initiation, development and structure of root nodules in *Elaeagnus angustifolia* L. (Elaeagnaceae). *Protoplasma*, *128*, 107-119.

Miller I. M., and Baker D. D. (1986). Nodulation of actinorhizal plants by *Frankia* strains capable of both root hair infection and intercellular penetration. *Protoplasma*, *131*, 82-91.

Mullin, B. C., and Dobritsa, S. V. (1996). Molecular analysis of actinorhizal symbiotic systems: Progress to date. *Plant Soil*, *186*, 9-20.

Markham, J., and Chanway, C. P. (1998). *Alnus rubra* (Bong.) nodule spore type distribution in southwestern British Columbia. *Plant Ecol.*, *135*, 197.

Nair, P. K. R. (1990). *The prospects for agroforestry in the tropics*. Technical Paper Number 131. Washington, D.C.: The World Bank.

Newton, M., El Hassen, B. A., and Zavitkovski, J. (1968). Role of red alder in western Oregon forest succession. In J. M. Trappe et al. (Eds.), *Biology of alders*. (pp. 73-84). Portland, OR: U.S. Department of Agriculture.

Paschke, M. W., and Dawson, J. O. (1992). Avian dispersal of *Frankia*. *Can. J. Bot.*, *71*, 1128-1131.

Parrota, J. A., Baker, D. D., and Maurice, M. (1994). Application of ^{15}N-enrichment methodologies to estimate nitrogen fixation in *Casuarina equisetifolia. Can. J. Forest Res.*, *24*, 201-207.
Paschke M. W. (1997). Actinorhizal plants in rangelands of the western United States. *Journal of Range Management*, *50*, 62-72.
Perry, T. O. (1989). Tree roots: Facts and fallacies. *Arnoldia (Magazine of the Arnold Arboretum)*, 49, 3-21.
Pinyopusarerk, K., Tyrnbull, J. W., and Midgley, S. J. (Eds.) (1996). *Recent* Casuarina *research and development. Proceedings of the Third International Casuarina Workshop. Da Nang, Vietnam.* (249p.). Canberra, Australia: CSIRO, Forestry and Forest Products.
Prin, Y., Duhoux, E., Diem, H. G., Roederer, Y., and Dommergues, Y. (1991a). Nitrogen-fixing aerial nodules on the trunks of *Casuarina cunninghamiana. Nitrogen Fixing Tree Research Reports*, *9*, 100-101.
Prin, Y., Duhoux, E., Diem, H. G., Roederer, Y., and Dommergues, Y. (1991b). Aerial nodules in *Casuarina cunninghamiana*. *Appl. Environ. Microbiol.*, *57*, 871-874.
Provorov, N. A., Borisov, A. Y., and Tikhonovich, I. A. (2002). Developmental genetics and evolution of symbiotic structures in nitrogen-fixing nodules and arbuscular mycorrhiza. *J. Theor. Biol.*, *214*, 215-232.
Quispel, A, and Burggraff, A. J. P. (1981). *Frankia* the diazotrophic endophyte from actinorhizas. In A. H. Gibson and W. E. Newton (Eds.), *Current perspectives in nitrogen fixation.* (pp. 229-236). Canberra, Australia: Australian Academy of Sciences.
Rose, S. L. (1980). Mycorrhizal associations of some actinomycete nodulated nitrogen-fixing plants. *Can. J. Bot.*, *58*, 1449-1454.
Rose, S. L., and Trappe, J. M. (1980). Three new endomycorrhizal *Glomus* spp. associated with actinorhizal shrubs. *Mycotaxon.*, *10*, 413-420.
Rose, S. L., and Youngberg, C. F. (1981). Tripartite associations in snowbrush (Ceanothus velutinus): Effect of vesicular-arbuscular mycorrhizae on growth, nodulation and nitrogen fixation. *Can. J. Bot.*, *59*, 34-39.
Russo, R. O. (1989). Evaluating *Alnus acuminata-Frankia* -mycorrhizae interactions. I. Acetylene reduction in seedlings inoculated with *Frankia* strain ArI3 and *Glomus intraradices* under three different phosphorus levels. *Plant Soil*, *118*, 151-155.
Russo, R. O. (1990). Evaluating *Alnus acuminata* as a component in agroforestry systems. *Agroforestry Systems*, *10*, 241-252.
Russo, R. O. (1995). *Alnus acuminata* ssp. *arguta* (Schectendal) Furlow: A valuable resource for neotropic Highlands. *Nitrogen Fixing Tree Research Reports*, Special Issue, 156-163.
Russo, R. O., and Berlyn, G. P. (1989). The effect of a new growth biostimulant on acetylene reduction in nodulated seedlings of *Alnus ascuminata. Abstracts of 12th North American Symbiotic Nitrogen Fixation Conference,* Ames, Iowa.
Russo, R. O., Gordon, J. C., and Berlyn, G. P. (1993). Evaluating *Alnus acuminata-Frankia*-mycorrhizae interactions. Growth response of *Alnus acuminata* seedlings to inoculation with *Frankia* strain ArI3 and *Glomus intraradices*, under three phosphorus levels. *J. Sustainable Forestry*, *1*, 93-110.
Sampson, A. W., and Jespersen, B. S. (1963). *California range brushlands and browse plants.* (162 p.). Berkeley, CA: University of California, California Agricultural Experiment Station.
Schwintzer, C. R. (1990). Spore-positive and spore-negative nodules. In C. R. Schwintzer, and J. D. Tjepkema (Ed.), *The biology of* Frankia *and actinorhizal plants.* (pp. 177–193). New York, NY: Academic Press, Inc.
Schwintzer, C. R., and Tjepkema, J. D. (Eds.). (1990). *The biology of* Frankia *and actinorhizal plants.* (408 p.). New York, NY: Academic Press, Inc.
Sharma, G., Sharma, E., Sharma, R., and Singh, K. K. (2000). Performance of an age series of *Alnus*–Cardamom plantations in the Sikkim Himalaya: Productivity, energetics and efficiencies. *Ann. Bot.*, *89*, 261-272.
Sicco Smit, G. (1971). Notas silviculturales sobre *Alnus jorullensis* de Caldas, Colombia. *Turrialba, 21*, 83-88.
Simonet, P., Navarro, E., Rouvier, C., Reddell, P., Zimpfer, J., Dommergues, Y., *et al.* (1999). Co-evolution between *Frankia* populations and host plants in the family Casuarinaceae and consequent patterns of global dispersal. *Environ. Microbiol.*, *1*, 525-534.
Simpson, M. J. A., MacIntosh, D. F., Cloughley, J. B., and Stuart, A. E. (1996). Past, present and future utilisation of *Myrica gale* (Myricaceae). *Econ. Bot.*, *50*, 122-129.

Singer, R., and Gomez, L. D. (1984). The basidiomycetes of Costa Rica. III. The genus *Phylloporus* (Boletaceae). *Brenesia*, *22*, 163-181.

Skene, K. R. (1998). Cluster roots: Some ecological considerations. *J. Ecol.*, *86*, 1060–1064.

Swensen, S. M., and Mullin, B. C. (1997a). Phylogenetic relationships among actinorhizal plants: The impact of molecular systematics and implications for the evolution of actinorhizal symbioses. *Physiol. Plant.*, *99*, 565-573.

Swensen, S. M. and Mullin, B. C. (1997b). The impact of molecular systematics on hypotheses for the evolution of root nodule symbioses and implications for expanding symbioses to new host plant genera. *Plant Soil*, *194*, 185-192.

Tjepkema, J. D., Schwintzer, C. R., Burris, R. H., Johnson, G. V., and Silvester, W. B. (2000). Natural abundance of ^{15}N in actinorhizal plants and nodules. *Plant Soil*, *219*, 285-289.

Torrey, J. G. (1990). Cross-Inoculation Groups within *Frankia*. In D. D. Baker, J. D. Tjepkema, and C. R. Schwintzer (Eds.), *The biology of* Frankia *and actinorhizal plants*. San Diego, CA: Academic Press Inc.

Torrey, J. G., and Tjepkema, J. D. (1979). Symbiotic nitrogen fixation with actinomycete-nodulated planta. *Botanical Gazette*, *140*, supplement, i-ii.

U.S. Department of Agriculture, Forest Service - USFS. (1937). *Range plant handbook*. Washington, D.C.: U.S. Forest Service.

Valdés, M., and Cruz-Cisneros, R. (1996). Root and stem nodulation of *Casuarina* in Mexico. *Forest, Farm, and Community Tree Research Reports, 1*, 61-65.

Vercoe, T. K. (1993). Australian trees on tour: A review of the international use of Australian forest genetic resources. In *Proceedings of 15th Biennial Conference of the Institute of Foresters of Australia*. Canberra, Australia: CSIRO.

Warcup, J. H. (1980). Ecotomycorrhizal associations of Australian indigenous plants. *New Phytol., 85*, 531 –535.

Watt, M., and Evans J. R. (1999). Proteoid roots: Physiology and development. *Plant Physiol.*, *121*, 317–323.

Warembourg, F. R. (1993). Nitrogen fixation in soil and plant systems. In R. Knowles, and T. H. Blackburn (Eds.), *Nitrogen isotope techniques*. (pp. 127-156). New York, NY: Academic Press, Inc.

Watanabe, I. (2000). Lecture at Cantho University, Vietnam. Retrieved June 6, 2002, from http://www.asahi-net.or.jp/~it6i-wtnb/BNF.html#ch4.

Webster, S. R., Youngberg, C. T., and Wollum, A. G. (1967). Fixation of nitrogen by bitterbrush (*Purshia tridentata* (Pursh)D.C.). *Nature, 216*, 392-393.

Werner, D., and Müller, P. (Eds.) (1990). *Fast growing trees and nitrogen fixing trees* (pp. 3-8). Stuttgart, Germany: G. Fischer Verlag.

Yang, Y. (1995). The effect of phosphorus on nodule formation and function in the *Casuarina-Frankia* symbiosis. *Plant Soil*, *176*, 161-169.

Youngberg, C. T., and Wollum, A. G. (1976). Nitrogen accretion in developing *Ceanothus velutinus* stands. *Soil Sci. Soc. Am. J.*, *40*, 109-112.

Zavitkovski, J., and Newton, M. (1968). Ecological importance of snowbrush *Ceanothus velutinus* in the Oregon *Cascades. Ecology*, *49*, 1134-1145.

Zhong Chonglu (2000). A New Record on Casuarina Blister Bark Disease in Southern China. *NFT News Improvement and Culture of Nitrogen Fixing Trees*, *3*, 6-7.

CHAPTER 9

MOLECULAR ECOLOGY OF N_2-FIXING MICROBES ASSOCIATED WITH GRAMINEOUS PLANTS: HIDDEN ACTIVITIES OF UNKNOWN BACTERIA

T. HUREK AND B. REINHOLD-HUREK
Allgemeine Mikrobiologie, Fachbereich Biologie/Chemie der Universität Bremen, Postfach 33 04 40, 28334 Bremen, Germany

1. INTRODUCTION

Nitrogen is essential for all living organisms. It controls important biogeochemical processes, such as decomposition, nutrient cycling, and atmospheric chemistry. The availability of this element often limits plant growth in both terrestrial and aquatic ecosystems, affecting both the productivity and species composition of plant communities and ecosystem processes at all levels. Nitrogen is unique among the other essential elements because its most common form, N_2 gas, must be fixed from the atmospheric reservoir into a biologically usable form. The majority of fixed nitrogen is provided biochemically by biological nitrogen fixation (BNF), an energetically costly process, exclusively carried out by prokaryotes that possess the enzyme nitrogenase. BNF is of great biogeochemical significance. It is estimated to provide worldwide 200-300 million tons of biologically usable (fixed) nitrogen per year. Inputs of fixed nitrogen range from 90-130 Tg N per year for terrestrial ecosystems (Galloway *et al.*, 1995) to 100-200 Tg N per year for marine ecosystems (Karl *et al.*, 2002).

Usually, diazotrophic prokaryotes have to be mineralized before biologically fixed nitrogen becomes available for use by other bacteria or eukaryotes. But organisms, like some animals or plants that are either in close association or form a symbiosis with N_2-fixing microbes, may also profit from biologically fixed nitrogen directly and contribute huge amounts of biologically available nitrogen into the nitrogen cycle. One of the best understood nitrogen-fixing interactions, the nodule

D. Werner and W. E. Newton (eds.), Nitrogen Fixation in Agriculture, Forestry, Ecology, and the Environment, 173-198.

symbiosis between rhizobia and legumes, provides at least 70 million tonnes of N per year worldwide (Brockwell *et al.*, 1995). Nitrogen is required in substantial amounts, in order to support crop growth and feed the growing world population. In agriculture, mineral nitrogen fertilizer is the most widely used commercial external resource with a still increasing demand worldwide. For example, in Asia, the use of synthetic nitrogen fertilizer has increased more than 30-fold from 1.5 million tons in 1961 to almost 47 million tons in 1996 (Dawe, 2000). The human-induced dramatic increase of biologically available nitrogen by approximately 150 million tons per year (Mosier *et al.*, 2002) has altered the global nitrogen cycle, caused ecosystem changes, and incurred considerable costs to society (Vitousek *et al.*, 1997a; 1997b; Mosier *et al.*, 2002).

Despite the crucial importance of fixed nitrogen for essential ecological and biogeochemical processes, information on the biology of important diazotrophs in many environments is still very limited. This situation is especially true for some grass-dominated ecosystems, which have a shortfall in the nitrogen budget. Members of the grass family (*Poaceae*, formerly *Gramineae*) do not naturally form specialized symbiotic structures and, like maize, wheat, and rice, represent the most important crops worldwide. Gramineous crops, such as certain sugar cane cultivars in Brazil, can obtain a substantial part of plant nitrogen from BNF (Lima *et al.*, 1987; Boddey, 1995). The amount of newly fixed nitrogen provided by BNF lies in the medium range of experimental estimates for many legumes (Peoples *et al.*, 1995). However, in contrast to either nodular or cyanobacterial plant symbioses, information on key diazotrophs in gramineous plants is still limited due to the failure of researchers to isolate bacteria that fulfill Koch's postulates. So far, the major nitrogen-fixing diazotrophs, which contribute biologically fixed nitrogen to support plant growth in some uninoculated gramineous plants, have not been identified. This review will focus on the molecular ecology of nitrogen-fixing bacteria that are associated with gramineous plants in their natural environment and provide some functional insights into their populations.

2. THE PROBLEM OF IDENTIFYING KEY DIAZOTROPHIC BACTERIA IN GRAMINEOUS PLANTS: THE CLASSICAL APPROACH

The extent to which potentially N_2-fixing bacteria in the rhizosphere of gramineous plants provide fixed N to the plant is a long-standing question that dates back at least to Hiltner's (1899) report of supposed nitrogen fixation in darnel (*Lolium temulentum*), which could be not confirmed later (McLennan, 1920). It is now well established that more than 150 kg N ha^{-1} y^{-1} (up to 70% of plant nitrogen) may be added to the N-budget of several uninoculated outdoor-grown Brazilian sugar cane cultivars by heterotrophic N_2 fixation (Boddey, 1995; Urquiaga *et al.*, 1992) and that up to 50–100 kg N ha^{-1} y^{-1} may be similarly added to the N-budget of flooded rice (*Oryza*) (Cassman *et al.* 1995). There is increasing evidence that BNF may be a key source for new plant nitrogen in other gramineous plants as well, especially those which thrive in periodically or continually flooded environments, where primary production is probably nitrogen limited. Good candidates are smooth cordgrass (*Spartina alterniflora*) (McClung *et al.*, 1983; Lovell *et al.*, 2000; Brown *et al.*, 2003), Kallar grass (*Leptochloa fusca*) (Reinhold, *et al.*, 1986; Malik *et al.*, 1986; Hurek *et al.*, 2002), and purple moor-grass (*Mollinia coerulea*) (Hamelin *et al.*, 2002), which, like rice, are all flood-tolerant plants that can grow into extensive monoculture stands in, respectively, the salt-marsh ecosystems of North America, the subtropical wetland soils in Asia, and the nutrient-poor fens or bogs in Europe.

Plants relocate up to 40% of their photosynthetic production to roots and into the adjacent soil (Van Veen *et al.*, 1991; Whipps, 1984). This rhizodeposition is an important source of energy for microorganisms (Van Veen *et al.*, 1989). Because BNF is energetically costly and sensitive to O_2 as well, it is reasonable to assume that this process takes place in close proximity to the plant, preferably in the rhizosphere, where diazotrophic microorganisms have an advantage in substrate competition over other soil organisms. Diazotrophs are generally enriched in the rhizosphere compared to the non-rhizosphere soil (Balandreau and Knowles, 1978). However, it has been long known that considerable numbers of diazotrophs, which are affiliated to many distantly related genera, occur in the rhizospheres of tropical and subtropical grasses (Döbereiner, 1961; Haahtela *et al.*, 1983; McClung *et al.*, 1983; Rennie, 1980), and that frequently several species can be isolated from one plant (Patriquin *et al.*, 1983; Ruschel and Vose, 1984). Which N_2-fixing bacteria are the important ones?

Before molecular phylogenetic methods became available for environmental studies in the 1990's (*e.g.*, DeLong *et al.*, 1989; Giovannoni *et al.*, 1990, Amann *et al.*, 1995), the isolation of bacteria was a prerequisite for identifying key diazotrophic microbes in gramineous plants just like it had been for the successful isolations from the rhizobial (Beijerinck, 1888) and actinorhizal (Callaham *et al.*, 1978) nodular symbioses long ago. Diazotrophs that colonize the interior of gramineous plant roots were a priority target in cultivation efforts for many years because it was reasonably assumed that, similar to nodular symbioses, either the gramineous plant's internal tissues or niches that are proximal to living plant tissues are the site of biological nitrogen fixation. The colonization of such sites is likely to result in an advantage in substrate competition over non-diazotrophs in the natural environment (Reinhold *et al.*, 1986; Reinhold, 1988). Accordingly, N_2-fixing bacteria, which penetrate deeply into the root interior (endorhizosphere), should be more important for the supply of fixed-nitrogen to the plant than diazotrophs, which primarily colonize the root surface (rhizoplane) (Reinhold *et al.*, 1986), like, *e.g.*, *Azospirillum* (Aßmus *et al.*, 1995; Steenhoudt and Vanderleyden, 2000).

Root environments in wetlands may be particularly favorable for BNF (Engelhard *et al.*, 2000) because they probably constitute a suitable low-O_2-environment in an anoxic soil. Flood-tolerant plants form longitudinally interconnected intercellular gas spaces in the cortex of roots. These so-called aerenchyma have a high potential to aerate roots by longitudinal O_2 transport and to oxidize the rhizosphere by radial O_2 loss *in situ* (De Simone *et al.*, 2002). Such an environment may provide anoxic microniches for the high nitrogenase activities of aerobic and anaerobic diazotrophic bacteria, if suitable biological (Hurek *et al.*, 1995; Hurek and Reinhold-Hurek, 1998) or chemical (Armstrong and Boatman, 1967; Armstrong, 1978; Begg *et al.*, 1994) O_2 sinks are present.

3. THE SIGNIFICANCE OF DIAZOTROPHIC GRASS ENDOPHYTES

Not completely unexpectedly, distinct diazotrophs have been cultured out in very high numbers (up to 10^8 bacteria per g root dry weight) from internal root tissues of flood-tolerant plants, such as smooth cord (McClung *et al.*, 1983), Kallar grass (Reinhold *et al.*, 1986), and rice (Barraquio *et al.*, 1997). Similar results were obtained for sucrose-rich members of the grass-family, such as sugarcane (Cavalcante and Döbereiner, 1988; Li and MacRae, 1992; Dong *et al.*, 1994), sweet potato (*Ipomoea batatas*), and Cameroon grass (*Pennisetum purpureum*) (Caballero-Mellado *et al.*, 1995), where diazotrophic bacteria were also recovered from stems, leaves, and tubers.

These observations almost immediately prompted investigations into the taxonomy and ecology of these bacteria. It turned out that the most promising candidates for a key role in N_2 fixation with grasses - those which could be repeatedly isolated from internal plant tissues - are affiliated to the distantly related, previously unknown, ß-proteobacterial genera, *Herbaspirillum* (Baldani *et al.*, 1986; 1996) and *Azoarcus sensu lato* (Reinhold *et al.*, 1993b; Reinhold-Hurek and Hurek, 2000), and to the new α-proteobacterial species, *Acetobacter diazotrophicus* (Gillis *et al.*, 1989), which was recently renamed *Gluconacetobacter diazotrophicus* (Yamada *et al.*, 1997; 1998). All are capable of penetrating deeply into plants (Figure 1a, b) (Hurek *et al.*, 1991; Hurek *et al.*, 1994; James, *et al.*, 1997; Olivares *et al.*, 1997; James and Olivares, 1998; James *et al.*, 2001; James *et al.*, 2002) and, as was suspected from their isolation but shown with certainty, all spend most of their life cyle inside plant tissues without causing symptoms of plant damage. They are, therefore, referred to as endophytes (Döbereiner *et al.*, 1993; Hurek *et al.*, 1994) according to a definition given by Quispel (1992).

In general, all diazotrophic grass endophytes are apoplastic colonizers and do not form an endosymbiosis in living plant cells (Reinhold-Hurek and Hurek, 1998b; James, 2000). In contrast to other plant-colonizing diazotrophic bacteria, such as *Rhizobium* or *Azospirillum*, they do not survive well in, and often cannot be isolated from, root-free soil (Döbereiner *et al.*, 1988; Boddey *et al.*, 1995; Olivares *et al.*, 1996; Reinhold-Hurek and Hurek, 1998b), probably because soil populations are not self-sustaining. Diazotrophic grass endophytes may be too specialized to persist in soil and are maintained by input from the plant or other sources (Hurek *et al.*, 1997a). The few data available on the genetic structure of their populations (Caballero-Mellado and Martinez-Romero, 1994; Caballero-Mellado *et al.*, 1995) (Figure 2) suggest that the genetic diversity of these bacteria is limited which might be related to their endophytic lifestyle. Similar to nonendophytic bacteria, such as *Azospirillum* (Christiansen-Weniger *et al.*, 1992; Vande Broek *et al.*, 1993; Arsène *et al.*, 1994), they can express the iron protein of nitrogenase (*nifH*) in the rhizosphere (Figure 1c) (Egener *et al.*, 1999; Hurek and Reinhold-Hurek, 1998; Gyaneshwar *et al.*, 2002; James *et al.*, 2002) and are able to supply fixed-nitrogen to inoculated gramineous plants (Sevilla *et al.*, 2001; Hurek *et al.*, 2002; Gyaneshwar *et al.*, 2002; James *et al.*, 2002; for excellent reviews on nonendophytic bacteria, see: Giller and Day, 1985; Okon and Labandera-Gonzalez, 1994; Bashan and Holguin, 1997).

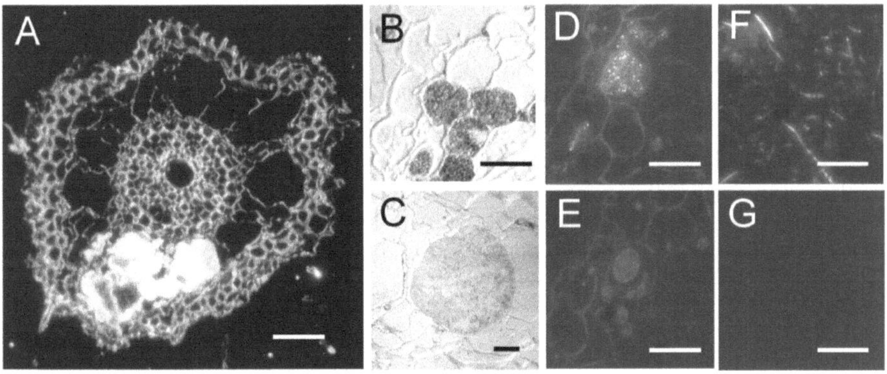

Figure 1. Colonization of rice roots in gnotobiotic culture (A-C) and nifH *expression in Kallar grass (*Leptochloa fusca*) roots grown in soil (D-G).*
*Dark field (A) or bright field (B, C) micrographs of semi-thin sections of rice seedling roots 11-14 days after inoculation with an isogenic mutant (*nifH::gus*) of* Azoarcus *sp. strain BH72. Subsequent sections were immunogold-silver stained using either antiserum directed against BH72 (A, B) or antiserum against the iron protein of nitrogenase (C).*
Detection of nitrogenase gene expression of Azoarcus *sp. by* in situ *hybridization (D-G). Epifluorescence micrographs of transversal semi-thin sections of resin-embedded root of uninoculated, soil-grown Kallar grass (D, E) or of cells of strain BH72 grown on N_2 (F, G)., hybridized with either a fluorescing antisense (D, F) or sense (E, G) probe to nifH.*
Bars 100 µm (A), 20 µm (B-E) or 5 µm (F, G).
Pictures are from Hurek and Reinhold-Hurek (2003) and were originally published by Egener et al.*, (1999) (A, C) and Hurek* et al.*, (1991), (1997b) (B, D-G).*

Clearly, endophytic as well as nonendophytic bacteria are able to fix N_2 in the rhizosphere of inoculated gramineous plants, but there is increasing evidence that diazotrophic grass endophytes that are affiliated with the genera, *Azoarcus* or *Herbaspirillum*, are much better equipped for growth in internal tissues of non-sucrose-rich plants than are other bacteria. For microorganisms of the genus, *Azoarcus*, this seems to hold even for N_2-dependent growth. In many cases, both the growth and N_2 fixation of microorganisms that colonize the rhizosphere of inoculated gramineous plants seems to be carbon-limited. Some bacteria, like either *Klebsiella pneumoniae* (Chelius and Triplett, 2000) or *Serratia marcescens* (Gyaneshwar *et al.*, 2001), and *Azorhizobium caulinodans* (van Nieuwenhove *et al.*, 2000), apparently require the addition of huge amounts of external carbon sources (> 1g l^{-1} Na-malate, sucrose, *etc.*) to plant cultivation media either to express *nifH* in the rhizosphere or to infect the internal tissues of inoculated maize (*Zea mays*) or rice seedlings, respectively. Others, like *Azospirillum brasilense* on wheat (*Triticum aestivum*), express *nifH* only epiphytically either with (Christiansen-Weniger *et al.*, 1992; Vande Broek *et al.*, 1993) or, to a lesser extent, without (Arséne *et al.*, 1994) carbon amendment. In contrast, endophytic microorganisms, such as *Azoarcus* sp. BH72, are able to express *nifH* abundantly and also to translate *nifH* transcripts into protein in roots of certain rice varieties without extra nutrients, albeit to a lesser extent than in the presence of externally supplied carbon sources (Egener *et al.*, 1999; see also Figure 6 of Reinhold-Hurek and Hurek, 1998a). Likewise, a heavy infection of the internal tissue of an Al-tolerant variety of *Oryza sativa* by the diazotrophic grass endophyte, *Herbaspirillum seropedicae* Z67, was accompanied by a considerable level of nitrogen fixation on the rhizoplane, when no external carbon source was added (Gyaneshwar *et al.*, 2002). These results did not depend on whether plants were grown in a semi-solid medium (e.g. Chelius and Triplett, 2000) or whether a possible artefactual situation caused by the use of agar-solidifying plant media (McCully, 2001) was avoided (*e.g.*, Gyaneshwar *et al.*, 2001; 2002; Egener *et al.*, 1999; van Nieuwenhove *et al.*, 2000; Gyaneshwar *et al.*, 2002).

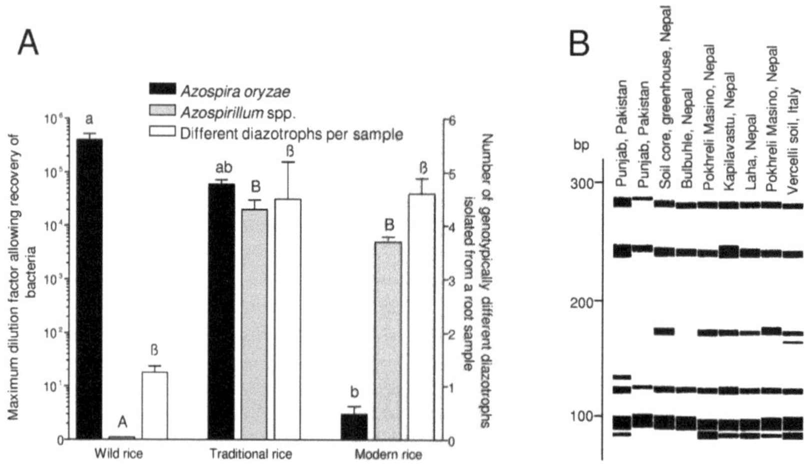

Figure 2. Distribution and genetic diversity of culturable diazotrophic grass endophytes in rice. (A) Differences in populations of culturable diazotrophs associated with the root interior of 2 different wild rice species, 4 different land races, and 3 different modern cultivars of Oryza sativa. All plants were grown in waterlogged soil. Columns that are different from each other at P<0.05 are headed by different letters. (B) Genomic fingerprints of Azospira oryzae *isolated from Kallar grass, one wild rice species, 3 different land races, and one modern cultivar of rice from locations in Asia and Europe.*
Picture modified from Engelhard et al., (2000).

The consensus of these studies is that diazotrophic grass endophytes are much less subject to carbon limitation than other plant-colonizing bacteria when they grow in internal plant tissues. Probably these bacteria have access to carbon sources that are not available to other microorganisms. One possibility is that endophytes might elicit a K^+ efflux / H^+ influx, like an exchange response (XR; Atkinson *et al.*, 1993; He *et al.*, 1994; Alfano and Collmer, 1996), in internal plant tissues by partial degeneration of the host cell membrane (Hallmann, 2001). An XR would cause an alkalinization of the intercellular fluid, which would stimulate both leakage of sucrose (and other compounds) from cells into the apoplast and bacterial growth (Atkinson and Baker, 1987a; 1987b). The abilitity to elicit a XR-like process in undiseased plants may be an important mechanism through which diazotrophic grass endophytes acquire nutrients for apoplastic colonization, so avoiding the putative carbon limitation of endophytic N_2 fixation. Efficient entry into plant tissues (Reinhold-Hurek *et al.*, 1993a; Dörr *et al.*, 1998) and metabolic adaptations to available nutrients (Hurek and Reinhold-Hurek, 1998) may also contribute to the capacity of these bacteria to establish considerable endophytic or endorhizospheric populations in various situations. Altogether, these skills are likely to provide diazotrophic grass endophytes with an important competitive advantage over other nonpathogenic microorganisms in the natural environment and so identify them as candidates of choice for an important role in gramineous plants.

However, whether bacteria are endophytic or not, the amount of biologically fixed nitrogen supplied by them to inoculated plants is, in all plant-bacteria combinations examined so far, not enough to explain the amount of BNF observed in the natural environment. Is the assumption wrong that the key organisms for BNF in gramineous plants are primarily recruited from the pool of specialized diazotrophic grass endophytes? Are several distinct microbes required to achieve the levels of BNF recorded in some grass-dominated environments while one is not sufficient?

4. THE SIGNIFICANCE OF CULTURE-INDEPENDENT METHODS

The above questions can only be answered by including molecular ecological studies of the activity of these or other diazotrophs in the natural environment. Culturing techniques have been widely used to identify the most active diazotrophic bacterium(a). However, if important diazotrophs are free-living or involved in inconspicious relationships with unknown partners, Koch`s postulates, which are designed to distinguish the true agent of infection from a secondary invader, are not applicable for separating important from unimportant microbes. In this case, primary diazotrophs can only be identified by employing culture-independent methods. Moreover, culture-dependent approaches can lead to misinterpretations of the microbial diversity and cause biased results because, as Ward *et al.*, (1990) pointed out, "a microorganism can be cultivated only after its physiological niche is perceived and duplicated experimentally". Culture-independent analyses, based on comparisons of 16S rRNA sequences, have revolutionized our knowledge of microbial diversity. An important conclusion from these studies is that the vast majority of microorganisms remains undiscovered. However, 16S rRNA is not a good marker to evaluate functional microbial diversity, *i.e.*, the functional differences among microbial species, because taxonomic affiliation and biochemical, physiological, or ecological capabilities of bacteria are often not correlated. Protein-coding genes are much better suited for this purpose because they can be used for surveying a particular capability or process and sometimes can be helpful for identification as well.

5. THE SIGNIFICANCE OF *NIFH*-TARGETED METHODS

The iron-protein gene (*nifH*, *anfH*, or *vnfH*, here collectively called "*nifH*") of the evolutionary conserved nitrogenase protein complex (Howard and Rees, 1996) is the marker of choice for monitoring biological N_2 fixation. This nitrogenase protein complex is distributed in more than 100 prokaryotic genera over most of the major phyla (for reviews, see Young, 1992; and in Volume 3 of this series, *Genomes and Genomics of Nitrogen-Fixing Organisms*) and is represented by three highly related nitrogenases; a molybdenum-based (*nif*), an iron-only based (*anf*), and a vanadium-based (*vnf*) nitrogenase. Although most N_2-fixing prokaryotes carry only *nif* genes, some organisms, like *Azotobacter vinelandii*, are able to synthesize any of the three nitrogenases, depending on the growth conditions (Bishop and Premakumar, 1992).

Various sets of universal *nifH* primers are available (*e.g.*, Zehr and McReynolds, 1989; Ueda *et al.*, 1995; Okhuma *et al.*, 1996; Piceno *et al.*, 1999; Widmer *et al.*, 1999), which amplify successfully fragments of distantly related diazotrophs, but only those of Zehr and McReynolds, (1989) have been shown to amplify highly diverged *nifH* genes with equal efficiencies (Tan *et al.*, 2003). These primers do not target the O_2-tolerant nitrogenase system of *Streptomyces thermoautotrophicus* (Ribbe *et al.*, 1997), where *nifH* is replaced by an unrelated gene that encodes a superoxide oxidoreductase. This unique nitrogenase requires the formation of superoxide, provided through the oxidation of CO by a CO dehydrogenase, to reduce N_2 to NH_3. Although it would be interesting to know how widely this nitrogenase is distributed, the apparent requirement of high concentrations of CO for biological nitrogen fixation in *Streptomyces thermoautotrophicus* suggests that this system is restricted to unusual CO-rich environments, from which this bacterium was isolated. Therefore, it appears to be unlikely that this nitrogenase system is relevant to BNF in gramineous plants and is not further considered here.

Whereas *nif* DNA retrieval from the natural environment (*e.g.*, Hamelin *et al.*, 2002; Engelhard *et al.*, 2000, Lovell, *et al.*, 2000; Ohkuma and Kudo, 1996, Ueda *et al.*, 1995) only shows the presence of nitrogenase genes or the diversity of diazotrophs, surveys of *nif*-mRNA expression can also evaluate whether the bacteria studied are of any importance for BNF in their natural environment. Plant endophytes, which are known to be highly related taxonomically, can be very diverse in their genetic and biochemical capacities (Stierle *et al.*, 1993; Young and Haukka, 1996). For example, a N_2-fixing isolate may be affiliated with the same genus or even species as the key diazotroph in the natural environment but might contribute only marginally to BNF in a gramineous plant. Using reverse-transcription-polymerase chain reaction (RT-PCR) to study the expression of nitrogenase genes would not only evaluate the actual activity of a particular microbe within a pool of other microbes in the natural environment, but would also allow identification of the primary diazotrophic bacteria by a comparison of sequences retrieved from the environment with sequences from cultivated organisms. For example, a match of >99.7% nucleotide identity between an environmentally transcribed *nifH* sequence and a *nifH* sequence from an isolate, *e.g.*, *Azoarcus* sp. BH72, would indicate that the cultivated bacterium (*Azoarcus* sp. BH72) is active in situ. The match is at or below the divergence introduced by PCR error (99.7%) and well above the strain- (~99%) or species (<95%) level divergence of highly related *nifH* genes within the genus *Azoarcus* (Hurek *et al.*, 2002).

6. LIMITATIONS OF *NIFH*-TARGETED METHODS

Matches below 95% nucleotide identity would be difficult to assign because proteobacterial lineages may show conflicting *nifH* and rRNA phylogenies at the genus level (Hurek *et al.*, 1997b). For example, a 336bp-long *nifH* fragment from *Azoarcus tolulyticus* (Genbank accession U97122) groups with *nifH* sequences from α- instead of ß-proteobacteria. Furthermore, it has 87% nucleotide identity with the corresponding *nifH* fragment from *Burkholderia vietnamensis* (Genbank accession AJ512206) compared to only 77% with that from *Azoarcus* sp. BH72 (Genbank accession Y12545). Also, the evolutionary histories of several *nifH* genes from *Bacteria* and *Archaea* in clusters II-IV are not in accord with those from ribosomal RNA (Figure 3). This incongruency of *nifH* and rRNA phylogenies in several organisms may reflect substantial horizontal descent of nitrogenase genes (Hennecke *et al.*, 1985; Normand and Bousquet, 1989) and hampers phylogenetic assignments considerably.

The rationale for using *nifH* mRNA-targeted techniques is that there is usually a tight relationship between nitrogenase activity and *nifH* transcription in diazotrophs (Merrick and Edwards, 1995), which holds for liquid cultures, soil (Bürgmann *et al.*, 2003), and the guts of termites (Noda *et al.*, 1999) as well. If nitrogen fixation is repressed, because either combined nitrogen is present or O_2 concentrations are unfavourable, the *nif* genes are not transcribed (Merrick, 1992). For example, in either the soil bacterium, *Azotobacter vinelandii*, or the diazotrophic grass endophyte, *Azoarcus* sp. BH72, no *nifH* transcripts can be detected in Northern blots in the presence of either 15 mM NH_4^+ (Blanco *et al.*, 1993) or 10 mM NH_4^+ (Egener *et al.*, 2001), respectively. Correspondingly, *nifH*-reporter-gene transcriptional-fusion assays in either *Azotobacter vinelandii* or *Azoarcus* sp. BH72 revealed that the activities measured at either 15 mM NH_4^+ (Bali *et al.*, 1992) or 0.5 mM NH_4^+ (Egener *et al.*, 1999), respectively, are only 3% of that obtained in the absence of ammonium. Consistent with these results, no PCR products are obtained with our protocol (Hurek *et al.*, 2002) in *nifH*-specific RT-PCRs with RNA preparations from cultures of *Azoarcus* sp. BH72 grown in a complex medium with 10 mM NH_4^+ (Hurek, unpublished). However, Bürgmann *et al.*, (2003) were able to detect by RT-PCR (background-level) *nifH* expression of *Azotobacter vinelandii* at 9 mM NH_4NO_3, when nitrogenase activity (acetylene reduction) was undetectable. Either their RNA-extraction protocol has a higher RNA-extraction efficiency and is producing lesser RT-PCR inhibitors than our method or the *nifH* mRNA of *Azotobacter vinelandii* is more stable than that of *Azoarcus* sp. BH72.

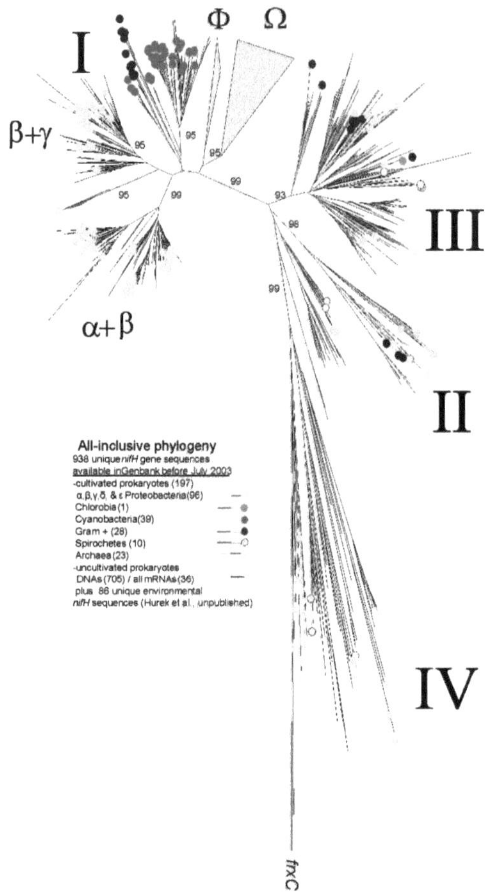

Figure 3. Unrooted phylogenetic tree of NIFH protein sequences (129 sequence positions) based on a ClustalX alignment (Thompson et al., *1997) as described previously (Hurek* et al., *1997b). Clusters I-IV as defined by Chien and Zinder (1994); frxC, chlorophyllide reductase, Genbank accession D00665 was added and which had served as an outgroup for previous phylogenetic analysis (*e.g., *Zehr* et al., *1998). Two groups (Ω, Φ) of entirely uncultured prokaryotes are detailed separately. The Ω (76 sequences) and Φ clusters (10 sequences) are named after* ***O****ryza (1) and* ***f****resh water, respectively, where sequences were first obtained (from Hurek* et al., *2002; Hurek* et al., *unpublished). The phylogenetic tree was inferred by a minimum evolution analyses of all unique 938 known sequences available up to July 2003 in Genbank plus 86 environmental sequences (Hurek* et al., *unpublished). Interior branch test support ≥ 93% is given for major clusters. Terminals representing* nifH *sequences from cultivated organisms are labelled with a circle according to taxonomic affiliation.*

Another possible discrepancy between nitrogenase activity and *nifH* transcription is the rapid inactivation of nitrogenase activity upon addition of ammonium, which leads to an immediate cessation in nitrogen fixation although mRNA is still present (Egener *et al.*, 2001). Because mRNAs in bacteria underlie a rapid turnover and rates of transcription can be reasonably assumed to be adjusted accordingly in the equilibrium situations in the natural environment, such a conflict between N_2 fixation and nitrogenase expression should not arise. Consistent with this assumption, both nitrogenase activity (measured as acetylene-reduction activity) and *nifH* expression decrease below the level of detection in both pot-grown *Orza* plants (Hurek *et al.*, unpublished data) and for the symbiotic microbial community in the gut of the termite, *Neotermes koshunensis* (Noda *et al.*, 1999), after application of a nitrogen source. Clearly, RT-PCR followed by comparative sequence analysis of *nifH* cDNAs is the method of choice to recognize the most active microbes if it is certain that the RT-PCR products do not represent background levels of *nifH* expression. One way to meet this condition is to sample RNA only when a parallel enzyme-based assay, such as either $^{15}N_2$ incorporation (Zehr *et al.*, 2001) or acetylene reduction (Noda *et al.*, 1999; Bürgmann *et al.*, 2003), indicates that nitrogenase is active. Another possibility is to ascertain, by quantitative RT-PCR (Hurek *et al.*, 2002), that the level of *nifH* transcription is far above background.

7. MANY DIAZOTROPHS DEFY CULTIVATION

RT-PCR-based techniques have been widely used to detect mRNA of several genes, including *nifH*, in the natural environment (*e.g.*, Brown *et al.*, 2003; Hurek *et al.*, 2002; Zani, *et al.*, 2000; Noda *et al.*, 1999; Kowalchuk *et al.*, 1999). Molecular ecological studies based on sequence analysis (*e.g.*, Hurek *et al.*, 2002; Zani *et al.*, 2000), denaturing gradient gel electrophoresis (DGGE) (Lovell *et al.*, 2000), PCR restriction length polymorphism (RFLP) (Poly *et al.*, 2001), and fluorescently labelled terminal restriction fragment length polymorphism (T-RFLP) (Moeseneder *et al.*, 1999; Tan *et al.*, 2003) of either *nifH* DNA or mRNA fragments have revealed a wide range of mostly (as yet) uncultured bacteria that occur in natural habitats, including those dominated by gramineous plants (see Zehr *et al.*, 2003 for a review).

As many as 938 unique *nifH* sequences (<99.7 nucleotide identity) of an approx. 318-nucleotides long stretch of sequence from both isolates and from environments as diverse as the Termite gut and the oceans were available up to July 2003 in Genbank. Of these sequences, 79% belong to uncultured and only 21% to cultured bacteria (Figure 3). All environmental *nifH* sequences (79%) were obtained with universal *nifH*-targeted primers. A comparison of these sequences reveals that only nine environmental *nifH* phylotypes (1.2%) are from cultivated N_2-fixing microorganisms, if strain-level divergence (99-100% nucleotide identity) is used as an exclusion criterion (Table 1). From these bacteria, only three are represented by *nifH* transcripts in the natural environment and they are suspected of playing an important role to the nitrogen economy there. These are: (i) *Synechocystis* sp. WH8501, as a representative of nanoplankton in the oceans (Zehr *et al.*, 2001; Falćon *et al.*, 2002); (ii) *Spirochaeta aurantia*, as a spirochetal representative of the termite gut microflora (Lilburn *et al.*, 2001; Okhuma *et al.*, unpublished; see also Noda *et al.*, 2003); and (iii) *Azoarcus* sp. BH72, as a representative of diazotrophic grass endophytes in the rhizosphere of gramineous plants (Hurek *et al.*, 2002).

Table. 1 Cultivated diazotrophic bacteria represented by environmental *nifH* phylotypes from the 938 unique *nifH* sequences[1] available up to July 2003 in Genbank.

Cultivated organism		Environmental *nifH* phylotype		
Species affiliation	Accession No.	DNA / mRNA	Accession No.	Retrieved from
Azospirillum lipoferum	AF216882	DNA	AF414623	Seabed grass
Azospirillum brasilense	M64344	DNA	AF389742	dead biomass of *Spartina alterniflora*
Klebsiella sp.	M63691	DNA	U43445	Tomales Bay
Rhizobium fredii	L16503	DNA	AY159590	Sweet potato
Rhizobium etli	U80928	DNA	AY159592	Sweet potato
Sinorhizobium meliloti	J01781	DNA	AY159587	Sweet potato
Azoarcus sp. BH72	AF200742	mRNA	AY231507	Kallar grass
Spirochaeta aurantia	AF325791	mRNA	AB083575	Termite
Synechocystis sp. WH8501	AF300829	mRNA	AY100520	North Atlantic Ocean

[1]*From sequences with >99.7 % DNA sequence identity, all sequences but one were deleted.*

Up to now, only *Azoarcus* sp. BH72 is represented by *nifH* transcripts in the roots of gramineous plants in the natural environment (see Figure 1D, E, F, G) and so presumably this bacterium also fixes N_2 there. *NifH* transcripts from other plant-colonizing diazotrophic bacteria, such as *Herbaspirillum* spp., *Pseudomonas* sp., *Azospirillum* spp., *etc.*, were not detected in the environments examined. Further, a recent survey of the activity of diazotrophs in the rhizosphere of *Spartina alterniflora* could not ascribe any phylotype to a cultivated bacterium (Brown *et al.*, 2003). None of the *nifH* mRNA sequences retrieved, using universal primers, from the environment (AY098493-AY098507) have more than 90% nucleotide identity to *nifH* sequences from cultured bacteria in Genbank. Because all matches can be considered well above species-level divergence (95% nucleotide identity), sequences can be not assigned with confidence to any taxon because of possible horizontal descent (see above).

8. DIAZOTROPHIC GRASS ENDOPHYTES AS KEY ORGANISMS FOR BNF IN GRAMINEOUS PLANTS

8.1. Important Diazotrophs in Gramineous Plants may have Small Culturable Populations

One plausible explanation for the failure of identifying primary source organisms in grasses would be that the most active nitrogen-fixing bacteria have almost completely escaped cultivation up to now. Recent results on Kallar grass and its nitrogen-fixing endophytes on a salt-affected field in Pakistan support this assumption and suggest that important microbes have been rarely cultivated because their culturable populations in natural environments are usually very small (Hurek *et al.*, 2002). In this study, the contribution of the diazotrophic grass endophyte, *Azoarcus* sp. BH72, to the fixed-nitrogen supply of the original host plant, Kallar grass, was evaluated. *Azoarcus* sp. BH72 originates from surface-sterilized roots of Kallar grass, where it was cultured out in huge numbers (*ca.* 10^8 cells g^{-1} dry weight) 20 years ago (Reinhold *et al.*, 1986). This bacterium could never be isolated from that location again and this might indicate that *Azoarcus* sp. BH72 is a random isolate and is not indigenous to that environment. But *in-situ* hybridizations with a homologous antisense *nifH* probe identified *nifH* mRNA highly related to *nifH* mRNA from *Azoarcus* sp. BH72 in roots of Kallar grass (cortex region; see Figure 1D, E, F, G). This last result suggested that the bacterium was present and metabolically active but could not be cultivated. It also fuelled the speculation that *Azoarcus* sp. BH72 plays a key role in BNF there.

The loss of culturability in metabolically active cells is a typical feature of many pathogenic or endosymbiotic microorganisms (Douglas, 1995) and may have many reasons. In *Azoarcus* sp. BH72, it may be related to a global change of the proteome as the result of differential gene expression when these bacteria confront certain rhizospheric fungi (Hurek *et al.*, 1995; Karg and Reinhold-Hurek, 1996; Hurek and Reinhold-Hurek, 1998; Hurek *et al.*, 2002). So far, no sequence information was available to support the assumption that the bacterium was active in the natural environment and has lost culturability there. It was also not clear what the level of *nifH* transcription of strain BH72 might be there in relation to other diazotrophic microbes. Previous studies, using GFP-gene fusions, had shown beyond doubt that strain BH72 can express nitrogenase genes in plant roots (Reinhold-Hurek and Hurek, 1998b; Egener *et al.*, 1999). However, these previous experiments were done in microbiologically controlled pure cultures and not under *in situ* conditions. Only RT-PCR and not the use of either reporter-gene fusions or *in-situ* hybridizations can evaluate the actual contribution to nitrogen fixation of a particular microbe within a pool of other diazotrophs. This also applies to inoculation experiments, where plants might get contaminated by other diazotrophs, so raising the question of who is responsible for the contributed fixed-nitrogen, the inoculum or other diazotrophs; and is even more applicable to plants in soil cores.

To provide such data, plants that were grown in the greenhouse as well as uninoculated Kallar grass plants, which originated from the natural monoculture stand in the Punjab of Pakistan, were examined. In this environment, BNF is a key source of fixed-nitrogen. As much as 20-40 metric tons of Kallar grass hay, amounting to more than 100 kg N, could be harvested ha^{-1} $year^{-1}$ on a low-fertility wetland soil without addition of nitrogenous fertilizer for many years (Malik, 1986). Both N-balance and isotopic measurements, in combination with the retrieval of either polyadenylated (Figure 4) or PCR-amplified *nifH*-transcripts resembling the sequence of the inoculum (strain BH72), confirmed that this endophyte contributed a substantial amount of biologically fixed nitrogen to inoculated plants. Not completely unexpectedly, these bacteria could not be reisolated after 3 months. As mentioned before, they could also not be isolated from plants grown in soil cores collected from the natural side in Pakistan.

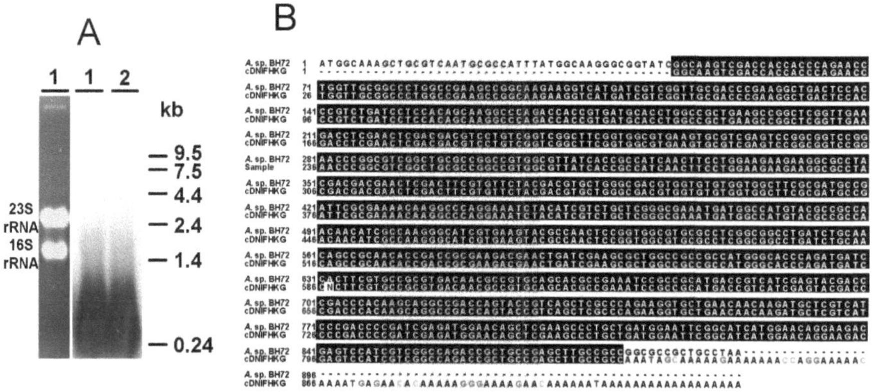

Figure 4. Polyadenylation of Azoarcus *sp. BH72 mRNA.*
(A). Agarose gel (left) and corresponding Northern blot (right) of mRNA from Azoarcus *sp. strain BH72 hybridized with a 27-base poly(dT) oligo at 54°C. Lane 1, total RNA; lane 2, total RNA purified with oligo(dT) cellulose.*
(B). Alignment of the nifH *DNA sequence of* Azoarcus *sp. strain BH72 (Genbank accession AF200742) with polyadenylated* nifH *mRNA from the same bacterium. The cDNA clone was retrieved from a library of polyadenylated mRNA which had been synthesized from roots of Kallar grass inoculated with* Azoarcus *sp. strain BH72. Identical nucleotides are boxed in black, bases other than A in the poly (A) tail are highlighted.*
Picture modified from Hurek et al., (2002).

8.2. Important Diazotrophs in the Natural Environment may contribute to BNF in Inoculated Plants

Extrapolated to an annual hectare basis, the amount of fixed N supplied by BNF corresponded to 34 kg N, which is less than one-third of the amount assumed to be provided to Kallar grass in the natural environment. The retrieval of polyadenylated *nifH* mRNA from *Azoarcus* sp. BH72 in root preparations of inoculated Kallar grass

(Figure 4) without PCR amplification and critical DNA removal step indicated unambiguously that this bacterium was metabolically active and probably was fixing nitrogen. It did not show whether other bacteria and how many of them did the same. Although RNA in prokaryotes may be polyadenylated as it is in eukaryotes, only a small portion of the mRNA population seems to be polyadenylated at any time and it is not known how widespread polyadenylation in bacteria is (Sarkar, 1997). Bacteria might have transcribed *nifH* genes without polyadenylating them. Comparative sequence analyses of three randomly selected clones after RT-PCR with universal primers for the *nifH* gene ruled this possibility out. All three clones carried inserts whose sequences were almost identical (99.7 % nucleotide identity) to the *nifH* gene from *Azoarcus* sp. BH72. This result was confirmed by hybridization analysis, which showed that the RT-PCR product hybridized with a *nifH* probe from strain BH72 at high stringency (hybridized at 70°C; washed in 0.1xSSC, 0.1 % SDS at 70°C). This indicated that the RT-PCR product had more than 95% nucleotide sequence identity to the corresponding *nifH* fragment from strain BH72. Clearly, unculturable *Azoarcus* sp. BH72 was the primary *nifH*-expressing bacterium and, therefore, also the most important N_2-fixing microorganism in inoculated plants.

It turned out to be the most important diazotroph in uninoculated Kallar grass as well. Once more, all possible *nifH* mRNAs were targeted with the universal *nifH* primers from Zehr and McReynolds (1989), plus a universal *nifH* primer for reverse transcription, which targets *nifH* mRNA of distantly related diazotrophs (Hurek *et al.*, unpublished). Comparative sequence analyses of six randomly selected clones showed that all clones carried inserts, whose sequences were almost identical (99.7 % nucleotide identity) to the *nifH* gene from *Azoarcus* sp. BH72. This result could be confirmed by hybridization analysis too. Compared to strain BH72, the level of BNF by all other microorganisms is presumably only small, because *nifH* transcripts from other bacteria were not abundant. Consistent with these results, 16S DNA fragments that were identical to a corresponding sequence stretch from strain BH72 could be retrieved from roots of uninoculated Kallar grass plants. The consensus of all these studies is that *Azoarcus* sp. BH72 is a constant member of the natural microbial community and is probably responsible for BNF with Kallar grass at the monoculture stand in the Punjab of Pakistan.

8.3. NifH Expression Equivalents of Diazotrophs in the Rhizosphere of Gramineous Plants come close to the Amount of Bacteroids in the Rhizobium-Legume Symbiosis

Additional experimental evidence for the above conclusion comes from quantification of *nifH* transcription. In order to be able to compare *nifH* expression of *Azoarcus* sp. BH72 in both inoculated and un-inoculated Kallar grass and to evaluate the expression level, a most-probable-number (MPN) RT-PCR was carried out (Figure 5). MPN-PCR assays have been widely used to measure DNA in soil, sediment, and laboratory experiments (Picard *et al.*, 1992; Degrange and Bardin, 1995; Murakami *et al.*, 1997; Michotey *et al.* 2000; Fredslund *et al.*, 2001).

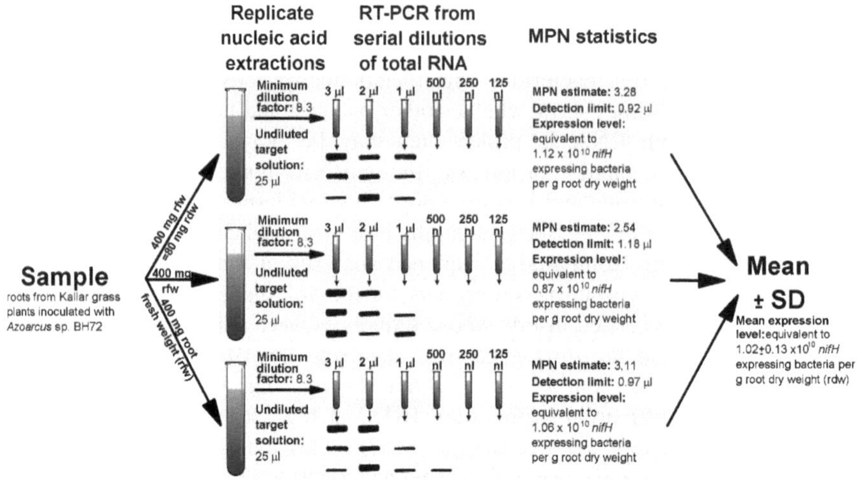

Figure 5. Graphically represented example of a most-probable-number reverse transcription PCR (MPN-RT- PCR) with replicate samplings according to Hurek et al., (2002). A mean estimate ± standard deviation of the level of nifH *expression in the rhizosphere of inoculated Kallar grass was calculated relative to the MPN-RT-PCR standard, which is sufficiently precise for most ecological and applied purposes. The expression level was calculated by: Dilution factor at detection limit x Standard x Correction for dry weight; the dilution factor of the target solution at the detection limit is given by: Undiluted target solution x Detection limit* $^{-1}$*; the detection limit is given by: Undiluted target solution x Minimum dilution factor* $^{-1}$ *x MPN estimate* $^{-1}$*. The standard is the detection limit of the method and corresponds to the number of* N_2*-fixing cells of* Azoarcus *sp. BH72 (3.3 x* 10^7*), which can be detected.*

Apart from the current study, MPN-PCR assays have been also applied recently to quantify virus-RNA in shellfish (Le Guyader *et al.*, 2003). MPN-PCR assays, like probably all quantitative PCR techniques, are not suitable for absolute quantification in an environmental context (Ferre, 1992; Rongson and Liren, 1997; Chandler, 1998). MPN-PCR procedures provide a conservative estimate which is limited by the efficiency of nucleic-acid extraction, target dilution, and presence of PCR inhibitors, which co-purify with nucleic acids (Chandler, 1998). In comparison to quantitative competitive PCR (cPCR), MPN-PCR techniques are much more robust and less subject to violation of underlying assumptions. Quantification by MPN-PCR does not rely upon addition of an internal standard. A problem, like a preferential amplification of target over the standard template, which would impair an enumeration in cPCR (Lee *et al.*, 1996), does not exist in a MPN-PCR assay. Also, the MPN-PCR does not have to be run within the exponential phase of amplification to provide a quantifiable estimate of the target concentration as in either cPCR (Diaco, 1995) or real-time quantitative PCR. Rather, quantification by MPN-PCR depends on a simple all-or-none determination at the endpoint of PCR when the terminal plateau has been reached. Consequently, this technique is capable to handle a relatively wide range of different amplification efficiencies without compromising the enumeration of the targets in solution (Sykes and Morley, 1995). It is likely to outperform cPCR and real-time quantitative PCR, when target concentrations are low, the molecular sampling error is high, and the PCR is inhibited by environmental contaminants, which occur when targets are extracted from soil, sediment, or plants, *etc.* Therefore, MPN-PCR appears to be suitable for quantification of *nifH* expression in the environment and was used here.

As mentioned above, the level of BNF in the inoculated set-up was only one-third or less than that of the un-inoculated system, *i.e.*, N_2 fixation by *Azoarcus* sp. BH72 probably contributed 34 kg N ha^{-1} $year^{-1}$ to inoculated Kallar grass plants but more than 100 kg N ha^{-1} $year^{-1}$ to the natural monoculture stand in Pakistan. Therefore, *nifH* expression should also differ a factor of >3 in both situations. Moreover, the level of *nifH* expression should be far above background in both cases. Because un-inoculated plants had been not sampled directly in the field, but rather after cultivation for 3 months in a phytotron, a factor of >3 is probably an overestimate. It would have been astonishing if the full N_2-fixation potential of *Azoarcus* sp. BH72 in the natural Kallar grass stand in Pakistan could have been maintained in the phytotron without time-consuming optimization of environmental parameters. Not unexpected, the difference found in *nifH* mRNA transcription was only 2-fold, but more surprising were the levels of iron-protein gene expression. The relative amount of *nifH* transcripts per g root dry weight in these un-inoculated and inoculated plants corresponded to the amount of *nifH* transcripts in 2×10^{10} and 1×10^{10} N_2-fixing cells of strain BH72, respectively (see Figure 5). These values were not only far above background but, based on data provided by Bergersen (1987) and Bergersen *et al.* (1985), also indicate a close approach to the number of soybean bacteroids in the *Rhizobium*-legume symbiosis, indicating a very high potential for *Azoarcus* sp. BH72-dependent nitrogen fixation in Kallar grass. Bergersen (1987) calculated that field-grown soybean crops, which fixed either 3.8 or 6.0 kg N ha^{-1} d^{-1}, had 100 and 165 mg dry mass of bacteroids per plant, respectively. Using an estimated dry weight of an *E. coli* cell as 83-1172 fg (Loferer-Krößbacher *et al.*, 1998), these amounts of nodule endosymbiont translate into 2×10^{13} to 8.5×10^{11} bacteroids per plant. With the dry weight of roots ranging from 2.3 to 8.9 g per soybean plant (Bergersen *et al.*, 1985), there must be between 9×10^{12} to 1×10^{11} bacteroids per g root dry weight.

9. SUMMARY AND OUTLOOK

Direct targeting of *nifH* mRNA is the tool of choice to identify primary nitrogen-fixing organisms in habitats where major diazotrophs are inconspicuous. The very high *nifH*-transcript levels detected in preparations of roots from Kallar grass confirm that, as has been discussed before, roots are probably the site of N_2-fixation in gramineous plants and suggest that nitrogen fixation is not carbon-limited there. These high levels are presumably under-estimates rather than over-estimates because MPN estimates are conservative and false negatives would have only increased the values. This approach also shows that, similar to other plant-microbe interactions, such as endomycorrhizae, very high microbial activities may be reached in roots by one particular microorganism without readily discernible changes of the root system like those that occur in nodule symbioses. Although in endomycorrhizae, a root can be clearly identified as either mycorrhizal or not after staining and microscopic examination, the endorhizospheres of Kallar grass and probably also of other gramineous plants continue to be habitats, where primary diazotrophs and sites of nitrogen fixation are well concealed. This situation is highly unsatisfactory. We calculate that roots of Kallar grass can accommodate 1 x 10^9 bacteria per g root dry weight (Hurek *et al.*, 2002). In combination with the observation that the *nifH*-expression level of bacteria inside the root can be more than a magnitude higher than in pure cultures, *nifH*-expression equivalents of 1-2 x 10^{10} can still be explained. But what about *nifH* transcription in roots of gramineous plants that is more than 1-2 orders of magnitude higher?

By comparing nitrogenase-gene transcription in two different natural environments (Hurek *et al.*, 2002) and more recently in many others, including swampy and dry grasslands, we found that un-culturability of the actively nitrogen-fixing microflora associated with grasses is the rule rather than the exception. Consistent with this conclusion, in a survey of N_2-fixing bacteria in the rhizosphere of *Spartina alterniflora*, no *nifH* transcripts were retrieved which could be assigned to a cultivated bacterium (Brown *et al.*, 2003). We also realized that the level of nitrogenase-gene transcription can be much higher than previously observed with Kallar grass. However, *nifH*-gene expression by *Azoarcus* sp. BH72 was below detection level in all samples examined, probably because the Kallar grass field in Pakistan constitutes a unique habitat and the populations of diazotrophic endophytes associated with grasses may be highly dynamic (Figure 6).

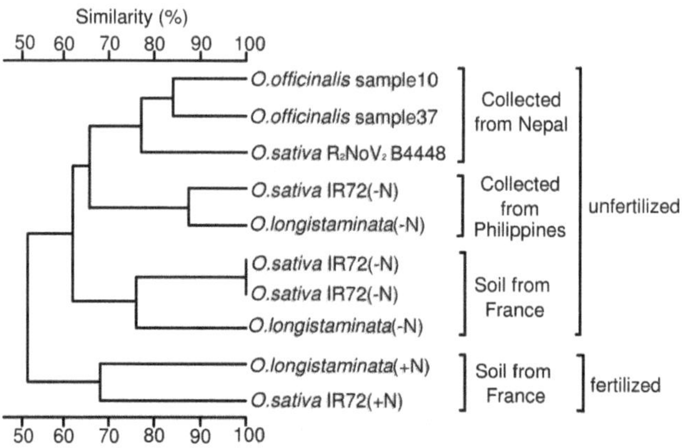

Figure 6. Cluster analysis by UPGMA (unweighted pair group with mathematical averages) method of T-RFLP patterns of nifH*-gene fragments amplified from different root samples of* Oryza *species grown at different places and fertilized (+N) or not fertilized (-N) with combined nitrogen. Picture taken from Tan et al. (2003).*

Instead, we found a phylogenetically well-supported group of, as yet, entirely uncultured microbes, which were detected in almost all samples (Figure 3; Hurek *et al.*, unpublished). This group was named the Ω group by us because sequences frequently occur and have been first detected in *Oryza* spp. (Hurek *et al.*, 2002).

Future research should focus on the population structures and sites of N_2-fixation in the most active nitrogen-fixing gramineous plant-microbe communities. In parallel, the microbes involved, probably diazotrophic grass endophytes, should be identified and their actual contribution of fixed nitrogen to the plant determined. Only by using this combination can we be certain that the high levels of *nif*-gene products in the vicinity of the plant do indeed indicate a high benefit of the plant from biological nitrogen fixation.

REFERENCES

Alfano, J. R., and Collmer, A. (1996). Bacterial pathogens in plants: Life up against the wall. *Plant Cell, 8*, 1683-1698.

Amann, R. I., Ludwig, W., and Schleifer, K. (1995). Phylogenetic identification and *in situ* detection of individual microbial cells without cultivation. *Microbial Rev., 9*, 143-169.

Armstrong, W. (1978). Root aeration in the wetland condition. In D. D. Hook and R. M. M. Crawford (Eds.), *Plant life in anaerobic environments* (pp. 269-298). Ann Arbor, Michigan: Ann Arbor Science.

Armstrong, W., and Boatman, D. J. (1967). Some field observations relating the growth of bog plants to conditions of soil aeration. *J. Ecol., 55*, 101-10.

Arsène, F., Katupitiya, S., Kennedy, I.R., and Elmerich, C. (1994). Use of *lacZ* fusions to study expression of *nif* genes of *Azospirillum brasilense* in association with plants. *Mol. Plant-Microbe Interact., 7*, 748-757.

Aßmus, B., Hutzler, P., Kirchhof, G., Amann, R., Lawrence, J. R., and Hartmann, A. (1995). *In situ* localization of *Azospirillum brasilense* in the rhizosphere of wheat with fluorescently labeled, rRNA-targeted oligonucleotide probes and scanning confocal laser microscopy. *Appl. Environ. Microbiol., 61*, 1013-1019.

Atkinson, M. M. (1993). Molecular mechanisms of pathogen recognition by plants. In J. H. Andrews and I. C. Tommerup (Eds.), *Advances in plant pathology* (pp. 35-64). Vol. 10. New York, NY: Academic Press, Inc.

Atkinson, M. M., and Baker, C. J. (1987a). Alteration of plasmalemma sucrose transport in *Phaseolus vulgaris* by *Pseudomonas syringae* pv. *syringae* and its association with K^+ / H^+ exchange. *Phytopathology, 77*, 1573-1578.

Atkinson, M. M., and Baker, C. J. (1987b). Association of host plasma membrane K^+ / H^+ exchange with multiplication of *Pseudomonas syringae* pv. *syringae* in *Phaseolus vulgaris. Phytopathology, 77*, 1273-1279.

Balandreau, J., and Knowles, R. (1978). The rhizosphere. In Y. R. Dommergues and S. V. Krupa (Eds.), *Interactions between non-pathogenic soil organisms and plants* (pp. 243-268). Amsterdam, The Netherlands: Elsevier/North-Holland Publishing Co.

Baldani, J. I., Baldani, V. L. D., Seldin, L., and Döbereiner, J. (1986). Characterization of *Herbaspirillum seropedicae* gen. nov., sp. nov., a root-associated nitrogen-fixing bacterium. *Int. J. Syst. Bacteriol., 36*, 86-93.

Baldani, J. I., Pot, B., Kirchhof, G., Falsen, E., Baldani, V. L. D., Olivares, F. L., Hoste, B., *et al.* (1996). Emended description of *Herbaspirillum*; inclusion of [*Pseudomonas*] rubrisubalbicans, a mild plant pathogen, as *Herbaspirillum rubrisubalbicans* comb. nov.; and classification of a group of clinical isolates (EF group 1) as *Herbaspirillum* species 3. *Int. J. Syst. Bacteriol., 46*, 802-810.

Bali, A., Blanco, G., Hill, S., and Kennedy, C. (1992). Excretion of ammonium by a *nifL* mutant of *Azotobacter vinelandii* fixing nitrogen. *Appl. Environ. Microbiol., 58*, 1711-1718.

Barraquio, W. L., Revilla, L., and Ladha, J. K. (1997). Isolation of endophytic diazotrophic bacteria from wetland rice. *Plant Soil, 194*, 15-24.

Bashan, Y., and Holguin, G. (1997). *Azospirillum*-plant relationships: Environmental and physiological advances (1990-1996). *Can. J. Microbiol., 43*, 103-121.

Begg, C. B. M., Kirk, G. J. D., Mackenzie, A. F., and Neue, H.-U., (1994). Root-induced iron oxidation and pH changes in the lowland rice rhizosphere. *New Phytologist, 128*, 469-477.

Beijerinck, M. W. (1888). Die Bakterien der Papilionaceen-Knöllchen. *Botanische Zeitung, 46*, 725-735, 741-750, 757-771, 781-790, 797-804.

Bergersen, F. J., Turner, G. L., Gault, R. F., Chase, D. L., and Brockwell, J. (1985). The natural abundance of ^{15}N in an irrigated soybean criop and its use for calculation of nitrogen fixation. *Austr. J. Agric. Res., 36*, 411-423.

Bergersen, F. G. (1987). Concluding remarks. *Phil. Trans. R. Soc. Lond., B, 317*, 295-297.

Bishop, P. E., and Premakumar, R. (1992). Alternative nitrogen fixation systems. In G. Stacey, R. H. Burris, H. J. Evans (Eds.), *Biological nitrogen fixation* (pp. 736–762). New York, NY: Chapman and Hall.

Blanco, G., Drummond, M., Woodley, P., and Kennedy, C. (1993). Sequence and molecular analysis of the *nifL* gene of *Azotobacter vinelandii. Mol. Microbiol., 9*, 869-879.

Boddey, R. M. (1995). Biological nitrogen fixation in sugar cane: a key to energetically viable biofuel production. *Crit. Rev. Plant Sci. 14*, 263-279.

Boddey, R. M., De Oliveira, O. C., Urquiaga, S., Reis, V. M., De Olivares, F. L., Baldani, V. L. D., and Döbereiner, J. (1995). Biological nitrogen fixation associated with sugar cane and rice: Contributions and prospects for improvement. *Plant Soil, 174*, 195-209.

Brockwell, J. P., Bottomley J., and Thies, J. E. (1995). Manipulation of rhizobia microflora for improving legume productivity and soil fertility: A critical assessment. *Plant Soil, 174*, 143-180.

Brown, M. M., Friez, M. J., and Lovell, C. R. (2003). Expression of *nifH* genes by diazotrophic bacteria in the rhizosphere of short form *Spartina alterniflora. FEMS Microbiol. Ecol. 43*, 411-417.

Bürgmann H., Widmer F., Sigler, W. V., and Zeyer, J. (2003). mRNA extraction and reverse transcription-PCR protocol for detection of *nifH* gene expression by *Azotobacter vinelandii* in soil. *Appl. Environ. Microbiol., 69*, 1928-1935.

Caballero-Mellado, J., and Martinez-Romero, E. (1994). Limited genetic diversity in the endophytic sugarcane bacterium *Acetobacter diazotrophicus. Appl. Environ. Microbiol., 60*, 1532-1537.

Caballero-Mellado, J., Fuentes-Ramírez, L. E., Reis, V. M., and Martínez-Romero, E. (1995). Genetic structure of *Acetobacter diazotrophicus* populations and identification of a new genetically distant group. *Appl. Environ. Microbiol., 61*, 3008-3013.

Callaham, D., Del Tredici, P., and Torrey, J. G. (1978). Isolation and cultivation *in vitro* of the actinomycete causing root nodulation in *Comptonia. Science, 199*, 899-902.

Cassman, K. G, De Datta, S. K., Olk, D. C, Alcantara, J. M., Samson, M. I., Descalsota, J. P., and Dizon, M. A. (1995). Yield decline and the nitrogen conomy of long-term experiments on continuous, irrigated rice systems in the tropics. In R. Lal and B. A. Stewart (Eds.), *Soil management: Experimental basis for sustainability and environmental quality* (pp 181–222). Boca Raton, FL: Lewis/CRC Publishers.

Cavalcante, V. A., and Döbereiner, J. (1988). A new acid-tolerant nitrogen-fixing bacterium associated with sugarcane. *Plant Soil, 108*, 23-32.

Chandler, D. P. (1998). Redefining relativity: Qquantitative PCR at low template concentrations for industrial and environmental microbiology. *J. Industrial Microbiol. Biotechnol., 21*, 128-140.

Chelius, M. K., and Triplett, E. W. (2000). Immunolocalization of dinitrogenase reductase produced by *Klebsiella pneumoniae* in association with *Zea mays* L. *Appl. Environ. Microbiol., 66*, 783-787.

Chien, Y.-T., and Zinder, S. H. (1994). Cloning, DNA sequencing, and characterization of a *nifD*-homologous gene from the archaeon *Methanosarcina barkeri* 227 which resembles *nifD1* from the eubacterium *Clostridium pasteurianum. J. Bacteriol., 176*, 6590–6598.

Christiansen-Weniger, C., Groneman, A. F., and van Veen, J. A. (1992). Associative N_2 fixation and root exudation of organic acids from wheat cultivars of different aluminium tolerance. *Plant Soil, 139*, 167-174.

Dawe, D. (2000). The potential role of biological nitrogen fixation in meeting future demand for rice and fertilizer. In J. K. Ladha and P. M. Reddy (Eds.), *The quest for nitrogen fixation in rice* (pp. 1-9). Makati City, The Philippines: International Rice Research Institute.

Degrange, V., and Bardin, R. (1995). Detection and counting of *Nitrobacter* populations in soil by PCR. *Appl. Environ. Microbiol., 61*, 2093-2098.

DeLong, E. F., Schmidt, T. M., and Pace, N. R., (1989). Analysis of single cells and oligotrophic picoplancton populations using 16S rRNA sequences. In T. Hattori, Y. Ishida, Y. Maruyama, R. Y. Morita and A. Uchida (Eds.), *Recent advances in microbial ecology* (pp. 697-700). Tokyo, Japan: Japanese Science Society Press.

Diaco, R. (1995). Practical considerations for the design of quantitative PCR assays. In M. A. Innis, D. H. Gelfand and J. J. Sninsky (Eds.), *PCR strategies* (pp. 84-108). San Diego, CA: Academic Press, Inc.

Döbereiner, J. (1961). Nitrogen-fixing bacteria of the genus *Beijerinckia* Derx in the rhizosphere of sugar cane. *Plant Soil, 15*, 211-216.

Döbereiner, J., Reis, V. M., and Lazarini, A. C. (1988). New N_2 fixing bacteria in association with cereals and sugar cane. In H. Bothe, F. J. de Bruijn, and W. E. Newton (Eds.), *Nitrogen fixation hundred years after* (pp. 717-722). Stuttgart, Germany: Gustav Fischer.

Döbereiner, J., Reis, V. M., Paula, M. A., and Olivares, F. (1993). Endophytic diazotrophs in sugar cane, cereals and tuber plants. In R. Palacios, J. Mora, and W. E. Newton (Eds.), *New horizons in nitrogen fixation* (pp. 671-676). Dordrecht, The Netherlands: Kluwer Academic Publishers.

Dörr, J., Hurek, T., and Reinhold-Hurek, B. (1998). Type IV pili are involved in plant-microbe and fungus-microbe interactions. *Mol. Microbiol., 30*, 7-17.

Dong, Z., Canny, M. J., McCully, M. E., Roboredo, M. R., Cabadilla, C. F., Ortega, E., and Rodés. R. (1994). A nitrogen-fixing endophyte of sugarcane stems. A new role for the apoplast. *Plant Physiol., 105*, 1139-1147.

Douglas, A. E., (1995). The ecology of symbiotic microorganisms. *Adv. Ecol. Res., 26*, 69-103.

Egener, T., Hurek, T., and Reinhold-Hurek, B. (1999). Endophytic expression of *nif* genes of *Azoarcus* sp. strain BH72 in rice roots. *Mol. Plant-Microbe Interact., 12*, 813-819.

Egener, T., Martin, D. E., Sarkar, A., and Reinhold-Hurek, B. (2001). Role of a ferredoxin gene cotranscribed with the *nifHDK* operon in N_2 fixation and nitrogenase "switch-off" of *Azoarcus* sp. strain BH72. *J. Bacteriol., 183*, 3752-3760.

Engelhard, M., Hurek, T., and Reinhold-Hurek, B. (2000). Preferential occurrence of diazotrophic endophytes, *Azoarcus* spp., in wild rice species and land races of *Oryza sativa* in comparison with modern races. *Environ. Microbiol., 2*, 131-141.

Fálcon, L. I., Cipriano, F., Chistoserdov, A. Y., and Carpenter, E. J. (2002). Diversity of diazotrophic unicellular cyanobacteria in the tropical North Atlantic Ocean. *Appl. Environ. Microbiol., 68*, 5760-5764.

Ferre, F. (1992). Quantitative or semi-quantitative PCR: Reality versus myth. *PCR Methods Applications, 2*, 1-9.

Fredslund, L., Ekelund, F., Jacobsen, C. S., and Johnsen, K. (2001). Development and application of a Most-Probable-Number-PCR assay to quantify flagellate populations in soil samples. *Appl. Environ. Microbiol., 67*, 1613-1618.

Galloway, J. N., Schlesinger, W. H., Levy, H. I., Michaels, A. F., and Schnoor, J. L. (1995). Nitrogen fixation: Anthropogenic enhancement-environmental response. *Global Biogeochemical Cycles, 9*, 235-252.

Giller, K. E., and Day, J.M. (1985). Nitrogen fixation in the rhizosphere: Significance in natural and agricultural systems. In A. H. Fitter, D. Atkinson, D. J. Read, and M. Busher (Eds.), *Ecological interactions in soil,* (pp. 127-147). Oxford, UK: Blackwell Scientific Publications.

Gillis, M., Kersters, K., Hoste, B., Janssens, D., Kroppenstedt, R. M., Stephan, M. P., Teixeira, K. R. S., Döbereiner, J., and De Ley, J. (1989). *Acetobacter diazotrophicus* sp. nov., a nitrogen-fixing acetic acid bacterium associated with sugarcane. *Int. J. Syst. Bacteriol., 39*, 361-364.

Giovannoni, S. J., Britschgi, T. B., Moyer, C. L., and Field, K. G. (1990). Genetic diversity in Sagasso Sea bacterioplankton. *Nature, 345*, 60-63.

Le Guyader, F. S., Neill, F. H., Dubois, E., Bon, F., Loisy, F., Kohli, E., Pommepuy, M., and Atmar, R. L. (2003). A semiquantitative approach to estimate Norwalk-like virus contamination of oysters implicated in an outbreak. *Int. J. Food Microbiol., 87*, 107-112.

Gyaneshwar, P., James, E. K., Mathan, N., Reddy, P. M., Reinhold-Hurek, B., and Ladha, J. K. (2001). Endophytic Colonization of Rice by a Diazotrophic Strain of *Serratia marcescens*. *J. Bacteriol., 183*, 2634-2645.

Gyaneshwar, P., James, E. K., Reddy, P. M., and Ladha, J. K. (2002). *Herbaspirillum* colonization increases growth and nitrogen accumulation in aluminium-tolerant rice varieties. *New Phytol., 154*, 131-145.

Haahtela, K. Wartioraara, T., Sundmann, V., and Skujins, J. (1981). Root-associated N_2 fixation (acetylene reduction) by *Enterobacteriaceae* and *Azospirillum* strains in cold-climate spodosols. *Appl. Environ. Microbiol., 41*, 203-206.

Hallmann, J. (2001). Plant interactions with endophytic bacteria. In M. J. Jeger and N. J. Spence (Eds.), *Biotic interactions in plant-pathogen associations.* (pp. 87-119). Wallingford, UK: CABI Publishing.

Hamelin, J., Fromin, N., Tarnawski, S., Teyssier-Cuvelle, S., and Aragno, M. (2002). *NifH* gene diversity in the bacterial community associated with the rhizosphere of *Molinia coerulea*, an oligonitrophilic perennial grass. *Environ. Microbiol., 4*, 477–481.

He, S. Y., Bauer, D. W., Collmer, A., and Beer, S. V. (1994). Hypersensitive response elicited by *Erwinia amylovora* harpin requires active plant metabolism. *Mol. Plant-Microbe Interact., 7*, 289-292.

Hennecke, H., Kaluza, K., Thony, B., Fuhrmann, M., Ludwig, W., Stackebrandt, E., (1985). Concurrent evolution of nitrogenase genes and 16S rRNA in *Rhizobium* species and other nitrogen fixing bacteria. *Arch. Microbiol., 142*, 342-348.

Hiltner, L. (1899). Über die Assimilation des freien atmosphärischen Stickstoffs durch in oberirdischen Pflanzenteilen lebende Mycelien. *Zentralblatt für Bakteriologie und Parasitenkunde. Abteilung II., 5*, 831-837.

Howard, J. B., and Rees, D. C. (1996). Structural basis of biological nitrogen fixation. *Chem. Rev., 96*, 2965–2982.

Hurek, T., Reinhold-Hurek, B., Van Montagu, M., and Kellenberger, E. (1991). Infection of intact roots of Kallar grass and rice seedlings by *Azoarcus.* In M. Polsinelli, R. Materassi, and M. Vinzenzini (Eds.), *Nitrogen fixation.* (pp. 235-242). Dordrecht, The Netheralnds: Kluwer Academic Publishers.

Hurek, T., Reinhold-Hurek, B., Van Montagu, M. and Kellenberger, E. (1994). Root colonization and systemic spreading of *Azoarcus* sp. strain BH72 in grasses. *J. Bacteriol., 176*, 1913-1923.

Hurek, T., Van Montagu, M., Kellenberger, E., and Reinhold-Hurek, B. (1995). Induction of complex intracytoplasmic membranes related to nitrogen fixation in *Azoarcus* sp. BH72. *Mol. Microbiol., 18*, 225-236.

Hurek, T., Wagner, B., and Reinhold-Hurek, B. (1997a). Identification of N_2-fixing plant- and fungus-associated *Azoarcus* species by PCR-based genomic fingerprints. *Appl. Environ. Microbiol., 63*, 4331-4339.

Hurek, T., Egener, T., and Reinhold-Hurek, B. (1997b). Divergence in nitrogenases of *Azoarcus* spp., *Proteobacteria* of the ß-subclass. *J. Bacteriol., 179*, 4172-4178.

Hurek, T., Handley, L., Reinhold-Hurek, B., and Piché, Y. (2002). *Azoarcus* grass endophytes contribute fixed nitrogen to the plant in an unculturable state. *Mol. Plant-Microbe Interact., 15*, 233-242.

Hurek, T., and Reinhold-Hurek, B. (1998). Interactions of *Azoarcus* sp. with rhizosphere fungi. In A. Varma and B. Hock (Eds.), *Mycorrhiza*, 2nd Edit. (pp. 595-614). Berlin, Germany: Springer Verlag.

Hurek, T., and Reinhold-Hurek, B. (2003). *Azoarcus* sp. strain BH72 as a model for nitrogen-fixing grass endophytes. *J. Biotechnol.*, in press.

James, E. K. (2000). Nitrogen fixation in endophytic and associative symbiosis. *Field Crops Res., 65*, 197-199.

James, E. K., Olivares, F. L., Baldani, J. I., and Döbereiner J. (1997). *Herbaspirillum*, an endophytic diazotroph colonizing vascular tissue in leaves of *Sorghum bicolor* L. Moench. *J. Exp. Bot., 48*, 785–797.

James, E. K., and Olivares, F. L. (1998). Infection and colonization of sugar cane and other graminaceous plants by endophytic diazotrophs. *Crit. Rev. Plant Sci., 17*, 77-119.

James, E. K., Olivares, F. L., de Oliveira, A. L. M., dos Reis, Jr., F. B., da Silva, L. G., and Reis, V. M. (2001). Further observations on the interaction between sugar cane and *Gluconacetobacter diazotrophicus* under laboratory and greenhouse conditions. *J. Exp. Bot., 52*, 747-760.

James, E. K., Prasad, G., Mathan, N., Barraquio, W. L., Reddy, P. M., Iannetta, P. P. M., Olivares, F. L., and Ladha, J. K., (2002). Infection and colonization of rice seedlings by the plant growth-promoting bacterium *Herbaspirillum seropedicae* Z67. *Mol. Plant-Microbe Interact., 15*, 894-906.

Karg, T., and B. Reinhold-Hurek (1996). Global changes in protein composition of N_2-fixing *Azoarcus* sp. strain BH72 upon diazosome formation. *J. Bacteriol., 178*, 5748-5754.

Karl, D., Bergman, B., Capone, D., Carpenter, E., Letelier, R., Lipschultz, F., Paerl, H., Sigman, D., and Stal, L. (2002). Dinitrogen fixation in the world's oceans. *Biogeochemistry, 57/58*, 47-98.

Kowalchuk, G. A., Naoumenko, Z. S., Derikx, P. J. L., Felske, A., Stephen, J. R., and Arkhipchenko, I. A. (1999). Molecular analysis of ammonia-oxidizing bacteria of the beta subdivision of the class *Proteobacteria* in compost and composted materials. *Appl. Environ. Microbiol., 65*, 396-403.

Lee, S. Y., Bollinger, J., Bezdicek, D., and Ogram, A. (1996). Estimation of the abundance of an uncultured soil bacterial strain by a competitive quantitative PCR method. *Appl. Environ. Microbiol., 62*, 3787-3793.

Li, R., and MacRae, I. C. (1992). Specfic identification and enumeration of *Acetobacter diazotrophicus* in sugarcane. *Soil Biol. Biochem., 24*, 413-419.

Lilburn,T. G., Kim, K. S., Ostrom, N. E., Byzek, K. R., Leadbetter, J. R., and Breznak,J. A. (2001). Nitrogen fixation by symbiotic and free-living spirochetes. *Science, 292*, 2495-2498.

Lima, E., Boddey, R. M., and Döbereiner, J. (1987). Quantification of biological nitrogen fixation associated with sugar cane using a ^{15}N-aided nitrogen balance. *Soil Biol. Biochem., 19*, 165-170.

Loferer-Krößbacher, M., Klima, J., and Psenner, R. (1998). Determination of bacterial cell dry mass by transmission electron microscopy and densitometric image analysis. *Appl. Environ. Microbiol., 64*, 688-694.

Lovell, C. R., Piceno, Y. M., Quattro, J. M., and Bagwell, C. E. (2000). Molecular analysis of diazotrophic diversity in the rhizosphere of the smooth cordgrass, *Spartina alterniflora. Appl. Environ. Microbiol., 66*, 3814-3822.

Malik, K. A., Aslam, Z., and Naqvi, M., (1986). *Kallar grass: A plant for saline land*. Faisalabad, Pakistan: Nuclear Institute for Agriculture and Biology.

Merrick, M. J. (1992). Regulation of nitrogen fixation genes in free-living and symbiotic bacteria. In G. Stacey, R. H. Burris, and H. J. Evans (Eds.), *Biological nitrogen fixation*. (pp. 835–876). New York, NY: Chapman and Hall.

Merrick, M. and Edwards, R. (1995). Nitrogen control in bacteria. *Microb. Rev., 59*, 604-622.

McClung, C. R., Van Berkum, P., Davis, R. E., and Sloger, C. (1983). Enumeration and localization of N_2-fixing bacteria associated with roots of *Spartina alterniflora* Loisel. *Appl. Environ. Microbiol., 54*, 1914-1920.

McCully, M. E. (2001). Niches for bacterial endophytes in crop plants: A plant biologist's view. *Austr. J. Plant Physiol., 28*, 983-990.

McLennan, E. (1920). The endophytic fungus of *Lolium*, Part I *Proceedings of the Royal Society of Victoria, 32*, 252-285.

Michotey, V., Méjean, V., and Bonin, P. (2000). Comparison of methods for quantification of cytochrome cd_1-denitrifying bacteria in environmental marine samples. *Appl. Environ. Microbiol., 66*, 1564-1571.

Moeseneder, M. M., Arrieta, J. M., Muyzer, G., Winter, C., and Herndl G. J. (1999). Optimization of terminal-restriction fragment length polymorphism analysis for complex marine bacterioplankton communities and comparison with denaturing gradient gel electrophoresis. *Appl. Environ. Microbiol., 65*, 3518-3525.

Mosier, A. R., Bleken, M. A., Chaiwanakupt, P., Ellis, E. C., Freney, J. R., Howarth, R. B., Matson, P. A., Minami, K., Naylor, R., Weeks, K. N., and Zhu, Z. L.(2002). Policy implications of human-accelerated nitrogen cycling. *Biogeochemistry, 57/58*, 477–516.

Murakami, K., Kanzaki, K., Okada, K., Matsumoto, S., and Oyaizu, H. (1997). Biological control of *Rhizoctonia solani* AG2-2 IIIB on creeping bentgrass using an antifungal *Pseudomonas fluorescens* HP72 and its monitoring in fields. *Ann. Phytopathol. Soc. Japan, 63*, 437-444.

Van Nieuwenhove, C., van Holm, L., Kulasooriya, S. A., and Vlassak, K. (2000). Establishment of *Azorhizobium caulinodans* in the rhizosphere of wetland rice (*Oryza sativa* L.). *Biol. Fertil. Soils, 31*, 143-149.

Noda, S., Ohkuma, M., Usami, R., Horikoshi, K., and Kudo, T. (1999). Culture-independent characterization of a gene responsible for nitrogen fixation in the symbiotic microbial community in the gut of the termite *Neotermes koshunensis*. *Appl. Environ. Microbiol., 65*, 4935-4942.

Noda, S., Ohkuma, M., Yamada, A., Honhoh, Y., and Kudo, T., (2003). Phylogenetic position and *in situ* identification of ectosymbiotic spirochetes on protists in the termite gut. *Appl. Environ. Microbiol., 65*, 625-633.

Normand, P., and Bousquet, J. (1989). Phylogeny of nitrogenase sequences in *Frankia* and other nitrogen-fixing microorganisms. *J. Mol. Evol. 29*, 436-447.
Ohkuma, M., and Kudo, T. (1996). Phylogenetic divsersity of the intestinal bacterial community in the termite *Reticulitermes speratus*. *Appl. Environ. Microbiol., 62*, 461-468.
Okon, Y., and Labandera-Gonzalez, C. A. (1994). Agronomic applications of *Azospirillum* - an evaluation of 20 years worldwide field inoculation. *Soil Biol. Biochem., 26*, 1591-1601.
Olivares, F. L., Baldani, V. L. D., Reis, V. M., Baldani, J. I., and Döbereiner, J. (1996). Occurrence of the endophytic diazotrophs *Herbaspirillum* spp. in roots, stems and leaves predominantly of Gramineae. *Biol. Fertil. Soils, 21*, 197–200.
Olivares, F. L., James, E. K., Baldani, J. I., and Döbereiner J. (1997). Infection of mottled stripe disease susceptible and resistant varieties of sugar cane by the endophytic diazotroph *Herbaspirillum*. *New Phytol., 135*, 723–737.
Patriquin, D. G. Döbereiner, J., and Jain, D. K. (1983). Sites and processes of association between diazotrophs and grasses. *Can. J. Microbiol., 29*, 900-915.
Peoples , M. B., Herridge, D. F., and Ladha, J. K. (1995). Biological nitrogen fixation: An efficient source of nitrogen for sustainable agricultural production? *Plant Soil, 174*, 3-28.
Picard, C., Ponsonnet, C., Paget, E., Nesme, X., and Simonet, P. (1992). Detection and enumeration of bacteria in soil by direct DNA extraction and polymerase chain reaction. *Appl. Environ. Microbiol., 58*, 2717-2722.
Piceno, Y. M., Noble, P. A., and Lovell, C. R. (1999). Spatial and temporal assessment of diazotroph assemblage composition in vegetated salt marsh sediments using denaturing gradient gel electrophoresis analysis. *Microb. Ecol., 38*, 157-167.
Poly, F., Ranjard, L., Nazaret, S., Gourbière, F., and Monrozier L. J. (2001). Comparison of gene pools in soils and soil microenvironments with contrasting properties. *Appl. Environ. Microbiol., 67*, 2255-2262.
Quispel, A. (1992). A search of signals in endophytic microorganisms. In D. P. S. Verma (Ed.), *Molecular signals in plant-microbe communications.* (pp. 471-491). Boca Raton, FL: CRC Press.
Reinhold, B., (1988). Systemanalyse einer natürlichen Assoziation von diazotrophen Bakterien mit Gramineen am Beispiel von *Leptochloa fusca* (L.) Kunth. Dissertation, University of Hannover, Germany.
Reinhold, B., Hurek, T., Niemann, E.-G., and Fendrik., I. (1986). Close association of *Azospirillum* and diazotrophic rods with different root zones of Kallar grass. *Appl. Environ. Microbiol., 52,* 520-526.
Reinhold-Hurek, B., Hurek, T., Claeyssens, M., and Van Montagu, M. (1993a). Cloning, expression in *Escherichia coli*, and characterization of cellulolytic enzymes of *Azoarcus* sp., a root-invading diazotroph. *J. Bacteriol., 175*, 7056-7065.
Reinhold-Hurek, B., Hurek, T., Gillis, M., Hoste, B., Vancanneyt, M., Kersters, K., and De Ley, J. (1993b). *Azoarcus* gen. nov., a nitrogen-fixing proteobacteria associated with roots of Kallar grass (*Leptochloa fusca* (L.) Kunth) and description of two species *Azoarcus indigens* sp. nov. and *Azoarcus communis* sp. nov. *Int. J. Syst. Bacteriol., 43*, 574-584.
Reinhold-Hurek, B., and Hurek, T. (1998a). Interactions of gramineous plants with *Azoarcus* spp. and other diazotrophs: Identification, localization and perspectives to study their function. *Crit. Rev. Plant Sci. 17*, 29-54.
Reinhold-Hurek, B., and Hurek, T. (1998b). Life in grasses: Diazotrophic endophytes. *Trends in Microbiol., 6*, 139-144.
Reinhold-Hurek, B., and Hurek, T. (2000). Reassessment of the taxonomic structure of the diazotrophic genus *Azoarcus* sensu lato and description of three new genera and species, *Azovibrio restrictus* gen. nov., sp. nov., *Azospira oryzae* gen. nov., sp. nov, and *Azonexus funguphilus* gen. nov., sp. nov. *Int. J. Syst. Bacteriol., 50*, 649-659.
Rennie, R. J. (1980). Dinitrogen-fixing bacteria: Computer-assisted identification of soil isolates. *Can. J. Microbiol., 26*, 1275-1283.
Ribbe, M., Gadkari, D., and Meyer, O. (1997). N_2 fixation by *Streptomyces thermoautotrophicus* involves a molybdenum-dinitrogenase and a manganese-superoxide oxidoreductase that couple N_2 reduction to the oxidation of superoxide produced from O_2 by a molybdenum-CO dehydrogenase. *J. Biol. Chem. 42*, 26627-26633.
Rongson, S., and Liren, M. (1997). Review on quantitative PCR assay. *J. Radioanal. Nucl. Chem., 221*, 137-142.

Ruschel, A. P., and Vose, P. B. (1984). Biological nitrogen fixation in sugar cane. In N. S. Subba Rao (Ed.), *Current developments in biological nitrogen fixation* (pp. 219-235). London, UK: Edward Arnold (Publishers) Ltd.

Sarkar, N. (1997). Polyadenylation of mRNA in prokaryotes. *Ann. Rev. Biochem., 66*, 173-197.

Sykes, P. J. and Morley, A. A. (1995). Limiting dilution polymerase chain reaction. In J. W. Larrick and P. D. Siebert (Eds.), *Reverse transcriptase PCR.* (pp. 150-165). New York, NY: Ellis Horwood.

Sevilla, M., Burris, R. H., Gunapala, N., and Kennedy, C. (2001). Comparison of benefit to sugarcane plant growth and $^{15}N_2$ incorporation following inoculation of sterile plants with *Acetobacter diazotrophicus* wild-type and nif- mutant strains. *Mol. Plant-Microbrobe Interact., 14*, 358-366.

De Simone, O., Haase, K., Müller, E., Junk, W.J., Gonsior, G., and Schmidt, W. (2002). Impact of root morphology on metabolism and oxygen distribution in roots and rhizophere from two Central Amazon floodplain tree species. *Functional Plant Biology, 29*, 1025-1035.

Steenhoudt, O., and Vanderleyden, J. (2000). *Azospirillum*, a free-living nitrogen-fixing bacterium closely associated with grasses: Genetic, biochemical and ecological aspects. *FEMS Microbiol. Rev., 24*, 487-506.

Stierle, A., Strobel, G., and Stierle, D. (1993). Taxol and taxane production by *Taxomyces andreanae*, an endophytic fungus of Pacific yew. *Science, 260*, 214-216.

Tan, Z., Hurek, T., and Reinhold-Hurek, B. (2003). Effect of N-fertilization, plant genotype and environmental conditions on *nifH* gene pools in roots of rice. *Environ. Microbiol., 2*, 1009-1015.

Thompson, J. D., Gibson, T. J., Plewniak, F., Jeanmougin, F., and Higgins, D. G. (1997). The CLUSTAL_X windows interface: Flexible strategies for multiple sequence alignment aided by quality analysis tools. *Nucleic Acids Res., 25*, 4876-4882.

Ueda, T., Suga, Y., Yahiro, N., and Matsuguchi, T. (1995). Remarkable N_2-fixing bacterial diversity detected in rice roots by molecular evolutionary analysis of *nifH* gene sequences. *J. Bacteriol., 177*, 1414-1417.

Urquiaga, S., Cruz, K. H. S., and Boddey, R. M. (1992). Contribution of nitrogen fixation to sugar cane: Nitrogen-15 and nitrogen balance estimates. *Soil Sci. Soc. Am. J., 56*, 105-114.

Vande Broek, A.., Michiels, J., Van Gool, A., and Vanderleyden, J. (1993). Spatial-temporal colonization patterns of *Azospirillum brasilense* on the wheat root surface and expression of the bacterial *nifH* gene during association. *Mol. Plant-Microbe Interact., 6*, 592-600.

Van Veen, J. A., Merckx, R., and Van de Geijn, S. C. (1989). Plant and soil related controls of the flow of carbon from roots through the soil microbial biomass. *Plant Soil, 115*, 179–188.

Van Veen, J. A., Liljeroth, E., Lekkerkerk, L. J. A., and Van de Geijn, S. C. (1991). Carbon fluxes in plant-soil systems at elevated atmospheric CO_2. *Ecological Applications, 1*, 175–181.

Vitousek, P. M., Mooney, H. A., Lubchenco, J., and Melillo, J. M. (1997a). Human domination of earth's ecosystems. *Science, 277*, 494-499.

Vitousek, P. M., Aber, J. D., Howarth, R. H., Likens, G. E., Matson, P. A., Schindler, D. W., Schlesinger, W. H., and Tilman. D. G. (1997b). Technical report: Human alteration of the global nitrogen cycle: Sources and consequences. *Ecological Applications, 7*, 737-750.

Ward, D. M., Weller, R., Bateson, M. M. (1990). 16S rRNA sequences reveal numerous uncultured microorganisms in a natural community. *Nature, 345*, 63-65.

Whipps, J. M. (1984). Environmental factors affecting the loss of carbon from the roots of wheat and barley seedlings. *J. Exp. Bot., 35*, 767–773.

Widmer, F., Shaffer, B. T., Porteous, L. A., and Seidler, R. J. (1999). Analysis of *nifH* gene pool complexity in soil and litter at a Douglas fir forest site in the Oregon Cascade Mountain Range. *Appl. Environ. Microbiol., 65*, 374–380.

Yamada, Y., Hoshino, K., and Ishikawa, T. (1997). The phylogeny of acetic acid bacteria based on the partial sequences of 16S ribosomal RNA: The elevation of the subgenus *Gluconoacetobacter* to generic level. *Biosci. Biotechnol. Biochem., 61*, 1244-1251.

Yamada, Y., Hoshino, K.-I., and Shikawa, T. (1998). *Gluconacetobacter* nom. corrig. (*Gluconoacetobacter* [sic]). In validation of publication of new names and new combinations previously effectively published outside the IJSB, List no. 64. *Int. J. Syst. Bacteriol., 48*, 327-328.

Young, J. P. W. (1992). Phylogenetic classification of nitrogen-fixing organisms. In G. Stacey, R. H. Burris, and H. J. Evans (Eds.), *Biological nitrogen fixation.* (pp. 43-86). New Yok, NY: Routledge, Chapman and Hall, Inc.

Young, J. P. W., and Haukka, K. E. (1996). Diversity and phylogeny of rhizobia. *New Phytol., 133*, 87-94.

Zani, S., Mellon, M.T., Collier, J. L., and Zehr, J. P. (2000). Expression of *nifH* genes in natural microbial assemblages in Lake George, New York, detected by reverse transcriptase PCR. *Appl. Environ. Microbiol., 66*, 3119-3124.

Zehr, J. P., and McReynolds, L. A. (1989). Use of degenerate oligonucleotides for amplification of the *nifH* gene from the marine cyanobacterium *Trichodesmium* spp. *Appl. Environ. Microbiol., 55*, 2522–2526.

Zehr, J. P., Mellon, M. T., and Zani, S. (1998). New nitrogen-fixing microorganisms detected in oligotrophic oceans by amplification of nitrogenase (*nifH*) genes. *Appl. Environ. Microbiol., 64*, 3444-3450.

Zehr, J. P., Waterbury, J., Turner, P. J., Montoya, J., Omoregie, E., Steward, G. F., Hansen, A., and Karl, D. M. (2001). Unicellular Cyanobacteria fix N_2 in the subtropical North Pacific Ocean. *Nature, 412*, 635-638.

Zehr, J. P, Jenkins, B. D., Short, S. M., and Steward, G. F. (2003). Nitrogenase gene diversity and microbial community structure: A cross-system comparison. *Environ. Microbiol., 5*, 539-554.

Chapter 10

INTERACTIONS OF ARBUSCULAR MYCORRHIZA AND NITROGEN-FIXING SYMBIOSIS IN SUSTAINABLE AGRICULTURE

J. M. BAREA[1], D. WERNER[2], C. AZCÓN-GUILAR[1], AND R. AZCÓN[1]

[1]*Departamento de Microbiología del Suelo y Sistemas Simbióticos, Estación Experimental del Zaidín, CSIC, Profesor Albareda 1, 18008-Granada, Spain; and* [2]*Fachbereich Biologie, Fachgebiet Zellbiologie und Angewandte Botanik, Philipps-Universität Marburg, D-35032 Marburg, Germany*

1. INTRODUCTION

Current public concerns about the side effects of agrochemicals on human health have led to research efforts being aimed at developing alternative agricultural strategies that would both optimise the rate of turnover and recycling of organic matter and plant nutrients and encourage the use of biological control agents, so reducing the use of chemical fertilizers and biocides. These efforts then directly lead to so-called sustainable developments in agriculture (Altieri, 1994). Sustainability-based agricultural practices emerge, therefore, as a response to social expectations with regard to the production of high-quality food and the preservation of the environment. Considering the fundamental role of soil microbes in nutrient cycling and their involvement in controlling certain plant diseases (Barea *et al.*, 1997), management of microbial activities and their effects on plant fitness and soil quality has been proposed as a strategy to help sustainable agricultural developments (Kennedy and Smith, 1995). Consequently, there is an increasing interest in improving our understanding of the diversity, dynamics, and significance of microbial populations in soil, and the mechanisms that need to be activated in the rhizosphere (Werner, 1998; Kennedy, 1998; Bowen and Rovira, 1999). This

D. Werner and W. E. Newton (eds.), Nitrogen Fixation in Agriculture, Forestry, Ecology, and the Environment, 199-222.

interest has facilitated the manipulation of selected microbial groups with regard to sustainability issues (Gianinazzi *et al.*, 2002).

Certain bacteria and fungi are known to play pre-eminent roles in sustainable agriculture. The main microbial types include both saprophytes and plant-symbionts, which may be either detrimental or beneficial to the soil-plant system (Barea, 1997). Detrimental microbes comprise the major plant pathogens as well as minor parasitic and non-parasitic deleterious rhizosphere organisms (Weller and Thomashow, 1994; Nehl *et al.,* 1996). Beneficial microorganisms are known to play fundamental roles in the sustainability of both agro-ecosystems and natural ecosystems, and some of them can be used as inoculants to benefit plant growth and health through either activating nutrient cycling or by controlling plant pathogens (Bowen and Rovira, 1999; Barea *et al.,* 2002a; 2002b). Saprophytic bacteria and fungi colonize below-ground plant surfaces, where a subset of the total rhizosphere bacterial community, termed "rhizobacteria", is known to display a specific ability for root colonization (Kloepper, 1994; 1996). The beneficial root-colonizing rhizosphere bacteria, the so-called "plant growth-promoting rhizobacteria" (PGPR), carry out important activities at the root/soil interfaces (Bashan, 1999; Dobbelaere *et al.*, 2001; Probanza *et al.*, 2002).

Several plant symbiotic bacteria and fungi, either pathogens or mutualistics, have been well studied. Arbuscular mycorrhizal (AM) fungi and some nitrogen-fixing (NF) bacteria, such as *Rhizobium* spp. and *Frankia* spp., are the main mutualistic microbial symbionts. AM fungi are known to play key roles in plant nutrition and health and in soil quality (Barea, 1991), whereas NF symbiotic bacteria, by cycling N to the biosphere from the atmosphere, represent a key input of fixed-N into plant productivity (Vance, 2001). Very important crop plants, *i.e.*, legumes species, are able to form dual symbiosis with both AM fungi and *Rhizobium* spp. (Barea and Azcón-Aguilar, 1983), a fact of great ecological and/or agricultural importance with regard to sustainability (Jeffries and Barea, 2001).

2. PURPOSE OF REVIEW

The co-existence of the AM and NF symbioses in most legume species of agricultural interest (including horticulture) has been the subject of many experimental studies (see Barea and Azcón-Aguilar, 1983; Mosse, 1986; Hayman, 1986; Azcón-Aguilar and Barea, 1992; and Barea *et al.,* 1992 as reference sources for the pioneering literature. Further developments on the ecology, physiology, biochemistry, molecular biology, and biotechnology of microbe-plant relationships have given new insights to the understanding of the formation and functioning of the AM and NF symbioses and their interactions in legumes. This situation, together with the current interest in legumes in the context of "sustainable agriculture", prompted us to analyse recent advances on this research area.

Many definitions of and conceptual approaches to "sustainable agriculture" have been given (see Barea and Jeffries (1995) for some examples). In sustainable systems, modifications of existing strategies, such as reduced agrochemical inputs, are proposed, so promoting the maximum use of beneficial soil microbiota. The formation and functioning of the tripartite legume-AM fungi-rhizobia will be discussed here, taking into account current sustainability-based principles in agricultural practices. It is generally accepted that sustainability-based approaches must also consider natural ecosystems, where the aims are to provide and/or maintain an appropriate vegetation cover, thus reducing the susceptibility of soils to erosion, while simultaneously preserving the structure and diversity of natural plant communities (Jeffries and Barea, 2001). Therefore, the significance of the interactions of the legumes-AM fungi-rhizobia symbioses in revegetation programs to restore degraded natural ecosystems will also be considered in this review.

However, before discussing the interactions between AM fungi and rhizobia, some brief comments on each of these symbiotic systems will be provided. Because this series of volumes is focussed on nitrogen fixation, detailed information concerning NF-symbioses is available elsewhere both in other Chapters of this volume and in other volumes of the series. Therefore, only a short description will be given for NF-related symbiosis. With regard to the AM associations, relevant current information will be condensed to achieve an up-to-date presentation of this subject and to create a conceptual background for non-specialist readers.

3. NITROGEN-FIXING SYMBIOSES

NF is a key factor in biological productivity; it is generally accepted that more than 60% of the N-input to plants has a biological origin. Half of this input is due to the activity of symbiotic plant-bacteria systems. Legume-rhizobia associations are the most important for the incorporation of N into pasture and agricultural systems. The significance of this symbiosis is strengthened by the growing interest in legumes for food production, forage, green manure, and horticulture (Postgate, 1998). Legumes also play a fundamental role in natural ecosystems (Jeffries and Barea, 2001). The Leguminosae (or Fabaceae) is one of the largest and possibly the most diverse families of angiosperms. Species range from small herbaceus plants to giant rainforest trees. Legumes have played an important role as an integral part of sustainable agriculture in the past and they are likely to become even more important in the future. Many of these plants are able to associate with bacteria to form nodules (usually on the roots) wherein atmospheric nitrogen is fixed. According to evolution studies (Parniske, 2000; Provorov *et al.,* 2002), it appears that the common ancestor of all these plants was not a nitrogen-fixing system but an AM-fungal association. The ability to nodulate probably evolved independently in this family and could even have been lost in several genera, when the availability of mineral nitrogen increased. The bacteria that are able to establish symbiotic relationship with legume species belong to the genera, *Rhizobium, Sinorhizobium, Bradyrhizobium, Mesorhizobium,* and *Azorhizobium*; they are often collectively termed as *Rhizobium* or rhizobia. They interact with legume roots leading to formation of N_2-fixing nodules (Spaink *et al.*, 1998). *Rhizobium* strains have also been described as colonizing the rhizosphere of non-legume hosts, where they establish positive interactions with AM fungi (Galleguillos *et al.*, 2000).

The agronomic significance of the symbiotic N_2 fixation by legume systems has been analysed through many experiments (as reviewed by Vance, 2001). The genetic and signalling programs in the rhizobia-legume symbiosis have also been reviewed recently (Long 2001) and several molecular approaches for diversity analysis of natural rhizobia populations have been developed (Flores *et al.*, 2000; Vinuesa *et al.*, 2000).

4. ARBUSCULAR MYCORRHIZA

Most plant species live in association with certain soil fungi, which colonize their roots by establishing what is known as *mycorrhiza*. The soil-borne mycorrhizal fungi, after the biotrophic colonization of the root cortex, develop an external mycelium, which is a bridge connecting the root with the surrounding soil microhabitats. The mycorrhizal (fungal-root) symbioses are critical for improving both plant fitness and soil quality through key ecological processes (Smith and Read, 1997; Van der Heijden and Sanders, 2002). The mycorrhizal associations are mutualistic symbioses because of the highly interdependent and mutually beneficial relationship established between both partners. The host plant receives mineral nutrients *via* the fungal mycelium (mycotrophism), whereas the heterotrophic fungus obtains carbon compounds from the host´s photosynthesis (Harley and Smith, 1983).

Mycorrhiza can be found in almost any kind of soil and ecosystems worldwide. All but a few vascular plant species belonging mainly to the families Cruciferae, Chenopodiaceae, Cyperaceae, Caryophyllaceae and Juncaceae are able to form mycorrhiza. The universality of this symbiosis implies a great diversity in the taxonomic features of the fungi and the plants involved. There are, in fact, considerable differences in the morphology of the mycorrhizal types and this is reflected in the resulting eco-physiological relationships. At least five types of mycorrhiza can be recognized, whose structural and functional features have been detailed elsewhere (see review by Smith and Read, 1997), thus, only a brief outline to differentiate these groups will be given here.

About 3% of the higher plants, mainly forest trees in the Fagaceae, Betulaceae, Pinaceae, *Eucalyptus,* and some woody legumes, form "*ectomycorrhiza*". The fungi involved are mostly higher Basidiomycetes and Ascomycetes, which colonize the cortical root tissues. A lack of intracellular penetration is characteristic. In general, the fungus develops a sheath or mantle around the feeder roots. Three other types of mycorrhiza can be grouped as *endomycorrhiza*, in which the fungus can colonize the root cortex intracellularly. One of these types is restricted to some species in the Ericaceae and is called "*ericoid*" mycorrhiza; the second type, called "*orchid*" mycorrhiza, is restricted to the Orchidaceae; and the third type, the "*arbuscular*" mycorrhiza, is by far the most widespread type. There is a fifth group, the "*ectendomycorrhiza*", that involves plant species in families other than Ericaceae within the Ericales and in the Monotropaceae. They have sheaths and produce intracellular penetrations ("arbutoid" and "monotropoid" mycorrhiza). Methods to study and manipulate mycorrhizal fungi have been described in a comprehensive and detailed book (Brundrett *et al.*, 1996).

Most of the major plant families are able to form mycorrhiza, with the arbuscular mycorrhizal (AM) associations being the most common mycorrhizal type involved in agricultural systems (Barea *et al.,* 1993). The AM fungi, which were formerly included in the order Glomales, Zygomycota (Morton and Benny, 1990; Morton *et al.*, 1995; Redecker *et al.*, 2000), have recently been moved to the new phylum Glomeromycota (Schüßler, 2001a and b). Identification of AM fungi has traditionally relied on the morphological and developmental characteristics of their large multinucleate spores (Morton and Benny, 1990). However, there is increasing evidence that morphology-based identification of AM fungi has a limited use in ecological settings because the low morphological diversity of AM spores may not reflect their physiological and genetic plasticity (Bachmann, 1998).

To circumvent this problem, biochemical and molecular approaches are being incorporated to define and relate taxa within AM fungi (Graham *et al.*, 1995; Madan *et al.*, 2002; Schüßler, 2001a). Several molecular techniques for the characterization of AM fungi have been applied (Redecker *et al.*, 1997a and b; Sanders *et al.*, 1995; Simon *et al.*, 1992; van Tuinen *et al.*, 1998) and show a considerable degree of variation among the copies of ribosomal DNA within single spores. However, recent reports indicate that small-subunit ribosomal-RNA (SSU rRNA) sequence analysis is a suitable tool to infer phylogenetic relationships of these AM fungi. It can be applied, for example, in diversity analysis of natural AM populations (Redecker *et al.*, 2000; Schüßler, 2001a; 2001b). The AM symbiosis evolved from the Ordovician-Devonian period (Redecker *et al.*, 2000) and the AM fungi are obligate symbionts, which are unable to complete their life cycle without colonizing a host plant (Azcón-Aguilar *et al.*, 1998).

During the process of AM formation (Giovannetti, 2000), the plant "accepts" the fungal colonization without any significant rejection reaction (Dumas-Gaudot *et al.*, 2000) and a series of root-fungus interactions lead to the integration of both organisms. The establishment of the symbiosis is the result of a continuous molecular 'dialogue' between plant and fungus through the exchange of both recognition and acceptance signals (Gianinazzi-Pearson *et al.*, 1996; Vierheilig and Piché, 2002). Cloning and identification of plant genes involved in AM formation and functioning, and the understanding of the signal-transduction pathways involved, are topics of current interest (Harrison *et al.*, 2000; Franken and Requena, 2001; Gollote *et al.*, 2002).

The AM symbiosis influences several aspects of plant physiology, such as plant rooting, closing of the nutrient cycles, nutrient acquisition, and plant protection (Kapulnik and Douds, 2000). The primary effect of the AM symbiosis is to increase the supply of mineral nutrients to the plant, particularly those nutrients whose ionic forms have a poor mobility rate or those present in low concentration in the soil solution. This situation mainly concerns phosphate, ammonium, zinc and copper (Barea, 1991). The processes of nutrient transport in AM systems have been reviewed recently (Smith *et al.*, 2001; Ferrol *et al.*, 2000). It has also been recognized that AM colonization affects a wide range of morphological parameters in developing root systems, with greater root branching as the most commonly described effect (Atkinson *et al.*, 1994; Berta *et al.*, 2002). AM formation changes, both quantitatively and qualitatively, the microbial populations in the rhizosphere (Azcón-Aguilar and Barea, 1992) and establishes the so-called mycorrhizosphere (Barea *et al.*, 2002b). As Bianciotto *et al.* (2001) reported, rhizobia can adhere to the hyphae of the AM fungi, where extra-cellular polysaccharides appear to be involved in the attachment of these bacteria, and then use the hyphae as a vehicle for root colonization.

In cooperation with other soil organisms, the external AM mycelium forms water-stable aggregates necessary for good soil tilth (Miller and Jastrow, 2000; Jeffries and Barea, 2001; Jeffries *et al.*, 2002). The AM associations also improve plant health through increased protection against biotic and abiotic stresses, indicating possible applications in both biocontrol of plant soil-borne microbial pathogens and in bioremediation of polluted soils (Bethlenfalvay and Linderman, 1992; Gianinazzi and Schüepp, 1994; Gianinazzi *et al.,* 2002).

It is important to point out that the enhancement of root resistence/tolerance to pathogen attack in AM plants is not exerted with the same effectivity by all AM fungi, is not applicable to all pathogens, and is not expressed in all substrata or in all environmental conditions. Nevertheless, there are examples which demonstrate that prior colonization by selected AM fungi protects plants against pathogenic fungi, such as *Phytophthora, Gaeumanomyces, Fusarium, Thielaviopsis, Pythium, Rhizoctonia, Sclerotium, Verticillium* and *Aphanomyces*, or nematodes, such as *Rotylenchus, Pratylenchus* and *Meloidogyne* (Linderman, 1994; Azcón-Aguilar and Barea, 1996; Azcón-Aguilar *et al.*, 2002; Pozo *et al.*, 2002; Werner *et al.*, 2002). The mechanisms that have been suggested to explain the protective action of AM symbiosis include: the improvement of plant vigor; damage compensation; competition with the pathogen for photosynthates or for colonization/infection sites; induction of changes in the morphology/anatomy of the root system; induction of changes in mycorrhizosphere populations; and activation of plant-defence mechanisms (Azcón-Aguilar *et al.*, 2002; Pozo *et al.*, 2002).

AM fungi also enable plants to cope with abiotic stress by alleviating nutrient deficiencies, improving drought tolerance, overcoming the detrimental effects of salinity, and enhancing tolerance to pollution by organic compounds or heavy metals (Augé, 2001; Miller and Jastrow, 2000; Jeffries *et al.*, 2002; Turnau and Haselwandter, 2002; Leyval *et al.*, 2002). AM fungi are known to improve the adaptation of sterile micro-propagated plants to un-sterile substrata and to field conditions (Azcón-Aguilar and Barea, 1997; Vestberg *et al.*, 2002). The interactions of AM fungi and other microbial components of rhizosphere (or mycorrhizosphere) populations have been studied in many experimental trials because of the repercussions for plant nutrition and health (Azcón-Aguilar and Barea, 1992; Barea, 1997; 2000; Barea *et al.*, 1997; 2002a and b).

Because the AM symbiosis can benefit plant growth and health, there is an increasing interest in ascertaining its effectiveness in particular plant-production systems and, consequently, in manipulating them, when feasible, so that they can be incorporated into production practices. Evidence is accumulating to show that indigenous and/or introduced AM fungi can benefit annual crops, such as cereals and legumes, vegetable crops, temperate fruit trees or shrubs, tropical plantation crops, ornamentals, spices, *etc.* (Azcón-Aguilar and Barea, 1997). Selection of the appropriate AM fungi (Estaún *et al.,* 2002), the production of quality inocula (von Alten et al. 2002), and the analysis of the ecological considerations of AM inoculation (Vosatka and Dodd, 2002; Feldmann and Grotkass, 2002) are critical issues for the application of mycorrhizal technology in agriculture.

5. INTERACTIONS BETWEEN AM FUNGI AND RHIZOBIA TO IMPROVE LEGUME PRODUCTIVITY IN AGRICULTURE

A century ago Janse (1896) was the first to report the coexistence of certain bacteria and fungi colonizing the root system of legume plants. The bacteria produce nodular structures on the root and the fungi establish a type of "root infection" that was later recognized as mycorrhizal development (Jones, 1924). Further, Asai (1944) stated that root nodulation by the bacteria can be dependent on the formation of mycorrhiza. Nowadays, both the widespread presence of the AM symbiosis in nodulated legumes and the role of AM in improving both nodulation and *Rhizobium* activity within the nodules are universally recognized (Vance, 2001; Provorov *et al.*, 2002; Barea *et al.*, 2002a; 2002b) and a great deal of work has been carried out on the tripartite symbiosis among legume-mycorrhiza-rhizobia. Relevant current information that has arisen since our previous reviews (Barea *et al.,* 1992; Barea *et al.,* 1997), will be summarized here by discussing: (1) the fundamental studies of the formation and functioning of the tripartite symbiosis; (2) strategic studies related to agricultural systems; (3) strategic studies related to the revegetation of degraded ecosystems; (4) biotechnological developments for integrated management; and (5) future trends for this research area.

5.1. The Formation and Functioning of the Tripartite Symbiosis

With the exception of *Azorhizobium caulinodans*, which also forms nodules on stems, the rhizobia form root nodules and coexist with AM fungi either as root symbionts or as free-living rhizosphere microorganisms. Coexistence commonly means interactions and these actually occur at the level of colonization and/or at the functionality (nutritional) level (Barea and Azcón-Aguilar, 1983). As Hayman (1986) and Mosse (1986) described, because of the relatively high P demand for nodule formation, it is obvious that a major benefit of AM on the symbiotic role of *Rhizobium* must be the P supplied by the fungus. However, nutrients other than P, such as Zn, Cu, Mo, Ca, *etc*., can affect both the infectivity and the symbiotic effectiveness of *Rhizobium.* Therefore, the enhanced uptake of these elements by the AM symbiosis may also be involved in the interactions. Conversely, fixed-N supplied by N_2 fixation by rhizobial activity, can be critical for maintaining a balanced physiological status in the plant, which in turn is important for AM formation and functioning. In addition, there is a high requirement for fixed-N by the AM fungi to synthesize chitin, the main constituent of its walls. Therefore, nodulation and AM formation appear to be mutually supportive.

Relevant information from studies carried out over the last years will be summarized under the following categories: (i) genetic and molecular aspects of AM fungi-rhizobia interactions; (ii) eco-physiological relationships in symbiosis formation and functioning; (iii) use of ^{15}N to ascertain the AM role in N_2 fixation by legume-rhizobia associations; (iv) studies involving genetically modified (GM) rhizobia; and (v) interactions involving other soil microorganisms.

5.1.1. Genetic and Molecular Aspects of AM Fungi-Rhizobia Interactions

Evolutionary timing and interaction patterns of microbe-plant symbioses, including both mutualistic, either N_2-fixing or mycorrhizal, and pathogenic associations, have suggested a common developmental program for all of these compatible microbe-plant associations (Parniske, 2000). A common ancestral plant-fungal interaction has been proposed and, because rhizobia-legume symbiosis appears much latter than AM associations (Sprent 1994; Redecker *et al.,* 2000; Provorov *et al.*, 2002), it has been hypothesised that the cellular and molecular events occurring during legume nodulation may have evolved from those already established in the AM symbiosis (Gianinazzi-Pearson, 1997). In fact, legume-rhizobia symbiosis seems to have evolved from a set of pre-adaptations during co-evolution with AM fungi (Provorov *et al.*, 2002). The possibility that some plant genes can modulate both types of legume symbiosis is a research field of current interest (Ruiz-Lozano *et al.*, 1999b). Recently, a plant receptor like kinase, required for both symbiosis, has been identified (Stracke *et al.*, 2002).

The use of mycorrhiza-defective legume mutants (Myc^-) has allowed critical advances in the understanding of common cellular and genetic programs for the legume-root symbioses (Gollotte *et al.,* 2002). Duc *et al.* (1989) first showed that the expression of the Myc^- character in model legumes was associated with that of the Nod^- character (non-nodulating plants), suggesting that the establishment of both types of symbiosis depends on the expression of some common gene(s) through which the plant controls early steps in the infection processes. Further, as reviewed by Gollote *et al.* (2002), work with a number of legume mutants, which are altered in their interaction with *Rhizobium*, indicated 12 symbiosis-related genes that possessed pleiotropic effects on both nodulation and AM formation. The expression of defence-related genes is also enhanced in Myc^- model legume mutants inoculated with either AM fungi or rhizobia (Gollotte *et al.*, 2002; Werner *et al.,* 2002). In particular, an accumulation of salicylic acid was described in the roots of these mutants when inoculated with either of their microbial symbionts (Blilou *et al.*, 1999; Ruiz-Lozano *et al.,* 1999a).

The use of Myc^- legumes has also contributed to a better understanding of the signalling processes involved in the formation of microbe-legume symbiosis. It has been suggested that both AM formation and nodulation share a common signal-transduction pathway (Hirsch and Kapulnik, 1998). Current advances in this area have been recently reviewed (Gollotte *et al.,* 2002).

5.1.2. Eco-physiological Relationships in Symbiosis Formation and Functioning

Between the pioneering work by Asai (1944) and recent publications (Vance 2001), a great number of reports detail the approaches made to elucidate the physiological and biochemical basis of AM fungi-*Rhizobium* interaction. In summary (see Barea *et al.,* 1992), it is clear the main cause for such interactions is the supply of P by the AM fungi to satisfy the high P demand of nodule formation. The role of AM fungi is a generalized influence on plant nutrition but more localized effects, as exerted either at the root, nodule, or bacteroid levels, have also been described. Because the nodules usually have 2- to 3-times more P than the root on which they are formed

(Mosse, 1986), it might be that nodules and rhizobial bacteroids actually have a "special demand" for P. This concept has been confirmed by time-course experiments, which showed that, early in their development, the nodules seem to have priority for P (as a consequence of their higher P-dependency) relative to other plant organs (Smith *et al.*, 1979; Asimi *et al.*, 1980).

In natural conditions, AM fungi and *Rhizobium* colonize the root almost simultaneously but the two endophytes do not seem to compete for infection sites. In certain cases, previous inoculation with one of the endophytes can depress the development of the other (Bethlenfalvay *et al.*, 1985). This has been mainly attributed to competition for carbohydrates when host photosynthesis is limited. When this occurs, AM fungi usually show a competitive advantage for carbohydrates over *Rhizobium* (Brown and Bethlenfalvay, 1988).

The energy cost of the tripartite symbiosis was investigated by using $^{14}CO_2$ (Pang and Paul, 1980; Kucey and Paul, 1982; Harris *et al.*, 1985; Ames and Bethlenfalvay, 1987). It was concluded that the CO_2-fixation rate, expressed as g C g^{-1} shoot dry matter h^{-1}, increased in the symbiotic plant. This is, in fact, a mechanism that enhances photosynthesis to compensate for the C cost of the symbioses.

Because AM colonization can help plants to cope with drought and salinity stresses (Augé 2001), the role of this symbiosis with legumes is particularly interesting. As Azcón *et al.* (1988) found, AM inoculation improved both nodulation and N_2 fixation at low levels of water potential. The negative effects of salinity on both nodulation and N_2 fixation can also be compensated for by AM (Ruiz-Lozano and Azcón, 1993).

More recent experiments have corroborated a positive effect of the interactions between AM fungi and nodulating rhizobia under drought conditions (Goicoechea *et al.*, 1997; 1998; 2000; Ruiz-Lozano *et al.*, 2001). AM inoculation was found not only to protect soybeans plants against the detrimental effects of drought, but also to help legume plants cope with the premature nodule senescense induced by drought stress (Ruiz-Lozano *et al.*, 2001). Alleviation of oxidative damage could be involved in AM protection against nodule senescense.

The impact of the application of pesticides on AM and nodulated legumes was investigated by Abd-Alla *et al.*, (2000). Several currently used pesticides negatively affected plant growth, AM formation, and nodulation, however, because non-inoculated controls were used, a possible AM effect on alleviating the negative effect of the pesticides could not be ascertained.

Other effects of AM fungi-rhizobia inoculation on eco-physiological aspects symbiotic development in legumes are discussed in Session 5.1.4 with reference to the use of GM rhizobia.

5.1.3. Use of ^{15}N to ascertain the Role of AM on N_2 Fixation by Legume-Rhizobia Associations

The addition of a small amount of ^{15}N-enriched inorganic fertilizer and an appropriate "non-fixing" reference crop is the basis for a direct method to distinguish among the relative contribution of the three fixed-N sources, *i.e.*, soil, atmosphere and fertilizer, for a fixing crop (Danso, 1988). Consequently, ^{15}N-based methodologies also offer the possibility of assessing the influence of any treatment on N_2 fixation distinct from soil or fertilizer N supply. Therefore, the isotope-based techniques have been used to measure the effect of AM inoculation on N_2 fixation. In fact, the use of the isotope ^{15}N made it possible to ascertain and quantify the amount of N which is actually fixed by legume-rhizobia consortia in a particular situation, and the contribution of the AM symbiosis to the process. A lower $^{15}N/^{14}N$ ratio in the shoots of *Rhizobium*-inoculated AM plants with respect to those achieved by the same *Rhizobium* strain in non-mycorrhizal plants was found. This result indicated an enhancement of the N_2-fixation rate (an increase in ^{14}N from the atmosphere) induced by the AM activity (Barea *et al.*, 1992). To corroborate this fiinding and to ascertain the AM role on N_2 fixation, the ^{15}N-based methodologies have been applied in both greenhouse and field studies (Azcón *et al.*, 1988; 1991; Barea *et al.*, 1997).

Isotopic techniques have been also used to measure N-transfer in mixed cropping systems, where legumes are usually involved (Zapata *et al.*, 1987). Because the AM mycelia can link different plant species growing nearby and so help overlap the pool of available nutrients for the intercropped plant species, the fixed-N that is released into the overlapping mycorrhizospheres by either legume-root exudation or by nodule decay can make fixed-N available to non-fixing plants (Haystead *et al.*, 1988). Following appropriate experimental designs using ^{15}N, the role of AM in this system has been assessed under both greenhouse and field conditions (van Kessel *et al.*, 1985; Haystead *et al.*, 1988; Barea *et al.*, 1989a; 1989b; Toro *et al.*, 1998; Barea *et al.*, 2002c). Two of these field experiments that used ^{15}N to study both N_2 fixation (Barea *et al.*, 1987) and fixed-N transfer (Barea *et al.*, 1989a), carried out in our Laboratory, will be described in Section 5.2.

5.1.4. Interactions involving Genetically Modified (GM) Rhizobia

Due to the importance of AM associations in the context of rhizosphere ecology, both the AM fungi and the AM symbiosis have been proposed as target bio-safety model systems, *i.e.*, as bio-markers, for the reliable application of microbial inoculants, when released at the soil-plant interfaces (Barea *et al.*, 1996; Tobar *et al.*, 1996). It is of particular importance to ascertain whether or not inocula produced from genetically modified microorganisms interfere with AM formation and/or functioning.

In relation to these bio-safety related purposes, a series of experiments compared the effects on AM formation and function of a wild type (WT) *Rhizobium meliloti* strain with those of its genetically modified (GM) derivative, which was developed to improve the nodulation competitiveness of the WT strain (Sanjuan and Olivares, 1991). This GM rhizobial strain was found not to interfere with AM formation by the representative AM fungus, *G. mosseae*. The parameters tested included spore germination, mycelial growth from the AM propagules, and AM "entry point" formation on the developing root system of the common host plant, *Medicago sativa* (Barea *et al.*, 1996). Indeed, the GM *Rhizobium* increased both the number of AM colonization units and the nutrient-acquisition ability in AM plants as compared to the WT rhizobial strain (Tobar *et al.,* 1996). The establishment of the symbiotic interactions also induced changes in root morphology, in particular, the degree of branching increased and the number of lateral roots was higher in AM plants inoculated with the GM *Rhizobium* strain (Barea *et al.*, 1996).

The differential effects of a WT and its GM derivative *Rhizobium meliloti* on the physiological/biochemical activity of the AM fungus *Glomus mosseae* colonizing alfalfa (*Medicago sativa* L.) roots were also tested. In one of these studies (Vazquez *et al.*, 2000), the effect of these two rhizobial strains on the fungal metabolic activity was evaluated using histochemical staining methods as markers for succinate dehydrogenase (SDH) activity. The GM strain did not adversely affect the development of the AM symbiosis and *ca.* 80% of the AM mycelium in plants inoculated with the GM *Rhizobium* was alive (had SDH activity), whereas only a 10-20% of the intra-radical mycelium remained alive in plants inoculated with the WT strain.

In another study (Vazquez *et al.*, 2002), the WT and its GM derivative *Rhizobium meliloti* were compared for their effects on plant growth, nitrate reductase activity (NR; EC 1.6.6.1), and protein content of *Medicago sativa* on co-inoculation with the AM fungus, *G. mosseae.* Varying distributions of NR activity were found in AM plant parts depending on the particular rhizobial strain used. The NR activity was enhanced in the root of plant inoculated with the GM strain, whereas the WT strain increased the NR activity in plant shoots. Protein content was substantially higher in AM plants. In general, AM legumes, which were nodulated by the GM rhizobial strain, displayed better physiological responses than those nodulated by the WT strain.

5.1.5. Interactions involving Other Soil Microorganisms

Because they share common micro-habitats at the root-soil interface, the AM fungi, rhizobia, and PGPR have to interact either during their colonization of the root or while functioning. Rhizobia and PGPR can influence AM formation and function and, conversely, the AM symbiosis can affect the rhizobial and PGPR population in the rhizosphere (Barea, 1997). In addition, PGPR can improve nodulation of legume roots (Staley *et al.,* 1992; Andrade *et al.,* 1998).

The well-known activities of rhizobia and phosphate-solubilizing microorganisms in improving the bioavailability of the major plant nutrients, N and P, are very much enhanced in the rhizosphere of AM plants, where synergistic interactions of such microorganisms with AM fungi occur. These interactions are the basis for the development of experimental approaches in the general context of rhizosphere biotechnology and their application to sustainable agriculture (Barea *et al.*, 2002a; 2002b).

Multi-microbial interactions, including AM fungi, *Rhizobium* and PGPR, have been tested with microorganisms that were isolated from a representative area of a desertification-threatened semi-arid ecosystem in the southeast of Spain (Requena *et al.,* 1997). *Anthyllis cytisoides*, an AM-dependent pioneer legume, which is dominant in the target mediterranean ecosystem, was the test plant. Microbial cultures from existing collections were also included in the screening process.

Several dual or triple microbial combinations were effective in improving plant development, nutrient uptake, N_2 fixation and/or root-system quality. These results indicate that selective and specific functional compatibility relationships exist among the microbial inoculants with respect to plant responses. Similar results were obtained by using *Medicago arborea*, a woody legume of interest for revegetation and biological reactivation of desertified semi-arid Mediterranean ecosystems (Valdenegro *et al.*, 2001).

In another experiment, a soil microcosm system was designed to evaluate the interactive effects of some multifunctional microbial-inoculation treatments and rock-phosphate (RP) application on N and P acquisition by alfalfa plants. The microbial inocula consisted of the wild type *Rhizobium meliloti* strain, its genetically modified derivative, which had enhanced competitiveness, the AM fungus (*Glomus mosseae*), and phosphate-solubilizing bacteria. Inoculated microorganisms established themselves in the root tissues and/or in the rhizosphere soil of the alfalfa plants. Measurements of the $^{15}N/^{14}N$ ratio in plant shoots indicated an enhancement of the N_2-fixation rate in *Rhizobium*-inoculated AM-plants over that achieved by the same *Rhizobium* strain in non-mycorrhizal plants. Regardless of the *Rhizobium* strain and of whether or not RP was added, AM-inoculated plants showed a lower specific activity ($^{32}P/^{31}P$) than did their comparable non-mycorrhizal controls, suggesting that the plant is using otherwise unavailable P sources (Toro *et al.*, 1998).

The effect of the phytostimulator bacteria of the genus *Azospirillum* has also been tested in interactions with the tripartite association, rhizobia-AM fungus-alfalfa. Positive effects of the multitrophic interactions both on symbiosis formation and micronutrient acquisition (Biró *et al.,* 2000) and on photosynthetic activity (Tsimilli-Michael *et al.,* 2000) were described.

Interactions, which involve either biocontrol *Pseudomonas* (Lesueur *et al.*, 2001; Kumar *et al.*, 2001) or plant growth-promoting yeasts (Vassilev *et al.*, 2001), also showed some beneficial effects on plant performance.

5.2. Strategic Studies related to Agricultural Systems

Only a few studies involving AM fungi-rhizobia interactions have been carried out under field conditions. One report (Azcón-Aguilar *et al.,* 1979) showed that the dual inoculation improved both the growth and nutrition of *Medicago sativa* in normal cultivation on an arable soil. Similarly, the interaction of rhizobia and AM fungi improved the performance of soybean plants under field conditions (Gonzalez-Mendez, 1990).

A ^{15}N-based technique was applied to estimate N_2 fixation by the forage legume, *Hedysarum coronarium,* and to ascertain the role of AM inoculation in plant-N nutrition throughout a growing season under field conditions. The absence of the *Rhizobium* specific for this forage legume in the test soil allowed the same legume to be used as the reference 'non-fixing' crop (Barea *et al.,* 1987). AM fungal inoculation enhanced dry matter yield, N concentration, and total N yield as compared to either phosphate-added or control plants. The use of ^{15}N distinguished between the sources of the acquired N and indicated that the AM fungi enhanced the amount of N derived from both soil and fixation. This result indicated that AM fungi acted both by a P-mediated mechanism to improve N_2 fixation and by enhancing N uptake from soil.

The ^{15}N isotope was also used to estimate both N_2 fixation by white clover and N-transfer from clover to perennial ryegrass. Pure and mixed stands of these pasture plants were established in a field soil (Barea *et al.*, 1989a; 1989b). The total N, P and dry matter yields in the grass/clover mixture were greater than in the monocultures. A lower ^{15}N enrichment of the grass growing in mixed stands compared to that growing alone suggested N-transfer. The AM colonization of the grass in the mixed grass/clover sward was significantly enhanced when compared with that of the grass in pure stands. Because the AM hyphae are known to be involved in NH_4^+ uptake, translocation and transfer to the host plants, this enhanced AM colonization may partly explain the improvement of pasture productivity when grass and mycotrophic legumes are grown together.

The interactive effects of phosphate-solubilizing bacteria, AM fungi and rhizobia were found to improve the agronomic efficiency of rock phosphate (RP) for legume crops under field conditions (Barea *et al.*, 2002c). Inoculation with AM fungi significantly increased plant biomass and both N and P accumulation in plant tissues. The phosphate-solubilizing rhizobacteria improved AM responses in soil receiving both RP and organic-matter amendments. The addition of organic matter favoured RP solubilization and this, together with a tailored microbial inoculation, increased the agronomic efficiency of RP in the test soil, which was Ca deficient at neutral pH.

A comprehensive experiment by Houngnandan *et al.*, (2000) tested the effect of rhizobial inoculation of macuna plants in 15 farmers′ fields in a savanna soil. They successfully established the plants, which in turn were colonization by AM fungi and a build up of AM spores. This technique is of interest for improving AM performance in cases where the availability of AM inoculum is scarce.

5.3. Strategic Studies related to the Revegetation of Degraded Ecosystems

Firstly, it must be stated that disturbance of natural plant communities is the first visible indication of a desertification process, but damage to the physical, chemical and biological properties of the soil is known to occur simultaneously. Such soil degradation limits the re-establishment of the natural plant cover. In particular, desertification disturbs plant-microbe symbioses, which are a critical ecological factor that help further plant growth in degraded ecosystems (Francis and Thornes 1990; Jeffries and Barea, 2001). However, only a few experiments have studied the effect of AM fungi-rhizobia-legume interactions in improving revegetation of degraded ecosystems (Jeffries and Barea, 2001).

In one of these studies, Herrera *et al.* (1993) reported the results of a 4-year field revegetation trial carried out in a semi-arid region of Spain. A number of woody species, common in revegetation programs in Mediterranean ecosystems, were used, including two native shrubs (*Anthyllis cytisoides* and *Spartium junceum*) and four tree legumes (*Acacia caven, Medicago arborea, Prosopis chilensis* and *Robinia pseudoacacia*). Plant species and microsymbionts were screened for appropriate combinations and a simple procedure to produce plantlets with an optimised mycorrhizal and nodulated status was developed. Results indicate that: (i) only the native shrub legumes were able to establish under the local environmental conditions; and (ii) inoculation with rhizobia and AMF improved plant survival and biomass development. Because these two shrub legumes are found in the natural plant community, they can be used to revegetate desertified areas. A reclamation strategy was proposed, using *Anthyllis cytisoides,* which is a particularly drought-tolerant legume species that is highly mycotrophic. This technique, involving the artificial acceleration of natural revegetation, could be accomplished by replanting randomly spaced groups of shrubs according to the natural pattern and structure of the undisturbed ecosystem (Francis and Thornes, 1990). *Anthyllis cytisoides* is very dependent on AM to reach optimal development in natural conditions (López-Sánchez *et al.*, 1992). It may have more widespread applications and might be used in areas contaminated with heavy metals because inoculation with certain AM fungi has been shown to reduce the toxicity effects of Pb and Zn (Díaz *et al.,* 1996).

In another example (Requena *et al.*, 2001), two long-term experiments in a desertified Mediterranean ecosystem indicated that inoculation with indigenous AM fungi, and with rhizobial nitrogen-fixing bacteria, not only enhanced establishment of key plant species, the legume *Anthyllis cytisoides,* but also increased soil fertility and quality. The dual symbiosis increased soil nitrogen content, organic matter and hydro-stable soil aggregates, and it enhanced N-transfer from N_2-fixing to non-fixing species associated within the natural succession. It was concluded that the introduction of target indigenous species of plants, which are associated with a managed community of microbial symbionts, is a successful biotechnological tool to aid the recovery of desertified ecosystems.

Dual inoculation with rhizobia and AM fungi also helped the establishment and biomass production of the woody legume, *Prosopis* sp., in wasteland in India (Bhatia *et al.*, 1998). This 6-year experiment clearly demonstrates the appropriate management of microbial symbiosis in legumes for restoration purposes.

A general conclusion from these experiments is that management of appropriate microsymbionts can help legumes to promote the stabilization of a self-sustaining ecosystem. The mycorrhizal shrubs, acting as a "fertile islands" (Azcón-Agular *et al.,* 2002), can serve as sources of inoculum for the surrounding area and could improve N nutrition for the non N_2-fixing vegetation in semi-arid ecosystems.

5.4. Biotechnological Developments for Integrated Management

An increasing demand for low-input agriculture has resulted in greater interest in the manipulation and use of soil microorganisms that are able either to benefit soil fertility or to improve plant nutrition and health. Because the appropriate management of these microbes can reduce the use of chemicals and energy in agriculture, this agro-biotechnology approach will lead to a more economical and sustainable production system, while minimizing environmental degradation. Consequently, some biotechnological inputs have been proposed concerning rhizosphere technology, a fact of particular interest for legume crops because the rhizosphere system of these plants harbours micro-symbionts and other associated microbes of great relevance for plant productivity. These biotechnological developments include the use of microbial inoculants. For legumes, the main target microbes are obviously AM fungi and rhizobia. However, PGPR microbes can also be included (2nd generation inoculants), particularly those either promoting AM formation and/or nodulation or used for the biological control of soil-borne plant pathogens (Barea *et al.*, 2002a; 2002b). The existence of some specificity in the host and/or microbial genotype relationships must be considered in selecting appropriate microbial consortia (Azcón *et al.*, 1991; Stanley *et al.*, 1992; Ruíz-Lozano and Azcón, 1993; Monzón and Azcón, 1996; Xavier and Germida, 2002). Therefore, microbial selection processes must consider "functional compatibility" criteria with regard to the microbial components, the host plant, and the environmental conditions.

Key issues in the production of microbial inoculants are inoculum technology, inoculum registration, bioethical and risk-assessment considerations (particularly for inocula based on genetically modified microorganisms), quality control, technology transfer, and commercialisation. The major concerns with respect to inoculant technology are: microbial selection and characterization; analysis of microbial competitiveness; inocula production (substrates and formulation), delivery (field release) and inoculation rate; monitoring the establishment of inoculated micro-organisms; and the fate and behaviour of inoculants after application (monitoring population dynamics, effect on target micro-organisms and microbial processes, effects on plant growth and health).

Although the technology for the production of rhizobial and free-living PGPR is commercially available, the production of inocula and the development of inoculation techniques have limited the manipulation of AM fungi. In fact, AM fungi technology deserves some particular comments. As obligate symbionts, the use of AM fungi in plant biotechnology differs from that of other beneficial soil microorganisms, like *Rhizobium*. Selection of the appropriate AM fungi is a key step (Estaún *et al.*, 2002) and specific procedures are required to multiply AM fungi and to produce high quality inocula (Azcón-Aguilar *et al.*, 2000; von Alten *et al.*, 2002). Multi-microbial inoculum production technologies are developed in cooperation with companies involved in the R and D activities, through the COST Action 8.38, which is a pan-European network that has been established to study "managing arbuscular mycorrhizal fungi for improving soil quality and plant health in agriculture." Another important issue is the analysis of the ecological conditions for the appropriate application of AM technology in agriculture. These studies have recently been reviewed (Vosatka and Dodd, 2002; Feldmann and Grotkas, 2002).

5.5. Future Trends in this Research Area

The application of selected microbial inoculants will become even more important in the near future because the use of agrochemicals must be reduced, and even avoided, to increase food quality, sustainable food production, and environmental protection. Because mycorrhizal and nodulated legumes are target crops for environmentally friendly food production, improvements in their use in sustainable agriculture is a challenge for future research strategies. To reach such a goal, it is important to explore the natural diversity of rhizobial and AM fungal populations in the rhizosphere of legumes, and to exploit selected microbes as a source of inocula. This is critical, in general, but particularly relevant for developing countries.

Molecular microbial ecology techniques are being used to explore the natural diversity of beneficial WT microbes either to be applied in plant production systems or to be used as a base for the use of improved GM strains. In this context, modern biotechnology, which has emerged as one of the most promising and crucial technologies for sustainable agricultural developments, must be considered. A research area of current interest is the use of genetically modified organisms (GMO). These include both novel GM crops improved for characteristics, such as resistance to pathogens, and new environmentally friendly GM microbial inoculants, which can both protect plants from disease and increase plant growth. These new products are expected to lead to a reduction in the use of biocides and chemical fertilizers and provide new options for the control of crop diseases, while reducing, or even avoiding, the application of these agrochemicals. Bio-safety issues must be considered for the release of GMOs to the environment.

Genetic manipulation of AM fungi is a matter of current research but the GM rhizobial strains are already available with an increased competitiveness, capability for environmental stress resistance, and synergy with low N-fertilizer regimes. These GM microbial inoculants must be released and tested under commercial field conditions, following *ad hoc* International Directives for the deliberate release of these GMOs, and the data concerning the effectiveness of WT or GM microbial inoculants on crop production, soil biomass, and soil fertility collected. It will also be critical to assess the impact of GM inoculants on key ecosystem processes, and the biological perturbations that they can induce in the rhizosphere.

6. CONCLUSIONS

The great agricultural and environmental importance of legumes, plus the ability of their rhizosphere system is able to harbour symbionts and other associated microbes of great relevance to plant productivity, make legumes target crops in sustainable agriculture. Current developments in the ecology, physiology, biochemistry, molecular biology, and biotechnology of microbe-plant relationships have given new insights into understanding the formation and functioning of the tripartite arbuscular-mycorrhizal and nitrogen-fixing symbioses of legumes and their interactions with PGPR. Although the technology for the production of rhizobial and free-living PGPR is commercially available, the production of AM fungi inocula and the development of inoculation techniques have limited the manipulation of AM fungi. However, current biotechnology practices now allow the production of efficient AM-fungal inoculants. Therefore, an appropriate management of selected AM fungi, rhizobia, and PGPR is now an available technique for exploiting the benefits of these microorganisms in agriculture, horticulture, and in revegetation of degraded ecosystems.

ACKNOWLEDGEMENTS

This study is included in the framework of the following Research Projects: INCO-DEV (ICA4-CT-2001-10057-"Soybean BNF + MYC") UE, CICyT (REN2000-1506- "Ecosymbionts"), Spain, and ENVIR-TN (EVK2-2001-00254- "Consider") UE. We thank Prof. J. Olivares, EEZ, CSIC, for comments and suggestions. D. Werner also thanks the German Science Foundation for support in the Schwerpunktprogramm "Molekulare Grundlagen der Mykorrhiza-Symbiosen".

REFERENCES

Abd-Alla, M. H., Omar, S. A., and Karanxha, S. (2000). The impact of pesticides on arbuscular mycorrhizal and nitrogen-fixing symbioses in legumes. *Appl. Soil Ecol., 14,* 191-200.

Altieri, M. A. (1994). Sustainable Agriculture. *Encylopedia of Agriculture Science, 4,* 239-247.

Ames, R. N., and Bethlenfalvay, G. J. (1987). Mycorrhizal fungi and the integration of plant and soil nutrient dynamics. *J. Plant Nutr., 10,* 1313-1321.

Andrade, G., de Leij, F. A. A. M., and Lynch, J. M. (1998). Plant mediated interactions between *Pseudomonas fluorescens, Rhizobium leguminosarum* and arbuscular mycorrhizae on pea. *Letters in Appl. Microbiol., 26,* 311-316.

Asai, T. (1944). Die bedeutung der mikorrhiza für das pflanzenleben. *Japanese J. Bot., 12,* 359-408.

Asimi, S., Gianinazzi-Pearson, V., and Gianinazzi, S. (1980). Influence of increasing soil-phosphorus levels on interactions between vesicular arbuscular mycorrhizae and *Rhizobium* in soybeans. *Can. J. Bot., 58,* 2200-2205.

Atkinson, S., Berta, G., and Hooker, J. E. (1994). Impact of mycorrhizal colonisation on root architecture, root longevity and the formation of growth regulators. In S. Gianinazzi and H. Schüepp (Eds.), *Impact of Arbuscular Mycorrhizas on Sustainable Agriculture and Natural Ecosystems* (pp. 47-60). Basel, Switzerland: Birkhäuser Verlag.

Augé, R. M. (2001). Water relations, drought and vesicular-arbuscular mycorrhizal symbiosis. *Mycorrhiza, 11,* 3-42.

Azcón, R., El-Atrash, F., and Barea, J. M. (1988). Influence of mycorrhiza vs soluble phoshate on growth, nodulation, and N_2 fixation (^{15}N) in alfalfa under different levels of water potential. *Biol. Fertil. Soils, 7,* 28-31.

Azcón, R., Rubio, R., and Barea, J. M. (1991). Selective interactions between different species of mycorrhizal fungi and *Rhizobium* meliloti strains, and their effects on growth, N_2 fixation (^{15}N) and nutrition of *Medicago sativa* L. *New Phytol., 117,* 399-404.

Azcón-Aguilar, C., Azcón, R., and Barea, J. M. (1979). Endomycorrhizal fungi and *Rhizobium* as biological fertilizers for *Medicago sativa* in normal cultivation. *Nature, 279,* 325-327.

Azcón-Aguilar, C., and Barea, J. M. (1992). Interactions between mycorrhizal fungi and other rhizosphere microorganisms. In M. J. Allen (Ed.), *Mycorrhizal Functioning. An Integrative Plant-Fungal Process Routledge* (pp. 163-198). New York, NY: Chapman and Hall Inc.

Azcón-Aguilar, C., and Barea, J. M. (1996). Arbuscular mycorrhizas and biological control of soil-borne plant pathogens - An overview of the mechanisms involved. *Mycorrhiza, 6,* 457-464.

Azcón-Aguilar, C., and Barea, J. M. (1997). Applying mycorrhiza biotechnology to horticulture, significance and potentials. *Scientia Horticulturae, 68,* 1-24.

Azcón-Aguilar, C., Bago, B., and Barea, J. M. (1998). Saprophytic growth of arbuscular-mycorrhizal fungi. In B. Hock and A. Varma (Eds.), *Mycorrhiza: Structure, Function, Molecular Biology and Biotechnology* (pp. 391-408). Heidelberg, Germany: Springer-Verlag.

Azcón-Aguilar, C., Palenzuela, E. J., and Barea, J. M. (2000). Substrato para la producción de inóculos de hongos formadores de micorrizas. Patent no. 9901814, Spain, CSIC.

Azcón-Aguilar, C., Jaizme-Vega, M. C., and Calvet, C. (2002). The contribution of arbuscular mycorrhizal fungi to the control of soil-borne plant pathogens. In S. Gianinazzi, H. Schüepp, J. M. Barea and K. Haselwandter (Eds.), *Mycorrhiza Technology in Agriculture, from Genes to Bioproducts* (pp. 187-197). Heidelberg, Germany: ALS Birkhäuser Verlag.

Azcón-Aguilar, C., Palenzuela, J., Roldan, A., Bautista, S., Vallejo, R., and Barea, J. M. (2002). Analysis of the mycorrhizal potential in the rhizosphere of representative plant species from desertification-threatened Mediterranean shrublands. *Appl. Soil Ecol., 21,* 1-9.

Bachmann, K. (1998). Species as units of diversity: An outdated concept. *Theory Bioscience, 117,* 213-230.

Barea, J. M. (1991). Vesicular arbuscular mycorrhizae as modifiers of soil fertility. In B. A. Stewart (Ed.), *Advances in Soil Science. Vol 15* (pp. 1-39). New York, NY: Springer-Verlag.

Barea, J. M. (1997). Mycorrhiza/bacteria interactions on plant growth promotion. In A. Ogoshi, L. Kobayashi, Y. Homma, F. Kodama, N. Kondon and S. Akino (Eds.), *Plant Growth-Promoting Rhizobacteria, Present Status and Future Prospects* (pp. 150-158). Paris, France: OCDE.

Barea, J. M. (2000). Rhizosphere and mycorrhiza of field crops. In J. P. Toutant, E. Balazs, E. Galante, J. M. Lynch, J. S. Schepers, D. Werner and P. A. Werry (Eds.), *Biological Resource Management, Connecting Science and Policy (OECD)* (pp. 110-125). Heidelberg, Germany: INRA, Editions and Springer.

Barea, J. M., Azcón, R., and Azcón-Aguilar, C. (1989a). Time-course of N_2-fixation (^{15}N) in the field by clover growing alone or in mixture with ryegrass to improve pasture productivity, and inoculated with vesicular-arbuscular mycorrhizal fungi. *New Phytol., 112,* 299-404.

Barea, J. M., Azcón, R., and Azcón-Aguilar, C. (1992). Vesicular-arbuscular mycorrhizal fungi in nitrogen-fixing systems. In J. R. Norris, D. Read and A. Varma (Eds.), *Methods in Microbiology. Vol 24. Techniques for the Study of Mycorrhizae* (pp. 391-416). London, UK: Academic Press.

Barea, J. M., Azcón, R., and Azcón-Aguilar, C. (1993). Mycorrhiza and crops. In I. Tommerup (Ed.), *Advances in Plant Pathology. Vol. 9. Mycorrhiza, A synthesis* (pp. 167-189). London, UK: Academic Press.

Barea, J. M., Azcón, R., and Azcón-Aguilar, C. (2002a). Mycorrhizosphere interactions to improve plant fitness and soil quality. *Antoine van Leeuwenhoek, 81,* 343-351.

Barea, J. M., and Azcón-Aguilar, C. (1983). Mycorrhizas and their significance in nodulating nitrogen fixing plants. *Adv. Agron., 36,* 1-54.

Barea, J. M., Azcón-Aguilar, C., and Azcón, R. (1987). Vesicular-arbuscular mycorrhiza improve both symbiotic N_2-fixation and N uptake from soil as assessed with a ^{15}N technique under field conditions. *New Phytol., 106,* 717-721.

Barea, J. M., Azcón-Aguilar, C., and Azcón, R. (1997). Interactions between mycorrhizal fungi and rhizosphere microorganisms within the context of sustainable soil-plant systems. In A. C. Grange and V. K. Brown (Eds.), *Multitrophic Interactions in Terrestrial Systems* (pp. 65-77). Cambridge, UK: Blackwell Science.

Barea, J. M., El-Atrach, F., and Azcón, R. (1989b). Mycorrhiza and phosphate interactions as affecting plant development, N_2 fixation, N-transfer and N uptake from soil in legume grass mixtures by using a ^{15}N dilution technique. *Soil Biol. Biochem., 21,* 581-589.

Barea, J. M., Gryndler, M., Lemanceau, Ph., Schüepp, H., and Azcón, R. (2002b). The rhizosphere of mycorrhizal plants. In S. Gianinazzi, H. Schüepp, J. M. Barea and K. Haselwandter (Eds.), *Mycorrhiza Technology in Agriculture, from Genes to Bioproducts* (pp. 1-18). Basel, Switzerland: Birkhäuser Verlag.

Barea, J. M., and Jeffries, P. (1995). Arbuscular mycorrhizas in sustainable soil plant systems. In A. Varma and B. Hock (Eds.), *Mycorrhiza, Structure, Function, Molecular Biology and Biotechnology* (pp. 521-559). Heidelberg, Germany: Springer-Verlag.

Barea, J. M., Tobar, R. M., and Azcón-Aguilar, C. (1996). Effect of a genetically modified *Rhizobium meliloti* inoculant on the development of arbuscular mycorrhizas, root morphology, nutrient uptake and biomass accumulation in *Medicago sativa. New Phytol., 134,* 361-369.

Barea, J. M., Toro, M., Orozco, M. O., Campos, E., and Azcón, R. (2002c). The application of isotopic ^{32}P and ^{15}N-dilution techniques to evaluate the interactive effect of phosphate-solubilizing rhizobacteria, mycorrhizal fungi and *Rhizobium* to improve the agronomic efficiency of rock phosphate for legume crops. *Nutrient Cycling in Agroecosystems, 63,* 35-42.

Bashan, Y. (1999). Interactions of *Azospirillum* spp. in soils: A review. *Biology and Fertility of Soils, 29,* 246-256.

Berta, G., Fusconi, A., and Hooker, J. E. (2002). Arbuscular mycorrhizal modifications to plant root systems: scale, mechanisms and consequences. In S. Gianinazzi, H. Schüepp, J. M. Barea and K. Haselwandter (Eds.), *Mycorrhiza Technology in Agriculture, from Genes to Bioproducts* (pp. 71-85). Basel, Switzerland: Birkhäuser Verlag.

Bethlenfalvay, G. J., Brown, M. S., and Stafford, A. E. (1985). *Glycine-Glomus Rhizobium* symbiosis. II. Antagonistic effects between mycorrhizal colonization and nodulation. *Plant Physiol., 79,* 1054-1058.

Bethlenfalvay, G. J., and Linderman, R. G. (1992). *Mycorrhizae in Sustainable Agriculture.* SAS Special publication no 54 Madison, Wisconsin.

Bhatia, N. P., Adholeya, A., and Sharma, A. (1998). Biomass production and changes in soil productivity during longterm cultivation of *Prosopis juliflora* (Swartz) DC inoculated with VA mycorrhiza and *Rhizobium* spp. in a semi-arid wasteland. *Biol. Fertil. Soils, 26,* 208-214.

Bianciotto, V., Andreotti, S., Balestrini, R., Bonfante, P., and Perotto, S. (2001). Extracellular polysaccharides are involved in the attachment of *Azospirillum brasilense* and *Rhizobium leguminosarum* to arbuscular mycorrhizal structures. *Eur. J. Histochem., 45,* 39-49.

Biró, B., Köves-Péchy, K., Vörös, I., Takács, T., Eggenberger, P., and Strasser, R. J. (2000). Interrelations between *Azospirillum* and *Rhizobium* nitrogen-fixers and arbuscular mycorrhizal fungi in the rhizosphere of alfalfa in sterile, AMF-free or normal soil conditions. *Appl. Soil Ecol., 15,* 159-168.

Blilou, I., Ocampo, J. A., and García-Garrido, J. M. (1999). Resistance of pea roots to endomycorrhizal fungus or *Rhizobium* correlates with enhanced levels of endogenous salicylic acid. *J. Exp. Bot., 50,* 1663-1668.

Bowen, G. D., and Rovira, A. D. (1999). The rhizosphere and its management to improve plant growth. *Adv. Agron., 66,* 1-102.

Brown, M. S., and Bethlenfalvay, G. J. (1987). *Glycine-Glomus-Rhizobium.* 6. Photosynthesis in nodulated, mycorrhizal or N-fertilized and P-fertilized soybean plants. *Plant Physiol., 85,* 120-123.

Brown, M. S., and Bethlenfalvay, G. J. (1988). The *Glycine-Glomus-Rhizobium* symbiosis. 7. Photosynthetic nutrient use efficiency in nodulated, mycorrhizal soybeans. *Plant Physiol., 86,* 1292-1297.

Brundrett, M., Bougher, N., Dell, B., Gove, T., and Malajczuk, N. (1996). *Working with Mycorrhizas in Forestry and Agriculture.* Canberra, Australia: ACIAR.

Danso, S. K. A. (1988). The use of ^{15}N enriched fertilizers for estimating nitrogen fixation in grain and pasture legumes. In D. P. Beck and L. Materon (Eds.), *Nitrogen Fixation by Legumes in Mediterranean Agriculture* (pp. 345-358). ICARDA.

Díaz, G., Azcón-Aguilar, C., and Honrubia, M. (1996). Influence of arbuscular mycorrhizae on heavy metal (Zn and Pb) uptake and growth of *Lygeum spartum* and *Anthyllis cytisoides. Plant Soil, 180,* 241-249.

Dobbelaere, S., Croonenborghs, A., Thys, A., Ptacek, D., Vanderleyden, J., Dutto, P., Labandera-Gonzalez, C., Caballero-Mellado, J., Aguirre, J. F., Kapulnik, Y., Brener, S., Burdman, S., Kadouri, D., Sarig, S., and Okon, Y. (2001). Response of agronomically important crops to inoculation with *Azospirillum. Aust. J. Plant Physiol., 28,* 1-9.

Duc, G. Trouvelot, A., Gianinazzi-Pearson, V., and Gianinazzi, S. (1989). First report of non-mycorrhizal plant mutants (Myc-) obtained in pea (*Pisum sativum* L) and fababean (*Vicia faba* L). *Plant Science, 60,* 215-222.

Dumas-Gaudot, E., Gollotte, A., Cordier, C., Gianinazzi, S., and Gianinazzi-Pearson, V. (2000). Modulation of host defence systems. In Y. Kapulnick and D. D. Douds, Jr. (Eds.), *Arbuscular Mycorrhizas: Physiology and Functions* (pp. 121-140). Dordrecht, The Netherlands: Kluwer Academic Publishers.

Estaún, V., Camprubí, A., and Joner, E. J. (2002). Selecting arbuscular mycorrhizal fungi for field application. . In S. Gianinazzi, H. Schüepp, J. M. Barea and K. Haselwandter (Eds.), *Mycorrhiza Technology in Agriculture, from Genes to Bioproducts* (pp. 249-259). Basel, Switzerland: Birkhäuser Verlag.

Feldman, F., and Grotkass, C. (2002). Directed inoculum production – shall we be able to design populations of arbuscular mycorrhizal fungi to achieve predictable symbiotic effectiveness?. In S. Gianinazzi, H. Schüepp, J. M. Barea and K. Haselwandter (Eds.), *Mycorrhiza Technology in Agriculture, from Genes to Bioproducts* (pp. 261-296). Basel, Switzerland: Birkhäuser Verlag.

Ferrol, N., Barea, J. M., and Azcón-Aguilar, C. (2000). The plasma membrane H-ATPase genes family in the arbuscular mycorrhizal fungus *Glomus mosseae. Curr. Genet., 37,* 112-118.

Flores, M., Mavingui, P., Perret, X., Broughton, W. J., Romero, D., Hernández, G., Davila, G., and Palacios, R. (2000). Prediction, identification, and artificial selection of DNA rearrangements in *Rhizobium*: Toward a natural genomic design. *Proc. Natl. Acad. Sci. USA, 97,* 9138-9143.

Francis, D. F., and Thornes, J. B. (1990). Matorral erosion and reclamation. Soil Degradation and Rehabilitation in Mediterranean Environmental Conditions. In J. Albaladejo, M. A. Stocking and E. Díaz (Eds.), *Soil Degradation and Rehabilitation in Mediterranean Environmental Conditions* (pp. 87-115). Murcia, Spain: CSIC.

Franken, P., and Requena, N. (2001). Analysis of gene expression in arbuscular mycorrhizas: New approaches and challenges. *New Phytol., 150,* 517-523.

Galleguillos, C., Aguirre, C., Barea, J. M., and Azcón, R. (2000). Growth promoting effect of two *Sinorhizobium* meliloti strains (a wild type and its genetically modified derivative) on a non-legume plant species in specific interaction with two arbuscular mycorrhizal fungi. *Plant Sci., 159,* 57-63.

Gianinazzi, S., and Schüepp, H. (1994). *Impact of Arbuscular Mycorrhizas on Sustainable Agriculture and Natural Ecosystems.* Basel, Switzerland: ALS Birkhäuser Verlag.

Gianinazzi, S., Schüepp, H., Barea, J. M., and Haselwandter, K. (2002). *Mycorrhizal Technology in Agriculture, from Genes to Bioproducts.* Basel, Switzerland: Birkhäuser Verlag.

Gianinazzi-Pearson, V. (1997). Have common plant systems co-evolved in fungal and bacterial root symbioses? In A. Legocki, H. Bothe and A. Pühler (Eds.), *Biological Fixation of Nitrogen for Ecology and Sustainable Agriculture* (pp. 322-324). Berlin and Heidelberg, Germany: Spinger-Verlag.

Gianinazzi-Pearson, V., Dumas-Gaudot, E., Gollotte, A., Tahiri-Alaoui, A., and Gianinazzi, S. (1996). Cellular and molecular defence-related root responses to invasion by arbuscular mycorrhizal fungi. *New Phytol., 133,* 45-57.

Giovannetti, M. (2000). Spore germination and pre-symbiotic mycelial growth. In Y. Kapulnik and D. D. Douds, Jr. (Eds.), *Arbuscular Mycorrhizas: Physiology and Function* (pp. 3-18). Dordrecht, The Netherlands: Kluwer Academic Publsihers.

Goicoechea, N., Antolín, M. C., and Sánchez-Díaz, M. (1997). Influence of arbuscular mycorrhizae and *Rhizobium* on nutrient content and water relations in drought stressed alfalfa. *Plant Soil, 192,* 261-268.

Goicoechea, N., Antolín, M. C., and Sánchez-Díaz, M. (2000). The role of plant size and nutrient concentrations in associations between *Medicago* and *Rhizobium* and/or *Glomus. Biol. Plant., 43,* 221-226.

Goicoechea, N., Szalai, G., Antolín, M. C., Sánchez-Díaz, M., and Paldi, E. (1998). Influence of arbuscular Mycorrhizae and *Rhizobium* on free polyamines and proline levels in water-stressed alfalfa. *J. Plant Physiol., 153,* 706-711.

Gollotte, A., Brechenmacher, L., Weidmann, S., Franken, P., and Gianinazzi-Pearson, V. (2002). Plant genes involved in arbuscular mycorrhiza formation and functioning. In S. Gianinazzi, H. Schüepp, J. M. Barea and K. Haselwandter (Eds.), *Mycorrhiza Technology in Agriculture, from Genes to Bioproducts* (pp. 87-102). Basel, Switzerland: Birkhäuser Verlag.

González-Méndez, S. B. (1990). Ecofisiología y biotecnología de las micorrizas VA en leguminosas. Ph. D. Thesis. Universidad de Granada. Facultad de Ciencias (Sección de Biología).

Graham, J. H., Hodge, N. C., and Morton, J. B. (1995). Fatty acid methyl ester profiles for characterization of *Glomalean* fungi and *Endomycorrhizae. Appl. Environ. Microbiol., 61,* 58-64.

Harley, J. L., and Smith, S. E. (1983). *Mycorrhizal Symbiosis.* New York, NY: Academic Press.

Harris, D., Pacovsky, R. S., and Paul, E. A. (1985). Carbon economy of soybean-*Rhizobium-Glomus* associations. *New Phytol., 101,* 427-440.

Harrison, M. J., Liu, J., Dewbre, G. R., Blaylock, L. A., and Zhao, L. (2000). Toward an understanding of the development and functioning of an arbuscular mycorrhiza: Molecular and genetic approaches. In H. C. Weber, S. Imhof and D. Zeuske (Ed.), *Proceedings of the Third International Congress on Symbiosis* (pp. 84). Marburg, Germany.

Hayman, D. S. (1986). Mycorrhizae of nitrogen-fixing legumes. *MIRCEN Journal, 2,* 121-145.

Haystead, A., Malajczuk, N., and Grove, T. S. (1988). Underground transfer of nitrogen between pasture plants infected with vesicular arbuscular mycorrhizal fungi. *New Phytol., 108,* 417-423.

Herrera, M. A., Salamanca, C. P., and Barea, J. M. (1993). Inoculation of woody legumes with selected arbuscular mycorrhizal fungi and rhizobia to recover desertified mediterranean ecosystems. *Appl. Environ. Microbiol., 59,* 129-133.

Hirsch, A. M., and Kapulnik, Y. (1998). Signal transduction pathways in mycorrhizal associations: Comparisons with the *Rhizobium*-legume symbiosis. *Fungal Genet. Biol., 23,* 205-212.

Houngnandan, P., Sanginga, N., Woomer, P., Vanlauwe, B., and van Cleemput, O. (2000). Response of *Mucuna pruriens* to symbiotic nitrogen fixation by rhizobia following inoculation in farmers' fields in the derived savanna of Benin. *Biol. Fertil. Soils, 30,* 558-565.

Janse, J. M. (1896). Les endophytes radicaux des quelques plantes Javanaises. *Annales du Jardin Botanique Buitenzorg, 14,* 53-212.

Jeffries, P., and Barea, J. M. (2001). Arbuscular mycorrhiza - a key component of sustainable plant-soil ecosystems In B. Hock (Ed.), *The Mycota. Vol. IX. Fungal Associations* (pp. 95-113). Berlin and Heidelberg, Germany: Springer-Verlag.

Jeffries, P., Craven-Griffiths, A., Barea, J. M., Levy, Y., and Dodd, J. C. (2002). Application of arbuscular mycorrhizal fungi in the revegetation of desertified Mediterranean ecosystem. In S. Gianinazzi, H. Schüepp, J. M. Barea and K. Haselwandter (Eds.), *Mycorrhiza Technology in Agriculture, from Genes to Bioproducts* (pp. 151-174). Heidelberg, Germany: ALS, Birkhäuser Verlag.

Jones, F. R. (1924). A mycorrhizal fungus in the roots of legumes and some other plant. *Journal of Agriculture Research, 29,* 459-470.

Kapulnik, Y., and Douds, D. D. Jr. (2000). *Arbuscular Mycorrhizas: Physiology and Function.* Dordrecht, The Netherlands: Kluwer Academic Publsihers.

Kennedy, A. C. (1998). The rhizosphere and spermosphere. In D. M. Sylvia, J. J. Fuhrmann, P. G. Hartel and D. A. Zuberer (Eds.), *Principles and Applications of Soil Microbiology* (pp. 389-407). Upper Saddle River, NJ: Prentice Hall.

Kennedy, A. C., and Smith, K. L. (1995). Soil microbial diversity and the sustainability of agricultural soils. *Plant Soil, 170,* 75-86.

Kloepper, J. W. (1994). Plant growth-promoting rhizobacteria (other systems). In Y. Okon (Ed.), *Azospirillum / plant associations* (pp. 111-118). Boca Raton, FL: CRC Press.

Kloepper, J. W. (1996). Host specificity in microbe-microbe interactions. *BioScience, 46,* 406-409

Kucey, R. M. N., and Paul, E. A. (1982). Carbon flow. photosynthesis, and N_2 fixation in mycorrhizal and nodulated faba beans (*Vicia faba* L). *Soil Biol. Biochem., 14,* 407-412.

Kumar, B. S. D., Berggren, I., and Martensson, A. M. (2001). Potential for improving pea production by co-inoculation with fluorescent *Pseudomonas* and *Rhizobium*. *Plant Soil, 229,* 25-34.

Lesueur, D., Ingleby, K., Odee, D., Chamberlain, J., Wilson, J., Manga, T. T., Sarrailh, J. M., and Pottinger, A. (2001). Improvement of forage production in *Calliandra calothyrsus*: Methodology for the identification of an effective inoculum containing *Rhizobium* strains and arbuscular mycorrhizal isolates. *J. Biotechnol., 91,* 269-282.

Leyval, C., Joner, E. J., del Val, C., and Haselwandter, K. (2002). Potential or arbuscular mycorrhizal fungi for bioremediation. In S. Gianinazzi, H. Schüepp, J. M. Barea and K. Haselwandter (Eds.), *Mycorrhiza Technology in Agriculture, from Genes to Bioproducts* (pp. 175-186). Basel, Switzerland: Birkhäuser Verlag.

Linderman, R. G. (1994). Role of VAM fungi in biocontrol. In F. L. Pfleger and R. G. Linderman (Eds.), *Mycorrhizae and Plant Health* (pp. 1-25). St Paul, MN: APS Press.

Long, S. R. (2001). Genes and signals in the *Rhizobium*-legume symbiosis. *Plant Physiology, 125,* 69-72.

López-Sánchez, M. E., Díaz, G., and Honrubia, M. (1992). Influence of vesicular-arbuscular mycorrhizal infection and P addition on growth and P nutrition of *Anthyllis cytisoides* L. and *Brachypodium retusum* (Pers.) Beauv. *Mycorrhiza, 2,* 41-45.

Madan, R., Pankhurst, C., Hawke, B., and Smith, S. (2002). Use of fatty acids for identification of AM fungi and estimation of the biomass of AM spores in soil. *Soil Biol. Biochem., 34,* 125-128.

Miller, R. M., and Jastrow, J. D. (2000). Mycorrhizal fungi influence soil structure. In Y. Kapulnik and D. D. Douds, Jr. (Eds.), *Arbuscular mycorrhizas: Physiology and Functions* (pp. 3-18). Dordrecht, The Netherlands: Kluwer Academic Publishers.

Monzón, A., and Azcón, R. (1996). Relevance of mycorrhizal fungal origin and host plant genotype to inducing growth and nutrient uptake in *Medicago* species. *Agric. Ecosyst. Environ., 60,* 9-15.

Morton, J. B., and Benny, G. L. (1990). Revised classification of arbuscular mycorrhizal fungi (zygomycetes), a new order, *Glomales*, two new suborders, *Glomineae* and *Gigasporineae,* and two new families, *Acaulosporaceae* and *Gigasporaceae*, with an emendation of glomaceae. *Mycotaxon, 37,* 471-491.

Morton, J. B., Franke, M., and Bentivenga, S. P. (1995). Developmental foundations for morphological diversity among endomycorrhizal fungi in *Glomales*. In A. Varma and B. Hock (Eds.), *Mycorrhiza,*

Structure, Function, Molecular Biology and Biotechnology (pp. 669-683). Heidelberg, Germany: Springer-Verlag.
Mosse, B. (1986). Mycorrhiza in a sustainable agriculture. *Biol. Agric., 3,* 191-209.
Nehl, D. B., Allen, S. J. and Brown, J. F. (1996). Deleterious rhizosphere bacteria: An integrating perspective. *Appl. Soil Ecol.*, 5, 1-20.
Pang, P. C., and Paul, E. A. (1980). Effects of vesicular arbuscular mycorrhiza on C^{14} and ^{15}N distribution in nodulated fababeans. *Can. J. Soil, 60,* 241-250.
Parniske, M. (2000). Intracellular accommodation of microbes by plants: A common developmental program for symbiosis and disease. *Curr. Opinion Plant Biol., 3,* 320-328.
Postgate, J. R. (1998). *Nitrogen Fixation.* London, UK: Cambridge University Press.
Pozo, M. J., Slezack-Deschaumes, S., Dumas-Gaudot, E., Gianinazzi, S., Azcón-Aguilar, C. (2002). Plant defense responses induced by arbuscular mycorrhizal fungi. In S. Gianinazzi, H. Schüepp, J. M. Barea and K. Haselwandter (Eds.), *Mycorrhiza Technology in Agriculture, from Genes to Bioproducts* (pp. 103-111). Basel, Switzerland: Birkhäuser Verlag.
Probanza, A., Lucas García, J. A., Ruiz Palomino, M., Ramos, B., and Gutiérrez Mañero, F. J. (2002). *Pinus pinea* L. seedling growth and bacterial rhizosphere structure after inoculation with PGPR *Bacillus* (*B. licheniformis* CECT 5106 and *B. pumillus* CECT 5105). *Appl. Soil Ecol., 20,* 75-84.
Provorov N. A., Borisov A. Y., and Tikhonovich, I. A. (2002). Developmental genetics and evolution of symbiotic structures in N_2-fixing nodules and arbuscular mycorrhiza. *J. Theor. Biol., 214,* 215-232.
Redecker, D., Morton, J. B., and Bruns, T. D. (2000). Ancestral lineages of arbuscular mycorrhizal fungi (*Glomales*). *Molecular Phylogenetics and Evolution, 14,* 276-284.
Redecker, D., von Berswordt-Wallrabe, P., Beck, D. P., and Werner, D. (1997a). Influence of inoculation with arbuscular mycorrhizal fungi on stable isotopes of nitrogen in *Phaseolus vulgaris. Biol. Fertil. Soils, 24,* 344-346.
Redecker, D., Thierfelder, H., Walker, C., and Werner, D. (1997b) Restriction analysis of PCR-amplified internal transcribed spacers of ribosomal DNA as a tool for species identification in different genera of the order Glomales. *Appl. Environ. Microbiol., 63,* 1756-1761.
Requena, N., Jimenez, I., Toro, M., and Barea, J. M. (1997). Interactions between plant-growth-promoting rhizobacteria (PGPR), arbuscular mycorrhizal fungi and *Rhizobium* spp. in the rhizosphere of *Anthyllis cytisoides*, a model legume for revegetation in mediterranean semi-arid ecosystems. *New Phytol., 136,* 667-677.
Requena, N., Pérez-Solís, E., Azcón-Aguilar, C., Jeffries, P., and Barea, J. M. (2001). Management of indigenous plant-microbe symbioses aids restoration of desertified ecosystems. *Appl. Environ. Microbiol., 67,* 495-498.
Ruiz-Lozano, J. M., and Azcón, R. (1993). Specificity and functional compaibility of VA mycorrhizal endophytes in association with *Bradyrhizobium* strains in *Cicer arietinum. Symbiosis, 15,* 217-226.
Ruiz-Lozano, J. M., Collados, C., Barea, J. M. and Azcón, R. (2001). Arbuscular mycorrhizal symbiosis can alleviate drought-induced nodule senescence in soybean plants. *New Phytol., 151,* 493-502.
Ruiz-Lozano, J. M., Gianinazzi, S., and Gianinazzi-Pearson, V. (1999a). Genes involved in resistance to powdery mildew in barley differentially modulate root colonization by the mycorrhizal fungus *Glomus mosseae. Mycorrhiza, 9,* 237-240.
Ruiz-Lozano, J. M., Roussel, H., Gianinazzi, S., and Gianinazzi-Pearson, V. (1999b) Defense genes are differentially induced by a mycorrhizal fungus and *Rhizobium* sp. in wild-type and symbiosis-defective pea genotypes. *Mol. Plant-Microbe Interact., 12,* 976-984.
Sanders, I. R., Alt, M., Groppe, K., Boller, T., and Wiemken, A. (1995). Identification of ribosomal DNA polymorphisms among and within spores of the Glomales, Application to studies on the genetic diversity of arbuscular mycorrhizal fungal communities. *New Phytol., 130,* 419-427.
Sanjuan, J., and Olivares, J. (1991). Multicopy plasmids carrying the *Klebsiella pneumoniae* nifA gene enhance *Rhizobium meliloti* nodulation competitiveness on alfalfa. *Mol. Plant-Microbe Interact., 4,* 365-369.
Schüβler, A., Gehrig, H., Schwarzott, D., and Walker, C. (2001a). Analysis of partial Glomales SSU rRNA gene sequences, implications for primer design and phylogeny. *Mycol. Res., 105,* 5-15.
Schüβler, A., Schwarzott, D., and Walker, C. (2001b). A new fungal phylum, the Glomeromycota, phylogeny and evolution. *Mycol. Res., 105,* 1413-1421.
Simon, L., Lalonde, M., and Bruns, T. D. (1992). Specific amplification of 18S fungal ribosomal genes from vesicular-arbuscular endomycorrhizal fungi colonizing roots. *Appl. Environ. Microbiol., 58,* 291-295.

Smith, S. E., Dickson, S., and Smith, F. A. (2001). Nutrient transfer in arbuscular mycorrhizas: How are fungal and plant processes integrated? *Austr. J. Plant Physiol., 28,* 683-694.
Smith, S. E., Nicholas, D. J. D., and Smith, F. A. (1979). Effect of early mycorrhizal infection on nodulation and nitrogen fixation in *Trifolium subterraneum* L. *Austr. J. Plant Physiol., 6,* 305-316.
Smith, S. E., and Read, D. J. (1997). *Mycorrhizal Symbiosis.* San Diego, CA: Academic Press.
Spaink, H. P., Kondorosi, A., and Hooykaas, P. J. J. (1998). *The Rhizobiaceae.* Dordrecht, The Netherlands: Kluver Academic Publishers.
Sprent, J. I. (1994). Evolution and diversity in the legume-*Rhizobium* symbiosis: Chaos theory? *Plant Soil, 161,* 1-10.
Stanley, T. W., Lawrence, E. G., and Nance, E. L. (1992). Influence of a plant growth-promoting *Pseudomonas* and vesicular-arbuscular mycorrhizal fungus on alfalfa and birdsfoot trefoil growth and nodulation. *Biol. Fertil. Soils, 14,* 175-180.
Stracke, S., Kistner, C., Yoshida, S., Mulder, L., Sato, S., Kaneko, T., Tabata, S., Sandal, N., Stougaard, J., Szczyglowski, K. and Parniske, M. (2002) A plant receptor-like kinase required for both bacterial and fungal symbiosis. *Nature 417,* 959-962.
Tobar, R. M., Azcón-Aguilar, C., Sanjuán, J., Barea, J. M. (1996). Impact of a genetically modified *Rhizobium* strain with improved nodulation competitiveness on the early stages of arbuscular mycorrhiza formation. *Appl. Soil Ecol., 4,* 15-21.
Toro, M., Azcón, R., Barea, J.M. (1998). Use of isotopic dilution techniques to evaluate the interactive effects of *Rhizobium* genotype, mycorrhizal fungi, phosphate-solubilizing rhizobacteria and rock phosphate on nitrogen and phosphorus acquisition by *Medicago sativa. New Phytol., 138,* 265-273.
Tsimilli-Michael, M., Eggenberg, P., Biró, B., Köves-Pechy, K., Vörös, I., and Strasser, R. J. (2000). Synergistic and antagonistic effects of arbuscular mycorrhizal fungi and *Azospirillum* and *Rhizobium* nitrogen-fixers on the photosynthetic activity of alfalfa, probed by the polyphasic chlorophyll a fluorescence transient O-J-I-P. *Appl. Soil Ecol., 15,* 169-182.
Turnau, K., and Haselwandter, K. (2002). Arbuscular mycorrhizal fungi, an essential component of soil microflora in ecosystem restoration. In S. Gianinazzi, H. Schüepp, J. M. Barea and K. Haselwandter (Eds.), *Mycorrhiza Technology in Agriculture, from Genes to Bioproducts* (pp. 137-149). Basel, Switzerland: Birkhäuser Verlag.
Valdenegro, M., Barea, J. M., and Azcón, R. (2001). Influence of arbuscular-mycorrhizal fungi, *Rhizobium meliloti* strains and PGPR inoculation on the growth of *Medicago arborea* used as model legume for re-vegetation and biological reactivation in a semi-arid mediterranean area. *Plant Growth Regulation, 34,* 233-240.
Vance, C. P. (2001). Symbiotic nitrogen fixation and phosphorus acquisition. Plant nutrition in a world of declining renewable resources. *Plant Physiol., 127,* 390-397.
van der Heijden, M. G. A., and Sanders, I. R. (2002). *Mycorrhizal Ecology.* Berlin and Heidelberg, Germany: Springer-Verlag.
van Kessel, Ch., Singleton, P. W. and Hoben H. J. (1985). Enhanced N-transfer from soybean to maize by vesicular-arbuscular (VAM) fungi. *Plant Physiol., 79,* 562-563.
van Tuinen, D., Jacquot, E., Zhao, B., Gollotte, A., and Gianinazzi-Pearson, V. (1998). Characterization of root colonization profiles by a microcosm community of arbuscular mycorrhizal fungi using 25S rDNA-targeted nested PCR. *Mol. Ecol., 7,* 879-887.
Vassilev, N., Vassileva, M., Azcón, R., and Medina, A. (2001). Interactions of an arbuscular mycorrhizal fungus with free or co-encapsulated cells of *Rhizobium trifoli* and *Yarowia lipolytica* inoculated into a soil-plant system. *Biotechnol. Lett., 23,* 149-151.
Vázquez, M. M., Barea, J. M., and Azcón, R. (2002). Influence of arbuscular mycorrhizae and a genetically modified strain of *Sinorhizobium* on growth, nitrate reductase activity and protein content in shoots and roots of *Medicago sativa* as affected by nitrogen concentrations. *Soil Biol. Biochem., 34,* 899-905.
Vázquez, M. M., Bejarano, C., Azcón, R., and Barea, J. M. (2000). The effect of a genetically modified *Rhizobium meliloti* inoculant on fungal alkaline phosphatase and succinate dehydrogenase activities in mycorrhizal alfalfa plants as affected by the water status in soil. *Symbiosis, 29,* 49-58.
Vestberg, M., Cassells, A. C., Schubert, A., Cordier, C., and Gianinazzi, S. (2002). Arbuscular mycorrhizal fungi and micropropagtion of high value crops. In S. Gianinazzi, H. Schüepp, J. M. Barea and K. Haselwandter (Eds.), *Mycorrhiza Technology in Agriculture, from Genes to Bioproducts* (pp. 223-2233). Basel, Switzerland: Birkhäuser Verlag.

Vierheilig, H., and Piché, Y. (2002). Signalling in arbuscular mycorrhiza: Facts and hypotheses. In B. Buslig and J. Manthey (Ed.), *Flavonoids in cell functions* (pp. 23-29). New York, NY: Kluwer Academic/Plenum Publishers.

Vinuesa, P., Kurz, E.M., Thierfelder, H., Leon-Barrios, M. Thynn, M. Sicardi-Mallorca, M., Rademaker, J.L.W., Martinez-Romero, E., de Bruijn, F.J., Bedmar, E., Izaguirre-Mayoral, M.L. and Werner, D. (2000). Genotypic diversity of *Bradyrhizobium* strains of tropical and temperate origin and the identification of the new genomic species nodulating endemic woody legumes (Fabaceae: Genisteae) from the Canary Islands. In F. O. Pedrosa, M. Hungría, M. G. Yates and W. E. Newton (Eds.), *Nitrogen Fixation: From Molecules to Crop Productivity* (pp. 195-210). Dordrecht, The Netherlands: Kluwer Academic Publishers.

von Alten, H., Blal, B., Dodd, J. C., Feldmann, F., and Vosatka, M. (2002). Quality control of arbuscular mycorrhizal fungi inoculum in Europe. In S. Gianinazzi, H. Schüepp, J. M. Barea and K. Haselwandter (Eds.), *Mycorrhiza Technology in Agriculture, from Genes to Bioproducts* (pp. 281-296). Basel, Switzerland: Birkhäuser Verlag.

Vosatka, M., and Dodd, J. C. (2002). Ecological considerations for successful application of arbuscular mycorrhizal fungi inoculum. In S. Gianinazzi, H. Schüepp, J. M. Barea and K. Haselwandter (Eds.), *Mycorrhiza Technology in Agriculture, from Genes to Bioproducts* (pp. 235-247). Basel, Switzerland: Birkhäuser Verlag.

Weller, D. M., and Thomashow, L. S. (1994). Current challanges in introducing beneficial microorganisms into the rhizosphere. In F. O´Gara, D. N. Dowling and B. Boesten (Eds.), *Molecular Ecology of Rhizosphere Microorganisms Biotechnology and the Release of GMOs* (pp. 1-18). Weinheim, Germany: VCH.

Werner, D. (1998). Organic signals between plants and microorganisms. In R. Pinton, Z. Varanini and P. Nannipieri (Eds.), *The Rhizosphere: Biochemistry and Organic Substances at the Soil-Plant Interfaces*. New York, NY: Marcel Dekker.

Werner, D., Barea, J. M., Brewing, N. J., Cooper, J. E., Katinakis, P., Lindström, K., O'Gara, F., Spaink, H. P., Truchet, G., and Müller, P. (2002). Symbiosis and defence in the interaction of plants with microorganisms. *Symbiosis, 32,* 83-104.

Xavier, L. J. C., and Germida, J. J. (2002). Response of lentil under controlled conditions to co-inoculation with arbuscular mycorrhizal fungi and rhizobia varying in efficacy. *Soil Biol. Biochem., 34,* 181-188.

Zapata, F., Danso, S. K. A., Hardanson, G., and Fried, M. (1987). Nitrogen-fixation and translocation in field grown fababean. *Agron. J., 79,* 505-509.

Chapter 11

INOCULANT PREPARATION, PRODUCTION AND APPLICATION

M. HUNGRIA[1], M. F. LOUREIRO[2], I. C. MENDES[3], R. J. CAMPO[1] AND P. H. GRAHAM[4]

[1]Embrapa Soja, Cx. Postal 231, 86001-970, Londrina, PR, Brazil
[2]UFMT/FAMEV, Av. Fernando Correa s/n, Campus Universitário, 78000-900, Cuiabá, MT, Brazil
[3]Embrapa Cerrados, Cx. Postal 08223, 73301-970, Planaltina, DF, Brazil
[4]Department of Soil, Water, and Climate, University of Minnesota, 1991 Upper Buford Circle, St Paul, MN 55108, USA

1. INTRODUCTION

Progressive chemical and physical degradation of soil is a major factor affecting crop yield worldwide (Cassman, 1999). The situation is most serious in tropical regions, where soils are often structurally fragile, have low organic matter and nutrient content, and are frequently subject to erosion or inappropriate farm management. In these areas, nutrient depletion may be accentuated by the high cost of fertilizers, especially fixed-N sources, the majority of which are imported from developed countries (Hungria and Vargas, 2000; Giller, 2001). Thus, smallholders in Africa, for example, commonly apply less N, P, and K than is removed in the grain (Giller and Cadisch, 1995; Franzluebbers *et al.*, 1998; Sanchez, 2002), suggesting annual average depletion rates across 37 African countries of 22 kg N, 2.5 kg P and 15 kg K ha^{-1}. Because of such depletion, biological nitrogen (N_2) fixation is critical to the agricultural sustainability of these areas, but is often constrained by the absence in the soil of efficient and competitive rhizobia. There is an obvious need to improve the availability, quality, and delivery of such rhizobia for every cropped legume.

The practice of transferring soil from a field where legumes have been grown to new areas being planted to the same crop, dates back to ancient times. It became the recommended method of inoculation after Hellriegel´s report on the N nutrition of

D. Werner and W. E. Newton (eds.), Nitrogen Fixation in Agriculture, Forestry, Ecology, and the Environment, 223-253.

leguminous plants in 1886, and was followed soon thereafter by the first use of rhizobial inoculants (Voelcker, 1896; Fred *et al.*, 1932). However, after over a century of rhizobial inoculation, most of the inoculants produced in the world are still of relatively poor quality (FAO, 1991; Olsen *et al.*, 1994; 1996; Brockwell and Bottomley 1995; Lupwayi *et al.*, 2000; Stephens and Rask, 2000). In this chapter, we discuss some aspects related to inoculant production and inoculation. Complementary information can be obtained from other reviews (Smith, 1992; Brockwell and Bottomley, 1995; Brockwell *et al.*, 1995; Lupwayi *et al.*, 2000; Stephens and Rask, 2000; Catroux *et al.*, 2001; Date, 2001).

2. STRAIN SELECTION

Successful inoculation starts with the establishment of long-term programs of strain selection and the identification of elite strains for each legume host of interest. Emphasis in this selection program should be given to a high capacity for N_2 fixation with all commonly used cultivars of the legume in question, competitiveness with indigenous or naturalized rhizobia, tolerance to environmental constraints, and the ability to persist in soil. Selection for specific ecosystems, unusual soil physical or chemical constraints, specific environmental concerns (temperature or soil acidity), or specific local cultivars may also be important (*e.g.*, Jones and Hardarson, 1979; Hungria and Bohrer, 2000; Hungria and Vargas, 2000; Mpeperecki *et al.*, 2000; Stephens and Rask, 2000; Chen *et al.*, 2002; Mostasso *et al.*, 2002). Important characteristics not often considered in strain selection are performance in storage and culture (Balatti and Freire, 1996), genetic stability (FAO, 1991), and the ability to survive on seeds (Lowther and Patrick, 1995). These traits have also been considered by Burton (1981), Roughley (1970) and Keyser *et al.* (1993).

2.1. A Successful Approach: The Brazilian Strain Selection Program for Soybean and Common Bean

Soybean (*Glycine max* L. Merr.) was introduced to Brazil in 1882, with large-scale cultivation of bred cultivars initiated in the 1960s (Vargas and Hungria, 1997; Hungria and Bohrer, 2000). As Brazilian soils were originally devoid of bradyrhizobia that were effective with soybean (Vargas and Suhet, 1980; Hungria and Vargas, 2000; Ferreira and Hungria, 2002), inoculants were also introduced in the early 1960s, mainly from the United States. Strain selection both for locally adapted cultivars and for tolerance to the often acid-soils conditions started immediately (Döbereiner *et al.*, 1970; Peres and Vidor, 1980; Vargas *et al.*, 1992; Peres *et al.*, 1993; Hungria and Vargas, 2000) with outstanding results. However, as the national mean yield for soybean has increased from 1,166 kg ha^{-1} in 1968/69 to 2,765 kg ha^{-1} in 2002/2003, plant demand for N has also increased. Further, more than 90% of the areas cropped to soybean today have been previously inoculated and have established bradyrhizobial populations of at least 10^3 cells g^{-1} of soil. Both situations contribute to a need for more efficient and competitive strains (Hungria *et al.*, 2001b; 2002).

Strain-selection programs in Brazil initially emphasized elite strains from foreign countries, but have since changed to selection amongst adapted strains obtained from locally-grown soybeans several years after their introduction. Grain yield has always been the major factor considered, but other parameters used in the identification of superior strains have included plant vigor, N_2 (C_2H_2) reduction, total N accumulated in tissues, N harvest index, ureide content in tissues, and nodule occupancy (Peres and Vidor, 1980; Peres *et al.*, 1984; 1993; Neves *et al.*, 1985; Vargas *et al.*, 1992; Hungria *et al.*, 1998; Santos *et al.*, 1999; Hungria and Vargas, 2000). The four strains used in the production of commercial inoculants in Brazil today can each fulfill the crop´s need for N at yields greater than 4,000 kg ha^{-1} (Vargas *et al.*, 1992; Peres *et al.*, 1993; Vargas and Hungria, 1997; Hungria *et al.*, 2001b). This program continues (see Table 1) with soybean bradyrhizobia having both a higher capacity for N_2 fixation and improved competitiveness already available and soon to be released for commercial purposes.

Table 1. Nodulation, nodule occupancy, and yield of soybean cultivar BR 133 inoculated with parental and variant Bradyrhizobium japonicum strains. Experiments performed in oxisols of Londrina, State of Paraná, Brazil[1].

Treatment	Nodulation (mg pl^{-1})	Nodule occupancy by inoculated strain (% before/% after)	Increase in nodule occupancy (%)	Yield (kg ha^{-1})
C - N[2]	97 b[3]	-	-	1,928 c[3]
C + N[2]	13 c	-	-	3,444 a
SEMIA 566	134 a	23/59	156	2,723 b
Variant of 566	161 a	23/65	183	3,415 a
CB1809 (=SEMIA 586)	99 b	8/22	175	3,029 b
CB1809 variant	155 a	8/45	462	3,772 a

[1]After M. Hungria and R.J. Campo (unpublished).
[2]Non-inoculated control (C) without or with N-fertilizer (200 kg of N ha^{-1}, as urea, split twice - at sowing and at flowering time).
[3]Means of three field trials, performed in three crop seasons, each with six replicates. Within a column, values followed by the same letter are not statistically different (Duncan, $p \leq 0.05$).

As with soybean, Brazil is also the largest producer of common bean (*Phaseolus vulgaris* L.) in the world, with beans being the most important source of protein in the Brasilian diet. Average bean yields in Brazil have been very low, ca. 728 kg ha^{-1} in 2002/2003, mainly because of the limited technology used by small farmers. Lack of response to inoculation in this crop has been attributed to high populations of indigenous but ineffective common-bean rhizobia in soil (Graham, 1981; Buttery *et al.*, 1987; Ramos and Boddey, 1987; Hardarson, 1993), but soil temperature and acidity are also important in Brazil. As with soybean, the strain-selection program with beans has emphasized the isolation and selection of efficient strains from local bean soils.

Many of these strains belong to the species *Rhizobium tropici,* an organism not normally associated with beans in the centers of origin of this crop. Recently, this approach allowed the identification of the efficient, competitive and high-temperature tolerant *R. tropici* strain, PRF 81 (= SEMIA 4080). In field trials in soils already containing 10^4-10^5 bean rhizobia g^{-1}, inoculation with this strain increased both nodulation and N_2 fixation, and improved grain yield up to 900 kg ha^{-1}. PRF 81 has been recommended for use in commercial inoculants since 1998 (Hungria *et al.*, 2000a) and promotion of inoculation through active extension programs has increased by 25% the sale of bean inoculant in a three-year period. The search within local indigenous populations continues, with two other *R. tropici* strains from the Cerrado area (H12 and H20; Mostasso *et al.*, 2002; Hungria *et al.*, 2003) also contributing to significant yield increases in beans (Figure 1). Searching for strains within a naturalized population is a time-consuming process involving thousands of plates, and many greenhouse and field trials, but a further advantage is that the strains obtained are not genetically modified, avoiding legal and socioeconomic problems.

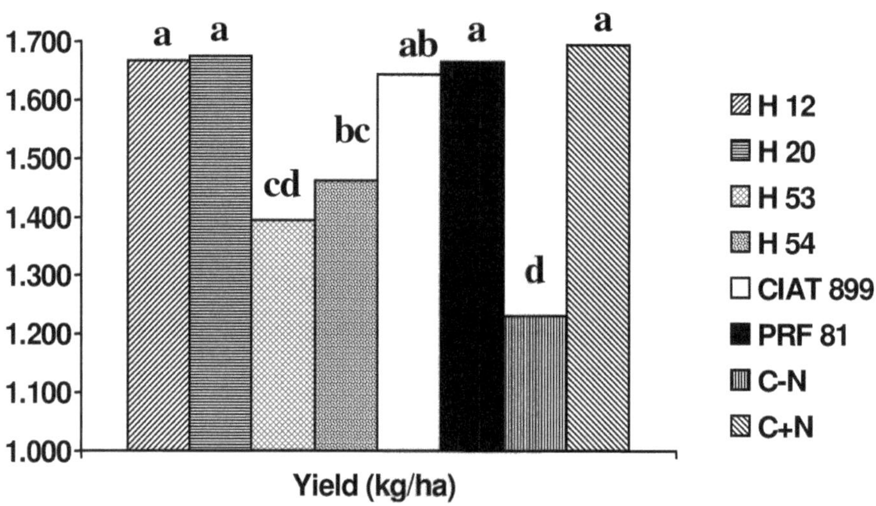

Figure 1. Effects of inoculation with Rhizobium tropici strains on yield of common bean. Uninoculated control treatments received either no N or 60 kg of N ha^{-1}, split between sowing and flowering. Mean of six field experiments performed in oxisols of Londrina, State of Paraná, Brazil, each with six replicates. Values followed by the same letter are not significantly different (Duncan, $p \leq 0.05$). After Hungria et al. *(2003).*

Because of the importance of soybean to the Brazilian economy and of beans to the country's nutrition, research on N_2 fixation in these crops has been well supported. As a result, the eleven inoculant manufacturers in Brazil sold a total of 14 million doses of inoculant for bean and soybean in 2001/2002, an increase of 16% over the previous two years. The situation is similar in Argentina where 10.3 million doses of inoculant were sold in 2001/2002. Unfortunately the financial support for strain selection and extension activities with other important legume crops is low. Although an enormous effort has been made by some government institutions, with more than 150 strains now recommended for the 90+ legume species used in grain crop or green manure production, pastures, and agroforestry (Hungria and Araujo, 1995), inoculants for these species still represent less than 1% of the overall market.

The favorable situation for both soybean and bean inoculation in Brazil differs from that in many other regions of the world. Brockwell and Bottomley (1995) suggest that 90% of all inoculants used provide no practical benefits. Furthermore, Karanja *et al.* (2000) note declining inoculant production in several regions of Africa, with only 12% of farmers using inoculants. Both poor inoculant quality and extension, in those areas of the world where inoculation is ineffective or little used, need to be addressed (Hall and Clark, 1995; Marufu *et al.*, 1995).

2.2. Selection of Fast-growing Strains for Soybean: Differences among Ecosystems

Fast-growing rhizobia that were able to effectively nodulate soybean were first isolated in 1982 from soils and nodules from the People´s Republic of China, which is the center of origin and diversity of this legume (Keyser *et al.*, 1982). Today, they are classified as *Sinorhizobium fredii* (Scholla and Elkan, 1984) and *S. xinjiangensis* (Chen *et al.*, 1988). Fast-growing strains belonging to other rhizobial species have also been isolated from nodulated soybean in Brazil and Paraguay (Chen *et al.*, 2000; 2002; Hungria *et al.*, 2001a; 2001c). Initially, it seemed that these fast-growing soybean rhizobia were only able to nodulate unimproved genotypes (Keyser *et al.*, 1982), but several modern soybean cultivars have now been reported as effectively nodulated by these strains (Balatti and Pueppke, 1992; Chueire and Hungria, 1997), which raises the possibility of their use in inoculants.

Among the advantages of using fast-growing strains are a shorter time for production of inoculant, a lower probability of contamination during the industrial process, easier establishment in the soil, and easier manipulation of genes (Chatterjee *et al.*, 1990; Cregan and Keyser, 1988). However, although high rates of N_2 fixation have been achieved in single inoculation experiments, fast-growing strains have proved to lack competitiveness against *Bradyrhizobium* isolates (McLoughlin *et al.*, 1985; Cregan and Keyser, 1988; Chueire and Hungria, 1997; Hungria *et al.*, 2001a). This limited competitiveness appears to be a function of low soil pH (Hungria *et al.*, 2001a). Buendia-Claveria *et al.* (1994) reported greater success with *S. fredii* as a soybean inoculant under alkaline soil conditions in Spain. This result reinforces the importance of strain selection under local conditions.

2.3. Use of Strains in Commercial Inoculants

In many countries, including the United States, microorganisms can be patented, with interpretation of the law often covering both artificially modified and "purified or isolated" preparations of newly discovered naturally occurring microbes. For natural microorganisms, a limitation can be that patents cover only a specific strain and its derivatives, but a broader protection may be possible for genetic modified organisms (Keyser *et al.*, 1993). Clearly, without strong patent protection, companies will not invest in strain selection and product development. A problem in countries, such as Brazil and Mexico, is that only genetically modified organisms can be patented, but these have problems in obtaining permission for field release.

The first North-American patent for pure cultures of rhizobia, and artificial inoculation, was obtained in 1896 by Nobbe and Hiltner and covered pure cultures of the desired *Rhizobium*, which were grown in flat glass bottles containing only a small amount of gelatin medium (Smith, 1992). Today, genetically modified and patented strains include a USDA *B. japonicum* strain, which is claimed to increase yield by 5-7%, and *S. meliloti* strain RMBPC-2, which is modified for NifA expression; both are commercialized by Urbana Laboratories (2002). Improvements in our knowledge of rhizobial genetics increase the potential for obtaining genetically modified rhizobia with superior symbiotic performance (Maier and Triplett, 1996; Sessitsch *et al.*, 2002). Additional genetically modified strains are likely to be released soon as products containing such rhizobia have been approved for field trials in several countries. The patenting of inoculant strains can only reduce the comparative testing of different inoculant-quality rhizobia.

Countries also differ in their policies concerning recommendation of strains for commercial inoculants. In countries such as the United States, each company determines which strains will be used in their products. In other countries, such as those belonging to Mercosur (Brazil, Argentina, Paraguay and Uruguay), commercial inoculants must contain the strains recommended by an official committee of rhizobiologists.

Commercial inoculants may contain one or more strains. Multistrain products may be important and recommended for several different hosts, *e.g.*, for both clover (*Trifolium* spp.) and alfalfa (*Medicago sativa*) (Roughley, 1970; Keyser *et al.*, 1993), or for African acacias (Sutherland *et al.*, 2000), but they may also be used for a single host (Roughley, 1970; Keyser *et al.*, 1993). Strains for such mixed inoculants should be grown in separate fermentors before being mixed into the carrier. Even then, it is difficult to ensure either a balanced growth among the strains in the inoculant (Roughley, 1970; Frankenberg *et al.*, 1995) or that strains perform similarly in terms of nodule occupancy. This situation probably explains why the N_2-fixation rates achieved with multistrain inoculants are often lower than those achieved with the single most effective strain (Bailey, 1988; Somasegaran and Bohlool, 1990). Benefits of single-strain inoculants would include avoidance of antagonistic effects between strains in the mixed inoculant, easier diagnosis of loss of effectiveness, and greater facility in quality control (Thompson, 1980). Today, the tendency is to use single-strain inoculants in countries with strong inoculant quality-control programs as well as in those with a tendency to recommend specific strains for each ecosystem. These countries include Australia, France, New Zealand, South Africa, and Uruguay (Date, 2001). Multiple-strain inoculants can pose special problems for legume species of lesser importance, where the manufacturer may not be able to justify economically the testing of strains in the mixture on a regular basis.

2.4. Persistence of the Strains on the Soils

Several studies have followed the persistence of exotic rhizobia introduced into sterile or non-sterile soil. In some studies, population numbers decline rapidly at a rate that varies with the environmental conditions, soil characteristics, or rhizobial strain used, among others (Gibson *et al.*, 1976; Keyser *et al.*, 1993). In other studies, the introduced inoculant strains still dominate in soil 5-15 years after introduction (Diatloff, 1977; Brunel *et al.*, 1988: Lindström *et al.*, 1990; see also Figure 2).

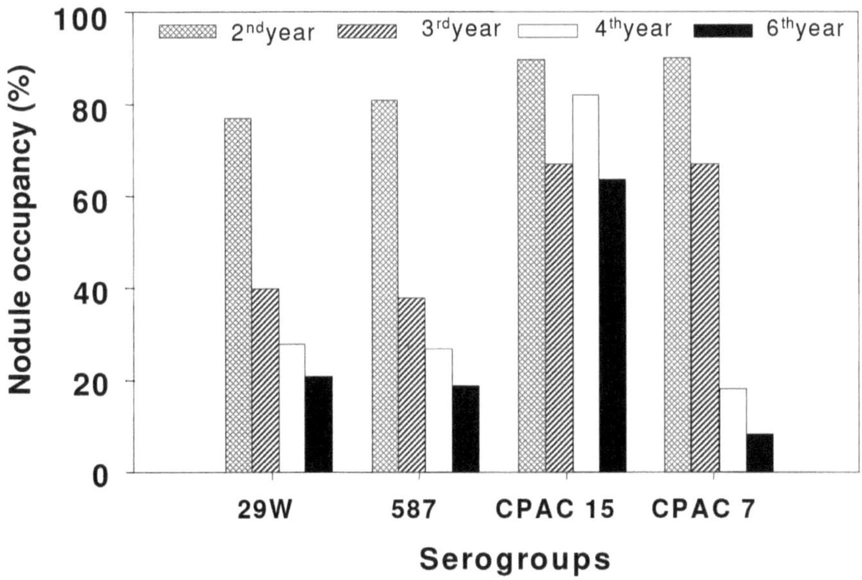

Figure 2. Dynamics of nodule occupancy by four Bradyrhizobium japonicum/B. elkanii strains for six years after their introduction into a Cerrados oxisol originally void of soybean bradyrhizobia. Data represent the mean values of four replicates. Modified from Mendes et al. *(2000).*

Saprophytic capacity is a desirable feature of inoculant rhizobia when one is sure of both their superior N_2-fixation capacity and their genetic stability. Replacing persistent strains with more efficient ones can be difficult as is evident from numerous studies with *B. japonicum* USDA 123 in the USA (Ham *et al.*, 1971). Figure 2 shows differences in the establishment of four soybean bradyrhizobia from Brazilian commercial inoculants in a Cerrados oxisol. In this experiment, the displacement of CPAC 15 (= SEMIA 5079 and belonging to the same serogroup as USDA 123) by other strains required annual and massive reinoculation.

Similar data for the USA is provided by Dunigan *et al.* (1984). van Elsas and Heijnen (1990) have suggested that the ideal situation would be one in which the inoculant organisms could be eliminated from the environment after completing their task. However, in some countries, the need for annual reinoculation might then significantly increase production costs. The molecular tools available today should be used to better follow the introduction, movement, and persistence of inoculant strains, and to determine the factors most important for strain persistence.

3. INOCULANT PRODUCTION

Inoculant production involves choosing and processing the carrier, culture maintenance and growth at increasing scales of production, and a final product of good quality, with a profitable benefit/cost ratio.

3.1. The Carrier

Desirable properties of a good inoculant carrier have been listed before (Keyser *et al.*, 1993; Walter and Paau, 1993; Balatti and Freire, 1996; Stephens and Rask, 2000), but can be summarized as: (i) readily available, uniform in composition and cheap in price; (ii) non-toxic to rhizobia; (iii) high water-retaining capacity; (iv) easily sterilized; (v) readily corrected to a final pH of 6.5 to 7.3; (vi) permitting good initial growth of the target organism; and (vii) maintaining high cell numbers during storage.

Peat has been the most suitable carrier for inoculant production because it usually meets these requirements. Another possible advantage of peat as a carrier is its adsorbent properties, which could reduce the effect of toxins that are built up during growth in fermentors and also lower the impact of bradyoxetin, which is a non-homoserine lactone signal molecule involved in quorum sensing that is produced by stationary-phase cultures and can inhibit *nod*-gene expression by bradyrhizobia (Loh *et al.*, 2002). Different peat sources vary in their capacity to support rhizobial multiplication and survival (Roughley and Vincent, 1967; Roughley, 1970; Somasegaran, 1985; Balatti and Freire, 1996). Among the best sources are peats from Argentina (Tierra del Fuego) and Canada, each with an organic matter content of 40-50%. For other sources, the quality of the peat can be improved by addition of humus.

If the harvested peat is wet, it should first be drained, then sieved to remove coarse material. The peat is then dried to a moisture content of about 5%, with the temperature kept below 100ºC to avoid the generation of toxic substances (Roughley and Vincent, 1967; Roughley, 1970). The peat is then ground because coarse particles adhere poorly to the seed coat. Burton (1967) and Roughley (1970) proposed that the peat be milled to pass a 0.20-0.25 mm sieve, whereas Strijdom and Deschodt (1976) recommended that 50% should pass through a 0.075 mm sieve. As many peat deposits are acid, pH should be corrected, as needed, to 6.5-7.0, usually with finely ground $CaCO_3$ (Roughley, 1970; Cattelan and Hungria, 1994).

Sterilization of the peat prior to inoculation is recommended but, unfortunately, there are still products available manufactured with non-sterile peat. Such inoculants may contain up to 1,000-fold fewer rhizobia than those made with sterilized peat, so reducing shelf life. The problem is even greater for slow-growing rhizobia (Roughley and Vincent, 1967, Roughley, 1970; Date and Roughley, 1977; Somasegaran, 1985; Lupwayi *et al.*, 2000; Stephens and Rask, 2000). Sterilization of the peat also reduces the frequency and level of contamination and, thus, the risk of introducing and disseminating plant, animal, and human pathogens (Lupwayi *et al.*, 2000, Catroux *et al.*, 2001). Difficulties associated with the sterilization of the peat include the identification of an appropriate, but low cost, methodology with high-capacity throughput, the need to follow aseptic procedures during culture addition to the pre-sterilized packaged carrier, and the difficulty in detecting contaminated packages (Smith, 1992; Balatti and Freire, 1996). In countries with high-quality products, such as Argentina, Australia, Canada, Czechoslovakia, France, The Netherlands, New Zealand, South Africa, and Uruguay, sterile products are either the general rule or are mandated by legislation.

For smaller quantities of peat, sterilization by autoclaving may be used with bulk peat autoclaved in either polyethylene bags or autoclave trays covered with foil at 121°C for a period of 1-3 hours (Somasegaran, 1991; Somasegaran and Hoben, 1994; Balatti and Freire, 1996). Other less-frequently used methods include fumigation with ethylene oxide or methyl bromide (Smith, 1992) and microwave radiation (Ferriss, 1984). Each of these methods has significant drawbacks. For large quantities of peat, either nuclear or γ-irradiation has been the sterilization method of choice because a more uniform final-product quality is obtained and rhizobial numbers are usually greater than obtained by autoclaving (Roughley and Vincent, 1967; Stephens and Rask, 2000). The level of γ-irradiation normally employed (5 Mrad) usually produces a sterile product (Parker and Vincent, 1981; Smith, 1992), but contaminants are consistently encountered at even higher levels (Yardin *et al.*, 2000). Smith (1992) suggests that survivors from treatment at 5 Mrad do not seriously affect rhizobial growth and survival and that higher-dosage rates could result in peat toxicity as well as raising production costs (Parker and Vincent, 1981). In Brazil, where current legislation demands that dilution counts from inoculants must be devoid of contaminants at the 10^{-5} dilution, 7 million doses per year are γ-irradiated at an average cost of U$ 0.10 $dose^{-1}$, which is less than 10% of the final price; however, 7 Mrad have to be used for some peat inoculants. A more recent non-nuclear sterilization technique used in Canada uses electron beam acceleration to generate 10^7 eV, and sterilization of pre-packaged peat takes only seconds, whereas γ irradiation takes hours (Stephens and Rask, 2000).

Peat carrier is usually packaged in polyethylene or polypropylene bags, with a thickness of 0.06-0.38 mm. Bags must be resistant to sterilization, allow safe transportation of inoculant, and be readily sealable to prevent contamination. They must retain moisture, so the peat does not dry out, but allow gas exchange; both of these factors are important in retaining rhizobial viability (Roughley, 1970; Keyser *et al.*, 1993). Package sizes vary considerably, generally from 40 g to 2.8 kg.

It is not always possible to use seed-applied peat as a carrier. Many soybean farmers, for instance, complain that peat-based products are time-consuming to use, especially when planting big areas. Some countries may not have natural deposits of peat, the peat available may not be suitable, or environmental regulations may prevent its harvest because many peat reserves are located alongside rivers and streams. A number of materials with the characteristics of a good carrier have been used as alternatives with different degrees of success. These include: vegetable oils (Kremer and Peterson, 1982); mineral oils (Chao and Alexander, 1984); plant materials, such as bagasse, silk cocoon waste (Marufu *et al.*, 1995; Jauhri *et al.*, 1989), sawdust, rice husk (Khatri *et al.*, 1973), and corncob (McLeod and Roughley, 1961); various clays, including vermiculite (Graham-Weiss *et al.*, 1987) perlite mixed with humus (Ballati and Freire, 1996), and diatomaceous earth (Sparrow and Ham, 1983); dehydrated sludge wastewater (Rebah *et al.*, 2002); polyacrylamide (Dommergues *et al.*, 1979) or cellulose gels (Jawson *et al.*, 1989); lignite and derivatives; coal; filter mud; and charcoal-bentonite (Keyser *et al.*, 1993; Marufu *et al.*, 1995). In contrast, survival was poor on a number of materials including diesel oil, some mineral oils, and kerosene (Faria *et al.*, 1985; Peres *et al.*, 1986).

Liquid or gel-based preparations also constitute a significant percentage of the inoculant market. They are of varying composition (magnesium silicate, potassium acrylate-acrylamide, grafted starch, hydroxyethyl cellulose products) and usually include additional proprietary substances to protect the rhizobia. Several studies report good performance of these preparations when compared to peat-treated seeds (Burton and Curley, 1965; Jawson *et al.*, 1989; Hynes *et al.*, 1995). However, in spite of good cell numbers in plate and most probable number (MPN) counts, many of them have failed to reproduce this performance under field conditions, especially under environmental stress in Brazil (Campo and Hungria, 2000b; Hungria *et al.*, 2001b; see Table 2), Uruguay and Canada.

Table 2. Soybean nodulation and yield in a Brazilian Cerrados oxisol as a result of different commercial inoculants. Experiment performed in a soil void of soybean bradyrhizobia and all inoculants contained the same strains[1].

Treatment[2]	Cells g^{-1} or ml^{-1} of inoculant	Dose	Cells $seed^{-1}$	Nodulation at flowering ($mg\ pl^{-1}$)	Yield ($kg\ ha^{-1}$)
C-N[3]				77 efgh	3,202 c
C+N[3]				24 h	3,334 bc
Traditional Peat Inoc.	$7.5x10^8$	$500g.50kg^{-1}$	$3.08x10^8$	232 a	4,226 a
Liquid 1-#1	$2.0x10^8$	$400ml.50kg^{-1}$	$2.80x10^3$	59 fgh	3,461 bc
Liquid 1-#2	$2.0x10^8$	$800ml.50kg^{-1}$	$4.95x10^4$	146 bcde	3,420 bc
Liquid 2	$1.0x10^7$	$200ml.50kg^{-1}$	$1.86x10^3$	43 gh	3,458 bc
Liquid 3	$1.6x10^9$	$150ml.50kg^{-1}$	$9.34x10^4$	133 cdef	3,363 bc
Peat 1	$2.0x10^9$	$20g.50kg^{-1}$	$2.80x10^4$	168 abcd	3,626 bc
Peat 2	$1.0x10^9$	$200g.50kg^{-1}$	$4.67x10^3$	103 defg	3,267 bc

[1]After I. C. Mendes (unpublished data). Within a column, values followed by the same letter are not statistically different (Duncan, $p \leq 0.05$).
[2]Liquid and peat inoculants are from different companies and used as recommended by manufacturers.
[3]Non-inoculated controls without or with 200 kg of N ha^{-1}, split twice - at sowing and at flowering stage.

Addition of proprietary cell protectants, sowing immediately after inoculation (Burton and Curley, 1965), and avoiding the use of fungicides with these products (Campo and Hungria, 2000a; 2000b) increases the probability of success.

Inoculants containing dried (either lyophilized or freeze-dried) and frozen (concentrated) cultures require more complex equipment for production and maintenance (Date, 2001); they are mixed with either a liquid or gel formulation at sowing (Walter and Paau, 1993). Tests performed with either dried or frozen inoculants in Brazil have shown that nodulation is usually lower than with inoculants containing bacteria prepared in the conventional way, probably because the rate of cell growth is lower than in other carriers. However, lyophilized cells may show a better survival in granular inoculants (Fouilleux *et al.*, 1994). Inoculants containing polymer microcapsules, beads, or clay pellets impregnated with rhizobia, and polyacrylamide, alginate, xanthan and carob gums have also been used as inoculants (Jung *et al.*, 1982; Bashan, 1986; Smith, 1992; Walter and Paau, 1993), but survival under dry conditions is often poor (Date, 2001).

3.2. *The Cultures*

3.2.1. *Strain Maintenance*

Keeping a pure strain alive with no variation or mutation is critical to inoculant manufacture. Strains may lose desirable properties either in storage or on repeated subculture. Careful maintenance of stock cultures and periodic testing of symbiotic efficiency are essential. Lapinskas (1990) and Lupwayi *et al.* (2000) urge periodic passage through the host under field conditions to maintain symbiotic efficiency. In countries such as the United States, where strains can be patented and individual inoculant manufacturers decide which strain(s) will be used, the maintenance of cultures is mainly left to the manufacturer. Where the use of particular strains is regulated by legislation, strains are usually kept at a central facility and forwarded to the industry as needed. Several methods for short- and long-term storage have been described (see Table 3). Most long-term storage is by cryopreservation (ultra-cold conditions) or freeze-drying (lyophilization) (Vincent, 1970; Somasegaran and Hoben, 1994; Balatti and Freire, 1996; Lupwayi *et al.*, 2000).

There are several sources of effective rhizobial cultures for research or inoculant production. They include the Microbial Resources Centre Network (MIRCEN), supported by the United Nations Education and Science Council (UNESCO) in Brazil (Porto Alegre), Kenya (Nairobi), Senegal (Dakar) and USA (Niftal, Hawaii (www.unesco.org/science/mircen_centres.html), and the USDA culture collection in Beltsville. The large CSIRO Tropical Pastures collection, previously maintained by Drs. D.O. Norris, R.A. Date and H.V.A. Bushby, is now maintained by Dr A. McInnes at the University of Western Sydney (a.mcinnes@uws.edu.au).

3.2.2. *Culture and Inoculant Production*

Product finishing is essential for good quality inoculants and it involves the steps of culture multiplication, aseptic injection of broth culture into the peat, proper maturation, and adequate packing. Either small or large fermentors made from glass or stainless steel can be used to grow cultures with growth conditions evaluated and optimized for each strain. Many media formulations have been described (Burton, 1967; Vincent, 1970; Roughley, 1970; Somasegaran and Hoben, 1994; Balatti and Freire, 1996), however, Stephens and Rask (2000) note that most of these were developed for general laboratory practice. They recommend a less nutrient-rich medium that is still able to support counts either at or exceeding 10^9 cells mL^{-1}. Several industrial by-products have also been used as carbon and/or nitrogen sources. They include corn steep liquor, proteolysed pea husks, malt extract, cheese derivatives, yeast extract, molasses, and casein hydrolysates. A common problem is of continuous supply and quality with these sources (Keyser *et al.*, 1993; Walter and Paau, 1993; Stephens and Rask, 2000).

Table 3. Method of maintenance, main characteristics, and cell viability related to each method.

Method	Main characteristics	Viability
Agar medium	Medium usually with yeast, mannitol, and salts, kept at 5-6°C for periodic transfer; simple low-cost.	1 year
Agar medium	Covered with sterilized mineral or paraffin oil, kept at 5-6°C; simple and low-cost	2 years
Porcelain beads	Dry suspension of cells on sterilized porcelain beads, kept in a tube with dehydrated silica	2 years
Soil, peat or clays	Preferentially with high water capacity, ground, corrected for chemical properties, and sterilized	2-4 years
Paper strips	Paper strip or disk saturated with a bacterial suspension and dried, kept in the refrigerator.	6 months
Freezing	With temperatures ranging from -70°C to -190°C, in deep freeze or liquid nitrogen. Viability depends on the culture medium, freezing speed, freezing temperature, and type of cryoprotectant used; good viability has been shown in a number of collections after 15-20 years.	From months to several years
Lyophili-zation	Viability depends on the physiological state of the culture, cell concentration, medium, and lyophilization rate; can be kept at room temperature for years, but not much information is available.	From months to several years

An initial inoculum of 0.2-1% or more is used with most fermentors to ensure sufficient growth while reducing the possibility of contamination. Transfers between fermentors may be necessary to obtain the final volume of broth needed. More details about fermentation processes and types of fermentors can be obtained elsewhere (Walter and Paau, 1993; Balatti and Freire, 1996). The factors usually controlled in culture production are the medium, temperature, agitation, pH, and aeration; inoculant batches should be checked for contamination at all steps. Under proper conditions, cell densities of 10^9 to 10^{10} mL^{-1} are usually obtained with dilution possible before injection into pre-sterilized peat carrier (Somasegaran, 1985; Keyser *et al.*, 1993; Balatti and Freire, 1996). Before injection into the carrier, cultures must be tested for pH change, Gram-stain reaction, and the number of viable cells, and contamination must be assessed (Balatti and Freire, 1996). Injection into pre-sterilized peat must be done under aseptic conditions and, when thousands of doses are being manufactured, electronic injection is desirable. In peat inoculants, cultures are usually mixed to establish a 45-60% moisture content on a wet weight basis (Roughley, 1970). For non-sterile peat, either a rotating bowl or a concrete mixer is used to facilitate mixing during broth addition. For pure-culture peat, the inoculated bags are either manually or mechanically agitated to distribute the inoculum and to remove lumps.

The importance of a period of storage (for maturing and curing) after inoculation has been recognized for some time. Burton and Curley (1965) showed that rhizobia, which were allowed to grow and colonize the peat particles after inoculation, were able to survive on seeds in greater number and for longer periods of time than rhizobia freshly adsorbed in peat. Inoculants are usually held at warm room temperature to stimulate multiplication. Materon and Weaver (1985) noted ten-times more growth of *R. leguminosarum* bv. trifollii when peat carriers were stored for at least four weeks after inoculation before being utilized. After curing, the inoculant is usually maintained at 4°C, however, the temperature used should be individually determined because the viability of some strains may decline at this temperature (Somasegaran, 1985). The storage conditions will also influence shelf life, so affecting cell viability, physiological characteristics (such as sensitivity to drying), and the time for colony and nodule appearance (Roughley, 1970; Burton, 1975; Revellin *et al.*, 2000; Catroux *et al.*, 2001). Inoculant labeling should include product registration information and batch numbers, the strain(s) of rhizobia and other microbes included, if genetically modified microorganisms are included, the number of cells guaranteed by the manufacturer, and instructions for use.

3.2.3. Inoculant Quality Control

The quality control of inoculants prepared in pre-sterilized peat is easier because the enumeration of rhizobia can be done by simple plate count methods (Vincent, 1970). For non-sterile carriers, most contaminants will grow faster than the rhizobia and sometimes appear similar to them, which complicates the counting. For non-sterile carriers, selective media that contain antibiotics, heavy metals, bacteriocides and/or fungicides have been described (Vincent, 1970; Tong and Sadowsky, 1994; Gomez *et al.*, 1995), but they are not equally effective for all strains. Plant infection, by MPN counts, should be carefully undertaken with rigid attention to detail. Results may vary not only with the number of serial dilutions, number of replicates, and the volume applied in each replicate, but also with the physiological state of the cells, the concentration of the inoculant, and the time allowed for plant growth (Lupwayi *et al.*, 2000; Catroux *et al.*, 2001). MPN counts are both time- and space-consuming and require about 30 days for plant growth and adequate greenhouse or growth chamber space. For specific strains, methods based on serological properties may be used (Keyser *et al.*, 1993; Lupwayi *et al.*, 2000), but other sophisticated methods to evaluate cell viability have been proposed (Catroux *et al.*, 2001).

At a national level, quality control varies with country. It can be left to the discretion of the manufacturer, as in the United States and United Kingdom, or evaluated through an organization in which the manufacturers participate voluntarily, as in South Africa, New Zealand, and the Australian Inoculant Research and Control Service, or regulated through a governmental institution, as in Brazil (through the Ministry of Agriculture) and Uruguay. Both rhizobial concentration and the level of contaminants are important and the standards vary with the country. In Australia, Canada, Czechoslovakia, France, India, Kenya, New Zealand, Rwanda, South Africa, Russia, Thailand, The Netherlands, and Zimbabwe, inoculants must contain 10^7-10^9 cells g^{-1} at manufacturing or mL^{-1} for shelf life, depending on the country. In these countries, the inoculants must also either be void of contaminants or with no contaminants at the 10^{-6} dilution (Smith, 1992; Lupwayi *et al.*, 2000; Stephens and Rask, 2000). In the countries of Mercosur, the manufacturers can be legally charged if the inoculants contain less than 10^8 cells g^{-1} or mL^{-1} of inoculant (for shelf life), with no contaminants permitted at 10^{-5} dilution.

It is well known that large inocula favor survival in greater numbers (Burton, 1976) and, as the retained inoculum may be as low as 5-10%, it is recommended that inoculant standards are based on numbers delivered per seed (FAO, 1991). Figure 3 shows the relationship between the number of cells applied per seed and the number of nodules produced in a field trial performed in Brazil.

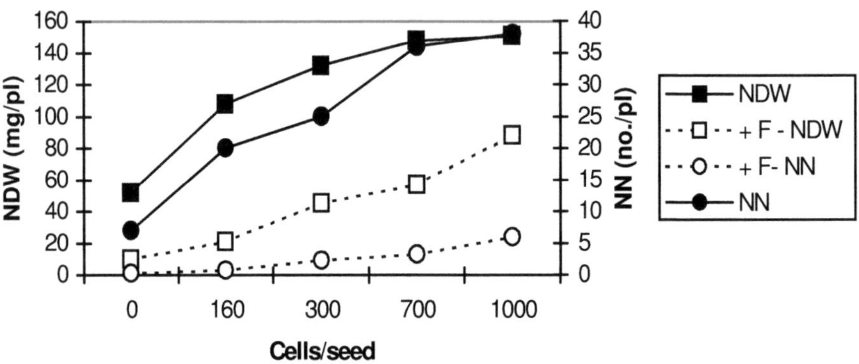

Figure 3. Nodule number (NN) and dry weight (NDW) of soybean cv. BR 16 inoculated with different concentrations of Bradyrhizobium elkanii SEMIA 587 with or without fungicide (F, thiram + thiabendazole).Experiment performed in an oxisol of Ponta Grossa, State of Paraná, Brazil, with less than 10 cells g^{-1}.
After M. Hungria and R.J. Campo (unpublished).

Standards usually dictate a minimum of 10^3 rhizobia $seed^{-1}$ for small-seeded legumes (such as clovers, *Trifolium* spp. and alfalfa, *Medicago sativa*), 10^4 rhizobia $seed^{-1}$ for medium-sized seeds (such as pigeonpea, *Cajanus cajan*), and 10^5 rhizobia $seed^{-1}$ for larger-seeded species (such as soybean, peas (*Vicia* spp.) and beans) (Thompson, 1980; Keyser *et al.*, 1993; Smith, 1992; Lupwayi *et al.*, 2000). In France, with a long tradition of inoculant legislation, a minimum of 10^6 cells $seed^{-1}$ is required for larger-seeded species; a similar level has been proposed in Canada (Lupwayi *et al.*, 2000). In Brazil, where the requirement has been slowly changing over time, the concentration recommended for large seeds is of 600,000 rhizobia $seed^{-1}$. The importance of cell number in the inoculant, when either conditions are adverse or the soil already contains ineffective rhizobia, has often been reported. For soybean, Weaver and Frederick (1974) indicated that cells of the added inoculant needed to outnumber resident rhizobia by 1,000-fold. Furthermore, in nine field sites in New Zealand, clover nodulation in pasture increased from 5% to 66% with an increase in the inoculation rate from $0.2x10^3$ to $260x10^3$ cells of *R. leguminosarum* bv. trifolii per seed (Patrick and Lowther, 1995).

4. INOCULANT APPLICATION

A number of different methods of inoculation are used by farmers, but not all are equally effective. In particular, the practice of mixing peat inoculant with the seeds in the planter box, although popular because of the ease with which it can be accomplished, is not recommended. Most of the inoculant will not stick to the seeds, resulting in non-uniform distribution; inoculant left on the box can gradually plug seeding tubes, so delaying sowing (Cattelan and Hungria, 1994). The sprinkle method, in which seeds are first sprinkled with water and then with dry inoculant powder, is little better. The dry seeds quickly absorb the water, not much is left as an adhesive for the peat, and most of the inoculant does not adhere to the seeds.

4.1. Slurry method

The slurry method is recommended for seed inoculation is the slurry method. First, the inoculant is mixed with a solution containing adhesive, then this slurry is applied to and mixed with the seeds until a uniform coverage is achieved. Seeds are then allowed to dry under cool conditions before sowing. Adhesives ("stickers") commonly used include sucrose (usually as a 10% solution), fungicide- and bactericide-free gum arabic (usually at 40%), carboxy-, methyl-ethylcellulose and methyl-hydroxypropyl cellulose (about 2-4%), or home-made gums prepared from cassava (*Manihot* spp.), starch, or wheat (*Triticum* spp.) flour (Roughley, 1970; Elegba and Rennie, 1984; Faria *et al.*, 1985; Cattelan and Hungria, 1994; Horikawa and Ohtsuka, 1996a). Either carpenter's glue or wallpaper adhesive have also been used, but some of these products contain fungicides that can kill the rhizobia. Slurry volume, in the case of soybean, should not exceed 300 ml per 50 kg of seeds. Peat inoculant is applied at rates of 1-10g kg^{-1} seed, depending on the cell concentration.

Table 4 shows the importance of sucrose solution, the most popular adhesive in Brazil. Without it, almost 50% of the peat inoculant was lost as the seeds dried.

Table 4. Slurry method of inoculation: Effect of different sucrose solution concentrations in the adherence of a peat inoculant onto the seeds and on grain yield of soybean cv. BR-37. The adhesive and the inoculant (10^8 cells g^{-1}) were applied at a dose of 300 mL of sucrose solution per 500 g of inoculant per 50 kg of seeds. Experiments carried out in Londrina and Ponta Grossa, State of Paraná, Brazil, in soils with established populations of soybean bradyrhizobia[1].

Sucrose concentration	Inoculant adherence (%)	Inoculant adherence (g)	Grain yield (kg ha^{-1}) Londrina	Grain yield (kg ha^{-1}) Ponta Grossa
0	48.2 b[2]	241.0	2,692 ab[2]	2,312 a[2]
10%	91.5 a	457.4	2,952 a	2,290 a
15%	92.0 a	460.0	2,568 b	2,460 a
20%	88.0 a	440.0	2,680 ab	2,393 a
25%	80.9 a	404.5	2,710 ab	2,363 a
CV (%)	13.0	13.0	10.7	13.7

[1]After Brandão Junior and Hungria (2000).
[2]Data for each site represent the means of two experiments, each with six replicates, and when followed by the same letter, within the same column, do not show statistical difference (Duncan, $p \leq 0.05$).

In this experiment, it is also important to emphasize that the use of sucrose did not result in either seed diseases or in changes of seed vigor (Brandão Junior and Hungria, 2000). Sucrose may also act as a cell protectant, increasing cell viability, and allowing the storage of the inoculated seed under dry, cool conditions for periods of up to a week before sowing (Burton, 1975; Peres *et al.*, 1986). One important factor not often considered is the quality of the water used to make the slurry. Many farmers use containers previously used for fertilizers or fungicides, both of which are toxic to rhizobia. A second problem, also evident in Table 2, is that, although the peat inoculant may be of high quality, the rate of application may be too low. As a result, the number of cells applied seed^{-1} may be inadequate and some of the benefits of inoculation lost.

Because of the time consumed in slurry inoculation at sowing, low-cost machines, which have separate compartments for the peat inoculant and slurry, have been developed in Brazil. The seeds may also receive liquid fungicides and micronutrients (Hungria *et al.*, 2001b). These machines are designed for 60 bags of seeds (50 kg bag^{-1}) h^{-1}.

4.2. Seed Pelleting

Seed pelleting is used where conditions at sowing are either less than desirable (either high temperature or low soil pH) or the seed is sown from the air. In this procedure, the seed is slurry inoculated using a strong adhesive, such as 40% gum arabic, it is then rolled in very finely ground calcium carbonate or rock phosphate or clay to form the pellet (Roughley, 1970; Burton, 1975; Smith, 1992; Thompson and Stout, 1992; Horikawa and Ohtsuka; 1996b). In Australia, specific micronutrients were also added to the seed-coating material and usually showed good results in acid soils; however, negative results were obtained with Mo and Co added to seed (see Table 6). The amount of seed-coating material and adhesive depends on the seed size. Another variation in this method is the use of mineral microgranules amended with nutrients and inoculated with either peat or liquid inoculants. Fouilleux *et al.* (1996) reported a significant increase in both early nodulation and N grain content, and better survival of *Bradyrhizobium* in soil undergoing desiccation, using this procedure.

4.3. Pre-inoculation

Although rhizobial numbers are usually greater on freshly inoculated seeds than on seeds that have been preinoculated for subsequent sale (Rice *et al.*, 2001), pre-inoculation of seed can be useful either when sowing large areas in a short time or when the weather is unstable. For forage legumes, such as alfalfa (*Medicago sativa*), cell viability can be maintained even when pre-inoculation is performed several months before planting (Smith, 1992; Rice *et al.*, 2001). However, only highly reputable seed companies should be considered for the purchase of pre-inoculated seed because rhizobial numbers seed^{-1} can be dramatically reduced in pre-inoculated seed that has been improperly stored (Thompson *et al.*, 1975). An alternative, which has been used to increase cell viability, is pre-inoculation for sowing within, at most, 10 days of seeding. Even here, the quality of the product can be erratic (Brockwell and Bottomley, 1995). Vacuum processing and the use of adhesives with alfalfa can also markedly decrease both the viability of cells and nodulation, even when stored at cool conditions (Horikawa and Ohtsuka, 1996a; 1996b).

4.4. Soil Inoculation

For leguminous crops such as soybean, dry beans, and particularly peanut, a disadvantage of seed inoculation (with either peat or liquid inoculants) is the incompatibility between inoculant strains and seed-applied fungicides, insecticides, or micronutrient preparations. Another constraint is the limited number of cells that can be applied to small-seeded legumes, particularly under difficult seeding conditions. It is well known that inoculants should be placed as close as possible to the seeds because the movement of rhizobia in soil is limited (McDermott and Graham, 1989). Chamblee and Warren (1990) reported a lateral movement of rhizobia of only 15 cm in an 11-month period.

To overcome the limitations described above, the inoculant can be applied directly to the seed furrow in the soil as granules, peat, or liquid, separated from the seeds at planting time. The inoculants are not mixed with fertilizer, which can be injurious to the rhizobia, but separately banded into the soil. A disadvantage of this procedure is the higher cost because the quantity of inoculant used is higher than that used for seed inoculation. For peat inoculants, the rate of application is usually either 6-20 kg ha^{-1} or 1 g m^{-1} of row (FAO, 1984; CIAT, 1988) and usually the cost limits its applications (Walter and Paau, 1993). Gault *et al.* (1982) mentioned a variation of the standard seed-applied peat-powder method, which consists of suspending peat in water, screening the suspension to remove large peat fibers, diluting with water in a spray tank, and spraying the slurry into the furrow. The inoculation of soybean grown in a first-year area through irrigation with a peat slurry at rates of 2, 4, 6 and 8 kg ha^{-1} also resulted in good nodulation (Smith, 1992).

Granular inoculants may use peat preparations milled and sieved to provide particles between 0.35 mm and 1.18 mm in size. These absorb the culture rapidly and, after being cured, flow uniformly through a granular applicator. Such granular peat preparations deliver at least 10^{11} cells ha^{-1} and perform comparably to seed-applied inoculants (Smith, 1992; Lupwayi *et al.*, 2000).

Broth inoculants usually packaged in dispenser bottles can also be delivered directly into the seed bed, using an inoculant tank, a pump, a manifold, and capillary tubes to deliver the liquid culture. The equipment is not expensive and is in use in Australia, the United States, and Brazil (Hely *et al.*, 1976; Brockwell and Bottomley, 1995; Campo and Hungria, 2002b). The depth at which the liquid inoculant is placed is also important, *e.g.*, soybean nodulation was superior when the inoculant was applied either to the seed in the furrow or 2.5 cm below the seed as compared to application at both 5.0 cm and 7.5 cm below the seed (Smith, 1992). In Brazil, to avoid toxicity from seed-applied micronutrients and fungicides, Campo and Hungria (2002b) had to use eight-times more liquid inoculant in the seed bed, and up to ten-times more is recommended by Urbana Laboratories (2002), indicating that the procedure is useful only under specific conditions.

5. FACTORS AFFECTING THE SUCCESS OF INOCULATION

5.1. Effects of Inoculation with Selected Strains in the Presence of an Established Population

We will not discuss the soil and environmental factors that can affect the success of inoculation. Some of those factors have been reviewed (*e.g.*, Cattelan and Hungria, 1994; Hungria and Vargas, 2000) and will be discussed elsewhere in this volume.

There are numerous reports in which the use of inoculant-quality strains, which were applied to legumes in soils with a low level of soil N and few indigenous rhizobia, resulted in measurable benefits in terms of nodulation, N accumulation, plant biomass, and grain yield. Benefits are much less common, however, in soils having either indigenous or established rhizobial populations. Populations of soil rhizobia as low as 20-100 cells g^{-1} of soil can limit the response of both soybean and common bean to inoculation (Dunigan *et al.*, 1984; Singleton and Tavares, 1986; Thies *et al.*, 1991; Nazih and Weaver, 1994). When soils are devoid of rhizobia and inoculation is needed to ensure adequate nodulation and N supply, farmers tend to inoculate and follow recommended procedures (Smith, 1992; Hall and Clark, 1995).

However, when the soil contains established rhizobial populations and inoculation is used as insurance, attention to proper inoculation practice can be limited.

A lack of benefit from inoculation in soils containing established populations of root-nodule bacteria should not be taken for granted. There are numerous reports of positive responses to inoculation of both soybean and common bean in Brazil in soils containing high numbers of indigenous rhizobia (Vargas *et al.*, 1992; Peres *et al.*, 1993; Nishi *et al.*, 1996; Hungria *et al.*, 1998; Hungria and Vargas, 2000; Hungria *et al.*, 2000a; 2000b; 2001b; 2002). When 13 experiments performed with soybean in several Brazilian states were analyzed, re-inoculation increased nodule number and dry weight, and nodule occupancy by the inoculated strain in the majority of studies. In nine of the thirteen experiments, re-inoculation significantly increased yield and total N content of grains (Hungria *et al.*, 2000b; see Table 5). When the results of field trials performed in other seasons and sites were added to this data set, the national mean increase in grain yield was estimated at 4.5% (Hungria *et al.*, 2001b). It is interesting to note that, with both common bean (Hungria *et al.*, 2000a; Mostasso *et al.*, 2002) and soybean (Campo and Hungria, 2000a), a further increase in both nodulation and yield was obtained by the re-inoculation with the same selected strains in the second year of application. More studies in this area are warranted.

Table 5. Mean and maximum percentage increases in yield (kg ha^{-1}) and total N in grains (kg N ha^{-1}) due to the inoculation with the combination of strains Bradyrhizobium elkanii SEMIA 587 and B. japonicum *CPAC 7 (=SEMIA 5080), when compared to the non-inoculated control. The increases were obtained in thirteen experiments performed in two Brazilian Regions, in soils with established population of soybean bradyrhizobia*[1,2].

Region	Grain yield (% increase)		Total N in grains (% increase)	
	Mean	Maximum	Mean	Maximum
Central-West	7.8	23	8.1	25
South	3.8	20	4.3	24

[1]After Hungria *et al.* (2000b).
[2]Each experiment was performed with four to six replicates and, after the multivariate analysis, the data presented in this table were statistically significant (Duncan, $p \leq 0.05$).

Differences between North American and Brazilian reports on response to inoculation in soybean and bean are intriguing. Several conditions in the tropics, including high soil temperature, acid pH, limited soil moisture, and perhaps even micronutrient availability, could affect the physiological properties and activity of soil rhizobia and explain, at least partially, the more positive results to inoculation evident in these regions (Hungria and Vargas, 2000). Differences could also be related either to the higher competitiveness of the established population of soybean bradyrhizobia in North American soils (Ham *et al.*, 1971; Weber *et al.*, 1989) or perhaps to differences in their distribution in soil (McDermott and Graham, 1989). However, as pointed out before (Santos *et al.* 1999; Hungria *et al.*, 2001b; Ferreira and Hungria, 2002), it is also possible that the selection program in Brazil has paid more attention to the search for more efficient, competitive, and adapted strains.

5.2. Seed Treatment with Fungicides and other Agrochemicals

Seed treatment with fungicides has been an increasing problem that affects inoculation success in beans and soybeans. Insecticides and herbicides applied at sowing can also inhibit nodulation, N_2 fixation and yield (De Polli *et al.*, 1986; Evans *et al.*, 1991; Cattelan and Hungria, 1994; Campo and Hungria, 2000a; 2000b). In Brazil, more than 90% of soybean seeds are treated with fungicides, and cell death rates of up to 70%, after only two hours of contact with fungicides, have been reported (Table 6). Reduction of nodulation under field conditions has also been reported (Table 7). These effects are most severe in first-year cropping areas, mainly in sandy soils, but also in areas with established populations. Selecting strains with higher tolerance to fungicides is not easy (Evans *et al.*, 1989) and only limited efforts made to use compounds less toxic to the rhizobia.

Table 6. Percentage of Bradyrhizobium japonicum cell death rate in inoculated soybean seeds two hours after the treatment with fungicides or micronutrients[1].

Fungicide	Death (%)	Micronutrient[3]	Death (%)
Control[2]	0	Control[2]	0
Benomyl + captan	62	Sodium molybdate	46
Benomyl + thiram	41	Ammonium molybdate	41
Carbendazin + captan	60	Molybdenum trioxide	37
Carbendazin + thiran	64	Molybdic acid	78
Thiabendazole + captan	28	Commercial product 1	97
Thiabendazole + thiran	24	Commercial product 2	28

[1]Adapted from Campo and Hungria (2000a).
[2] Peat inoculant applied as a slurry (10% sucrose).
[3]The chemical compounds were applied at the doses of 20 g of Mo and 5 g of Co ha^{-1} and the commercial products according to the manufacturer.

One possible reason for the success of *Rhizobium tropici* strains in both Brazil and north-central Minnesota (Estevez de Jensen *et al.*, 2002) could be the marked tolerance of Type IIB, but not IIA, strains to both streptomycin and captan (B. Tlusty and P.H. Graham, unpublished). Integrated root-disease management strategies, including inoculation with biological control agents (Estevez de Jensen *et al.*, 2002), are needed. In Canada, fungicide-treated seeds require 2-3 times the usual inoculation rate (Agriculture and Agri-Food Canada, 2002). Vincent (1958) pointed out that peat may shelter the rhizobia from toxic substances in the seed coat. Consistent with this observation, when liquid inoculants were tested in the presence of several fungicides in Brazil, the cell death rate was higher than in the presence of peat inoculant (Campo and Hungria, 2000a; 2000b).

Table 7. Effects of seed treatment with systemic and non-systemic fungicides on both nodulation (NN, nodule number per plant) and decrease in nodulation in relation to the non-treated seeds (%) in experiments performed in either first-year cropping areas (Terra Roxa and Vera Cruz, State of Paraná, <10 cells g^{-1}) or areas with an established population of soybean bradyrhizobia (Cristalina, State of Goiás, 10^5 cells g^{-1} soil)[1].

Treatment	New areas				Old area	
	Terra Roxa (sandy soil)		Vera Cruz (clay soils)		Cristalina (sandy soil)	
	NN	(%)	NN	(%)	NN	(%)
Non- inoculated	1	-	5	-	34	-
Inoculated (I, peat, $3x10^5$ cells $seed^{-1}$)	23	0	34	0	44	0
I + Benomyl + Captan	6	74	26	24	-	-
I + Benomyl + Thiram	5	78	27	21	-	-
I + Benomyl + Tolylfluanid	5	78	25	27	-	-
I + Carbendazin + Captan	11	52	33	3	-	-
I + Carbendazin + Thiram	5	78	28	18	38	14
I + Carbendazin + Tolylfluanid	4	83	26	24	-	-
I + Carboxin + Thiram	14	39	29	15	33	25
I + Difenoconazole + Thiram	13	43	30	12	-	-
I + Thiabendazole + Captan	3	87	25	27	-	-
I + Thiabendazole+ Thiram	7	70	23	32	-	-
I + Thiabendazole + Tolylfluanid	5	78	32	6	34	23
C.V (%)	53	-	30	-	-	-
LSD (5%)[1]	3.6	-	6.2	-	-	-

[1]After Hungria *et al.* (2001b).
[2]Means of six replicates and values followed by the same letter do not show statistical difference (Test "t", $p \leq 0.05$).

The use of micronutrients, mainly Mo and Co, at sowing is also increasing as a result of both the continuous cropping of soils and the depletion of nutrient reserves. The importance of these two micronutrients for nodulation and yield in Brazilian soils cropped for several years has been consistently demonstrated with yield increases of up to 82% (Campo and Hungria, 2000a; 2002a; Campo *et al.*, 2000; Hungria *et al.*, 2001b).

However, micronutrient preparations may also cause a dramatic decline in inoculant-cell survival on seeds (see Table 6 above; Date and Hillier, 1968; Graham *et al.*, 1974). Most commercial products tested in Brazil had a pH below 3.5 and toxic concentrations of nutrients, but some of the manufacturers used balanced formulas with neutral pH that decreased cell death (see Product 1, Table 6). A strategy to avoid this problem has been to produce seeds enriched in Mo, obtained by spraying seed nurseries with micronutrients after flowering (Campo and Hungria, 2002a).

5.3. Inoculation under Unfavorable Conditions

Under unfavorable conditions, peat is the most suitable carrier. It does not necessarily have to be applied to the legume for which it is finally intended. One approach has been to inoculate the previous crop and success was obtained when wheat (*Triticum aestivum*), rice (*Oryza sativa*), or maize (*Zea mays*) were inoculated in anticipation of subsequent seeding with soybean, green gram (*Vigna radiata*), or groundnut (*Arachis hypogeae*) (Diatloff, 1969; Gaur *et al.*, 1980; Peres *et al.*, 1989). In most instances, this inoculation of a surrogate host appears to have limited practical value. It could be important, however, where soil conditions mitigate against rhizobial survival on the intended host, *e.g.*, with fungicide-treated seed. With rhizobial cell numbers on alternate hosts, such as wheat, increased by 10-fold to 300-fold following inoculation, it is possible that initial strain establishment under harsh environmental conditions might be better on cereals than on the legume host, but this remains to be determined. One situation where benefits are likely is in the revegetation of disturbed landscapes in the northern USA and Europe. The practice in Minnesota is to seed and rake in the cover crop before the first frost, but to broadcast legume seed on the surface. Rhizobia on such seed have to persist through the harsh winter period, with germination of the host legume delayed as much as 8-9 months after seeding (P.H. Graham and B. Tlusty, unpublished).

Inoculation of seedlings in either a greenhouse or nursery can be very useful in forestry to guarantee a successful nodulation (Keyser *et al.*, 1993) and a combination of seed and soil inoculation may also be recommended for unfavorable conditions (Brockwell *et al.*, 1985; Cattelan and Hungria, 1994). Post-planting inoculation can be either intentional or remedial (to correct poor nodulation). Among the alternatives mentioned in the literature are inoculation through a center pivot irrigation (Smith, 1992), sub-surface granular applications, and surface spray application (Rogers *et al.*, 1982). A three-year experiment with soybean showed that, in addition to seed inoculation, a cover inoculation of the soil with irrigation water (at either the time of sowing or at the three-node V3 stage) with peat inoculant in suspension increased nodulation by 1.4-times to 2.4-times (Ciafardini and Lombardo, 1991). Both positive and negative results are mentioned in the literature as a consequence of post-planting inoculation and it seems that the success of this procedure is related not only to the method of application, but also to the timing of the application and the soil conditions during inoculant delivery. Given the life cycle of most crop hosts, it is not to be expected that post-emergent inoculation would be effective more than 30 days after sowing.

A major factor affecting the success of inoculation is the application of N-fertilizers. In high profit crops, such as soybean, there is an increasing pressure to sell N-fertilizers to the farmers. There are some reports of benefits due to the application of starter N (van Kessel and Hartley, 2000) but, in Brazil, doses as low as 20-40 kg of N ha^{-1} have substantially decreased both nodulation and N_2 fixation with no yield benefit (Crispino *et al.*, 2001; Hungria *et al.*, 2001b). van Kessel and Hartley (2000) analyzed 600 experiments performed over a 25-year period and concluded that the contribution of N_2 fixation to crop growth and yield decreased after 1985 for soybean and after 1987 for common bean. They suggested that a major factor was the increased use of N-fertilizers worldwide. Therefore, more effort should be put into the use of inoculants and not fertilizers in legume crops.

It is also important to remember that, although the success of inoculation is related to good soil and crop management, what is good management in one environment may not apply in another. Thus, although higher rates of N_2 fixation have been reported under no-tillage conditions in the tropics and subtropics (Campo and Hungria, 2000b; Ferreira *et al.*, 2000; Hungria and Vargas, 2000; van Kessel and Hartley, 2000), the cooler spring temperatures under no-till conditions in the northern USA and Canada can delay nodulation and perhaps inhibit *nod*-gene expression (Zhang and Smith, 1996). In places where production or utilization of inoculants is limited, *e.g.*, in Asia and Africa (Eaglesham, 1989), new approaches should be sought for the use of promiscuous soybean cultivars that are able to effectively nodulate with indigenous bradyrhizobia (Mpepereki *et al.*, 2000).

5.4. Inoculation and Co-inoculation with other Microorganisms

Azospirillum is another diazotrophic and plant growth-promoting bacteria tested in multiple inoculation trials (Okon and Labandera-González, 1994). In trials in Mexico, yield differences in maize inoculated with *A .brasilensis* ranged from 15% to 78% (Y.Okon, pers. comm.) and more than two million ha cropped with maize is now being inoculated. Important results have also been obtained with micro-propagated sugarcane (*Saccharum officinalis*) inoculated with *Gluconoacetobacter diazotrophicus* in Brazil. We will not detail experiments involving co-inoculation, but several papers point either to synergism between co-inoculated rhizobia and plant growth-promoting bacteria (Smith, 1992; Okon and Labandera-González, 1994; Burdman *et al.*, 1996) or to benefits from inoculation with species of *Bacillus* having biocontrol activity (Araújo and Hungria, 1999; Estevez de Jensen *et al.*, 2002) among several others (van Elsas and Heijnen, 1990; Walter and Paau, 1993). Rhizobial inoculation can also stimulate other microorganisms, *e.g.*, root colonization by mycorrhizal fungus (Xie *et al.*, 1995), seedling emergence, and grain and straw yields of lowland rice (*Oryza sativa* L.) (Biswas *et al.*, 2000). Undoubtedly, there is a need for a range of additional studies to integrate the use of these different organisms, either alone or in combination, in agriculture.

6. MAIN CONCLUSIONS

The advances made in recent years have shown that it is possible to obtain inoculants with high rhizobial counts, which are free of contaminants and with a longer shelf life. Alternative carriers and technologies of inoculation have also been identified. Used appropriately, inoculants prepared using these methodologies can be important to agricultural sustainability, particularly in those countries where leguminous plants play a key role in the economy. Brazil, for example, has benefited enormously from an emphasis on nodulation and nitrogen fixation in crop and pasture species. However, the potential benefits of inoculation are often limited by the poor quality of inoculants either available in the market or used by farmers. Further improvements in the technical requirements for improved inoculant production and in their quality control are needed. The transference of existing or improved technologies to different agro-ecosystems can also be limited by political decisions and burocracy in extension agencies. Better communication and interaction between scientists, extension agents, and farmers is also needed as pointed out by Hall and Clark (1995) for Thailand and by Marufu *et al.* (1995) for Africa. To convince politicians and governmental institutions of the benefits of inoculation, greater emphasis should also be given to economic studies (like those performed by Panzieri *et al.*, 2000) that, in the great majority of cases, will demonstrate the value of using inoculants compared to chemical fertilizers.

ACKNOWLEDGEMENT

The work in Brazil is partially supported by CNPq (PRONEX and 520396/96-0).

REFERENCES

Agriculture and Agri-Food Canada (2002). Retrieved November 06, 2002, from http://res2.agr.ca/lacombe/pdf/hilites/00ResHiSpr.pdf.

Araújo, F. F. de, and Hungria, M. (1999). Nodulação e rendimento de soja co-inoculada com *Bacillus subtilis* e *Bradyrhizobium japonicum/B. elkanii*, *Pesq. Agropec. Bras., 34*, 1633-1643.

Bailey, L. D. (1988). Influence of single strains and a commercial mixture of *Bradyrhizobium japonicum* on growth, nitrogen accumulation and nodulation of two early-maturing soybean cultivars. *Can. J. Plant Sci., 69*, 41-418.

Balatti, A. P., and Freire, J. R. J. (Eds.) (1996). *Legume Inoculants. Selection and Characterization of Strains. Production, Use and Management.* La Plata: Editorial Kingraf.

Balatti. P. A., and Pueppke, S. G (1992). Identification of North American soybean lines that form nitrogen-fixing nodules with *Rhizobium fredii* USDA 257, *Can. J. Plant Sci., 72*, 49-55.

Bashan, Y. (1986). Alginate beads as synthetic carriers for slow release of bacteria that affect plant growth. *Appl. Environ. Microbiol., 51*, 1089-1098.

Biswas, J. C., Ladha, J. K., Dazzo, F. B., Yanni, Y. G., and Rolfe, B. G. (2000). Rhizobial inoculation influences seedling vigor and yield of rice. *Agron. J., 92*, 880-886.

Brandão Junior, O., and Hungria, M. (2000). Efeito de concentrações de solução açucarada na aderência do inoculante turfoso às sementes, na nodulação e no rendimento da soja, *Rev. Bras. Ci. Solo, 24*, 515-526.

Brockwell, J., and Bottomley, P. J. (1995). Recent advances in inoculant technology and prospects for the future. *Soil. Biol. Biochem., 27*, 683-697.

Brockwell, J., Gault, R. R., Chase, D. L., Turner, G. L., and Bergersen, F. J. (1985). Establishment and expression of soybean symbiosis in a soil previously free of *Rhizobium japonicum. Aust. J. Agric. Res., 36*, 397-409.

Brockwell. J, Bottomley, P. J., and Thies, J. E. (1995). Manipulation of rhizobia microflora for improving legume productivity and soil fertility: A critical assessment. *Plant Soil, 174*, 143-180.

Brunel, B., Cleyet-Marel, J. C., Normand, P., and Bardin, R. (1988). Stability of *Bradyrhizobium japonicum* inoculants after introduction into soil. *Appl. Environ. Microbiol., 54*, 2636-2642.

Buendía-Clavería, A. M, Rodriguez-Navarro, D. N., Santamaría-Linaza, C., Ruíz-Saínz, J. E., and Temprano-Vera, F. (1994). Evaluation of the symbiotic properties of *Rhizobium fredii* in European soils. *Syst. Appl. Microbiol., 17*, 155-160.

Burdman, S., Sarig, S., Kigel, J., and Okon, Y. (1996). Field inoculation of common bean (*Phaseolus vulgaris* L.) and chickpea (*Cicer arietinum* L.) with *Azospirillum brasilense* strain CD. *Symbiosis, 21*, 41-48.

Burton, J. C. (1967). *Rhizobium* culture and use. In H. J. Peppler (Ed.), *Microbial Technology* (pp. 1-33). New York, NY: Van Nostrand-Reinhold.

Burton, J. C. (1975). Methods of inoculating seeds and their effect on survival of rhizobia. In P.S. Nutman (Ed.), *Symbiotic Nitrogen Fixation in Plants* (pp. 175-189), (International Biological Programme, 7). Cambridge, UK: Cambridge University Press.

Burton, J. C. (1976). Problems in obtaining adequate inoculation of soybeans. *World Soybean Res.* Sept, 170-179.

Burton, J. C. (1981). *Rhizobium* inoculants for developing countries. *Trop. Agric.*, *58*, 291-295.

Burton, J. C., and Curley, R. L. (1965). Comparative efficiency of liquid and peat-based inoculants on field-grown soybeans (*Glycine max*). *Agron. J.*, *57*, 379-381.

Buttery, B. R., Park S. J., and Findlay, W. J. (1987). Growth and yield of white bean (*Phaseolus vulgaris* L.) in response to nitrogen, phosphorus and potassium fertilizer and to inoculation with *Rhizobium. Can J. Plant Sci.*, *67*, 425-432.

Campo, R. J., Albino, U. B., and Hungria, M. (2000). Importance of molybdenum and cobalt in the biological nitrogen fixation. In F. O. Pedrosa, M. Hungria, M. G Yates and W. E. Newton (Eds.), *Nitrogen Fixation: From Molecules to Crop Productivity* (pp. 597-598). Dordrecht, The Netherlands: Kluwer Academic Press.

Campo, R. J., and Hungria, M. (2000a). *Compatibilidade do Uso de Inoculantes e Fungicidas no Tratamento de Sementes de Soja* (Boletim de Pesquisa, 4). Londrina, Brazil: Embrapa Soja.

Campo, R. J., and Hungria, M. (2000b). Inoculação da soja em plantio direto. In *Anais do Simpósio sobre Fertilidade do Solo e Nutrição de Plantas no Sistema Plantio Direto* (pp. 146-160). Ponta Grossa, Brazil: Associação dos Engenheiros Agrônomos.

Campo, R. J., and Hungria, M. (2002a). Importância dos micronutrientes na fixação biológica do nitrogênio. In O.F. Saraiva and C.B. Hoffman-Campo (Orgs.), *Perspectivas do Agronegócio da Soja* (pp. 355-366). Londrina, Brazil: Embrapa Soja.

Campo, R. J., and Hungria, M. (2002b). Método alternativo para aplicação de inoculantes na presença de micronutrientes e fungicidas. In *Resumos da Fertbio 2002* (CD Rom). Rio de Janeiro, Brazil: UFRRJ/SBCS.

Cassman, K. G. (1999). Ecological intensification of cereal production systems: Yield potential, soil quality and precision agriculture. *Proc. Natl. Acad. Sci. USA, 96*, 5952-5959.

Catroux, G., Hartmann, A., and Revellin, C. (2001). Trends in rhizobial production and use. *Plant Soil, 230*, 21-30.

Cattelan, A. J., and Hungria, M. (1994). Nitrogen nutrition and inoculation. In *Tropical Soybean Improvement and Production* (pp. 201-215). Rome, Italy: FAO.

Centro Internacional de Agricultura Tropical (CIAT) (1988). *The Legume-Rhizobium Symbiosis: Evaluation, Selection and Agronomic Mañagement.* Cali, Colombia: CIAT.

Chamblee, D. S., and Warren, Jr., R. D. (1990). Movement of rhizobia between alfalfa plants. *Agron. J.*, *82*, 283-286.

Chao, W. L., and Alexander, M. (1984). Mineral soils as carriers for rhizobium inoculants. *Appl. Environ. Microbiol.*, *47*, 94-97.

Chatterjee, A., Balatti, P. A., Gibbons, W., and Pueppke S. G. (1990). Interactions of *Rhizobium fredii* USDA 257 and nodulation mutants derived from it with the agronomically improved soybean cultivar McCall. *Planta, 180*, 303-311.

Chen, L. S., Figueredo, A., Pedrosa, F. O., and Hungria, M. (2000). Genetic characterization of soybean rhizobia in Paraguay. *Appl. Environ. Microbiol., 66*, 5099-5103.

Chen, L. S., Figueredo, A., Villani, H., Michajluk, J., and Hungria, M. (2002). Diversity and symbiotic effectiveness of rhizobia isolated from field-grown soybean in Paraguay. *Biol. Fert. Soils, 35*, 448-457.

Chen, W. X., Yan, G. H., and Li, J. L. (1988). Numerical taxonomic study of fast-growing soybean rhizobia and a proposal that *Rhizobium fredii* be assigned to *Sinorhizobium* gen. nov., *Int. J. Syst. Bacteriol.*, *38*, 392-397.

Chueire, L. M. O., and Hungria, M (1997). N_2-fixation ability of Brazilian soybean cultivars with *Sinorhizobium fredii* and *Sinorhizobium xinjiangensis, Plant Soil, 196*, 1-5.

Ciafardini, G., and Lombardo, G. M. (1991). Nodulation, dinitrogen fixation, and yield improvement in second-crop soybean cover inoculated with *Bradyrhizobium japonicum. Agron. J.*, *83*, 622-625.

Cregan, P. B., and Keyser, H. H. (1988). Influence of *Glycine* spp. on competitiveness of *Bradyrhizobium japonicum* and *Rhizobium fredii*. *Appl. Environ. Microbiol., 54*, 803-808.

Crispino, C. C., Franchini, J. C., Moraes, J. Z., Sibaldelle, R. N. R., Loureiro, M. F., Santos, E. N., *et al.* (2001). *Adubação Nitrogenada na Cultura da Soja* (Comunicado Técnico, 75). Londrina, Brazil: Embrapa Soja.

Date, R. A. (2001). Advances in inoculant technology: A brief review. *Aust. J. Exp. Agric., 41*, 321-325.

Date, R. A., and Hillier, G. R. (1968). Molybdenum application in the lime of lime-pelleted subterranean clover seed. *J. Aust. Inst. Agric. Sci., 34*, 171-172.

Date, R. A., and Roughley, R. J. (1977). Preparation of legume seed inoculants. In R. W. F. Hardy and A. H. Gibson (Eds.), *A Treatise on Dinitrogen Fixation, Section IV, Agronomy and Ecology* (pp. 243-275). New York, NY: John Wiley and Sons.

De Polli, H., Souto, M., and Franco, A. A. (1986). *Compatibilidade de Agrotóxicos com Rhizobium spp. e a Simbiose das Leguminosas* (Documento, 3). Seropédica, Brazil: Embrapa-Uapnpbs.

Diatloff, A. (1969). The introduction of *Rhizobium japonicum* to soil by seed inoculation of non-host legumes and cereals. *Austr. J. Exp. Agric. Anim. Husb., 9*, 357-360.

Diatloff, A. (1977). Ecological studies of root nodule bacteria introduced into field environments. 6. Antigenic and symbiotic stability in *Lotononis* rhizobia over a 12 year period. *Soil. Biol. Biochem., 9*, 85-88.

Döbereiner, J., Franco, A. A., and Guzmán, I. (1970). Estirpes de *Rhizobium japonicum* de excepcional eficiência. *Pesq. Agropec. Bras., 5*, 155-161.

Dommergues, Y. R., Diem, H. G., and Divies, C. (1979). Polyacrylamide-entrapped *Rhizobium* as an inoculant for legumes. *Appl. Environ. Microbiol., 37*, 779-781.

Dunigan, E. P., Bollich, P. K., Huchinson, R. L., Hicks, P. M., Zaunbrecher, F. C, Scott, S. G., and Mowers, R. P. (1984). Introduction and survival of an inoculant strain of *Rhizobium japonicum* in soil. *Agron. J., 76*, 463-466.

Eaglesham, A. R. J. (1989). Global importance of *Rhizobium* as an inoculant. In R. Campbell and R. M. McDonald (Eds.), *Microbial Inoculation of Crop Plants* (pp. 29-48). Oxford, UK: Oxford University Press.

Elegba, M. S., and Rennie, R. J. (1984). Effect of different inoculant adhesive agents on rhizobial survival, nodulation, and nitrogenase (acetylene-reducing) activity of soybeans (*Glycine max* (L.) Merrill). *Can. J. Soil. Sci., 64*, 631-636.

Estevez de Jensen, C., Percich, J. A., and Graham, P. H. (2002). Integrated management strategies of bean root rot with *Bacillus subtilis* and *Rhizobium* in Minnesota. *Field Crops Res., 74*, 107-115.

Evans, J., O'Connor, G. E., Griffith, G., and Howieson, J. (1989). Rhizobial inoculant for iprodione-treated lupin seed: Evaluation of an iprodione-resistant *Rhizobium lupini*. *Aust. J. Exp. Agric., 29*, 641-646.

Evans, J., Seidel, J., O'Connor, G. E., Watt, J., and Sutherland, M. (1991). Using omethoate insecticide and legume inoculant on seed. *Aust. J. Exp. Agric., 31*, 71-76.

Faria, S. M. D., De Polli, H., and Franco, A. A. (1985). Adesivos para inoculação e revestimento de sementes de leguminosas. *Pesq. Agropec. Bras., 20*, 169-176.

Ferreira, M. C., and Hungria, M. (2002). Recovery of soybean inoculant strains from uncropped soils in Brazil. *Field Crops Res., 79*, 139-152.

Ferreira, M. C., Andrade, D. S., Chueire, L. M. de O., Takemura, S. M., and Hungria, M. (2000). Effects of tillage method and crop rotation on the population sizes and diversity of bradyrhizobia nodulating soybean. *Soil Biol. Biochem., 32*, 627-637.

Ferriss, R. S. (1984). Effects of microwave oven treatment on micro-organisms in soil. *Phytopathology, 74*, 121-126.

Food and Agriculture Organization of the United Nations (FAO) (1984). *Legume Inoculants and Their Use*. Rome, Italy: FAO.

Food and Agriculture Organization of the United Nations (FAO) (1991). *Expert Consultation on Legume Inoculant Production and Quality Control*. Rome, Italy: FAO.

Fouilleux, G., Revellin, C., and Catroux, G. (1994). Short-term recovery of *Bradyrhizobium japonicum* during an inoculation process using mineral microgranules. *Can. J. Microbiol., 40*, 322-325.

Fouilleux, G., Revellin, C., Hartmann, A., and Catroux, G. (1996). Increase of *Bradyrhizobium japonicum* numbers in soils and enhanced nodulation of soybean (*Glycine max* (L.) Merr.) using granular inoculants amended with nutrients. *FEMS Microb. Ecol., 20*, 173-183.

Frankenberg, C. L. C., Freire, J. R. J., and Thomas, R. W. S. P. (1995). Growth competition between two strains of *Bradyrhizobium japonicum* in broth and peat-based inoculant: Dinitrogen fixation efficiency and competition for nodulation sites. *Rev. Bras. Microbiol., 26*, 211-218.

Franzluebbers, K., Hossner, L. R., and Juo, A. S. R. (1998). *Integrated Nutrient Management for Sustained Crop Production in Sub-Saharan Agriculture* (Tropical Soils TAMU Technology Bulletin 98-03). College Station, TX: Texas A & M University.

Fred, E. B., Baldwin, I. L., and McCoy, E. (1932). *Root Nodule Bacteria and Leguminous Plants.* Madison, WI: University of Wisconsin Press.

Gault, R. R., Chase, D. L., and Brockwell, J. (1982). Effects of inoculation equipment on the viability of *Rhizobium* spp. in liquid inoculant for legumes. *Aust. J. Exp. Agric. Anim. Husb., 22*, 299-309.

Gaur, Y. D., Sem, A. N., and Subba Rao, N. S. (1980). Improved legume-rhizobium symbiosis by inoculating preceding cereal crop with *Rhizobium. Plant Soil, 54*, 313-316.

Gibson, A. H., Date, R. A., Ireland, J. A., and Brockwell, J. (1976). A comparison of competitiveness and persistence amongst five strains of *Rhizobium trifolii. Soil. Biol. Biochem., 8*, 395-401.

Giller, K. E. (2001). *Nitrogen Fixation in Tropical Cropping Systems.* Wallingford, UK: CAB International.

Giller, K. E., and Cadisch, G. (1995). Future benefits from biological nitrogen fixation: An ecological approach to agriculture. *Plant Soil, 174*, 255-277.

Gomez, M., Revellin, C., Hartmann, A., and Catroux, G. (1995). Improved enumeration of *Bradyrhizobium japonicum* in commercial soybean inoculants using selective media. *Lett. Appl. Microbiol., 21*, 142-145.

Graham, P. H. (1981). Some problems of nodulation and symbiotic nitrogen fixation in *Phaseolus vulgaris* L.: A review. *Field Crops Res., 4*, 93-112.

Graham, P. H., Morales, V. M., and Zambrano, O. (1974). Seed pelleting of a legume to supply molybdenum. *Turrialba, 24*, 335-336.

Graham-Weiss, L., Bennett, M. L., and Paau, A. S. (1987). Production of bacterial inoculants by direct fermentation on nutrient-supplemented vermiculite. *Appl. Environ. Microbiol., 53*, 2138-2140.

Hall, A., and Clark, N. (1995). Coping with change, complexity and diversity in agriculture - the case of rhizobial inoculants in Thailand. *World Dev., 23*, 1601-1614.

Ham, G. E., Frederick, L. R., and Anderson, I. C. (1971). Serogroups of *Rhizobium japonicum* in soybean nodules samples in Iowa. *Agron. J., 63*, 69-72.

Hardarson, G. (1993). Methods for enhancing symbiotic nitrogen fixation. *Plant Soil, 152*, 1-17.

Hely, F. W., Hutchings, R. J., and Zorin, M. (1976). Legume inoculation by spraying suspensions of nodule bacteria into soil beneath seed. *J. Aust. Inst. Agric. Sci., 42*, 241-244.

Horikawa, Y., and Ohtsuka, H. (1996a). Effects of coating and adhesive on the inoculation of *Rhizobium meliloti* to alfalfa (*Medicago sativa* L.) seeds for nodulation and seedling growth. *Grass. Sci., 41*, 275-279.

Horikawa, Y., and Ohtsuka, H. (1996b). Storage conditions and nodule formation of coated alfalfa (*Medicago sativa* L.) seeds inoculated with *Rhizobium meliloti. Grass. Sci., 42*, 7-12.

Hungria, M., and Araújo, R. S. (1995). Relato da VI reunião de laboratórios para recomendação de estirpes de *Rhizobium* e *Bradyrhizobium.* In M. Hungria, E. L. Balota, A. Colozzi-Filho, D. S. Andrade (Eds.), *Microbiologia do Solo: Desafios para o Século XXI* (pp. 476-489). Londrina, Brazil: Iapar/Embrapa-CNPSo.

Hungria, M., and Bohrer, T. R. J. (2000). Variability of nodulation and dinitrogen fixation capacity among soybean cultivars. *Biol. Fert. Soils, 31*, 45-52.

Hungria, M., and Vargas, M. A. T. (2000). Environmental factors affecting N_2 fixation in grain legumes in the tropics, with an emphasis on Brazil. *Field Crops Res., 65*, 151-164.

Hungria, M., Andrade, D. S., Chueire, L. M. de O., Probanza, A., Guttierrez-Mañero, F. J., and Megías, M. (2000a). Isolation and characterization of new efficient and competitive bean (*Phaseolus vulgaris* L.) rhizobia from Brazil. *Soil Biol. Biochem., 32*, 1515-1528.

Hungria, M., Boddey, L. H., Santos, M. A., and Vargas, M. A. T. (1998). Nitrogen fixation capacity and nodule occupancy by *Bradyrhizobium japonicum* and *B. elkanii* strains. *Biol. Fert. Soils, 27*, 393-399.

Hungria, M., Campo, R. J., Chueire, L. M. O., Grange, L., and Megías, M. (2001a). Symbiotic effectiveness of fast-growing rhizobial strains isolated from soybean nodules in Brazil. *Biol. Fert. Soils, 33*, 387-394.

Hungria, M., Campo, R. J., and Mendes, I. C. (2001b). *Fixação Biológica do Nitrogênio na Cultura da Soja* (Circular Técnica, 13). Londrina, Brazil: Embrapa Soja/Embrapa Cerrados.

Hungria, M., Campo, R. J., and Mendes, I. C. (2002). Aspectos básicos e aplicados da fixação simbiótica do nitrogênio. In O. F. Saraiva and C. B. Hoffman-Campo (Org.), *Perspectivas do Agronegócio da Soja* (pp. 258-268). Londrina, Brazil: Embrapa-Soja.

Hungria, M., Campo, R. J., and Mendes, I. C. (2003). Benefits of inoculation of the common bean (*Phaseolus vulgaris*) crop with efficient and competitive *Rhizobium tropici* strains. *Biol Fert. Soils, 39*, 88-93.

Hungria, M., Chueire, L. M. de O., Coca, R. G., and Megías, M. (2001c). Preliminary characterization of fast growing strains isolated from soybean nodules in Brazil. *Soil Biol. Biochem., 33*, 1349-1361.

Hungria, M., Vargas, M. A. T., Araujo, R. S., Kurihara, C., Maeda, S., Sá, *et al.* (2000b). Brazilian trials to evaluate the effects of soybean reinoculation. In F. O. Pedrosa, M. Hungria, M. G. Yates and W. E. Newton (Eds.), *Nitrogen Fixation: From Molecules to Crop Productivity* (pp. 549). Dordrecht, The Neetherlands: Kluwer Academic Press.

Hynes, R. K., Craig, K. A., Covert, D., Smith, R. S., and Rennie, R. J. (1995). Liquid rhizobial inoculants for lentil and field pea. *J. Prod. Agric., 8*, 547-552.

Jauhri, K. S., Gupta, M., and Sadasivam, K. V. (1989). Agro-industrial wastes as carriers for bacterial inoculants. *Biol. Wastes, 27*, 81-86.

Jawson, M. D., Franzluebbers, A. J., and Berg, R. K. (1989). *Bradyrhizobium japonicum* survival in and soybean inoculation with fluid gels. *Appl. Environ. Microbiol., 55*, 617-622.

Jones, D. G., and Hardarson, G. (1979). Variation within and between white clover varieties in their preference for strains of *Rhizobium trifolii. Ann. Appl. Biol., 82*, 221-228.

Jung, G., Mugnier, J., Diem, H. G., and Dommergues, Y. R. (1982). Polymerase entrapped *Rhizobium* as an inoculant for legumes. *Plant Soil, 65*, 219-231.

Karanja, N., Freire, J., Gueye, M., and DaSilva, E. (2000). MIRCEN networking: Capacity building and BNF technology transfer in Africa and Latin America. *AgBiotechNet, 2*, ABN 043.

Keyser, H. H., Bohlool, B. B., Hu, T. S., and Weber, D. F. (1982). Fast-growing rhizobia isolated from root nodules of soybeans. *Science, 215*, 1631-1632.

Keyser, H. H., Somasegaran P., and Bohlool, B. B. (1993). Rhizobial ecology and technology. In F. B. Metting, Jr. (Ed.), *Soil Microbial Ecology, Applications in Agricultural and Environmental Management* (pp. 205-226). New York, NY: Marcel Dekker.

Khatri, A. A., Choksey, M., and D'Silva, E. (1973). Rice husk as the medium for legume inoculants. *Sci. Cult., 39*, 194-196.

Kremer, R. J., and Peterson, H. L. (1982). Effect of inoculant carrier on survival of *Rhizobium* on inoculated seed. *Soil Sci., 134*, 177-125.

Lapinskas, E. B. (1990). Effect of different methods of passage through a plant host on the effectiveness of clover nodule bacteria. *Mikrobiologia, 49*, 535-540.

Lindström, K., Lipsanen, P., and Kaijalainen, S. (1990). Stability of markers used for identification of two *Rhizobium galegae* inoculant strains after five years in the field. *Appl. Environ. Microbiol., 56*, 444-450.

Loh, J., Carlson, R. W., York, W. S., and Stacey, G. (2002). Bradyoxetin, a unique chemical signal involved in symbiotic gene regulation. *Proc. Natl. Acad. Sci. USA, 99*, 14446-14451.

Lowther, W., and Patrick, H. N. (1995). *Rhizobium* strain requirements for improved nodulation of *Lotus corniculatus. Soil Biol. Biochem., 27*, 721-724.

Lupwayi, N. Z., Olsen, P. E., Sande, E. S., Kayser, H. H., Collins, M. M., Singleton, P. W., and Rice, W. A. (2000). Inoculant quality and its evaluation. *Field Crops Res., 65*, 259-270.

Maier, R. J., and Triplett, E. W. (1996). Towards more productive, efficient and competitive nitrogen-fixing symbiotic bacteria. *Crit. Rev. Plant Sci., 15*, 191-234.

Marufu, L., Karanja, N., and Ryder, M. (1995). Legume inoculant production and use in east and southern Africa. *Soil Biol. Biochem., 27*, 735-738.

Materon, L. A., and Weaver, R. W. (1985). Inoculant maturity influences survival of rhizobia on seed. *Appl. Environ. Microbiol., 49*, 465-467.

McDermott, T. R., and Graham, P. H. (1989). *Bradyrhizobium japonicum* inoculant mobility, nodule occupancy. and acetylene reduction in the soybean root system, *Appl. Environ. Microbiol., 55*, 2493-2498.

McLeod, R. W., and Roughley, R. J. (1961). Freeze-dried cultures as commercial legume inoculants. *Aust. J. Exp. Agric. Anim. Husb., 1*, 29-33.

McLoughlin, T. J., Owens, P. A., and Scott, A. (1985). Competition studies with fast-growing *Rhizobium japonicum* strains. *Can. J. Microbiol., 31*, 220-223.

Mendes, I. C., Vargas, M. A. T., and Hungria, M. (2000). *Estabelecimento de Estirpes de Bradyrhizobium japonicum/B. elkanii e seus Efeitos na Reinoculação da Soja em Solos de Cerrado* (Documentos, 20). Planaltina, Brazil: Embrapa Cerrados.

Mostasso, L., Mostasso, F. L., Dias, B. G., Vargas, M. A. T., and Hungria, M. (2002). Selection of bean (*Phaseolus vulgaris* L.) rhizobial strains for the Brazilian Cerrados. *Field Crops Res., 73*, 121-132.

Mpeperecki, S., Javaheri, F., Davis, F., and Giller, K. E. (2000). Soybean and sustainable agriculture: promiscuous soybeans in Southern Africa. *Field Crops Res., 65*, 137-150.

Nazih, N., and Weaver, R. W. (1994). Numbers of clover rhizobia needed for crown nodulation and early growth of clover in soil. *Biol. Fert. Soils, 23*, 110-112.

Neves, M. C. P., Didonet, A. D., Duque, F. F., and Döbereiner, J. (1985). *Rhizobium* strain effects on nitrogen transport and distribution in soybeans. *J. Exp. Bot., 36*, 1179-1192.

Nishi, C. Y. M., Boddey, L. H, Vargas, M. A. T., and Hungria, M. (1996). Morphological, physiological and genetic characterization of two new *Bradyrhizobium* strains recently recommended as Brazilian commercial inoculants for soybean. *Symbiosis, 20*, 147-162.

Okon, Y., and Labandera-Gonzalez, C. A. (1994). Agronomic applications of *Azospirillum* - an evaluation of 20 years worldwide field inoculation. *Soil Biol. Biochem., 26*, 1591-1601.

Olsen, P. E., Rice, W. A., and Collins, M. M. (1994). Biological contaminants in North American legume inoculants. *Soil Biol. Biochem., 27*, 699-701.

Olsen, P. E., Rice, W. A., Bordeleau, L. M., Demidoff, A. H., and Collins, M. M. (1996). Levels and identities of non-rhizobial microorganisms found in commercial legume inoculant made with non-sterile peat carrier. *Can. J. Microbiol., 42*, 72-75.

Panzireri, M., Marchettini, N., and Hallam, T. G. (2000). Importance of the *Bradyrhizobium japonicum* symbiosis for the sustainability of a soybean cultivation. *Ecol. Model., 135*, 301-310.

Parker, F. E., and Vincent, J. M. (1981). Sterilization of peat by gamma radiation, *Plant Soil, 61*, 285-293.

Patrick, H. N., and Lowther, W. (1995). Influence of the number of rhizobia on the nodulation and establishment of *Trifolium ambiguum. Soil Biol. Biochem., 27*, 717-720.

Peres, J. R. R., and Vidor, C. (1980). Seleção de estirpes de *Rhizobium japonicum* e competitividade por sítios de infecção nodular em cultivares de soja. *Agron. Sul Riogr., 16*, 205-219.

Peres, J. R. R., Mendes, I. C., Suhet, A. R., and Vargas, M. A. T. (1993). Eficiência e competitividade de estirpes de rizóbio para a soja em solos de Cerrados. *R. Bras. Ci. Solo, 17*, 357-363.

Peres, J. R. R., Suhet, A. R., and Vargas, M. A. T. (1986). Sobrevivência de estirpes de *Rhizobium japonicum* na superfície de sementes de soja inoculadas. *Pesq. Agropec. Bras., 21*, 489-493.

Peres, J. R. R., Suhet, A. R., and Vargas, M. A. T. (1989). Estabelecimento de *Bradyrhizobium japonicum* em um solo de cerrado pela inoculação de sementes de arroz. *Rev. Bras. Ci. Solo, 13*, 35-39.

Peres, J. R. R., Vargas, M. A. T., and Suhet, A. R. (1984). Variabilidade na eficiência de fixar nitrogênio entre isolados de uma mesma estirpe de *Rhizobium japonicum. Rev. Bras. Ci. Solo, 8*, 193-196.

Ramos, M. L. G., and Boddey, R. M (1987). Yield and nodulation of *Phaseolus vulgaris* and the competitivity of an introduced *Rhizobium* strain: Effects of lime, mulch and repeated cropping. *Soil Biol. Biochem., 19*, 171-177.

Rebah, F. B, Tyagi, R. D., and Prévost, D. (2002). Wastewater sludge as a substrate for growth and carrier for rhizobia: The effect of storage conditions on survival of *Sinorhizobium meliloti*. *Biores. Technol., 83*, 145-151.

Revellin, C., Meunier, G., Giraud, J. J., Sommer, G., Wadoux, P., and Catroux, G. (2000). Changes in the physiological and agricultural characteristics of peat-based *Bradyrhizobium japonicum* inoculants after long-term storage. *Appl. Microbiol. Biotechnol., 54*, 206-11.

Rice, W. A., Olsen, P. E., Lupwayi, N. Z., and Clayton, G. W. (2001). Field comparison of pre-inoculated alfalfa seeds and traditional seed inoculation with inoculant prepared in sterile or non-sterile peat. *Commun. Soil Sci. Plant Anal., 32*, 2091-2107.

Rogers, D. D., Warren, R. D., and Chamblee, D. S. (1982). Remedial postemergence legume inoculation with *Rhizobium*. *Agron. J., 74*, 613-619.

Roughley, R. J. (1970). The preparation and use of legume seed inoculants. *Plant Soil, 32*, 675-701.

Roughley, R. J., and Vincent, J. M. (1967). Growth and survival of *Rhizobium* spp. in peat culture. *J. Appl. Bacteriol., 30*, 362-376.

Sanchez, P. A. (2002). Soil fertility and hunger in Africa. *Science, 295*, 2019-2020.

Santos, M. A., Vargas, M. A. T., and Hungria, M. (1999). Characterization of soybean bradyrhizobia strains adapted to the Brazilian Cerrados Region. *FEMS Microbiol. Ecol., 30*, 261-272.

Scholla, M. H., and Elkan, G. H. (1984). *Rhizobium fredii* sp. nov., a fast-growing species that effectively nodulates soybeans. *Int. J. Syst. Bacteriol., 34*, 484-486.

Sessitsch, A., Howieson, J. G., Perret, X., Antoun, H., and Martínez-Romero, E. (2002). Advances in *Rhizobium* research. *Crit. Rev. Plant Sci., 21*, 323-378.

Singleton, P. W., and Tavares, J. W. (1986). Inoculation response of legumes in relation to the number and effectiveness of indigenous *Rhizobium* populations. *Appl. Environ. Microbiol., 51*, 1013-1018.

Smith, R. S. (1992). Legume inoculant formulation and application. *Can. J. Microbiol., 38*, 485-492.

Somasegaran, P. (1985). Inoculant production with diluted liquid cultures of *Rhizobium* spp. and autoclaved peat: Evaluation of diluents, *Rhizobium* spp., peats, sterility requirements, storage, and plant effectiveness. *Appl. Environ. Microbiol., 50*, 398-405.

Somasegaran, P. (1991). *Inoculant Production with Emphasis on Choice of Carriers, Methods of Production and Reliability Testing/Quality Assurance Guidelines* (pp. 87-105). Rome, Italy: FAO.

Somasegaran, P., and Bohlool, B. B. (1990). Single-strain vs. multistrain inoculation: Effect of soil mineral N availability on rhizobial strain effectiveness and competition for nodulation on chickpea, soybean and drybean. *Appl. Environ. Microbiol., 56*, 3298-3303.

Somasegaran, P., and Hoben, H. J. (1994). *The Handbook for Rhizobia: Methods in Legume-Rhizobia Technology*. New York, NY: Springer Verlag.

Sparrow, S. D., and Ham, G. E. (1983). Survival of *Rhizobium phaseoli* in six carrier materials. *Agron. J., 75*, 181-184.

Stephens, J. H. G., and Rask, H. M. (2000). Inoculant production and formulation. *Field Crops Res., 65*, 249-258.

Strijdom, B. W., and Deschodt, C. C. (1976). Carriers of rhizobia and the effects of prior treatment on the survival of rhizobia. In P.S. Nutman (Ed.), *Symbiotic Nitrogen Fixation in Plants* (pp. 151-168). London, UK: Cambridge University Press.

Sutherland, J. M., Odee, D. W., Muluvi, G. M., McInroy, S. G., and Patel, A. (2000). Single and multi-strain rhizobial inoculation of African acacias in nursery conditions. *Soil Biol. Biochem. 32*, 323-333.

Thies, J. E, Singleton, P. W., and Bohlool, B. B. (1991). Influence of the size of indigenous rhizobial population on establishment and symbiotic performance of introduced rhizobia on field-grown legumes. *Appl. Environ. Microbiol., 57*, 19-28.

Thompson, D. J., and Stout, D. G. (1992). Influence of three commercial seed coatings on alfalfa seedling emergence, nodulation and yield. *J. Seed Techn., 16*, 9-16.

Thompson, J. A. (1980). Production and quality control of legume inoculants. In F.J. Bergersen (Ed.), *Methods for Evaluating Nitrogen Fixation* (pp. 489-533). New York, NY: John Wiley & Sons Inc.

Thompson, J. A., Brockwell, J., and Roughley, R. J. (1975). Preinoculation of legume seed. *J. Aust. Inst. Agric. Sci., 41*, 253-254.

Tong, Z., and Sadowsky, M. J. (1994). A selective medium for the isolation and quantification of *Bradyrhizobium japonicum* and *Bradyrhizobium elkanii* strains from soils and inoculants. *Appl. Environ. Microbiol., 60*, 581-586.

Urbana Laboratories (2002). Retrieved November 06, 2002, from www.urbana-labs.com.
van Elsas, J. D., and Heijnen, C. E. (1990). Methods for the introduction of bacteria into soil: A review. *Biol. Fert. Soils, 10*, 127-133.
van Kessel, C., and Hartley, C. (2000). Agricultural management of grain legumes: Has it led to an increase in nitrogen fixation? *Field Crops Res., 65*, 165-181.
Vargas, M. A. T., and Hungria, M. (1997). Fixação biológica do N_2 na cultura da soja. In M. A. T. Vargas and M. Hungria (Eds.), *Biologia dos Solos de Cerrados* (pp. 297-360). Planaltina, Brazil: Embrapa-CPAC.
Vargas, M. A. T., Mendes, I. C., Suhet, A. R., and Peres, J. R. R. (1992). *Duas Novas Estirpes de Rizóbio para a Inoculação da Soja* (Comunicado Técnico, 62). Planaltina, Brazil: Embrapa-CPAC.
Vargas, M. A. T., and Suhet, A. R (1980). Efeito de tipos e níveis de inoculantes na soja cultivada em um solo de cerrado. *Pesq. Agropec. Bras., 15*, 343-347.
Vincent, J. M (1970). *Manual for the Practical Study of Root Nodule Bacteria* (International Biological Programme, 15). Oxford, UK: Blackwell.
Vincent, J. M. (1958). Survival of the root nodule bacteria. In E. G. Hallsworth (Ed.), *Nutrition of the Legumes* (pp. 108-123). New York, NY: Academic Press.
Voelcker, J. A. (1896) "Nitragin" or the use of "pure cultivation" bacteria for leguminous crops. *J. Roy. Agr. Soc. 3rd Ser., 7*, 253-264.
Walter, R. W., and Paau, A. S. (1993). Microbial inoculant production and formulation. In F. B. Metting Jr (Ed.), *Soil Microbial Ecology* (pp. 579-594). New York, NY: Marcel Dekker Inc.
Weaver, R. W., and Frederick, L. R (1974). Effect of inoculum rate on competitive nodulation of *Glycine max* L. Merrill. I - Greenhouse studies. *Agron. J., 66*, 229-232.
Weber, D. F., Keyser, H. H., and Uratsu, S. L. (1989). Serological distribution of *Bradyrhizobium japonicum* from U.S. soybean production areas. *Agron. J., 81*, 786-789.
Xie, Z. P., Staehelin, C., Vierheilig, H., Wiemken, A., Jabbouri, S., Broughton, W. J., *et al.* (1995). Rhizobial nodulation factors stimulate mycorrhizal colonization of nodulating and nonnodulating soybeans. *Plant Physiol., 108*, 1519-1525.
Yardin, M. R., Kennedy, I. V., and Thies, J. E. (2000). Development of high quality carrier materials for field delivery of key microorganisms used as bio-fertilisers and bio-pesticides. *Rad. Phys. Chem., 57*, 565-568.
Zhang, F., and Smith, D. L. (1996). Genistein accumulation in soybean (*Glycine max* L. Merr.) root systems under suboptimum root zone temperatures. *J. Exp. Bot., 47*, 785-792.

Chapter 12

NITRIFYING BACTERIA

C. FIENCKE, E. SPIECK AND E. BOCK

Institut für Allgemeine Botanik, Abteilung Mikrobiologie, Universität Hamburg, D-22609 Hamburg, Germany

1. NITRIFICATION AS PART OF THE NITROGEN CYCLE

Nitrogen is one of the most important elements of life. In nature, several redox reactions of nitrogen are carried out almost exclusively by microorganisms (Figure 1). The most important nitrogen component on earth is dinitrogen (N_2), a highly stable gas that can only be used as a nitrogen source by a restricted group of microorganisms, the nitrogen-fixing bacteria (Figure 1, pathway 1). Biological nitrogen fixation is an important process because it provides ammonia (NH_3) for plants and animals. Nitrogen in form of ammonia can be assimilated into organic material (Figure 1, pathway 2). In modern agriculture, additional ammonia is supplied by mineral fertilizers derived from industrially fixed N_2. Conversely, ammonia is released from organic nitrogen compounds by microbial activity called ammonification (Figure 1, pathway 3). In soils, ammonia is mainly bound to clay particles primarily in the form of ammonium (NH_4^+), where it is available for utilization. Ammonia/ammonium is the most frequently found form of nitrogen in the biosphere and ammonia is transferred efficiently over long distances *via* volatilization.

Under oxic conditions, ammonia is not stable and is converted to nitrate by nitrifying bacteria in soil, freshwater, and marine environments (Figure 1, pathways 4-5). The microbial oxidation of ammonia to nitrite (NO_2^-) and nitrate (NO_3^-) is called nitrification. Nitrification is catalyzed by two physiological groups of bacteria. The aerobic ammonia-oxidizing bacteria produce nitrite (Figure 1, pathway 4), which is further metabolized by nitrite-oxidizing bacteria to nitrate (Figure 1, pathway 5). In contrast to ammonia/ammonium, nitrite is usually found only in trace amounts in aerobic habitats and it rarely accumulates at low partial

D. Werner and W. E. Newton (eds.), Nitrogen Fixation in Agriculture, Forestry, Ecology, and the Environment, 255-276.

pressures of O_2, *e.g.*, either in soils with high water potential or in alkaline environments. Due to the toxicity of nitrite for living organisms, the maintenance of a low nitrite concentration in aerobic habitats is essential. In general, nitrite is immediately consumed by nitrite oxidizers. The end product of nitrification, nitrate, is mobile and can readily leach into both ground water used for drinking and surface waters to cause eutrophication. Instead of or in combination with ammonia, some organisms can use nitrate as a nitrogen source for growth (Figure 1, pathway 6).

Nitrate also serves as a substrate for denitrification and is used for anaerobic respiration (Figure 1, pathways 7-9). The denitrifying bacteria generally produce N_2 *via* the intermediate, nitrite, and the greenhouse gases, nitric oxide (NO) and nitrous oxide (N_2O). Therefore, denitrification is involved in the destruction of stratospheric ozone, global warming, and the loss of ammonia fertilizer to the atmosphere as N_2. In addition to aerobic nitrification, anaerobic ammonia oxidation (Anammox) has been recently described (Jetten, 2001). The, so far, unisolated planctomycetes combine ammonia and nitrite to produce N_2 (Figure 1, pathway 10).

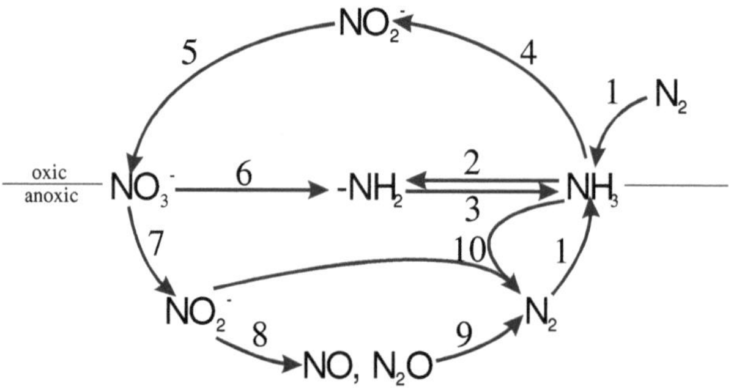

Figure 1. Nitrogen cycle. (1) Dinitrogen (N_2) fixation; (2) assimilation of ammonia (NH_3) into the amino group ($-NH_2$) of protein; (3) ammonification; (4) ammonia oxidation; (5) nitrite (NO_2^-) oxidation; (6) assimilation of nitrate (NO_3^-); (7, 8, 9) denitrification via *nitrite, nitric oxide (NO) and nitrous oxide (N_2O); (10) anaerobic ammonia oxidation.*

Nitrification, the biological oxidation of ammonia to nitrate *via* nitrite, occupies a central position within the nitrogen cycle (Figure 1, pathways 4-5). Nitrifying bacteria are the only organisms capable of converting the most reduced form of nitrogen, ammonia, to the most oxidized form, nitrate. The nitrification process has various direct and indirect implications for natural and man-made systems. It increases the loss of soil nitrogen due to leaching of nitrate, volatilization of nitrogen gases by denitrification, and chemodenitrification (loss of nitrite under acid conditions) and, therefore, influences the fixed nitrogen supply to plants. Furthermore, in unbuffered soils, the oxidation of ammonia to nitric and nitrous acid results in acidification, which allows the formation of aluminum ions, which are toxic to roots of trees, from insoluble aluminates (Mulder *et al.*, 1989). Consequently, nitrification is not beneficial in agricultural soils. Therefore, anhydrous ammonia is used as nitrogen fertilizer and chemicals are occasionally added to the fertilizer to inhibit the nitrification process. However, plants can benefit from nitrifying activity. In particular, many trees prefer nitrate instead of ammonia as a nitrogen source. The production of acids by nitrifiers also contributes to the bio-deterioration of calcareous stone, for example, those of historical buildings (Mansch and Bock, 1996; 1998).

In contrast, nitrification is highly desirable in the treatment of sewage for the efficient removal of ammonium (Painter, 1986). Nitrifiers oxidize ammonium, which is, together with urea, the most frequently found nitrogen compound in sewage, to nitrate. Nitrate can subsequently be removed by denitrifying bacteria *via* anaerobic respiration predominantly in form of dinitrogen gas. This treatment is an integral part of modern nutrient removal of waste-water treatment plants protecting environments from high amounts of ammonia. In addition, ammonia oxidizers have potential applications in the bioremediation of polluted soils and waters that are contaminated with chlorinated aliphatic hydrocarbons, *e.g.*, trichloroethylene (TCE) (Hyman et al., 1995), which is currently the most widely distributed carcinogenic organic groundwater pollutant.

The understanding of the metabolism of both ammonia and nitrite oxidizers will facilitate attempts both to reduce their negative effects and to make use of their beneficial effects.

2. TWO PHYSIOLOGICAL GROUPS OF BACTERIA CONTRIBUTE TO NITRIFICATION

Under oxic conditions, the most important group of organisms involved in nitrification are aerobic chemolithoautotrophic nitrifying bacteria, the ammonia and nitrite oxidizers. For these organisms, the oxidation of inorganic nitrogen compounds serves as their characteristic energy source. They can derive all cellular carbon from carbon dioxide (CO_2). No chemolithotroph is known that can carry out the complete oxidation of ammonia to nitrate. Under anoxic conditions, it has been recently indicated that representatives of the order Planctomycetes are involved in the chemolithotrophic nitrification (Strous et al., 1999). These organisms seem to combine ammonia and nitrite directly into dinitrogen gas (Figure 1, pathway 10). Unfortunately, pure cultures of the involved bacteria have not been obtained so far.

In addition to lithotrophic nitrification, various heterotrophic bacteria, fungi, and algae are capable of oxidizing ammonia to nitrate in the presence of O_2 (Kilham, 1986). However, in contrast to lithotrophic nitrification, heterotrophic nitrification is not coupled to energy generation. Consequently, the growth of heterotrophic nitrifiers is dependent on the oxidation of organic substrates. During heterotrophic nitrification, either ammonia or reduced nitrogen from organic compounds is co-oxidized. The rate of heterotrophic nitrification is much slower than that accomplished by the chemolithotrophic nitrifying bacteria and may not be ecological significant. In the following discussion, the aerobic lithoautotrophic nitrifying bacteria will be described exclusively. For more details about nitrifiers, see Bock and Wagner (2001).

Originally, the lithoautotrophic nitrifying bacteria were collected together within one family, named *Nitrobacteracea,* where ammonia oxidizers are characterized by the prefix *Nitroso-* and the nitrite oxidizers are characterized by the prefix *Nitro-* (Watson *et al.*, 1989). However, phylogenetic investigations have made it evident that ammonia and nitrite oxidizers are not closely related (Teske *et al.*, 1994; Woese *et al.*, 1984, 1985). Several genera are recognized on the basis of both morphology and gene-sequence analyses. So far, five genera of ammonia oxidizers, including *Nitrosomonas, Nitrosospira, Nitrosovibrio, Nitrosolobus*, and *Nitrosococcus*, and four genera of nitrite oxidizers, including *Nitrobacter, Nitrospira, Nitrospina*, and *Nitrococcus*, have been described.

According to comparative 16S rDNA sequence analyses, all recognized ammonia oxidizers are members of two lineages within the β- and γ-subclasses of the Proteobacteria (Koops and Pommerening-Röser, 2001). The marine species of the genus *Nitrosococcus* cluster together in the γ-subclass of Proteobacteria (Woese *et al.*, 1985). The four other genera of ammonia oxidizers, *Nitrosomonas* (including *Nitrosococcus mobilis*, which belongs phylogenetically to *Nitrosomonas*), *Nitrosospira, Nitrosovibrio*, and *Nitrosolobus*, form a monophyletic assemblage within the β-subclass of Proteobacteria (Woese *et al.*, 1984).

Nitrite oxidizers are phylogenetically more distinct. Among the nitrite oxidizers, the genera *Nitrobacter* and *Nitrococcus* were assigned to the α- and γ-subclass of Proteobacteria, respectively (Teske *et al.*, 1994). Nitrite oxidizers of the genus *Nitrospira* are affiliated with the recently described phylum *Nitrospira*e, which represents an independent line of decent within the domain Bacteria (Ehrich *et al.*, 1995; Spieck and Bock, 2001). Because of limited sequence data, *Nitrospina* was first aligned with the δ-subclass of Proteobacteria (Teske *et al.*, 1994), however, in the latest release of ribosomal database project, it is listed in its own subdivision (http://rdp.cme.msu.edu/cgis/treeview.cgi).

All genera of nitrifiers can be initially distinguished by their cell shape. They are rod-shaped, ellipsoidal, spherical, spirillar, vibrioid, or lobular. All species have a typical Gram-negative multilayered cell wall. Many of the nitrifying bacteria have complex intracytoplamic membranes, which may occur as either flattened lamellae or randomly arranged tubes. Cells may be either motile or non-motile and the flagella of motile cells are polar to lateral or peritrichous. Glycogen and polyphosphate inclusions have been observed. In some species, polyhedral cellular inclusions, called carboxysomes, occur and these contain the key enzyme of carbon-dioxide fixation. Cells occur singly, in pairs and, occasionally, in short chains. Extracelluar polymeric substances (EPS), which cause cell aggregation, have been observed both in cultures as well as in natural habitats.

3. ECOLOGY AND DETECTION OF NITRIFYING BACTERIA

Nitrifying bacteria are widely distributed in nature. They are found in most aerobic environments where organic matter is mineralized and, therefore, considerable amounts of ammonia are present. They occur both in lakes and streams, which receive inputs of ammonia, and at the thermocline of lakes, where both ammonia and O_2 are present. However, in accordance with their ecophysiological characteristics, several species have been observed to occur either predominantly or exclusively at special sites, *e.g.*, in rivers, freshwater lakes, salt lakes, oceans, brackish waters, sewage disposal plants, rocks or natural stone of historical buildings, and acid soils (de Boer *et al.,* 1991; Mansch and Bock, 1998; Koops and Pommerening-Röser, 2001). Some species have an obligate salt requirement; others prefer either eutrophic or oligotrophic environments and may tolerate high or low temperatures (Golovacheva, 1976; Koops *et al.*, 1991). Furthermore, nitrifying bacteria are found in niches with either a low O_2 concentration or under anoxic conditions, like in Antarctic lakes and at a depth of 60m in permafrost soils (Voytek *et al.*, 1999; Wagner *et al.*, 2001). Especially in aquatic environments, ammonia-oxidizing bacteria often occur attached as flock or a biofilm (Stehr *et al.,* 1995).

Enrichment cultures of nitrifying bacteria are obtained by using selective media containing either ammonia or nitrite as electron donor and bicarbonate as sole carbon source. Nitrifying bacteria are slow-growing organisms because their cell growth is inefficient. The shortest generation times, as measured in laboratory experiments, were *ca.* 7 hours for *Nitrosomonas* and 10 hours for *Nitrobacter.* For cell division in natural environments, most nitrifying species need from several days to weeks, depending on substrate and O_2 availability, as well as on the temperature and pH-value. Because of the inefficiency of growth of these organisms, visible turbidity will not develop even after extensive nitrification has occurred, so that the best means of monitoring growth is to assay for either the production or disappearance of nitrite.

The slow growth rates of nitrifiers have severely hampered cultivation-dependent approaches to investigate the number as well as the community composition and dynamics of nitrifiers in different environments. The number of nitrifiers in complex systems has been traditionally determined by the most-probable-number (MPN) technique (Matulewich *et al.,* 1975) and selective plating. However, these techniques are time-consuming, often underestimate the number of nitrifiers, and do not allow discriminations at the species level (Belser, 1979; Konuma *et al.*, 2001). Depending on the culture medium and incubation conditions, only a fraction of the total nitrifying community can ever be measured by viable counting methods. Therefore, other methods have been developed for *in situ* identification and enumeration based on the specificity of antibodies and nucleic-acid sequences in order to avoid the limitations of the MPN-technique.

Nitrifiers can be detected in environmental samples, independent of their culturability, by using either antibodies or 16S rRNA-targeted oligonucleotide probes. The fluorescent antibody technique can be applied for direct microscopic enumeration of nitrifiers in complex environmental samples (Belser and Schmidt, 1978; Stanley and Schmidt, 1981; Völsch et al., 1990; Sanden et al., 1994). But for antibodies, which recognize epitopes of the cell wall, the target cells have to be isolated first as pure cultures and the produced antibodies often recognize only a few strains of a given species. Recently, antibodies that target the key enzymes of ammonia oxidizers and nitrite oxidizers were developed (Aamand *et al.*, 1996; Pinck *et al.,* 2001), which overcome the problems of antibodies recognizing only epitopes of the cell wall. They were successfully applied for the detection of nitrifiers in environmental samples (Bartosch *et al.,* 1999, 2002; Fiencke, pers. com.) as well as for physiological studies of the key enzyme (Pinck *et al.*, 2001). Alternatively, nitrifiers can be detected in environmental samples by using a variety of different PCR-techniques, which in general use 16S rRNA (Degrange and Bardin, 1995; Hiorns *et al.*, 1995; McCaig *et al.*, 1999) as well as key enzyme sequence information (Rotthauwe *et al.*, 1997; Nold *et al.*, 2000). Recently, investigations demonstrated that fatty acid profiles of nitrite oxidizers reflect their phylogenetic heterogeneity and could, therefore, be used to for the differentiation and allocation of new isolates (Lipski *et al.*, 2001). Quantitative population structure analysis of nitrifying bacteria can be performed by applying oligonucleotide probes for fluorescence *in situ* hybridization (FISH) (Wagner *et al.*, 1995, Mobarry *et al.*, 1996; Voytek *et al.*, 1999).

4. METABOLISM OF NITRIFYING BACTERIA

Nitrifiers are chemolithotrophic bacteria that are characterized by the ability to grow on reduced inorganic nitrogen compounds. They are autotrophic bacteria, which can derive all cellular carbon from carbon dioxide (CO_2) that is fixed *via* the Calvin/Benson-Cycle. In the past, they were described as obligate chemolithoautotrophic. Nitrifiers seemed incapable of growth with other carbon or energy sources and organic compounds seemed to be toxic to them. This restricted capability is surprising because both ammonia and nitrite are poor energy sources (Hooper, 1989). However, it has now been demonstrated that several nitrifiers are able to grow mixotrophically with either ammonia or nitrite as electron donors and a combination of carbon dioxide and organic compounds as carbon source (Clark and Schmidt, 1966; Martiny and Koops, 1982). Compared to purely autotrophic growth, the addition of organic compounds stimulated cell growth and increased cell yield (Steinmüller and Bock, 1976; Krümmel and Harms, 1982). Furthermore, the nitrite oxidizer, *Nitrobacter*, can grow heterotrophically and organotrophically with pyruvate, formate, and acetate, which serve as both energy and carbon sources (Bock, 1976). However, for these organisms, organotrophic growth was always slower than lithotrophic growth.

Recently, heterotrophic growth of *Nitrosomonas* with fructose and pyruvate as sole carbon source has also been detected (Hommes *et al.*, 2003). But, in contrast to nitrite oxidizers, organotrophic growth of ammonia oxidizers with organic compounds as energy source and O_2 as electron acceptor has not been detected so far (Matin, 1978; Krümmel and Harms, 1982). The reasons proposed for obligate chemolithotrophy of ammonia oxidizers are: (i) the incompleteness of the tricarboxylic acid cycle; (ii) the inability to regulate the levels of enzymes to effectively utilize organic substrates; and (iii) the inability to oxidize NADH at a rate sufficient to utilize organic substrate (Hooper, 1969).

Nevertheless, the metabolism of nitrifiers is more heterogeneous than originally expected. Under O_2 limitation and anoxic conditions, *Nitrosomonas* and *Nitrobacter* are able to denitrify concomitantly producing nitric oxide, nitrous oxide, and dinitrogen (Freitag *et al.*, 1987; Stüven *et al.*, 1992; Bock *et al.*, 1995). Moreover, *Nitrosomonas* has been shown to grow anaerobically with nitrogen dioxide/dinitrogen tetraoxide (NO_2/N_2O_4) as oxygen donor instead of O_2 (Schmidt and Bock, 1997; 1998).

4.1. Biochemistry of Ammonia Oxidizers

Ammonia oxidizers derive their energy from the oxidation of ammonia to nitrite, which is catalyzed by two key enzymes, ammonia monooxygenase (AMO) and hydroxylamine oxidoreductase (HAO). AMO oxidizes NH_3 to the intermediate hydroxylamine (NH_2OH) (equation 3). This oxidation is subdivided into two parts (equations 1 and 2). The exergonic reduction of O_2 to water (equation 2) enables the endergonic ammonia oxidation (equation 1). Consequently, the oxidation of ammonia needs two exogenously supplied electrons (Wood, 1988).

$$NH_3 + \tfrac{1}{2} O_2 \rightarrow NH_2OH \qquad \Delta G_0' = +17\ kJ.mol^{-1} \qquad (1)$$
$$\tfrac{1}{2} O_2 + 2\,H^+ + 2e^- \rightarrow H_2O \qquad \Delta G_0' = -137\ kJ.mol^{-1} \qquad (2)$$
$$NH_3 + O_2 + 2\,H^+ + 2e^- \rightarrow NH_2OH + H_2O \qquad \Delta G_0' = -120\ kJ.mol^{-1} \qquad ^{*}(3)$$

*The $\Delta G_0'$ values of the reaction were calculated using the assumption that the reducing equivalents for AMO are energetically near the ubiquinone level (+110 mV) (Schmidt and Bock, 1998).

Hydroxylamine is further oxidized by periplasmatic HAO to nitrite (NO_2^-) (see equation 4).

$$NH_2OH + H_2O \rightarrow HNO_2 + 4\,H^+ + 4e^- \qquad \Delta G_0' = +23\ kJ.mol^{-1} \qquad (4)$$
$$\tfrac{1}{2} O_2 + 2\,H^+ + 2e^- \rightarrow H_2O \qquad \Delta G_0' = -137\ kJ.mol^{-1} \qquad (5)$$
$$NH_2OH + \tfrac{1}{2} O_2 \rightarrow HNO_2 + 2\,H^+ + 2e^- \qquad \Delta G_0' = -114\ kJ.mol^{-1} \qquad (6)$$

This second oxidation is also composed of several parts in which enzyme bound nitroxyl (NOH) and nitric oxide (NO) are formed (Hooper and Terry, 1979; Hooper and Balny, 1982). NOH can be released as nitrous oxide (N_2O) and NO (Anderson, 1965; Anderson and Levine, 1986). Two electrons from hydroxylamine oxidation (equation 4) are transferred to ammonia oxidation (equation 3) and the other two electrons pass to the respiratory chain for energy generation (equation 5). The overall reaction shows that biogenic ammonia oxidation causes nitrous acid production (equation 7).

$$NH_3 + 1\tfrac{1}{2} O_2 \rightarrow HNO_2 + H_2O \qquad \Delta G_0' = -234\ kJ.mol^{-1} \qquad (7)$$

Under anoxic conditions, *Nitrosomonas eutropha* can grow with nitrogen dioxide/dinitrogen tetraoxide (NO_2/N_2O_4) as oxygen source instead of O_2 for ammonia oxidation to hydroxylamine (Schmidt and Bock, 1997, 1998). Further, the isolated AMO can oxidize ammonia with NO_2 under anoxic conditions (Fiencke, unpublished data). In both cases, NO is formed. Based on these findings, it was postulated that NO_2/N_2O_4 might also be used as oxygen donor under oxic conditions and the produced NO reacts with O_2 to regenerate NO_2/N_2O_4 (Zart and Bock, 1998).

This speculative nitrogen-oxide cycle is supported by the formation of NO and NO_2/N_2O_4 during aerobic ammonia oxidation (Stüven and Bock, 2001).

Ammonia can be used directly as substrate or can made available by hydrolization of urea (Koops *et al.*, 1991). The capability of using urea as an ammonia source might be important for ammonia oxidizers in acid soils, where the actual ammonia concentration is low as a result of the low pH (de Boer and Kowalchuk, 2001).

4.1.1. Ammonia Monooxygenase (AMO)

Because AMO is an important enzyme of nitrification, efforts have been initiated to isolate it. However, AMO is not stable once isolated from the cells and many attempts to purify the enzyme have failed (Suzuki *et al.*, 1981; Ensign *et al.*, 1993). Only recently have some attempts succeeded (Fiencke, unpublished data). Therefore, little is known about its structure and function. Most of the breakthroughs in our understanding of the molecular properties of AMO have been deduced from studies using intact cells of *Nitrosomonas europaea*. Using $^{18}O_2$, AMO was shown to be a monooxygenase (Dua *et al.*, 1979; Hollocher *et al.*, 1981). The K_m values and pH optima point to ammonia, rather than ammonium, as substrate (Drozd, 1976). The low K_m values for both ammonia (2-61 μM) (Koops and Pommerening-Röser, 2001) and O_2 (0.5-7.5 μM) (Laanbroek and Gerards, 1993) allow ammonia oxidation under low substrate concentrations.

AMO has a broad range of substrates. Besides ammonia, a huge number of compounds can be oxidized; these include carbon monoxide (CO) (Tsang and Suzuki, 1982), methane (CH_4) (Jones and Morita, 1983), alkyl and acryl (Hyman *et al.*, 1988), halogenated (Vanelli *et al.*, 1990), and aromatic hydrocarbons (Hyman *et al.*, 1985). This conspicuously broad substrate range offers potential application for bioremediation of sites contaminated with chlorinated aliphatic hydrocarbons. Unfortunately, the oxidation of these alternative substrates cannot supply energy for growth. On the contrary, these substrates act as reversible inhibitors of AMO (Keener and Arp, 1993). The inhibition of AMO by nonpolar compounds points towards a nonpolar catalytic site on the enzyme.

In addition to the reversible inhibition, AMO can be inhibited irreversibly by either acetylene (Hynes and Knowles, 1982; Hyman and Wood, 1985) or trichloroethylene (Hyman *et al.*, 1995). Acetylene is only inhibitory when AMO is catalytically active and high concentrations of ammonium protect the cells (Hyman and Arp, 1992). Therefore, it is assumed that acetylene is oxidized at the catalytic site. he oxygen reactive intermediates bind irreversible to AMO. When cells of *Nitrosomonas europaea* are incubated with radioactive acetylene ($^{14}C_2H_2$), AMO activity is lost and a membrane bound 27-kDa polypeptide, called AmoA, is labeled (Hyman and Wood, 1985). Therefore, AmoA is thought to contain the catalytic site for NH_3 oxidation (Hyman and Arp, 1992).

Other inhibitors of AMO are nitrapyrin (2-chloro-6-trichloromethyl pyridine) (Vannelli and Hooper, 1993) and chelating agents, like thiourea and allylthiourea (Hooper and Terry, 1973). Because these chelating agents preferentially bind copper (Cu) and the ammonia-oxidizing activity of cell-free extracts increased in the presence of added Cu, it is suggested that Cu is a cofactor for AMO (Bédard and Knowles, 1989; Ensign *et al.,* 1993).

The AMO enzyme likely consists of three polypeptides. In addition AmoA, a second polypeptide, called AmoB, copurified with AmoA (McTavish *et al.*, 1993). The evidence for AmoC, a third polypeptide of AMO, is indicated by the mRNA of AmoC being co-transcribed with the mRNA of AmoA (Sayavedra-Soto *et al.*, 1998). None of the subunits have been purified in an active state yet but the corresponding genes, *amoA, amoB, amoC*, have been identified (McTavish *et al.*, 1993; Bergmann and Hooper, 1994; Klotz *et al.*, 1997). The gene *amoA* codes for a hydrophobic 32-kDa protein consisting of four-to-five transmembrane sequences (McTavish *et al.*, 1993; Hooper *et al.*, 1997). In the same operon and adjacent to *amoA* is *amoB*, which codes for a less hydrophobic 43-kDa protein consisting of only two-to-three transmembrane domains and a hydrophobic leader sequence (McTavish *et al.*, 1993; Bergmann and Hooper, 1994). The *amoC* gene is located upstream of *amoA* and *amoB* and encodes a hydrophobic 31-kDa protein with six transmembrane domains (Klotz *et al.*, 1997).

The number of copies of the *amo* operon may be genus specific. Two nearly identical copies are present in strains of both *Nitrosomonas* and *Nitrosovibrio* and three copies are found in strains of *Nitrosospira* and *Nitrosolobus* (Norton *et al.*, 1996; 2002; Klotz and Norton, 1998; Sayavedra-Soto *et al.*, 1998), whereas only a single copy could be detected in marine *Nitrosococcus* strains of the γ-subclass of *Proteobacteria* (Alzerreca *et al.*, 1999). The proteins, AmoA, AmoB and AmoC, of the γ-ammonia oxidizers clearly differ from the AMO proteins of the β-ammonia oxidizers. There is only 37-43% amino-acid sequence similarity for the two groups, whereas the amino-acid sequences of AMO proteins of the β-ammonia oxidizers are 73-85 % similar (Alzerreca *et al.*, 1999).

Polyclonal antibodies have been developed against AmoA and AmoB (Pinck *et al.*, 2001; Fiencke, unpublished data). Quantitative immunoblot analysis, using polyclonal antibodies against AmoB, revealed that total cell protein of *Nitrosomonas eutropha* consisted of *ca.* 6% AmoB, when cells were grown under standard conditions (Pinck *et al.*, 2001). The specific amount of AMO in cells of *Nitrosomonas eutropha* was regulated by the ammonium concentration. At high ammonium concentrations, less AMO was found than under ammonium-limiting conditions. Furthermore, AMO seems to be strongly protected from degradation. Cells that were starved for one year for ammonia still contained high amounts of AMO, although they showed far less ammonia-oxidation activity than growing cells. Hence, the amount of AMO does not directly correlate with the activity of ammonia oxidation.

AMO is similar to particulate methane monooxygenase (pMMO) in the methane-oxidizing bacteria in its putative subunit composition, the inhibitor profiles, the broad substrate range, and the DNA sequences of the genes encoding the proteins (Bédard and Knowles, 1989; Semrau et al., 1995). It is postulated that AMO and pMMO are phylogenetically related and have evolved divergently in correlation with their different physiological roles (Klotz and Norton, 1998). Recently, a third member of this class of monooxygenases, one which oxidizes butane, was recognized (Hamamura *et al.*, 2001).

4.1.2. Hydroxylamine Oxidoreductase (HAO)

In contrast to AMO, HAO has been isolated and characterized. This second key enzyme of ammonia oxidation is located in the periplasmic space, but might be anchored in the cytoplasmic membrane (Olson and Hooper, 1983; McTavish *et al.*, 1995). HAO is a multi-heme enzyme, a complex homodimer or homotrimer of identical 63-kDa subunits, which contain seven c-type hemes and a novel heme, P460, per monomer (Arciero and Hooper, 1993; Sayavedra-Soto *et al.*, 1994; Iverson *et al.*, 1998; Hendrich *et al.*, 2001). The P460 heme is probably the catalytic site of hydroxylamine oxidation (Hooper *et al.*, 1983; Arciero *et al.*, 1993). HAO constitutes *ca.* 40% of the c-type heme of *Nitrosomonas europaea* (DiSpirito *et al.*, 1985a). The crystal structure of HAO revealed the orientation of the hemes in each subunit and suggested potential pathways of electron flow through the enzyme (Igarashi *et al.*, 1997). The oxidation of hydroxylamine to nitrite was postulated to be a two-step reaction with enzyme-bound nitroxyl (HNO) and NO as intermediates (Andersson and Hooper, 1983).

The detailed coupling between ammonia oxidation and hydroxylamine oxidation is not established. Addition of hydroxylamine to ammonia-oxidizing cells in the lag-phase increases the growth yield, however, all attempts to grow ammonia oxidizers on hydroxylamine alone have failed (Nicholas and Jones, 1960; Böttcher and Koops, 1994). The toxicity of hydroxylamine in the absence of ammonia might be due to a surplus of electrons at the AMO under these conditions. With O_2 present, this surplus at the AMO might cause the formation of peroxynitrite (ONO_2), which damages the cells. This suicidal activity of ammonia oxidizers might also cause nitrification breakdown in waste-water treatment plants if plenty of organics is available as an additional electron donor.

The four electrons originating from the oxidation of hydroxylamine are probably channeled from HAO to cytochrome c_{554}, a 26-kDa tetraheme protein (Figure 2) (Andersson *et al.*, 1986; Arciero *et al.*, 1991; Bergmann *et al.*, 1994). Cytochrome c_{554} is a positively charged (pI 10.7) periplasmic enzyme, which can bind to the negatively charged membrane (Yamanaka and Shinra, 1974; McTavish *et al.*, 1995). This two-electron carrier might form ionic complexes with negatively charged electron donors and acceptors, like HAO (pI 3.4) (Iverson *et al.*, 1998) and the 10-kDa monoheme cytochrome c_{552} (pI 3.7) (Fujiwara *et al*, 1995). Cytochrome c_{554} represents a branching point in the electron-transport chain. Two of the four electrons coming from HAO are passed through cytochrome c_{554}, the membrane-associated tetraheme cytochrome c_{m552} (Cyt c_B), and then the ubiquinone-cytochrome bc_1 complex to AMO (Wood, 1986; Whittaker et al., 2000). In addition to tri- and tetramethylhydroquinone, both hydroxylamine and NADH could be probable electron donors to AMO (Suzuki *et al.*, 1981; Shears and Wood, 1986). The remaining electrons from HAO are channeled through cytochrome c_{554} *via* cytochrome c_{552} to either the terminal cytochrome oxidase aa_3 (DiSpirito *et al.*,

1986) or the periplasmic nitrite reductase (Miller and Nicholas, 1985). On the electron pathway, the ubiquinone-cytochrome bc_1 complex might be a mediator between cytochrome c_{554} and cytochrome c_{552} (Wood, 1986). Ubiquinone and membrane-bound cytochromes of the b and c types were detected in *Nitrosomonas europaea* (DiSpirito *et al.*, 1985 a; Whittaker *et al.*, 2000) and other electron carriers have been identified whose functions are unknown (Hooper *et al.*, 1997; Whittaker *et al.*, 2000).

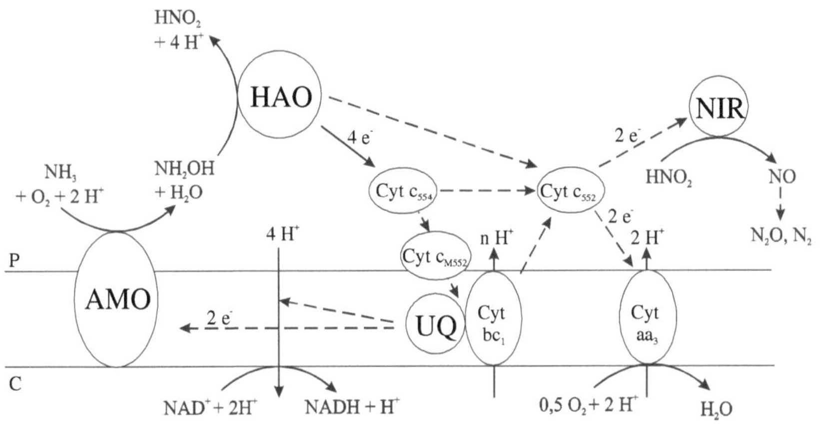

Figure 2. Model of electron transport chain of Nitrosomonas europaea modified from Whittaker et al. (2000) and Poughon et al. (2001).
AMO = ammonia monooxygenase; C = cytoplasmic side of the membrane; Cyt = cytochrome; HAO = hydroxylamine oxidoreductase; NIR = nitrite reductase; P = periplasmic side of the membrane; UQ = ubiquinone.

The first step of ammonia oxidation to hydroxylamine by AMO cannot be used for energy conservation. Thus, hydroxylamine is the real energy-conserving substrate. The oxidation of hydroxylamine to nitrite builds up a proton gradient across the cytoplasmic membrane (Hollocher *et al.*, 1982; Kumar and Nicholas, 1982). The gradient results in a proton motive force, which drives the synthesis of ATP (Drozd, 1976). The proton/oxygen (H^+/O) ratio depends on the substrate. The ratios are 3.4 for ammonia oxidation and 4.4 for hydroxylamine oxidation (Hollocher *et al.*, 1982). A portion of the electrons are used for the reduction of pyridine nucleotides (NAD(P)), which are necessary for carbon-dioxide fixation. Reducing power in form of NAD(P)H might be formed by reversed electron transport, using energy from the proton motive force (Aleem, 1966) (Figure 2). Fifty percent of the electrons circulating in the chain are used to oxidize ammonia, only 6% are involved in the reverse electron chain, and 44% reach the terminal oxidase (Poughon *et al.*, 2001).

In addition to the above-mentioned reactions, ammonia oxidizers show denitrifying activity. Instead of O_2, nitrite can be used as the terminal electron acceptor (Anderson *et al.*, 1993; Bock *et al*, 1995). Under these conditions, NO,

N_2O, and N_2 are produced (Poth and Focht, 1985). The nitrite reductase is induced under O_2 limitation (DiSpirito *et al.*, 1985 b; Miller and Nicholas, 1985). Neither an NO- nor a N_2O-reductase has been identified yet (Hooper *et al.*, 1997) although, in the genome of *Nitrosomonas europaea*, a probable gene for NO-reductase was detected (Whittaker *et al.*, 2000). Recently, anaerobic growth of pure cultures of *Nitrosomonas eutropha* with H_2 and pyruvate as energy source and nitrite as terminal electron acceptor has been described (Bock *et al.*, 1995).

Most studies have focused on *Nitrosomonas europaea*, which possesses a genome size of 2.8 Mbp, within which the genes of the electron-transport chain were identified (Chain *et al.*, 2003). These genes are frequently found in similar or identical copies, like the genes for AMO. In the genome, three nearly identical copies of genes for HAO (*hao*) were found (McTavish *et al.*, 1993; Hommes *et al.*, 2001). There is a copy of the gene that codes for cytochrome c_{554} (*cycA* or *hcy*) that is located downstream of each copy of *hao* (Sayavedra-Soto *et al.*, 1994; Hommes *et al.*, 1996). The membrane-bound tetraheme c-type cytochrome c_{m552} is encoded by genes contiguous with two of the three copies of *cycA* (Bergmann *et al.*, 1994). The genes for cytochrome c_{m552} are co-transcribed with the genes for cytochrome c_{554}.

4.2. Biochemistry of Nitrite Oxidizers

Nitrite oxidizers derive their energy from the oxidation of nitrite to nitrate (equation 8). Only a single step is involved in this oxidation, which is carried out by the enzyme nitrite oxidoreductase in *Nitrobacter*. In the genera *Nitrococcus, Nitrospina* and *Nitrospira*, the key enzyme is called the nitrite-oxidizing system. For the oxidation of nitrite, no O_2 is consumed because the oxygen atom in the nitrate molecule is derived from water. The two electrons released during oxidation are transported *via* cytochromes to O_2 for energy conservation (equation 9). There is no acid production when nitrite is oxidized to nitrate (equation 10).

$NO_2^- + H_2O$	$\rightarrow NO_3^- + 2\,H^+ + 2e^-$	$\Delta G_0' = +83$ kJ.mol^{-1}	(8)
$\frac{1}{2}\,O_2 + 2\,H^+ + 2e^-$	$\rightarrow H_2O$	$\Delta G_0' = -157$ kJ.mol^{-1}	(9)
$NO_2^- + \frac{1}{2}\,O_2$	$\rightarrow NO_3^-$	$\Delta G_0' = -74$ kJ.mol^{-1}	*(10)

*The $\Delta G_0'$ values of the reactions were calculated under the assumption that the reducing equivalents for the nitrite oxidoreductase are energetically near cytochromes of the a- and c-type (+/- 0 mV).

4.2.1. Nitrite Oxidoreductase (NOR)

Most biochemical investigations of nitrite oxidation have been performed with members of the genus *Nitrobacter* and, thus, cannot be generalized for the other genera of nitrite oxidizers. An active form of the membrane-bound nitrite oxidoreductase of *Nitrobacter* was isolated and characterized (Meincke *et al.*, 1992). It consists of 2-3 subunits depending on the isolation procedure (Sundermeyer-Klinger *et al.*, 1984). A membrane-associated α-subunit, NorA (115-130 kDa), and β-subunit, NorB (65 kDa), can be solubilized by heat treatment. An additional protein with a molecular mass of 32 kDa co-purified with the α- and β-subunits when using sodium deoxycholate and subsequent isolation by sucrose (Sundermeyer-Klinger *et al.*, 1984). This protein was described as the γ-subunit of nitrite oxidoreductase. Cytochromes of the a- and c-type were present when the enzyme was solubilized with Triton X-100 and purified by ion-exchange and size-exclusion chromatography (Tanaka et al., 1983). This purified nitrite oxidoreductase was composed of the three subunits of 55, 29, and 19 kDa. All preparations of NOR contained both molybdenum and iron-sulfur clusters (Meincke *et al.*, 1992). Molybdenum is essential for nitrite oxidation and occurs in form of molybdopterin.

NorA might contain the catalytic site of NOR and NorB might function as an electron-channeling protein between NorA and the membrane-integrated electron-transport chain. Using three monoclonal antibodies recognizing NorA and NorB of *Nitrobacter*, this key enzyme was found to be homologous in all four known species of *Nitrobacter* (Aamand *et al.*, 1996). Furthermore, the antibodies detected the NorB of the genera *Nitrococcus, Nitrospina* and *Nitrospira* (Bartosch *et al.*, 1999). The molecular masses of NorB of *Nitrobacter* and *Nitrococcus* were identical (65 kDa), whereas NorB of both *Nitrospina* (48 kDa) and *Nitrospira* (46 kDa) differed significantly.

The nitrite-oxidizing enzyme occurs as characteristic membrane-associated two-dimensional crystals in all nitrite oxidizers. In *Nitrobacter,* these regularly arranged particles, which appear as dimers with a size of 8-10 nm, are located at the inner surface of the cytoplasmic and intra-cytoplasmic membranes (Sundermeyer-Klinger *et al.*, 1984; Meincke et al., 1992; Spieck et al., 1996a). The integrity of a structure that extends between two neighboring particles was assumed to be necessary for conservation of activity (Tsien and Laudelout, 1968). The molecular weight of a single particle was 186 kDa suggesting that it represents a αβ-heterodimer (Spieck *et al.*, 1996a). The location of the particles correlated with immuno-gold labeling of the key enzymes, using antibodies against the NorA and NorB (Spieck *et al.*, 1996b). As in *Nitrobacter*, the nitrite-oxidizing system of *Nitrococcus* is located on the inner side of the cytoplasmic and intracytoplasmic membranes and the particles are arranged in rows (Watson and Waterbury, 1971). In cells of *Nitrospina* and *Nitrospira*, which do not possess intracytoplasmic membranes, the nitrite-oxidizing system was found in form of hexagonal particle patterns in the periplasmic space (Spieck et al., 1998). With regard to the location and molecular masses of the nitrite-oxidizing enzymes, those for the genera *Nitrobacter* and *Nitrococcus* are similar, whereas those for *Nitrospina* and *Nitrospira* are different and form a second coherent group reflecting the phylogeny of the genera.

The genes for both NorA and NorB (*norA* and *norB*) of *Nitrobacter hamburgensis* were identified and sequenced (Kirstein and Bock, 1993; Degrange, personal comm.). The genes cluster together with an additional open reading frame (*norX*) of unknown function in the order *norA*, *norX* and *norB*. Both subunits, NorA and NorB, show significant sequence similarities to several dissimilatory nitrate reductases of several chemoorganotrophic bacteria, *e.g.*, of *Escherichia coli*. NorA is similar to the catalytic α-subunits and NorB is similar to the β-subunits of dissimilatory nitrate reductases. NorB contains four cysteine clusters with striking homology to those of iron-sulfur centers of bacterial ferredoxins. Further, a close functional similarity between the nitrite oxidoreductase and dissimilatory nitrate reductases was suggested by Hochstein and Tomlinson (1988).

The pH optimum of the isolated enzyme was 8.0 and the K_m value for nitrite was 0.5-2.6 mM (Tanaka *et al.*, 1983) and 3.6 mM (Sundermeyer-Klinger *et al.*, 1984). Besides catalyzing nitrite oxidation to nitrate, the NOR of *Nitrobacter* can act as nitrate reductase under anoxic conditions and, therefore, also catalyze the reversible denitrifying process (Sundermeyer-Klinger *et al.*, 1984). During this type of growth, nitrate can be used as the acceptor for electrons derived form organic compounds. For the reduction of nitrate, a K_m value of *ca.* 0.9 mM was calculated.

In *Nitrobacter,* the NOR concentration varies with growth conditions and the enzyme is inducible by either nitrite or nitrate (Bock *et al.*, 1991). The nitrite-oxidizing activity of mixotrophically grown cells was higher than that of autotrophically grown cells (Milde and Bock, 1985). Although many strains of *Nitrobacter* are able to grow heterotrophically, growth is very inefficient and slow (Bock, 1976). During heterotrophic growth, NOR was repressed by more than 90% (Wiesche and Wenzel, 1998). Except for *Nitrobacter*, all other isolated nitrite oxidizers are obligate lithotrophs with nitrite serving as the only energy source. The K_m values for nitrite oxidation (15-270 μM NO_2^-) normally exceed nitrite concentrations in natural environments, where it rarely accumulates (Prosser, 1989). Because, at low pH, the oxidation of nitrite is inhibited by non-dissociated substrate (HNO_2), which is chemically unstable, nitrite enriches only in alkaline sites or when O_2 is limited.

The electrons from nitrite oxidation are transported to the electron-transport chain. They are probably channeled from the molybdopterin and the iron-sulphur clusters of NOR *via* cytochrome a_1 and c to the terminal cytochrome aa_3 oxidase (Figure 3; Hooper, 1989; Bock *et al.*, 1991).

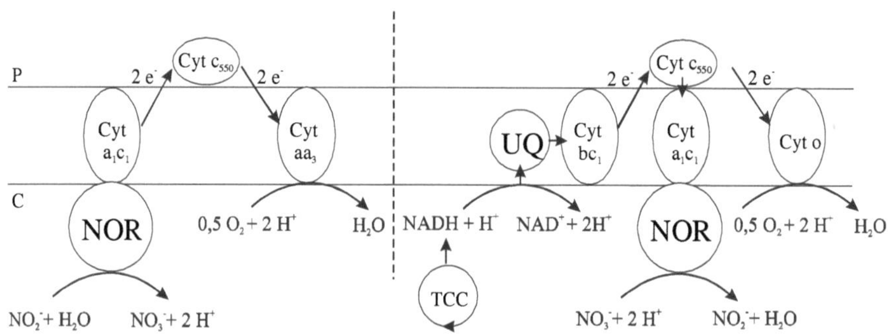

Figure 3. Model of electron-transport chain of Nitrobacter, with, on the left, lithotrophic growth with nitrite, and on the right, heterotrophic growth with organic substances. C = cytoplasmic side of the membrane; Cyt = cytochrome; NOR = nitrite oxidoreductase; P = periplasmic side of the membrane; TCC = tricarboxylic acid cycle; UQ = ubiquinone pool.

The reduction of cytochrome c is a thermodynamically unfavorable step because electrons derived from the high potential of the NO_3^-/NO_2^- redox couple (E_0' +420 mV) have to be transported to the low potential of the cytochrome c redox couple (E_0' +260 mV). A relatively high nitrite concentration would cause a lowering of the redox potential but, in natural habitats, high nitrite concentrations are rarely found. In fact, a highly active cytochrome aa_3 pushes nitrite oxidation by the removal of electrons from cytochrome c (O'Kelley *et al.*, 1970). At constant NOR content, nitrite-oxidation activity is regulated by the concentration of cytochrome aa_3. In contrast to the mitochondrial terminal oxidase, purified terminal oxidase aa_3 consists of two subunits with molecular masses of 40 and 27 kDa and contains magnesium but no zinc (Yamanaka *et al.*, 1981). Cells of *Nitrobacter* seem to possess different terminal oxidases depending on the growth conditions. During nitrite oxidation, cytochrome aa_3 was active, whereas a b-type cytochrome was used as terminal oxidase for heterotrophic growth (Kirstein *et al.*, 1986).

Nitrite oxidation has been reconstituted in proteoliposomes using isolated nitrite-oxidoreductase, cytochrome c oxidase, and nitrite as substrate. O_2 was consumed in the presence of membrane-bound cytochrome c-550 (Nomoto *et al.*, 1993). The electron-transport chain may differ between the genera because both *Nitrobacter* and *Nitrococcus* are rich in cytochromes c and a, whereas the other two genera *Nitrospina* and *Nitrospira* lack cytochromes of the a-type (Watson *et al.*, 1986; Ehrich *et al.*, 1995).

On heterotrophic growth of *Nitrobacter*, electrons pass from NADH via flavine mononucleotide and ubiquinone to a cytochrome bc_1-complex and finally to either the terminal oxidase with O_2 as electron acceptor or to the nitrate-reducing NOR with nitrate as electron acceptor (Figure 3; Kirstein *et al.*, 1986). Under certain conditions, NOR co-purifies with a nitrite reductase (NiR), which reduces nitrite to nitric oxide (Ahlers *et al.*, 1990). Because the activity of NiR is enhanced when the O_2 partial pressure is low, the enzyme is involved in denitrification. However, ATP generation has not been detected during nitrite reduction, which may, therefore, function as nitrite detoxification (Freitag and Bock, 1990).

Generation of membrane potential probably occurs for ATP synthesis and the reduction of pyridine nucleotides (NAD). However, the mechanisms of energy conservation are still unclear because proton translocation linked to the electron-transport chain has not been demonstrated yet (Hollocher *et al.*, 1982). A purified ATPase from *Nitrobacter* has been characterized (Hara *et al.*, 1991), but the primary energy product is NADH (Sundermeyer and Bock, 1981), which is used for ATP synthesis (Freitag and Bock, 1990). However, up to now, it is not clear how energy conservation occurs because the postulated reverse electron flow for the generation of NADH has not been confirmed. The reduction of cytochrome is thermodynamically unfavorable because of the high potential of the NO_3^-/NO_2^- redox couple, so the energy charge is extremely low (0.37) (Eigener, 1975). The low energy generation in *Nitrobacter* may be compensated by the large amount of nitrite oxidoreductase, which can represent 10-30% of total cell protein (Bock *et al.*, 1991).

REFERENCES

Aamand, J., Ahl, T., and Spieck, E. (1996). Monoclonal antibodies recognizing nitrite oxidoreductase of *Nitrobacter hamburgensis, N. winogradskyi,* and *N. vulgaris*. *Appl. Environ. Microbiol., 62*, 2352-2355.

Ahlers, B, König, W., and Bock, E. (1990). Nitrite reductase activity in *Nitrobacter vulgaris*. *FEMS Microbiol. Lett., 67*, 121-126.

Aleem, N. I. H. (1966). Generation of reducing power in chemosynthesis. II. Energy linked reduction of pyridine nucleotides in chemoautotroph *Nitrosomonas europaea*. *Biochim. Biophys. Acta, 113*, 216-224.

Alzerreca, J. J., Norton, J. M., and Klotz, M. G. (1999). The *amo* operon in marine ammonia-oxidizing γ-proteobacteria. *FEMS Microbiol. Lett., 180*, 21-29.

Anderson, J. H. (1965). Studies on the formation of nitrogenous gas from hydroxylamine by *Nitrosomonas*. *Biochim. Biophys. Acta, 97*, 337-339.

Anderson, I. C., and Levine, J. S. (1986). Relative rates of nitric oxide and nitrous oxide production by nitrifiers, denitrifiers, and nitrate respirers. *Appl. Environ. Microbiol., 59*, 3525-3533.

Anderson, I. C., Poth, M., Homstead, J., and Burdige, D. (1993). A comparison of NO and N_2O production by the autotrophic nitrifier *Nitrosomonas europaea* and the heterotrophic nitrifier *Alcaligenes faecalis*. *Appl. Environ. Microbiol., 51*, 938-945.

Andersson, K. K., Hooper, A. B. (1983). O_2 and H_2O are each the source of one O in NO_2^- produced from NH_4^+ by *Nitrosomonas*. ^{15}N-NMR evidence. *FEBS Lett., 164*, 36-240.

Andersson, K. K., Lipscomb, J. D., Valentine, M., Munck, E., and Hooper, A. B. (1986). Tetraheme cytochrome c554 from *Nitrosomonas europaea*: heme-heme interactions and ligand binding. *J. Biol. Chem., 261*, 1126-1138.

Arciero, D. M., Balny, C., and Hooper, A. B. (1991). Spectroscopic and rapid kinetic studies of reduction of cytochrome c-554 by hydroxylamine oxidoreductase from *Nitrosomonas europaea*. *Biochemistry, 30*, 11466-11472.

Arciero, D. M., and Hooper, A. B. (1993). Hydroxylamine oxidoreductase from *Nitrosomonas europaea* is a multimer of an octa-heme subunit. *J. Biol. Chem., 268*, 14645-14654.

Arciero, D. M., Hooper, A. B., Cai, M., and Timkovich, R. (1993). Evidence for the structure of the active site heme P460 in hydroxylamine oxidoreductase of *Nitrosomonas*. *Biochemistry, 32*, 9370-9378.

Bartosch, S., Wolgast, I., Spieck, E., and Bock, E. (1999). Identification of nitrite-oxidizing bacteria with monoclonal antibodies recognizing the nitrite oxidoreductase. *Appl. Environ. Microbiol., 65*, 4126-4133.

Bartosch, S., Hartwig, C., Spieck, E., and Bock, E. (2002). Immunological detection of *Nitrospira*-like bacteria in various soils. *Microbiol. Ecol., 43*, 26-33.

Bédard, C., and Knowles, R. (1989). Physiology, biochemistry, and specific inhibitors of CH_4, NH_4^+, and CO oxidation by methanotrophs and nitrifiers. *Microbiol. Rev.*, 53, 68-84.

Belser, L. W. (1979). Population ecology of nitrifying bacteria. *Annu. Rev. Microbiol., 16*, 309-333.

Belser, L. W., and Schmidt, E. L. (1978). Serological diversity within a terrestrial ammonia-oxidizing population. *Appl. Environ. Microbiol., 36*, 589-593.

Bergmann, D. J., and Hopper, A. B. (1994). Sequence of the gene, *amo*B, for the 43 kDa polypeptide of ammonia monooxygenase of *Nitrosomonas europaea*. *Biochim. Biophys. Res. Comm., 204*, 759-762.

Bergmann, D. J., Arciero, D. M., and Hooper, A. B. (1994). Organization of the hao gene cluster of *Nitrosomonas europaea*: genes for two tetraheme c cytochromes. *J. Bacteriol., 176*, 3148-3153.

Bock, E. (1976). Growth of *Nitrobacter* in the presence of organic matter. II. Chemoorganotrophic growth of *Nitrobacter agilis*. *Arch. Microbiol., 108*, 305-312.

Bock, E., Koops, H.-P., Harms, H., and Ahlers, B. (1991). The biochemistry of nitrifying organism. In J. M. Shively, and L. L. Barton (Eds.), *Variation in autotrophic life* (pp. 171-200). London, UK: Academic Press.

Bock, E., Schmidt, I., Stüven, R., and Zart, D. (1995). Nitrogen loss caused by denitrifying *Nitrosomonas* cells using ammonia or hydrogen as electron donors and nitrite as electron acceptor. *Arch. Microbiol., 163*, 16-20.

Bock, E., and Wagner, M. (2001). Oxidation of inorganic nitrogen compounds as an energy source. In E. Stackebrandt (Ed.), *The Prokaryotes* vol 3 (online). New York, NY: Springer-Verlag.

Böttcher, B., and Koops, H.-P. (1994). Growth of lithotrophic ammonia-oxidizing bacteria on hydroxylamine. *FEMS Microbiol. Lett.*, *122*, 263-266.

Chain, P., Lamerdin, J., Larimer, F., Regala, W., Lao, V., Land, *et al.* (2003). Complete genome sequence of the ammonia-oxidizing bacterium and obligate chemolithoautotroph *Nitrosomonas europaea*. *J. Bacteriol.* *185*, 2759-2773.

Clark, C., and Schmidt, E. L. (1966). Effect of mixed culture of *Nitrosomonas europaea* stimulated by uptake and utilization of pyruvate. *J. Bacteriol.*, *91*, 367-373.

de Boer, W., and Kowalchuk, G. A. (2001). Nitrification in acid soils: Micro-organisms and mechanisms. *Soil Biol. Biochem.*, *33*, 853-866.

de Boer, W., Klein, W., Gunnewiek, P. J. A., Veenhuis, M., Bock, E., and Laanbroek, H. J. (1991). Nitrification at low pH by aggregated chemolithotrophic bacteria. *Appl. Environ. Microbiol.*, *57*, 3600-3604.

Degrange, V., and Bardin, R. (1995). Detection and counting of *Nitrobacter* populations in soil by PCR. *Appl. Environ. Microbiol.*, *61*, 2093-2098.

DiSpirito, A. A., Taaffe, L. R., and Hooper, A. B. (1985a). Localization and concentration of hydroxylamine oxidoreductase and cytochromes c-552, c-554, c_m-552 and *a* in *Nitrosomonas europaea*. *Biochim. Biophys. Acta*, *806*, 320-330.

DiSpirito, A. A., Taaffe, L. R., Lipscomb, J. D., and Hooper, A. B. (1985b). A "blue" copper oxidase from *Nitrosomonas europaea*. *Biochim. Biophys. Acta*, *827*, 320-326.

DiSpirito, A. A., Lipscomb, J. D., and Hooper, A. B. (1986). A three-subunit cytochrome aa_3 from *Nitrosomonas europaea*. *J. Biol. Chem.*, *261*, 17048-17056.

Drozd, J. W. (1976). Energy coupling and respiration in *Nitrosomonas europaea*. *Arch. Microbiol.*, *110*, 257-262.

Dua, R. D., Bhandari, B., and Nicholas, D. J. D. (1979). Stable isotope studies on the oxidation of ammonia to hydroxylamine by *Nitrosomonas europaea*. *FEBS Lett.*, *106*, 401-404.

Ehrich, S., Behrens, D., Lebedeva, E., Ludwig, W., and Bock, E. (1995). A new obligately chemolithoautotrophic, nitrite-oxidizing bacterium, *Nitrospira moscoviensis* sp. nov. and its phyologenetic relationship. *Arch. Microbiol.*, *164*, 16-23.

Eigener, U. (1975). Adenine nucleotide pool variations in intact *Nitrobacter winogradskyi* cells. *Arch. Microbiol.*, *102*, 233-240.

Ensign, S. A., Hyman, M. R., and Arp, D. J. (1993). In vitro activation of ammonia monooxygenase from *Nitrosomonas* by copper. *J. Bacteriol.*, *175*, 1971-1998.

Freitag, A., Rudert, M., and Bock, E. (1987). Growth of *Nitrobacter* by dissimilatoric nitrate reduction. *FEMS Microbiol. Lett.*, *48*, 105-109.

Freitag, A., and Bock, E. (1990). Energy conservation in *Nitrobacter*. *FEMS Microbiol. Lett.*, *66*, 157-162.

Fujiwara, T., Yamanaka, T., and Fukumori, Y. (1995). The amino acid sequence of *Nitrosomonas europaea* cytochrome c-552. *Current Microbiol.*, *31*, 1-4.

Golovacheva, R. S. (1976). Thermophilic nitrifying bacteria from hot springs. *Microbiol.*, 45, 329-331.

Hamamura, N., Yeager, C., and Arp, D. J. (2001). Two distinct monooxygenases for alkane oxidation in *Nocarioides* sp. strain CF8. *Appl. Environ. Microbiol.*, *31*, 1-4.

Hara, T., Villalobos, A. P, Fukomori, Y., and Yamanaka, T. (1991). Purification and characterization of ATPase from *Nitrobacter winogradskyi*. *FEMS Microbiol. Lett.*, *82*, 49-54.

Hendrich, M. P., Petatis, D., Arciero, D. M., and Hooper, A. B. (2001). Correlations of structure and electronic properties from EPR spectroscopy of hydroxylamine oxidoreductase. *J. Am. Chem. Soc.*, *123*, 2997-3005.

Hiorns, W. D., Hastings, R. C., Head, I. M., McCarthy, A. J., Saunders, J. R., Pickup, R. W., and Hall, G. H. (1995). Amplification of 16S ribosomal RNA genes of autotrophic ammonia-oxidizing bacteria demonstrates the ubiquity of *Nitrosospira* in the environment. *Microbiol.*, *141*, 2793-2800.

Hochstein, L. I., and Tomlinson, G. A. (1988). The enzymes associated with denitrification. *Annu. Rev. Microbiol.*, *42*, 231-261.

Hollocher, T. C., Tate, M. E., and Nicholas, D. J. D. (1981). Oxidation of ammonia by *Nitrosomonas europaea*. Definitive ^{18}O-tracer evidence that hydroxylamine formation involves a monooxygenase. *J. Biol. Chem.*, *256*, 10834-10836.

Hollocher, T. C., Kumar, S., and Nicholas, D. J. D. (1982). Respiration-dependent proton translocation in *Nitrosomonas europaea* and its apparent absence in *Nitrobacter agilis* during inorganic oxidations. *J. Bacteriol.*, *149*, 1013-1020.

Hommes, N. G., Sayavedra-Soto, L. A., and Arp, D. J. (1996). Mutagenesis of hydroxylamine oxidoreductase in *Nitrosomonas europaea* by transformation and recombination. *J. Bacteriol., 178*, 3710-3714.

Hommes, N. G., Sayavedra-Soto, L. A., and Arp, D. J. (2001). Transcript analysis of multiple copies of *amo* (encoding ammonia monooxygenase) and *hao* (encoding hydroxylamine oxidoreductase) in *Nitrosomonas europaea. J. Bacteriol., 183*, 1096-1100.

Hommes, N. G., Sayavedra-Soto, L. A., and Arp, D. J. (2003) Chemolithoorganotrphic growth of *Nitrosomonas europaea* on fructose. *J. Bacteriol., 185*, 6809-6814.

Hooper, A. B. (1969). Biochemical basis of obligate autotrophy of *Nitrosomonas europaea. J. Bacteriol., 97*, 776-779.

Hooper, A. B. (1989). Biochemistry of the nitrifying lithoautotrophic bacteria. In H. G. Schlegel and B. Bowien (Eds.), *Autotrophic bacteria* (pp. 239-281). Madison, WI: Science Tech Publisher.

Hooper, A. B., and Balny, C. (1982). Reaction of oxygen with hydroxylamine oxidoreductase of *Nitrosomonas*: fast kinetics. *FEBS Lett., 144*, 299-303.

Hooper, A. B., and Terry, K. R. (1973). Specific inhibitors of ammonia oxidation in *Nitrosomonas. J. Bacteriol., 115*, 480-485.

Hooper, A. B., and Terry, K. R. (1979). Hydroxylamine oxidoreductase of *Nitrosomonas* – production of nitric oxide from hydroxylamine. *Biochim. Biophys. Acta, 571*, 12-20.

Hooper, A. B., Debey, P., Anderson, K. K., and Balny, C. (1983). Heme P460 of hydroxylamine oxidoreductase of *Nitrosomonas*. Reaction with CO and H_2O_2. *Eur. J. Biochem., 134*, 83-87.

Hooper, A. B., Vannelli, T., Bergmann, D. J., and Arciero, D. M. (1997). Enzymology of the oxidation of ammonia to nitrite by bacteria. *Ant. van Leeuwenhoek, 71*, 59-67.

Hyman, M. R., and Arp, D. J. (1992). $^{14}C_2H_2$- and $^{14}CO_2$-labelling studies of the de novo synthesis of polypeptides by *Nitrosomonas europaea* during recovery from acetylene and light inactivation of ammonia monooxygenase. *J. Biol. Chem., 267*, 1534-1545.

Hyman, M. R., and Wood, P. M. (1985). Suicidal inactivation and labelling of ammonia mono-oxygenase by acetylene. *Biochem. J., 227*, 719-725.

Hyman, M. R., Sansome-Smith, A. W., Shears, J. H., and Wood, P. M. (1985). A kinetic study of benzene oxidation to phenol by whole cells of *Nitrosomonas europaea* and evidence for the further oxidation of phenol to hydroquinone. *Arch. Microbiol., 143*, 302-306.

Hyman, M. R., Murton, I. B., and Arp, D. J. (1988). Interaction of ammonia monooxygenase from *Nitrosomonas europaea* with alkanes, alkenes, and alkynes. *Appl. Environ. Microbiol., 54*, 3187-3190.

Hyman, M. R., Russell, S. A., Ely, R. L., Williamson, K. J., and Arp, D. J. (1995). Inhibition, inactivation, and recovery of ammonia-oxidizing activity in cometabolism of trichloroethylene by *Nitrosomonas europaea. Appl. Environ. Microbiol., 61*, 1480-1487.

Hynes, R. K., and Knowles, R. (1982). Effect of acetylene on autotrophic and heterotrophic nitrification. *Can. J. Microbiol., 28*, 334-340.

Igarashi, N., Moriyama, H., Fujiwara, T., Fukumori, Y., and Tanaka, N. (1997). The 2.8 Å structure of hydroxylamine oxidoreductase from nitrifying chemoautotrophic bacterium, *Nitrosomonas europaea. Nature Struct. Biol., 4*, 276-284.

Iverson, T. M., Arciero, D. M., Hsu, B. T., Logan, M. S. P., Hooper, A. B., and Rees, D. C. (1998). Heme packing motifs revealed by the crystal structure of the tetra-heme cytochrome c554 from *Nitrosomonas europaea. Nature Struct. Biol., 5*, 1005-1012.

Jetten, M. S. M. (2001). New pathways for ammonia conversion in soil and aquatic systems. *Plant and Soil, 230*, 9-19.

Jones, R. D., and Morita, R. Y. (1983). Methane oxidation by *Nitrosococcus oceanus* and *Nitrosomonas europaea. Appl. Environ. Microbiol., 45*, 401-410.

Keener, W. K., and Arp, D. J. (1993). Kinetic studies of ammonia monooxygenase inhibition in *Nitrosomonas europaea* by hydrocarbons and halogenated hydrocarbons in an optimized whole-cell assay. *Appl. Environ. Microbiol., 59*, 2501-2510.

Kilham, K. (1986). Heterotrophic nitrification. In J. I. Prosser (Ed.), *Nitrification* (pp. 117-126). Oxford, UK: IRL Press.

Kirstein, K., and Bock, E. (1993). Close genetic relationship between *Nitrobacter hamburgensis* nitrite oxidoreductse and *Escherichia coli* nitrate reductase. *Arch. Microbiol., 160*, 447-453.

Kirstein, K. O., Bock, E., Miller, D. J., and Nicholas D. J. D. (1986). Membrane-bound b-type cytochrome in *Nitrobacter*. *FEMS Microbiol. Lett.*, *36*, 63-67.
Klotz, M. G., and Norton, J. M. (1998). Multiple copies of ammonia monooxygenase (*amo*) operons have evolved under biased AT/GC mutational pressure in ammonia-oxidizing autotrophic bacteria. *FEMS Microbiol. Lett.*, *168*, 303-311.
Klotz, M. G., Alzerreca, J., and Norton, J. M. (1997). A gene encoding a membrane protein exists upstream of the *amoA/amoB* genes in ammonia oxidizing bacteria: a third member of the *amo* operon? *FEMS Microbiol. Lett.*, *150*, 65-73.
Konuma, S., Satoh, H., Mino, T., and Matsu, T. (2001). Comparison of enumeration methods of ammonia-oxidizing bacteria. *Wat. Sci. Techn.*, *43*, 107-114.
Koops, H.-P., and Pommerening-Röser, A. (2001). Distribution and ecophysiology of nitrifying bacteria emphasizing cultured species. *FEMS Microbiol. Ecol.*, *1255*, 1-9.
Koops, H.-P., Böttcher, B., Möller, U. C., Pommerening-Röser, A., and Stehr, G. (1991). Classification of eight new species of ammonia-oxidizing bacteria: *Nitrosomonas communis* sp. nov., *Nitrosomonas ureae* sp. nov., *Nitrosomonas aestuarii* sp. nov., *Nitrosomonas marina* sp. nov., *Nitrosomonas nitrosa* sp. nov., *Nitrosomonas eutropha* sp. nov., *Nitrosomonas oligotropha* sp. nov. and *Nitrosomonas halophila* sp. nov. *J. Gen. Microbiol.*, *137*, 1689-1699.
Krümmel, A., and Harms, H. (1982). Effect of organic matter on growth and cell yield of ammonia-oxidizing bacteria. *Arch. Microbiol.*, *133*, 50-54.
Kumar, S., and Nicholas, D. J. D. (1982). A proton motive force-dependent adenosine-5′-triphosphate synthesis in speroplasts of *Nitrosomonas europaea*. *FEMS Microbiol. Lett.*, *14*, 21-25.
Laanbroek, H. J., and Gerards, S. (1993). Competition for limiting amounts of oxygen between *Nitrosomonas europaea* and *Nitrobacter winogradskyi* grown in mixed continuous cultures. *Arch. Microbiol.*, *159*, 453-459.
Lipski, A., Spieck, E., Makolla, A., and Altendorf, K. (2001). Fatty acid profiles of nitrite-oxidizing bacteria reflect their phylogenetic heterogeneity. *System. Appl. Microbiol.*, *24*, 377-384.
Mansch, R., and Bock, E. (1996). Simulation of microbial attack on natural and artificial stone. In E. Heitz, H.-C. Flemming, and W. Sand (Eds.), *Microbially influenced corrosion of materials* (pp. 167-186). Berlin, Heidelberg, Germany: Springer.
Mansch, R., and Bock, E. (1998). Biodeterioration of natural stone with special reference to nitrifying bacteria. *Biodegradation*, *9*, 47-64.
Martiny, H., and Koops, H.-P. (1982). Incorporation of organic compounds into cell protein by lithotrophic ammonia oxidizing bacteria. *Ant. van Leeuwenhoek*, *48*, 327-336.
Matin, A. (1978). Organic nutrition of chemolithotrophic bacteria. *Ann. Rev. Microbiol.*, 32, 311-318.
Matulewich, V. A., Strom, P. F., and Finstein, M. S. (1975). Length of incubation for enumerating nitrifying bacteria present in various environments. *Appl. Environ. Microbiol.*, *29*, 265-268.
McCaig, A. E., Philips, C. J., Stephen, J. R., Kowalchuk, G. A., Harvey, S. M., Herbert, R. A., Embley, T. M., and Prosser, J. I. (1999). Nitrogen cycling and community structure of proteobacterial beta-subgroup ammonia-oxidizing bacteria within polluted marine fish farm sediments. *Appl. Environ. Microbiol.*, *65*, 213-220.
McTavish, H., Fuchs, J. A., and Hooper, A. B. (1993). Sequence of the gene coding for ammonia-monooxygenase in *Nitrosomonas europaea*. *J. Bacteriol.*, *175*, 2436-2444.
McTavish, H., Arciero, D. M., and Hooper, A. B. (1995). Interactions with membranes of cytochrome c554 from *Nitrosomonas europaea*. *Arch. Biochem. Biophys.*, *324*, 53-58.
Meincke, M., Bock, E., Kastrau, D., and Kroneck, P. M. H. (1992). Nitrite oxidoreductase from *Nitrobacter hamburgensis*: Redox centers and their catalytic role. *Arch. Microbiol.*, *158*, 127-131.
Milde, K., and Bock, E. (1985). Comparative studies on membrane proteins of *Nitrobacter hamburgensis* and *Nitrobacter winogradskyi*. *FEMS Microbiol. Lett.*, *26*, 135-139.
Miller, D. J., and Nicholas, D. J. D. (1985). Characterization of a soluble cytochrome oxidase/nitrite reductase from *Nitrosomonas europaea*. *J. Gen. Microbiol.*, *131*, 2851-2854.
Mobarry, B. K., Wagner, M., Urbain, V., Rittmann, B. E., and Stahl, D. A. (1996). Phylogenetic probes for analyzing abundance and spatial organization of nitrifying bacteria. *Appl. Environ. Microbiol.*, *62*, 2156-2162.
Mulder, J., van Breemen, N., Rasmussen, L., and Driscoll, C. T. (1989). Aluminium chemistry of acidic sandy soils affected by acid atmospheric deposition in the Netherlands and in Denmark. In T. E. Lewis (Ed.), *Environmental chemistry and toxicology of aluminium* (pp. 171-194). Chelsea, MI: Lewis Publishing Inc.

Nold, S. C., Zhou, J., Devol, A. H., and Tiedje, J. M. (2000). Pacific northwest marine sediments contain ammonia-oxidizing bacteria in the β-subdivision of the *Proteobacteria*. *Appl. Environ. Microbiol.*, *66*, 4532-4535.

Nomoto, T., Fukumori, Y., and Yamanaka, T. (1993). Membrane-bound cytochrome c is an alternative electron donor for cytochrome aa_3 in *Nitrobacter winogradsky*. *J. Bacteriol.*, *175*, 4400-4404.

Norton, J. M., Low, J. M., and Klotz, M. G. (1996). The gene encoding ammonia monooxygenase subunit A exists in three identical copies in *Nitrosospira* sp. NpAV. *FEMS Microbiol. Lett.*, *139*, 181-188.

Norton, J. M., Alzerreca, J. J., Suwa, Y., Klotz, M. G. (2002). Diversity of ammonia monooxygenase operon in autotrophic ammonia-oxidizing bacteria. Arch. Microbiol., *177*, 139-149.

Nicholas, D. J. D., and Jones, O. T. G. (1960). Oxidation of hydroxylamine in cell-free extracts of *Nitrosomonas europaea*. *Nature*, *185*, 512-514.

O´Kelley, J. C., Becker, G. E., and Nason, A. (1970). Characterization of the particulate nitrite oxidase and its component activities from autotroph *Nitrobacter agilis*. *Biochim. Biophys. Acta*, *205*, 409-425.

Olson, T. C., and Hooper, A. B. (1983). Energy coupling in the bacterial oxidation of small molecules: an extracytoplasmatic dehydrogenase in *Nitrosomonas*. *Microbiol. Lett.*, *19*, 47-50.

Painter, H. A. (1986). Nitrification in the treatment of sewage and waste-water. In J. I. Prosser (Ed.), *Nitrification* (pp. 185-211). Oxford, UK: IRL Press.

Pinck, C., Coeur, C., Potier, P., and Bock, E. (2001). Polyclonal antibodies recognizing the AmoB protein of ammonia oxidizers of the β-subclass of the class *Proteobacteria*. *Appl. Environ. Microbiol.*, *67*, 118-124.

Poth, M., and Focht, D. D. (1985). ^{15}N kinetic analyses of N_2O production by *Nitrosomonas europaea*: An examination of nitrifier denitrification. *Appl. Environ. Microbiol.*, *49*, 1134-1141.

Poughon, L., Dussap, C.-G., and Gros, J.-B. (2001). Energy model and metabolic flux analysis for autotrophic nitifiers. *Biotechnol. Bioeng.*, *72*, 416-433.

Prosser, J. I. (1989). Autotrophic nitrification in bacteria. In A. H. Rose, and D. W. Tempest (Eds.), *Advances in microbial physiology*, Vol. 30 (pp. 125-181). London, UK: Academic Press.

Rotthauwe, J. H., Witzel, K. P., and Liesack, W. (1997). The ammonia monooxygenase structural gene *amoA* as a functional marker: Molecular fine-scale analysis of natural ammonia-oxidizing populations. *Appl. Environ. Microbiol.*, *63*, 4704-4712.

Sanden, B., Grunditz, C., Hansson, Y., and Dalhammar, G. (1994). Quantification and characterisation of *Nitrosomonas* and *Nitrobacter* using monoclonal antibodies. *Water Sci. Tech.*, *29*, 1-6.

Sayavedra-Soto, L. A., Hommes, N. G., and Arp, D. J. (1994). Characterization of the gene encoding hydroxylamine oxidoreductase in *Nitrosomonas europaea*. *J. Bacteriol.*, *176*, 504-510.

Sayavedra-Soto, L. A., Hommes, N. G., Alzerreca, J. J., Arp, D. J., Norton, J. M., and Klotz, M. G. (1998). Transcription of the *amoC*, *amoA* and *amoB* genes in *Nitrosomonas europaea* and *Nitrosospira* sp. NpAV. *FEMS Microbiol. Lett.*, *167*, 81-88.

Schmidt, I., and Bock, E. (1997). Anaerobic ammonia oxidation with nitrogen dioxide by *Nitrosomonas eutropha*. *Arch. Microbiol.*, *167*, 106-111.

Schmidt, I., and Bock, E. (1998). Anaerobic ammonia oxidation by cell-free extracts of *Nitrosomonas eutropha*. *Ant. van Leeuwenhoek*, *73*, 271-278.

Semrau, J. D., Chistoserdov, A., Lebron, J., Costello, A., Davagnino, J., and Kenna, E. (1995). Particulate methane monooxygenase genes in methanotrophs. *J. Bacteriol.*, *177*, 3071-3079.

Shears, J. H., and Wood, P. M. (1986). Tri- and tetra-methylhydroquinone as electron donors for ammonia monooxygenase in whole cells of *Nitrosomonas europaea*. *FEMS Microbiol. Lett.*, *33*, 281-284.

Spieck, E., and Bock, E. (2001). Genus *Nitrospira*. In D. R. Bone, R. W. Castenholz, and G. M. Garrity (Eds.), *Bergey´s Manual of Systematic Bacteriology*, 2nd ed, vol 1, The Archaea and the Deeply Branching and Phototrophic Bacteria (pp. 451-453). New York, NY: Springer-Verlag.

Spieck, E., Müller, S., Engel, A., Mandelkow, E., Patel, H., and Bock, E. (1996a). Two-dimensional structure of membrane-bound nitrite oxidoreductase *from Nitrobacter hamburgensis*. *J. Struct. Biol.*, *117*, 117-123.

Spieck, E., Aamand, J, Bartosch, S, and Bock, E. (1996b). Immunocytochemical detection and location of the membrane-bound nitrite oxidoreductase in cells of *Nitrobacter* and *Nitrospira*. *FEMS Mircobiol. Lett.*, *139*, 71-76.

Spieck, E., Ehrich, S., Aamand, J., and Bock, E. (1998). Isolation and immunocytochemical location of the nitrite-oxidizing system in *Nitrospira moscoviensis*. *Arch. Microbiol.*, *169*, 225-230.

Stanley, P. M., and Schmidt, E. L. (1981). Serological diversity of *Nitrobacter* spp. from soil and aquatic habitats. *Appl. Environ. Microbiol.*, *41*, 846-849.

Stehr, G., Böttcher, B., Dittberner, B., Rath, G., and Koops, H.-P. (1995). The ammonia-oxidizing nitrifying population of the River Elbe estuary. *FEMS Microbiol. Ecol.*, *17*, 177-186.

Steinmüller, W., and Bock, E. (1976). Growth of *Nitrobacter* in the presence of organic matter. *Arch. Microbiol.*, *108*, 299-304.

Strous, M., Fuerst, J. A., Kramer, E. H., Logemann, S., Muyzer, G., van de Pas-Schoonen, K. T., Webb, R., Kuenen, J. G., and Jetten, M. S. (1999). Missing lithotroph identified as new planctomycete. *Nature*, *400*, 446-449.

Stüven, R., and Bock, E. (2001). Nitrification and denitrification as a source for NO and NO_2 production in high-strength wastewater. *Wat. Res.*, *35*, 1905-1914.

Stüven, R., Vollmer, M., and Bock, E. (1992). The impact of organic matter on nitric oxide formation by *Nitrosomonas europaea*. *Arch. Microbiol.*, *158*, 439-443.

Sundermeyer, H., and Bock, E. (1981). Energy metabolism of autotrophically and heterotrophically grown cells of *Nitrobacter winogradskyi*. *Arch. Microbiol.*, 130, 2250-254.

Sundermeyer-Klinger, H., Meyer, W., Warninghoff, B., and Bock, E. (1984). Membrane-bound nitrite oxidoreductase of *Nitrobacter*: Evidence for a nitrate reductase system. *Arch. Microbiol.*, 140, 153-158.

Suzuki, I., Kwok, S.-C., Dular, U., and Tsang, D. C. Y. (1981). Cell-free ammonia-oxidizing system of *Nitrosomonas europaea*: general conditions and properties. *Can. J. Biochem.*, 59, 477-483.

Tanaka, Y., Fukumori, Y., and Yamanaka, T. (1983). Purification of cytochrome a_1c_1 form *Nitrobacter agilis* and characterization of nitrite oxidation system of the bacterium. *Arch. Microbiol.*, 135, 265-271.

Teske, A., Alm, E., Regan, J. M., Toze, S., Rittmann, B. E., and Stahl, D. A. (1994). Evolutionary relationship among ammonia- and nitrite-oxidizing bacteria. *J. Bacteriol.*, 176, 6623-6630.

Tsang, D. C. Y., and Suzuki, I. (1982). Cytochrome c554 as a possible electron donor in the hydroxylation of ammonia and carbon monoxide in *Nitrosomonas europaea*. *Can. J. Biochem.*, 60, 1018-1024.

Tsien, H.-C., and Laudelout, H. (1968). Minimal size of *Nitrobacter* membrane fragments retaining nitirite oxidizing activity. *Arch. Microbiol.* 61, 280-291.

Vannelli, T., and Hooper, A. B. (1993). Reductive dehalogenation of the trichloromethyl group of nitrapyrin by the ammonia-oxidizing bacterium *Nitrosomonas europaea*. *Appl. Environ. Microbiol.*, 59, 3597-3601.

Vannelli, T., Logan, M., Arciero, D. M., and Hooper, A. B. (1990). Degradation of halogenated aliphatic compounds by the ammonia-oxidizing bacterium *Nitrosomonas europaea*. *Appl. Environ. Microbiol.*, 56, 1169-1171.

Völsch, A., Nader, W. F., Geiss, H. K., Nebe, G., and Birr, C. (1990). Detection and analysis of two serotypes of ammonia-oxidizing bacteria in sewage plants by flow cytometry. *Appl. Environ. Microbiol.*, 140, 153-158.

Voytek, M. A., Priscu, J. C., and Ward, B. B. (1999). The distribution and relative abundance of ammonia-oxidizing bacteria in lakes of the McMurdo Dry Valley, Antarctica. *Hydrobiologia*, 401, 113-130.

Wagner, M., Rath, G., Amann, R., Koops, H.-P., and Schleifer, K. H. (1995). *In situ* identification of ammonia-oxidizing bacteria. *Syst. Appl. Microbiol.*, 18, 251-264.

Wagner, D., Spieck, E., Bock, E., and Pfeiffer, E.-M. (2001). Microbial life in terrestrial permafrost: Methanogenesis and nitrification in gelisols as potentials for exobiological processes. In G. Horneck and C. Baumstark-Khan (Eds.), Astrobiology - the quest for the conditions of life (pp. 143-159). Berlin, Germany: Springer-Verlag.

Watson, S. W., and Waterbury, J. B. (1971). Characteristics of two marine nitrite oxidizing bacteria, *Nitrospina gracilis* nov. gen. nov. sp. and *Nitrococcus mobilis* nov. gen. nov. sp. *Arch Microbiol.*, 77, 203-230.

Watson, S. W., Bock, E., Valois, F. W., Waterbury, J. B., and Schlosser, U. (1986). *Nitrospira marina* gen. nov. sp. nov.: A chemolithotrophic nitrite-oxidizing bacterium. *Arch. Microbiol.*, 144, 1-7.

Watson, S. W., Bock, E., Harms, H., Koops, H.-P., and Hooper, A. B. (1989). Nitrifying bacteria. In J. T. Stanley, M. P. Bryant, N. Pfennig, and J. G. Holt (Eds.), Bergey`s manual of systematic bacteriology, Vol. 3 (pp. 1808-1834). Baltimore MD: Williams and Wilkins Co.

Whittaker, M., Bergmann, D., Arciero, D., and Hooper, A. B. (2000). Electron transfer during the oxidation of ammonia by chemolithotrophic bacterium *Nitrosomonas europaea*. *Biochim. Biophys. Acta.*, 1459, 346-355.

Wiesche von der, M., and, Wetzel, A. (1998). Temporal and spatial dynamics of nitrite accumulation in the river Lahn. *Water Res.*, 32, 1653-1661.

Woese, C. R., Weisburg, W. G., Paster, B. J., Hahn, C. M., Tanner, R. S., Krieg, N. R., Koops, H.-P., Harms, H., and Stackebrandt, E. (1984). The phylogeny of purple bacteria: The beta subdivision. *Syst. Appl. Microbiol.*, 5, 327-336.

Woese, C. R., Weisburg, W. G., Hahn, C. M., Paster, B. J., Zablen, L. B., Lewis, B. J., Macke, T. J., Ludwig, W., and Stackebrandt, E. (1985). The phylogeny of purple bacteria: The gamma subdivision. *Syst. Appl. Microbiol.*, 6, 25-33.

Wood, P. M. (1986). Nitrification as a bacterial energy source. In J. I. Prosser (Ed.), Nitrification. Special publications of the society of general microbiology. Vol. 20 (pp. 39-62). Oxford, UK: IRL Press.

Wood, P. M. (1988). Monooxygenase and free radical mechanisms for biological ammonia oxidation. In J. A. Cole and S. Ferguson (Eds.), The nitrogen and sulfur cycles (pp. 217-243). Cambridge, UK: Cambridge University Press.

Yamanaka, T., and Shinra, M. (1974). Cytochrome c-552 and cytochrome c-554 derived from *Nitrosomonas europaea*. *J. Biochem.*, 75, 1265-1273.

Yamanaka, T., Kamita, Y., and Fukumori, Y. (1981). Molecular and enzymatic properties of cytrochrome aa_3-type terminal oxidase derived from *Nitrobacter agilis*. *J. Biochem.*, 89, 265-273.

Zart, D, and Bock, E. (1998). High rate of aerobic nitrification and denitrification by *Nitrosomonas eutropha* grown in a fermentor with complete biomass retention in the presence of gaseous NO_2 or NO. *Arch. Microbiol., 169*, 282-286.

Chapter 13

THE NITROGEN CYCLE: DENITRIFICATION AND ITS RELATIONSHIP TO N_2 FIXATION

R. J. M. VAN SPANNING[1], M. J. DELGADO[2] AND D. J. RICHARDSON[3]

[1]*Department of Molecular Cell Physiology, Free University of Amsterdam, De Boelelaan 1087, 1081 HV Amsterdam, The Netherlands*

[2]*Departamento de Microbiología del Suelo y Sistemas Simbióticos, Estación Experimental del Zaidín, CSIC, PO Box 419, 18080-Granada, Spain*

[3]*Centre for Metalloprotein Spectroscopy and Biology, School of Biological Sciences, University of East Anglia, Norwich NR4 7TJ, UK*

1. INTRODUCTION

All living cells require nitrogen for the synthesis of many of their biomolecules. The assimilation of nitrogen occurs via the incorporation of the ammonium ion. In Nature, however, nitrogen is present in many other oxidation states. The biologically most important compounds are nitrate, nitrite, nitric oxide, nitrous oxide, dinitrogen, and ammonium. The free concentration of each of these nitrogen compounds is mostly determined by the production and consumption rates of bacterial metabolic processes. Together, these processes drive the global nitrogen cycle and ensure a balanced recycling of the nitrogen compounds (Ferguson, 1998; Moreno-Vivian and Ferguson, 1998; Moura and Moura, 2001; Richardson and Watmough, 1999).

The fixation of atmospheric nitrogen is in part achieved by the chemical reaction of dinitrogen (N_2) and dioxygen (O_2) induced by lightning, which gives rise to nitric oxide (NO). In the O_2-rich atmosphere, NO is then oxidized to nitrogen dioxide (NO_2) and taken up in the oceans in the form of nitrate ions. Biological nitrogen fixation into ammonium as carried out by certain bacteria, however, is much more efficient and makes most of the nitrogen available to all living cells.

The production of gaseous N_2 is mainly carried out by denitrifying species. Denitrification is an anaerobic respiratory process carried out by many bacterial

D. Werner and W. E. Newton (eds.), Nitrogen Fixation in Agriculture, Forestry, Ecology, and the Environment, 277-342.

species and some fungi and yeasts whereby N-oxides substitute for O_2 as the terminal electron acceptor of respiration. In a sequence of four reactions, nitrate is reduced to N_2 via the intermediates, nitrite (NO_2^-), nitric oxide (NO), and nitrous oxide (N_2O), successively (Ferguson, 1994; Stouthamer, 1991; Stouthamer, 1992). The key reaction that distinguishes denitrification from nitrate respiration is the reduction of nitrite, which yields nitric oxide rather than ammonium. In the past few years, three-dimensional structures have been solved for the majority of these enzymes, revealing the architecture of the active metal sites as well as global structural and mechanistic aspects.

Bacteria that are able to denitrify are confronted with a paradox. On the one hand, the potential to denitrify enhances their metabolic flexibility because it allows them to grow in the absence of O_2. On the other hand, there is the potential risk that the free concentrations of the toxic intermediates, nitrite and nitric oxide, reach levels that are lethal to the cell. Due to the sequential order of reactions during denitrification, the reaction products of three of the four enzymes are substrates for the next enzyme. The toxicity of nitrite and nitric oxide imply that concentrations and activities of each of the enzymes should be well tuned in order to keep these steady-state concentrations in the cell low. It has now become evident that denitrifying organisms regulate the activity of denitrification both by means of the expression of each of the gene clusters involved (DNA level, long-term adaptation) and through the specific properties (K_m and k_{cat}) of the participating enzymes (protein level, short-term adaptation).

Denitrification and aerobic respiration have some features in common. These are: (i) cd_1-type nitrite reductase and *bc*-type nitric oxide reductase, which are capable of reducing O_2 to water; (ii) subunit I of the haem copper oxidases and the large subunit from nitric oxide reductase have similar topologies; (iii) a Cu_A site, which is active in electron transfer, is present in nitrous oxide reductase, in subunit II of aa_3-type cytochrome *c* oxidases, and in some quinol-oxidizing nitric oxide reductases; and (iv) subunit III of aa_3-type cytochrome *c* oxidases has homology with NorE, which is encoded by the nitric oxide reductase gene cluster. It is, therefore, tempting to speculate that some of the building blocks of the denitrification apparatus have been used and/or rearranged giving rise to the evolution of the aerobic respiratory system in a time where the O_2 concentration in the earth's atmosphere increased as a result of photosynthetic activity (Saraste, 1994; Saraste and Castresana, 1994).

Products of denitrification have manifold, mainly adverse, effects on the atmosphere, soils, and waters and thus have both agronomic and environmental impact. When nitrate is converted to gaseous nitrogen by denitrifying bacteria in agricultural soils, nitrogen is lost as an essential nutrient for the growth of plants. In contrast to ammonium, which is tightly bound in soil, nitrate is easily washed out and flows into the groundwater where it (and its reduction product, nitrite) adversely affects water quality. In addition, nitrogenous oxides released from soils and waters are, in part, responsible for the depletion of the ozone layer above the Antarctic and, in part, for the initiation of acid rain and global warming. Thus, the impact of products of denitrification in soils, waters, and the atmosphere is of extreme relevance for human welfare and makes a detailed knowledge of this process essential.

2. THE NITROGEN CYCLE

Ammonium is incorporated into central metabolism either through the combined action of glutamine synthetase and glutamate synthase (Reitzer, 1996) or through glutamate dehydrogenase. Mineralization processes on dead cells may release it again; this process is usually referred to as ammonification. The nitrogen cycle constitutes an oxidative phase, the conversion of ammonium to nitrate, and a reductive phase, the conversion of nitrate back to ammonium (Figure 1).

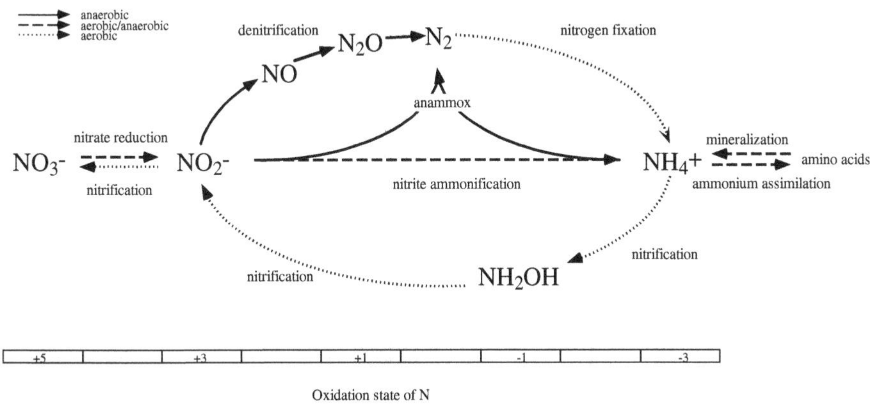

Figure 1. The biological nitrogen cycle.
Nitrate reduction involves respiratory membrane bound (Nar), periplasmic (Nap), and assimilatory (Nas) nitrate reductases. Nitrite ammonification involves either siroheme-type (NirB/NasB) or decaheme c-type (Nrf) nitrite reductases. Denitrification is the sequential reduction of nitrate/nitrite to dinitrogen by nitrite, nitric oxide and nitrous oxide reductases. Nitrogen fixation is catalyzed by nitrogenase. Nitrification reactions include ammonia oxidation to nitrite by ammonia monooxygenase and hydroxylamine oxidoreductase, and nitrite oxidation to nitrate by nitrite oxidase. Ammonia oxidation coupled to nitrite reduction is carried out by Anammox. Assimilatory reactions are carried out under both aerobic and anaerobic conditions. Respiratory nitrate reduction occurs sometimes aerobically, mostly anaerobically. Denitrification, respiratory nitrite ammonification, and Anammox are carried out under anaerobiosis. Nitrification requires the presence of O_2.

The oxidation of ammonium to nitrate is achieved by the sequential activities of ammonia oxidizers, the nitrosifiers, which oxidize ammonia to nitrite, and the nitrite oxidizers, the nitrifiers, which oxidize nitrite to nitrate. Together, these processes make up the nitrification pathway. Ammonia and nitrite oxidizers have an important role in the conversion of nitrogenous compounds both in their natural environment, *e.g.*, soils, sediments, and lakes, and in wastewater-treatment systems.

These habitats are very dynamic especially with regard to the availability of O_2. As a result of biological, climatological, or mechanical factors, the O_2 concentration is constantly fluctuating and requires subtle adaptive responses of its inhabitants to survive in the numerous microhabitats present in these environments.

The anaerobic oxidation of ammonium is catalyzed exclusively by some members of the Planctomycetes, like *Brocadia anammoxidans*, by a process called anammox (anoxic ammonia oxidation). The oxidation of ammonium in this process is coupled to the reduction of nitrite and yields N_2 and free energy for maintenance and growth. The biochemistry of the anammox process is still under investigation, but nitrite appears to be an electron acceptor and hydrazine an intermediate (Strous *et al.*, 1999; Van de Graaf *et al.*, 1995). Due to the toxicity of the intermediates, the key steps of the oxidation take place in a special membrane-enclosed compartment called the anammoxosome (Lindsay *et al.*, 2001). Carbon is fixed from CO_2 using nitrite as electron donor. The mechanism is not clear but it may be that the latter is mediated via an enzyme resembling the nitrite oxidoreductase found in nitrite oxidizers. The anammox process is of great ecological importance and allows the removal of ammonium from anaerobic sites both in natural environments and in man-made wastewater-treatment plants.

The initial step in the reductive phase of the nitrogen cycle is the two electron-reduction of nitrate to nitrite carried out by bacteria that have the enzyme nitrate reductase. There are three distinct types of bacterial nitrate reductase: a periplasmic enzyme (Nap); a membrane-bound enzyme with its catalytic site facing the cytoplasm (Nar); and a cytoplasmic one (Nas) (Moreno-Vivian and Ferguson, 1998; Richardson *et al.*, 2001; Stolz and Basu, 2002). The different nitrate reductases have structural and functional resemblance but different physiological roles. Some species have the genetic potential to express all three of them apparently enhancing their metabolic flexibility.

Nas has a role in the process called assimilatory nitrate/nitrite ammonification (Lin and Stewart, 1998). Nas provides the nitrite that is subsequently used for ammonium formation by a dedicated sirohaem-type nitrite reductase. The proteins are, therefore, referred to as assimilatory nitrate and nitrite reductases. They receive electrons from either NAD(P)H or ferredoxin (Blasco *et al.*, 1997; Gangeswaran *et al.*, 1993), their activities are tightly coupled in the cytoplasm, and the genes encoding them are usually located in a single *nas* operon (Lin *et al.*, 1994). They are found in many different types of organisms, in which they are co-ordinately expressed during conditions of ammonium depletion and nitrate availability. These conditions are sensed by global nitrogen-control systems (NtrC) and pathway specific nitrate/nitrite sensors (Wu *et al.*, 1999).

Nar and Nap are coupled to anaerobic respiratory processes that usually occur as alternative for O_2 respiration. Both enzymes receive electrons from quinol. They are, therefore, called respiratory nitrate reductases (Moreno-Vivian and Ferguson, 1998; Richardson *et al.*, 2001). In contrast to Nap, Nar is an electrogenic enzyme and as such it contributes to the generation of a proton-motive force during electron transfer from quinol to nitrate (Berks *et al.*, 1995). Nar is expressed when the O_2 concentration is low and nitrate is available. Due to the location of the catalytic subunit of Nar, nitrite is produced in the cytoplasm. In some nitrate-respiring species, Nar is co-expressed with a sirohaem-type nitrite reductase. This sirohaem-type nitrite reductase is structurally and functionally similar to the one involved in assimilatory functions, but its expression is brought under O_2 rather than under ammonium control (Cole, 1996; Tyson *et al.*, 1994; Wang and Gunsalus, 2000). Its role is to detoxify nitrite in a process called dissimilatory nitrite ammonification.

In yet other nitrate-respiring bacteria, but never in the same species, nitrite is transferred to the periplasm where it is further reduced to N_2 via the intermediate compounds nitric oxide (NO), nitrous oxide (N_2O), and N_2 (Berks *et al.*, 1995; Moura and Moura, 2001; Zumft, 1997). These reactions are carried out by dedicated N-oxide reductases, which receive electrons from respiratory components. This process is termed denitrification and serves to continue with respiration and free-energy transduction under anaerobic conditions. Expression of the enzymes is controlled by the O_2 status and the presence of N-oxides.

Some denitrifying species recruit Nap rather than Nar to initiate their denitrification mode but, because electron transfer through Nap is not electrogenic, the free-energy transduction in that reaction is less efficient. Nap has also a role in non-denitrifying species where its activity is coupled to yet another type of nitrite reductase, Nrf. Nrf is a decahaem-type nitrite reductase, which is, like Nap, located in the periplasm and receives electrons from quinol. Both Nap and Nrf appear to be co-ordinately expressed along with formate dehydrogenase and hydrogenase under conditions of O_2 depletion in some species (Cole, 1996; Simon, 2002). Together, they make up a respiratory system for the oxidation of formate and/or hydrogen and the concomitant reduction of nitrate *via* nitrite to ammonium. This respiratory process is called respiratory nitrate/nitrite ammonification. Free-energy transduction is driven by the electrogenic properties of the dehydrogenases and not by either Nap or Nrf. Both formate dehydrogenase and hydrogenase have the characteristic feature of enzymes that carry out electrogenic redox reactions, which give rise to the formation of a proton electrochemical gradient across the membrane in that they possess a trans-membrane "electron wire" composed of two haems *b* located at opposite sides of the membrane. During oxidation of their substrates in the periplasm, protons are released at the periplasmic site while the electrons move *via* the wire to a cytoplasmic site where they reduce quinone to quinol. Both Nap and Nrf receive electrons and protons from one side of the membrane, the periplasm, and are, therefore, non-electrogenic.

The N_2 that is liberated into the atmosphere by denitrifying species is recycled by nitrogen-fixing organisms in the process of nitrogen fixation. These species have the enzyme nitrogenase, which reduces N_2 to ammonium in a procees that also produces H_2. As such, they make sure that nitrogen becomes available again for all cellular life.

3. DENITRIFICATION

Zumft has defined denitrification as the dissimilatory reduction of either nitrate or nitrite to a gaseous N-oxide concomitant with free-energy transduction (Zumft,

1997). The process assembles nitrate respiration, nitrite respiration combined with NO reduction, and nitrous-oxide respiration. A scheme of the denitrification reactions as they occur in *Paracoccus denitrificans* is shown in Figure 2. Comprehensive reviews covering the physiology, biochemistry and molecular genetics of denitrification have been published elsewhere (Baker *et al.*, 1998; Hendriks *et al.*, 2000; Richardson *et al.*, 2001; Richardson and Watmough, 1999; Watmough *et al.*, 1999; Zumft, 1997). Here, we review the basic concepts of denitrification and the most recent developments in relation to other nitrogen-cycle processes. Tables 1 and 2 list the species that are known to have denitrification genes and denitrifying properties, respectively. Table 3 lists the species that do not have genes encoding the key enzymes of denitrification.

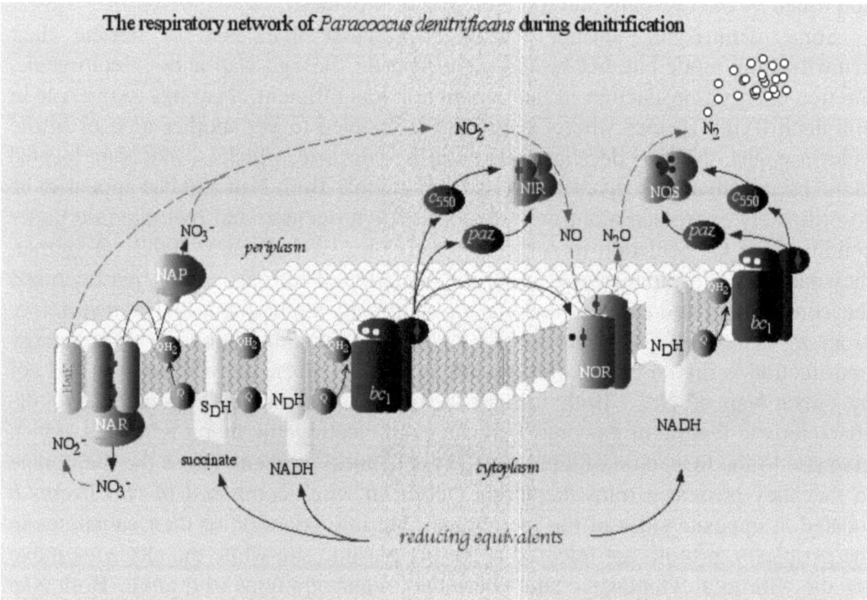

Figure 2. Scheme of the full denitrification process in Paracoccus denitrificans. Dashed arrows, N-oxide transport; straight arrows, electron transport. SDH, succinate dehydrogenase; NDH, NADH dehydrogenase; Q, quinone; bc_1, cytochrome bc_1 complex; c_{550}, cytochrome c; paz, pseudoazurin; NAR, membrane-bound nitrate reductase; NAP, periplasmic nitrate reductase; NIR, cd_1-type nitrite reductase; NOR, bc-type NO reductase; NOS, nitrous oxide reductase.

4. BACTERIAL RESPIRATORY NITRATE REDUCTASES

There are 2 types of respiratory nitrate reductase in bacteria; a membrane bound nitrate reductase (Nar), and a periplasmic one (Nap). Both types receive electrons from quinol and catalyse the two-electron reduction of nitrate to nitrite. Together with Nas, they belong to a family of mononuclear molybdenum enzymes, which also includes DMSO reductase (DMSOR), formate dehydrogenase H (Fdh), and

Table 1. The distribution of denitrification genes in Bacteria and Archaea

Bacteria	Nar	Nap	NirK	NirS	Nor	Nos
Alpha Proteobacteria						
Agrobacterium tumefaciens *	-	+	+	-	cNor	-
Aquaspirillum itersonii	?	?	-	w	?	?
Azospirillum brasilense SP7	?	+	-	+/	?	+/
Azospirillum halopraeferens	?	?	?	?	?	+/
Azospirillum lipoferum	?	?	-	+/	?	+/
Blastobacter denitrificans	?	?	+	-	?	?
Bradyrhizobium japonicum	-	+	+	-	cNor	+
Brucella melitensis *	+	-	+	-	cNor	+
Brucella suis *	+	-	+	-	cNor	+
Hyphomicrobium denitrificans	?	?	+	-	?	?
Hyphomicrobium zavarzinii	?	?	+	-	?	?
Magnetospirillum magnetotacticum	+	+	-	+	?	+
Ochrobactrum anthropi	?	?	?	+/	?	?
Paracoccus denitrificans	+	+	-	+	cNor	+
Paracoccus pantotrophus	+	+	-	+	cNor	+
Rhizobium sullae (hedysari)	?	?	+	-	-	-
Rhodobacter capsulatus AD2	+	+	-	-	?	?
Rhodobacter capsulatus BK5	+	-	(+)	-	cNor	?
Rhodobacter capsulatus E1F1	-	-	-	-	?	?
Rhodobacter sphaeroides2.4.1	-	+	-	-	cNor	+
Rhodobacter sphaeroides2.4.3	-	+	+	-	cNor	+
Rhodopseudomonas palustris	?	?	+	-	cNor	+
Roseobacter denitrificans	?	?	-	+	cNor	?
Sinorhizobium meliloti *	-	+	+	-	cNor	+
Beta Proteobacteria						
Achromobacter cycloclastes	?	?	+	-	cNor	+/
Acidovorax avenae	?	?	?	+/	?	?
Alcaligenes faecalis IAM1015	?	?	-	+	?	?
Alcaligenes faecalis S-6	?	?	+	-	cNor	+/
Alcaligenes xylosoxidans	?	?	+	-	?	+
Azoarcus evansii	?	?	-	+/	?	?
Azoarcus tolulyticus	?	?	-	+/	?	?
Burkholderia mallei	+	-	+	-	qNor	+
Burkholderia pseudomallei	+	-	+	-	qNor	+
Neisseria gonorrhoeae	?	?	+	-	qNor	+
Neisseria meningitidis *	-	-	+	-	qNor	-
Nitrosomonas europaea	-	-	+	-	cNor	-
Ralstonia eutropha	+	+	-	+	qNor	+
Ralstonia metallidurans	+	+	-	+	qNor	+
Ralstonia solanacearum *	+	-	+	-	qNor	+
Thauera aromatica	?	?	-	+/	?	?
Thauera chlorobenzoica	?	?	-	+/	?	?
Thauera mechernichensis	?	?	-	+/	?	?
Thauera selenatis	?	?	-	+/	?	?
Thiobacillus denitrificans	+	?	-	+/	?	?

Bacteria	Nar	Nap	NirK	NirS	Nor	Nos
Gamma Proteobacteria						
Halomonas halodenitrificans	+	?	-	+	cNor	?
Legionella pneumophila	?	+	?	?	qNor	?
Pseudomonas aeruginosa *	+	+	-	+	cNor	+
Pseudomonas aureofaciens	?	?	+	?	?	?
Pseudomonas denitrificans	?	?	w,p	?	?	+
Pseudomonas fluorescens	+	?	-	+	cNor	+
Pseudomonas G-179	-	+	+	-	cNor	?
Pseudomonas nautical	?	?	?	+	?	+
Pseudomonas putida	?	?	+/	?	?	?
Pseudomonas stutzeri	+	+	-	+	cNor	+
Delta Proteobacteria						
Geobacter metallireducens	+	?	?	?	qNor	?
Gram positive bacteria						
Bacillus azotoformans	?	?	w	?	qNor	?
Bacillus halodenitrificans	?	?	+	?	?	?
Bacillus stearothermophilus	+	-	+	-	qNor	?
Corynebacterium nephridii	?	?	w,p	?	?	?
Corynebacterium diphtheriae	+	?	+	?	qNor	?
Desulfitobacterium hafniense	?	?	?	?	cNor	+
Mycobacterium avium	?	?	?	?	qNor	?
Staphylococcus aureus (EMRSA-16)	?	?	+	?	qNor	?
Streptomyces thioluteus	?	?	+	?	?	?
Cyanobacteria						
Synechocystis PCC6803	-	-	-	-	qNor	-
Bacteroidetes						
Cytophaga hutchinsonii	?	?	?	?	qNor	?
Flavobacterium sp	?	?	?	w,p	?	?
Crenarchaeota						
Pyrobaculum aerophilum *	+	-	-	-	qNor	-
Sulfolobus solfataricus *	-	-	-	-	qNor	-
Euryarchaeota						
Haloarcula marismortui	+	?	+	-	?	?
Haloferax denitrificans	+	?	+	-	?	?

The table includes data derived from a BLAST analysis of 106 published and 134 unfinished bacterial and archeal genome sequences at the NCBI Blast Server (www.ncbi.nlm.nih.gov), 25 October 2002 release.

*Species that have their genome sequences published. +, present; -, absent; ?, unknown; w, p, +/, present as judged by Western analyses, gene probing, and PCR, respectively (Coyne *et al.*, 1989; Rosch *et al.*, 2002; Smith and Tiedje, 1992; Song and Ward, 2002).

Table 2. List of denitrifying species of which the genetics of denitrification is not known

Phylum or class	Species
Alpha Proteobacteria	*Paracoccus versutus*
	Pseudomonas carboxydohydrogena
	Rhodoplanes elegans
	Rhodoplanes roseus
Beta Proteobacteria	*Alicycliphilus denitrificans*
	Brachymonas denitrificans
	Burkholderia glumae
	Chromobacterium sp
	Comamonas denitrificans
	Eikenella sp
	Hydrogenophaga pseudoflava
	Kingella denitrificans
	Neisseria denitrificans
	Nitrobacter sp
	Sterolibacterium denitrificans
	Thermothrix thiopara
Gamma Proteobacteria	*Alteromonas sp*
	Beggiatoa sp
	Moraxella sp
	Oligella sp
	Pseudoalteromonas denitrificans
	Pseudomonades species (see Zumft, 1997)
	Shewanella denitrificans
	Shewanella putrefaciens??
	Thialkalivibrio denitrificans
	Thioploca sp
Epsilon Proteobacteria	*Arcobacter cryaerophilus*
	Thiomicrospira denitrificans
	Wolinella succinogenes
Bacteroidetes	*Empedobacter brevis*
	Flexibacter canadensis
	Sphingobacterium sp
Gram positive	*Geobacillus thermodenitrificans*
	Jonesia denitrificans
	Tsukamurella sp
Euryarchaeota	*Haloarcula vallismortui*

Table 3. Bacteria lacking denitrification genes in their genome sequences

Alpha Proteobacteria	**Aquificales**
Caulobacter crescentus	*Aquifex aeolicus*
Mesorhizobium loti	
Rickettsia conorii	**Chlamydiae**
Rickettsia prowazekii	*Chlamydia muridarum*
	Chlamydia trachomatis
Gamma Proteobacteria	*Chlamydophila pneumoniae*
Buchnera sp	
Buchnera aphidicola	**Chlorobi**
Escherichia coli	*Chlorobium tepidum TLS*
Haemophilus influenza	
Pasteurella multicoda	**Cyanobacteria**
Salmonella enterica typhi	*Nostoc sp. PCC 7120*
Shigella flexneri	*Thermosynechococcus elongatus*
Vibrio cholerae	
Wigglesworthia brevipalpis	**Deinococci**
Xanthomonas axonopodis	*Deinococcus radiodurans*
Xanthomonas campestris	
Xylella fastidiosa	**Fusobacteria**
Yersinia pestis	*Fusobacterium nucleatum*
Epsilon Proteobacteria	**Spirochaetes**
Campylobacter jejuni	*Borrelia burgdorferi*
Helicobacter pylori	*Leptospira interrogans*
	Treponema pallidum
	Thermotogae
	Thermotoga maritima
Gram positive bacteria	
Bacillus halodurans	*Mycoplasma pneumoniae*
Bacillus subtilis	*Mycoplasma pulmonis*
Bifidobacterium longum	*Oceanobacillus iheyensis*
Clostridium acetobutylicum	*Staphylococcus aureus MW2*
Clostridium perfringens	*Staphylococcus aureus Mu50*
Corynebacterium glutamicum	*Staphylococcus aureus N315*
Lactococcus lactis	*Streptococcus agalactiae*
Listeria innocua	*Streptococcus pneumoniae*
Listeria monocytogenes	*Streptococcus pyogenes*
Mycobacterium leprae	*Streptomyces coelicolor*
Mycobacterium tuberculosis	*Thermoanaerobacter tengcongensis*
	Ureaplasma urealyticum

None of the Archaea, except Pyrobaculum aerophilum *and* Sulfolobus solfataricus *(qNor), has either denitrification genes or other nitrate/nitrite reductase genes.*

biotin-*S*-oxide reductase (Hille, 1996). They share common architectural features within their catalytic subunit, especially with regard to the make up of their catalytic site, which contains two pterin cofactors that ligate a single molybdenum atom (bis-molybdopterin guanine dinucleotide cofactor, bis-MGD cofactor). In some members, *e.g.*, Nar, Nap and Fdh, the catalytic site also contains a [4Fe-4S] cluster, but this is absent in others, *e.g.*, DMSOR and TMAO reductase. Members of the family can also be distinguished by the nature of the amino-acid ligand to the Mo ion that is provided by the polypeptide; it which can be Cys (Nap), SeCys (Fdh) or Ser (DMSOR) (Dias *et al.*, 1999; Hilton and Rajagopalan, 1996; Schindelin *et al.*, 1996). Phylogenetic analyses revealed that the cytoplasmic Nas and periplasmic Nap proteins cluster together in a neighbour-joining phylogenetic tree distinct from the Nar group of proteins (Stolz and Basu, 2002). Nar and Nap differ in their subunit composition, the nature and number of cofactors involved in electron transport from quinol to the active site, their involvement in free-energy transduction, affinity for nitrate, induction patterns, and physiological role.

4.1. Membrane-bound Respiratory Nitrate Reductase (Nar)

Nar is found in both denitrifiers and non-denitrifiers of virtually all phyla of the bacterial kingdom. In non-denitrifiers, Nar is usually co-expressed with a cytoplasmic sirohaem-type nitrite reductase, which reduces nitrite to ammonium for detoxification. In denitrifying species, nitrite is translocated from the cytoplasm across the membrane to the periplasm, where it is further reduced *via* NO and N_2O to N_2 by the denitrification enzymes. In any case, Nar is expressed during O_2 depletion and nitrate availability.

4.1.1. Composition, Structure and Function of Nar

Nar is a three-subunit enzyme composed of NarGHI (Ballard and Ferguson, 1988). NarG, the α-subunit of *ca.* 140 kDa, contains the bis-MGD molybdopterin cofactor at its catalytic site and a [4Fe-4S] cluster. NarH, the β-subunit of *ca.* 60 kDa, contains four additional iron-sulphur centres, one [3Fe-4S] and three [4Fe-4S]. NarG and NarH are located in the cytoplasm and associate with NarI, the γ-subunit. The contact between NarGH and NarI is most likely mediated solely by the C-terminal part of NarH. NarI is an integral membrane protein of *ca.* 25 kDa with five trans-membrane helices and the N-terminus facing the periplasm. This subunit carries two haems *b*, one of low potential (haem b_l, with Em 8 = +20 mV, *E. coli* enzyme) located at the periplasmic site, and of one high potential (haem b_h, Em 8 = +125 mV) located at the cytoplasmic site (Rothery *et al.*, 1999). Histidine residues conserved in all known NarI proteins are supposedly involved in ligation of these haems (Berks *et al.*, 1995). The two haems *b* constitute a plane that is orthogonal to the membrane.

Nar receives electrons from quinol and is, therefore, linked to respiratory electron transfer. Oxidation of quinol occurs at the periplasmic site of NarI, where the protons are released and the two electrons are moved from the outside low-potential haem b_l to the inside high-potential haem b_h. This charge separation makes the enzyme electrogenic in that it contributes to the generation of a proton electrochemical gradient across the membrane (two charge separations during the transfer of two electrons to nitrate: 2q/2e-) (Berks *et al.*, 1995). Electrons from haem b_h of NarI are most likely donated to the [3Fe-4S] cluster of NarH (Rothery *et al.*, 2001). From there, they flow *via* the iron-sulphur clusters in NarH to the one in NarG, probably the direct electron donor to the bis-MGD cofactor containing catalytic site in NarG, where nitrate is reduced to nitrite (Rothery *et al.*, 1998). The precise pathway, however, has yet to be resolved. More details about the biochemistry of Nar from *E. coli* are described in a recent review (Blasco *et al.*, 2001).

Nar has not yet been crystallized, but much about its structure-function relationship can be learned from crystal analyses of related enzymes. Analyses of the structures of Nap (Dias *et al.*, 1999), and the soluble DMSO reductase of *Rhodobacter sphaeroides* (McAlpine *et al.*, 1998; Schindelin *et al.*, 1996) have recently been solved and a mechanism for the molybdenum-catalyzed two-electron transfer steps has been proposed. The structures suggest that the enzymes cycle between mono-oxo Mo(IV) and des-oxo Mo(V) states during turnover. Sequence similarities suggest that the mechanism of nitrate reduction in Nar proceeds in a similar fashion to that for DMSO reduction (Schindelin *et al.*, 1996; Stiefel, 1996).

The best model for the organization of NarGHI in the membrane comes from the recent structure of the nitrate-induced formate dehydrogenase (Fdh-N) of *Escherichia coli* (Jormakka *et al.*, 2002; Richardson and Sawers, 2002). This enzyme is related to Nar, especially with respect to the equivalent FeS-containing β-subunit and Mo-MGD catalytic α-subunit, and it also has a di-haem integral membrane subunit. The topology is, however, different because, in the formate dehydrogenase, both the FeS-containing and Mo-MGD-containing subunits are in the periplasm and the dehydrogenase catalyses the inward movement of electrons from the periplasmic side to the cytoplasmic side of the membrane, where menaquinone is reduced. It nevertheless resembles the nitrate reductase in constituting the electron-carrying arm of a redox loop that has a q^+/e ratio of 1 and it provides the first molecular insight into the architecture of such an electrogenic redox loop. In short, the structure shows that, in Fdh-N, the electrons extracted from formate at the periplasmic bis-MGD active site pass down a 90-Å "ladder" of eight redox centres and ultimately reduce menaquinone at the N-face of the cytoplasmic membrane. This redox ladder comprises the bis-MGD, five Fe-S clusters and two haems. Each is within 12 Å of its nearest neighbour ensuring rapid electron transfer.

In the FdhN β-subunit, the four iron-sulphur centres (all [4Fe-4S]) are arranged in two pairs in each of two domains. This arrangement was already predicted for Nar from detailed spectroscopic studies on *E. coli* NarH, the only difference being that one of the Nar FeS centres is a [3Fe-4S] cluster (Blasco *et al.*, 2001). In the case of NarH, the low redox potentials of two of the centres had raised the possibility that they are not directly involved in electron transfer between quinol and nitrate. However, consideration of the structure of the homologous Fdh-N β-subunit leaves no doubt that all four FeS centres will be directly involved in mediating electron transfer. Thus, it seems that a similar extended electron-transfer chain will serve in NarGHI and add to a growing number of cases of electron transfer complexes in bacteria that contain a mixture of endergonic and exergonic electron-transfer steps and which include the quinol-dependent nitrate electron-transport system to Nap discussed below.

4.1.2. Genetics of Nar

Nar proteins are encoded by genes of a *narGHJI* operon. The organization of this operon is conserved in most species that express Nar. The *narGHI* genes encode the structural subunits and *narJ* encodes a dedicated chaperone required for the proper maturation and membrane insertion of Nar (Blasco *et al.*, 1992). *E. coli* has a functional duplicate of the *narGHJI* operon, called *narZYWV*. The subunits of the two enzymes are interexchangeable (Blasco *et al.*, 1990). In many species, a *narK* gene encoding a nitrate/nitrite transporter precedes the *narGHJI* genes (Clegg *et al.*, 2002; Moir and Wood, 2001). In the case of *Paracoccus pantotrophus*, it has been shown that *narK* forms part of a *narKGHJI* operon.

A set of *narXL* genes encoding a two-component regulatory system, which modulates the expression of Nar in response to the nitrate/nitrite concentration, is found upstream of the *nar*-gene cluster in both *E. coli* and denitrifying *Pseudomonades* and downstream of the *nar*-gene cluster of *Ralstonia solanacearum* (Lissenden *et al.*, 2000; Philippot *et al.*, 2001; Rabin and Stewart, 1993; Stewart, 1993; Vollack and Zumft, 2001). The alpha-Proteobacteria, *Brucella suis*, *Brucella melitensis*, *Paracoccus denitrificans* and *Pa. pantotrophus*, have the *narXL* genes replaced by an *fnr*-like gene encoding an FNR homologue, which is designated as NarR (DelVecchio *et al.*, 2002; Wood *et al.*, 2001). This protein is a transcriptional regulator that modulates the expression of the *nar* operon in response to the intracellular nitrite concentration (Wood *et al.*, 2001). It remains to be seen if a comparable make-up of the *nar* operon is common to other species of this genus as well. Apart from a requirement of N-oxides, transcription of the *nar* operon is, in many cases, controlled by the O_2 status of the cell as sensed by Fnr-like proteins.

4.2. Periplasmic Nitrate Reductase (Nap)

Nap is widespread in all classes of denitrifying and non-denitrifying proteobacteria but as yet not found in other phyla of the bacterial superkingdom (Richardson, 2000). The best-studied Nap enzymes were isolated from *Pa. denitrificans* (Sears *et al.*, 1995), *Pa. pantotrophus* (Bell *et al.*, 1993; Berks *et al.*, 1994), *E. coli* (Grove *et al.*, 1996), *Rh. sphaeroides* (Richardson *et al.*, 1990), *Ralstonia eutropha* (Siddiqui *et al.*, 1993) and *Pseudomonas putida* (Carter *et al.*, 1995).

4.2.1. Composition, Structure and Function of Nap

Nap has three subunits called NapABC. The NapAB complex is located in the periplasm and associates with a trans-membrane NapC component. NapA (*ca.* 90

kDa) is the catalytic subunit with a [4Fe-4S] cluster (Em 7 = –160 mV) and a bis-MGD cofactor similar to the organization of cofactors in NarG. NapB (16 kDa) is a dihaem cytochrome c_{552}, with both haems being bis-histidinyl ligated (Em 7 = –15 mV and +80 mV) (Breton *et al.*, 1994).

Recently, the crystal structure of *Desulfovibrio desulfuricans* NapA (Dias *et al.*, 1999) has become available and diffracting crystals of a NapAB complex from *Rhodobacter capsulatus* have been reported (Pignol et al. 2001). These data allowed a prediction of the catalytic cycle (see the above part on Nar) that correlates with the molecular make up of Nap and related molybdenum enzymes (Dias *et al.*, 1999; McAlpine *et al.*, 1998). It is notable though that differences in the structure of the catalytic site may exist between NapA for *De. desulfuricans* (Dias *et al.*, 1999) and *Pa. pantotrophus*. In *De. desulfuricans,* the molybdenum amino-acid ligand is a cysteine residue. The Mo ion is additionally coordinated by four sulphur ligands provided by the two MGD moieties and a water/hydroxo ligand. Based on this oxidised structure, it was proposed that the enzyme cycles between des-oxo-Mo(IV) and mono-oxo-Mo(VI), which is then protonated to form the water/hydroxo ligand (Dias *et al.*, 1999). This scheme is different from that derived from solution X-ray absorption studies on the *Pa. pantotrophus* NapAB enzyme which pointed towards mono-oxo and di-oxo state interconversion during the catalytic cycle rather than des-oxo/mono-oxo interconversions. This difference may lie in the different experimental methods used to provide the data on which the catalytic cycle is proposed. However, based on primary-sequence analysis, Nap from *De. desulfuricans* is a rather divergent member of the Nap family. It is, for example, more closely related to Fdh from *E. coli* than is Nap from *Pa. pantotrophus* (Richardson *et al.*, 2001) and exhibits a formate-dehydrogenase activity that is absent in *Pa. pantotrophus* Nap (Bursakov *et al.*, 1997; Butler *et al.*, 1999). Thus, the possibility exists that differences between the two enzymes may account for different oxygen co-ordination at Mo during the catalytic cycle.

NapC (25 kDa) has a single N-terminal trans-membrane helix that anchors a globular domain with four *c*-type haems onto the periplasmic surface of the cytoplasmic membrane. All four haems have relatively low midpoint redox potentials (Em, 8.0 of –56 mV, –181 mV, –207 mV and –235 mV), are low spin, and apparently bis-histidinyl axially ligated (Roldan *et al.*, 1998). Likely candidates for the distal histidines (His81, His99, His174 and His194, using numbering of the *Pa. pantotrophus* NapC) were suggested from sequence alignment studies (Roldan *et al.*, 1998) and verified by mutagenesis studies (Cartron *et al.*, 2002). These studies also suggested that the soluble domain of NapC is made up of two sub-domains, each containing a di-haem pair with each haem-pair obtaining one distal histidine as haem ligand from its own domain and a second from the other domain.

NapC belongs to a large family of bacterial tetra-haem and penta-haem cytochromes that have been proposed to participate in electron transfer between the quinol/quinone pool and periplasmic redox enzymes, such as the TMAO reductase, DMSO reductase, fumarate reductase, nitrite reductases, and hydroxylamine oxidoreductase (Bergmann *et al.*, 1994; Jüngst *et al.*, 1991; Myers and Myers, 1997; Richardson, 2000; Shaw *et al.*, 1999; Simon, 2002). NapC can also be added to this list because its periplasmic domain can mediate electron transfer from quinols to

NapAB (Cartron *et al.*, 2002). On the basis of these features, the electron pathway from quinol to nitrate through Nap can be deduced. The enzyme receives electrons from membrane-embedded benzoquinols or napthoquinols, which are donated to the tetra-haem periplasmic domain of NapC. From there, they flow *via* the haems in NapB and the [4Fe-4S] cluster in NapA to the bis-MGD-containing catalytic site of NapA, where the two-electron reduction of nitrate to nitrite is catalyzed. As discussed above for Nar, this electron-transfer system will also contain a mixture of endergonic and exergonic electron-transfer steps but it is expected that none of the eight redox centres involved will be more than 14 Å apart, thus, ensuring rapid electron transfer. Electron transfer through Nap is non-electrogenic because electrons and protons required for the reduction of nitrate to nitrite are both taken up at the same side of the membrane, *i.e.*, the periplasm.

Except for the ability to reduce nitrate, Nap is catalytically distinct from Nar. Nar is able to reduce chlorate and bromate, whereas Nap is completely specific for nitrate as its substrate. Nar is sensitive to competitive inhibition by azide, whereas Nap is not. This undoubtedly reflects structural differences in the catalytic sites of the two enzymes that are also apparent from analysis of EPR spectra and primary-sequence analysis, which point towards Nar having a Ser rather than a Cys ligand to the molybdenum atom (Potter *et al.*, 2001). However, both nitrate reductases can use reduced viologens as electron donor. In whole cells, Nar activity can be distinguished from Nap activity, using viologen-linked assays. Nar, with its active site on the cytoplasmic side of the cytoplasmic membrane, is accessible to the membrane-permeable viologen, benzyl viologen, but not to the relatively membrane-impermeable methyl viologen. Nap is accessible to both viologens (Berks *et al.*, 1995).

4.2.2. Genetics of Nap

The NapABC proteins are encoded by the *napABC* genes, which make up part of a *nap* operon (Berks *et al.*, 1995; Grove *et al.*, 1996; Liu *et al.*, 1999; Richardson *et al.*, 2001; Vollack *et al.*, 1998). Except for *Shewanella putrefaciens* and *Campylobacter jejuni*, which lack *napC*, all operons studied thus far have the *napABCD* genes in common (Figure 3). Because NapC appears crucial in electron transfer between quinol and NapAB, it might well be that *Sh. putrefaciens* and *Ca. jejuni* recruit proteins with NapC activity from genes located elsewhere on the chromosome. One such homologue might be CymA of *Sh. putrefaciens*, which appeared to have a pleiotropic role in electron transport to periplasmic nitrate reductase, fumarate reductase, and iron reductase (Myers and Myers, 1997; Myers and Myers, 2000). A likely candidate in *Ca. jejuni* is NrfH, which is also a member of the NapC/NirT family of multi-haem cytochromes *c* (Einsle *et al.*, 2002). The gene is located in the *nrfAH* gene cluster.

The genes of the *nrfAH* cluster encode a nitrite-reductase complex involved in respiratory nitrite ammonification (Einsle *et al.*, 2002; Simon, 2002). The catalytic subunit is NrfA, which is a periplasmic decahaem-type cytochrome *c* (in the native dimeric form) and NrfH is the membrane-anchored subunit, which transfers electrons from quinol to NrfA. Whether NrfH is also used for periplasmic nitrate reduction in *Ca. jejuni* remains to be seen, but Nrf usually coexists and associates with Nap (Potter *et al.*, 2001). NapABC is expressed as proteins with a signal sequence that targets them to the membrane prior to translocation to the periplasm. The signal sequence of NapA contains a twin-arginine motif diagnostic for proteins that are translocated to the periplasm in a partially folded state *via* a special Tat (twin arginine translocon) system (Berks, 1996; Berks *et al.*, 2000). NapD is conserved in all *nap*-gene clusters. It has a likely role as a chaperone in the correct assemblage of Nap (Potter and Cole, 1999).

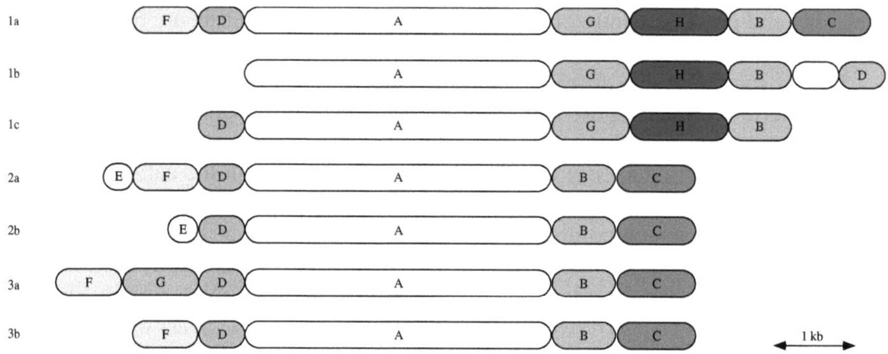

Figure 3. The organization of gene clusters encoding periplasmic nitrate reductase in different bacteria. The transcription direction is from left to right in all cases. Species in group 1a: E. coli, Ha. influenzae, Pa. multicoda, Sa. enterica (gamma proteobacteria), and Ma. magnetotacticum (alpha). Group 1b: Ca. jejuni (epsilon). Group 1c: Sh. putrefaciens (gamma). Species in group 2a: Rh. sphaeroides, Ag. tumefaciens, Si. meliloti (alpha proteobacteria), and Ps. aeruginosa, Ps. G179 (gamma). Group 2b: Pa. denitrificans, Pa. pantotrophus, and Br. japonicum (alpha). Species in group 3a: Vi. cholerae (gamma). Group 3b: Ye. pestis (gamma).

In addition to *napABCD*, *nap* operons may include one or more of *napKEFGH* genes, but their occurrence and position in the operons differ in different species (Figure 3). The *napE* and *napK* genes translate into small single helix integral membrane proteins of unknown function. Both NapF and NapG are 20-kDa proteins, which have four cysteine patterns diagnostic for binding [4Fe-4S] clusters. NapF is probably located in the cytoplasm, whereas NapG has a signal sequence for transport to the periplasm. NapG is structurally related to MauM, a protein of unknown function in methylamine metabolism and encoded by the *mau* operon of methylamine-utilizing bacteria. NapH has four transmembrane α-helices arranged such that the N- and C-termini face the cytoplasm. The C-terminal domain has four sets of cysteines that may be involved in ligation of four [4Fe-4S] clusters. It also has two conserved stretches of amino acids composed of CXXCP (where X is any amino acid), one of which located in the cytoplasmic loop after helix two and the other after helix four.

NapH belongs to a family of membrane-bound ferredoxin-like proteins, which are found in a wide variety of bacteria and archaea (Berks *et al.*, 1995). Their gene products share the basic structure of four trans-membrane helices with the four cysteine motifs at appropriate positions but do not otherwise show significant sequence similarity to one another. The conserved [4Fe-4S] clusters suggest an electron-transfer reaction for all of the members of this family and a role for NapGH in ubiquinol oxidation has been suggested from analysis of nitrate reduction in menaquinol-deficient strains of *E. coli* (Brondijk *et al.*, 2002). Representative members are: RdxA found in *Rh. sphaeroides* and involved in reduction of tellurite to tellurium metal; MauN, another protein encoded by the *mau* operon; and CcoG, which is essential for the maturation of the *cbb*3-type oxidase, a haem-copper oxidase that is recruited for respiration at low O_2 concentrations in many soil bacteria. The *ccoG* gene is part of an operon with a *copA*-like gene, which encodes a copper-transporting ATPase, suggesting that they have their role in copper uptake, processing, and insertion into the oxidase at low O_2 concentrations.

The family also includes NosR and NirI, which are found in *nos*- and *nir*-gene clusters of some denitrifying species, respectively (Berks *et al.*, 1995). The latter two have an N-terminal extension that folds into two additional α-helices and a large periplasmic loop in between them. NosR and NirI have a role in transcriptional regulation of the *nos*- and *nir*-gene clusters, respectively. Interestingly, many examples of this type of ferredoxin gene that have been recognised so far in the databases are part of operons, which encode proteins that require metals for their activity and/or proteins involved in metal transport. Only little is known about their function because mutations in either gene do not always display clear phenotypes. Perhaps the family members can substitute for each other in individual species depending on their genetic potential and its regulation.

The *nap* operons can be grouped into three types: the first encompasses *napGH* genes and lacks *napE*; the second includes *napE* but lacks the *napGH* couple; and the third lacks both the *napGH* couple and *napE*. Examples of first type are found as the *napFDAGHBC* operon in the alpha-proteobacterium, *Magnetospirillum magnetotacticum*, and in the gamma-proteobacteria, *E. coli*, *Haemophilus influenzae*, *Pasteurella multicoda*, and *Salmonella enterica*. *Sh. putrefaciens* (gamma) and *Ca. jejuni* (epsilon) have a truncated version of that operon in that they lack *napF* and *napC*, whereas the *Ca. jejuni* also has the *napD* gene located at the 3' part of the operon. Except for *Ma. magnetotacticum*, all of these species are non-denitrifiers, which reduce nitrite produced by Nap to ammonium with a respiratory decahaem-type nitrite reductase, Nrf. Some reports state that *Sh. putrefaciens* is also a denitrifier (Samuelsson, 1985), but this view may be challenged because the only nitrite reductase isolated from this organism thus far is a haem-*c*-containing enzyme of *ca.* 60 kDA, which is strongly associated with the membrane (Krause and Nealson, 1997). These data suggest that *Sh. putrefaciens* has Nrf just like the related species, *Shewanella oneidensis* (Simon, 2002). As far as currently known, Nrf and nitrite reductases involved in denitrification are never found together in the same cell. The biochemistry of nitrous-oxide production by *Sh. putrefaciens* remains to be elucidated. The second type of operon, the *napEFDABC* operon, is present in the alpha-proteobacteria *Rh. sphaeroides*, *Agrobacterium tumefaciens*, and *Sinorhizobium meliloti* and in the gamma species, *Pseudomonas aeruginosa* and *Pseudomonas G-179*. Similar operons, but devoid of *napF*, are found in the alpha-proteobacteria *Pa. denitrificans* and *Bradyrhizobium japonicum*. These species are all denitrifiers. The third type, the *napFGDABC* operon, is found in the gamma-proteobacterium, *Vibrio cholerae*. *Yersinia pestis* has a similar one but it lacks *napG*. These bacteria have no respiratory nitrite-reducing capacities.

A recent review on periplasmic nitrate reduction describes correlations between the organization of the *nap* genes, the physiology of the host, the conditions under which the *nap* genes are expressed, and even the fate of nitrite, the product of Nap activity (Potter *et al.*, 2001). It is evident that the distinctiveness of *nap*-operon organization does not show a phylogenetic pattern for the species in which they are found, which is suggestive for lateral transfer of the *nap* operons. It appears that there is a correlation between the type of operon present in a particular species and its mode of respiratory nitrite reduction (either respiratory nitrite ammonification or denitrification) if at all. Similarly, it might be that the absence of *nap* genes from particular operons is compensated for by gene functions located elsewhere on the genome.

Many bacteria recruit Nap either for anaerobic nitrate respiration (*Enterobacteriaceae*) or to promote denitrification (*Rhizobiaceae*). Transcription of their *nap*-gene clusters is under control of both O_2 depletion and nitrate/nitrite availability. In *E. coli*, Nap is maximally expressed at low nitrate concentrations under anaerobic conditions, and both Fnr and NarP are required for *nap*-gene expression (Darwin *et al.*, 1998). Nap expression in the denitrifiers *Pa. denitrificans* and *Pa. pantotrophus*, however, is not controlled by the O_2 and nitrate status of the cell. Rather it seems that their *nap*-gene clusters are transcribed in response to changes in the redox state of a respiratory component. Nap expression is optimal under aerobic conditions and when highly reduced substrates are used as carbon and free-energy sources, potentially resulting in an overshoot of electrons through the respiratory network. Because Nap is not electrogenic, it has been suggested that the enzyme from these bacteria serves as an electron sink to poise the redox state of the Q-pool under aerobic conditions whenever nitrate is available (Sears *et al.*, 2000; Sears *et al.*, 1997). Likewise, it may be that the enzyme facilitates the switch from aerobic respiration to denitrification as has been suggested for Nap from *Ralstonia* species (Siddiqui *et al.*, 1993; Warnecke-Eberz and Friedrich, 1993).

4.3. Nitrate and Nitrite Transporters

Because passive diffusion of N-oxyanions over the membrane is extremely slow and pH dependent (pKa of nitric acid and nitrous acid is –1.3 and 3.4, respectively), nitrate reducers that reduce nitrate to nitrite by Nar require special transporters not only to mediate the cellular influx of nitrate for reduction at the cytoplasmic face of Nar but also to facilitate the extrusion of its product nitrite in order to prevent the accumulation of this cytotoxic compound. Both functions are unified in a single protein in *E. coli*, designated NarK, which functions as a nitrate/nitrite antiporter (Demoss and Hsu, 1991; Noji *et al.*, 1989). Although this view has been challenged by Rowe et al, who suggested a role of NarK only in nitrite extrusion (Rowe *et al.*, 1994), others have recently provided convincing arguments for a role of *E. coli* NarK as a nitrate/nitrite antiporter (Clegg *et al.*, 2002; Wood *et al.*, 2002). Primary and secondary structure predictions show that NarK is a member of the major facilitator superfamily of transport proteins, which are integral membrane proteins with twelve alpha helices (Marger and Saier, 1993). The *E. coli narK* gene is located upstream the *narGHJI* gene cluster, and a copy of it, called *narU*, is upstream of the *narZYWV* operon. Orthologues of the *narK* gene have been encountered in many non-denitrifying nitrate-reducing species and in denitrifiers. They all share six conserved glycine residues, two conserved arginine residues in helices #2 and #8, a phenylalanine residue in helix #4, a tyrosine in helix #7, and an aspartate in the cytoplasmic loop in between helices #8 and #9. It has been suggested that the arginines in the helices have a key role in trans-membrane transport of the N-oxyanions, whereas aspartate controls their influx and efflux (Wood *et al.*, 2002).

A phylogenetic analysis revealed that NarK proteins cluster in two groups, the NarK1- and NarK2-types. NarK1-type proteins share four conserved tryptophan residues not found in the NarK2-types. These residues are located in the cytoplasm, two at the N-terminus, one in the loop between helices #2 and #3, and one at the C-terminus. NarK1-type proteins are most likely nitrate/proton symporters because their activity is dependent on the proton-motive force, whereas NarK2-type proteins have a nitrate/nitrite antiport activity that is independent of the proton-motive force (Clegg *et al.*, 2002; Moir and Wood, 2001; Wood *et al.*, 2002). It may well be that their differences in function are reflected by the presence or absence of the conserved tryptophan residues. Genes encoding NarK1-type proteins are found in species that reduce nitrate either by Nar (*Bacillus subtilis narK*, *Staphylococcus carnosus narT*, *Streptomyces coelicolor narK*) or by Nas (*Aquifex aeolicus nasA*, *Caulobacter crescentus nasA*). The latter enzyme provides the nitrite for a sirohaem-type nitrite reductase in the process of assimilatory nitrite ammonification. This observation suggests that NarK1 does not have a role in nitrite extrusion. NarK2-type proteins are those from *E. coli* (NarK and NarU) and orthologues are expected both in a few species closely related to *E. coli* and in *Mycobacterium tuberculosis* (three copies) and *Mycobacterium leprae* (three copies).

Most denitrifiers appear to have the potential to express both types of NarK protein. In *Ps. aeruginosa*, the *narK1* and *narK2* genes are clustered and located in front of the *narGHJI* gene cluster that encodes Nar. The same organization is found in *Ralstonia* and *Brucella* species, *Pseudomonas fluorescens*, *Pa. denitrificans* and in *Pa. pantotrophus*. The latter has the two genes that are fused and give rise to the synthesis of a single NarK1-NarK2 hybrid protein (Wood *et al.*, 2002). *Pseudomonas stutzeri* has a *narK2* gene, but has its *narK1* gene replaced by a *narC* gene, whose product is related to the pmf-driven high-affinity nitrate/proton symporters of yeast and plants (Hartig *et al.*, 1999). An old study of Boogerd *et al* suggested the operation of both a pmf-driven nitrate/proton symport and a nitrate/nitrite antiport system during denitrification in *Pa. denitrificans* (Boogerd *et al.*, 1983), and it now becomes evident that many, if not all, denitrifiers recruit both transport systems during denitrification.

It has been hypothesized that NarK1 initiates the uptake of nitrate at the expense of the membrane potential. Nitrate is subsequently reduced by Nar giving rise to an increasing intracellular nitrite concentration, which would then allow the operation of the nitrate/nitrite antiport system (Wood *et al.*, 2002). Fine tuning of their activities may be required both to allow high fluxes of nitrate into the cell while keeping the intracellular concentrations of nitrite below cytotoxic levels and to lower the activity of the energetically unfavourable nitrate/proton symport system once enough nitrite comes available for the antiporter. One of the environmental factors that might control the activity of NarK is the O_2 concentration. Denitrifying cells of *Pa. denitrificans* and *Ps. aeruginosa* immediately cease with nitrate reduction once they are exposed to either O_2 or other oxidizing compounds (Alefounder and Ferguson, 1980; Hernandez and Rowe, 1988). This cessation is apparently a consequence of a redox-inactivation of a nitrate-transport system rather than nitrate reductase itself because the addition of low concentrations of detergent restored nitrate reduction, and because inside-out vesicles prepared from *Pa. denitrificans* showed high nitrate-reductase activities (Alefounder and Ferguson, 1980). NarK from its relative *Pa. pantotrophus* has five cysteine residues, some of which may have a role in the proposed redox response (Moir and Wood, 2001).

5. NITRITE REDUCTASES

There are two types of respiratory nitrite reductase characterized in denitrifying bacteria, a homo-dimeric enzyme with haems *c* and d_1 (NirS, cd_1-type), and a homo-trimeric enzyme with copper atoms (NirK, copper-type). Both are periplasmic proteins, receive electrons from cytochrome *c* and/or a blue copper protein, pseudoazurin (Koutny *et al.*, 1999; Moir and Ferguson, 1994), and catalyse the one-electron reduction of nitrite to nitric oxide. Neither of the enzymes is electrogenic because both take up the electrons and protons required for nitrite reduction at the same side of the membrane, *i.e.*, the periplasm. NirK and NirS are never found together in a single species. The structural and functional characteristics of both enzymes have been reviewed (Cutruzzola, 1999; Watmough *et al.*, 1999).

5.1. cd_1-Type Nitrite Reductase

The cd_1-type nitrite reductase has been isolated from many denitrifying bacteria, including *Alcaligenes faecalis* (Iwasaki and Matsubara, 1971), *Azospirillum brasilense* (Danneberg *et al.*, 1986), *Halomonas halodenitrificans* (Mancinelli *et al.*, 1986), *Ma. magnetotacticum* (Yamazaki *et al.*, 1995), *Pa. denitrificans* (Timkovich *et al.*, 1982), *Pa. pantotrophus* (Moir *et al.*, 1993), *Ps. aeruginosa* (Walsh *et al.*, 1979; Wood, 1978), *Ps. stutzeri* (Weeg-Aerssens *et al.*, 1991), *Ps. nautica* (Besson *et al.*, 1995), *Ra. eutropha* (Sann *et al.*, 1994), *Roseobacter denitrificans* (Doi *et al.*,

1989), and *Thiobacillus denitrificans* (LeGall *et al.*, 1979). In addition to the reduction of nitrite to nitric oxide, cd_1-type nitrite reductase is also capable of catalysing the reduction of O_2 to water and, as such, it was long designated as cytochrome oxidase (Walsh *et al.*, 1979).

5.1.1. Composition, Structure and Function of cd_1-type Nitrite Reductase

The crystal structures of both oxidized and reduced cd_1-type nitrite reductase from *Pa. pantotrophus* (Fulop *et al.*, 1995; Williams *et al.*, 1997) and *Ps. aeruginosa* (Nurizzo *et al.*, 1998; Nurizzo *et al.*, 1997) have been solved. The enzymes are homodimers (*ca.* 120 kDa as the dimer), and each monomer consists of a small haem *c*-binding domain and a larger haem d_1-binding domain that harbours the active site. The haem d_1 moiety is markedly different from other types of haem. Two of the pyrroles are partly saturated and contain oxo and extra methyl groups which indicate that haem d_1 is actually a dioxoisobacteriochlorine (or dionehaem) (Chang *et al.*, 1986; Chang and Wu, 1986). Haem d_1 is related to sirohaem (an isobacteriochlorin), which is the cofactor of both nitrite and sulphite reductase from *E. coli* and *Salmonella typhymurium* (Goldman and Roth, 1993; Spencer *et al.*, 1993; Wu *et al.*, 1991). The haem *c*-binding domain is mostly α-helical, superficially resembling the folding of class I *c*-type cytochromes, but the threading and connectivities of the helices are different. Haem *c* of *Pa. pantotrophus* is bis-histidine liganded (His-17 (distal) and His-69 (proximal)) which was unexpected, because bis-histidine-liganded cytochromes normally have low redox potentials, and haem *c* in cd_1-type nitrite reductase is expected to have a redox potential of +200-300 mV in order to be able to accept electrons from its physiological electron donors, either cytochrome c_{550} or pseudoazurin and to donate them to the active site (Koutny *et al.*, 1999; Moir and Ferguson, 1994).

The haem d_1-binding domain of the *Pa. pantotrophus* enzyme is an eight-bladed β-propeller structure with the haem d_1 located in the centre. A comparable β-propeller domain is present in both nitrous oxide reductase and methanol dehydrogenase despite the enzymes sharing no common sequence motifs (Baker *et al.*, 1997). A histidine residue originating from the d_1-domain of one monomer and a tyrosine residue originating from the haem *c*-domain of the other ligate the haem iron positioned in the centre of the nitrite-reductase propeller. This domain cross-interaction was another unexpected feature of this enzyme. It is noteworthy that both unexpected features, the bis-histidine ligation of haem *c* and the tyrosine-ligation of haem d_1, are so far only present in the cd_1-type nitrite reductase of *Pa. pantotrophus*. It is intriguing that reduction of this crystal form resulted in loss of the tyrosine from the d_1 haem and, thus, the provision of a binding site on the iron for the substrate. Histidine-17 was also dissociated from the Fe of the *c*-type cytochrome domain and replaced by the sulphur of methionine-106, which had undergone a considerable spatial displacement following reduction of the enzyme (Williams *et al.*, 1997). This ligand switch would be predicted to raise the redox potential and to make it more favourable for electron transfer from either cytochrome c_{550} or pseudoazurin.

Magnetic circular dichroism (MCD) of *Ps. aeruginosa* enzyme showed that histidine and methionine residues ligate the haem *c*. Further, a counterpart of the ligating tyrosine could not be identified in the primary sequence of Nir from *Ps. stutzeri* (Cheesman *et al.*, 1997). Inspection of the crystal structures of the *Ps. aeruginosa* enzyme revealed that the conformational changes upon reduction of the oxidized enzyme are less extensive than those observed in the *Pa. pantotrophus* enzyme (Nurizzo *et al.*, 1998). These discrepancies indicate that there are structural diversities among the different cd_1-type nitrite reductases, although their mode of catalysis may be identical (Cheesman *et al.*, 1997). Possibly, either one or both enzymes as isolated are in a resting state rather than participating in catalysis. In this respect, when the fully reduced *Pa. pantotrophus* enzyme was oxidised by either O_2 or nitrite under rapid reaction conditions, the *c*-type cytochrome haem had spectroscopic properties consistent with ferric iron and histidine plus methionine coordination, rather than bis-histidinyl as observed for the ferric haem in the crystal structure (Allen *et al.*, 2000; George *et al.*, 2000; Koppenhofer *et al.*, 2000).

5.1.2. Genetics of cd_1-type Nitrite Reductase

The gene encoding the structural monomers of cd_1-type nitrite reductase, *nirS*, is part of a *nir*-gene cluster. The number and organization of the *nir* genes in these clusters differ in different species. The best-characterized clusters are from the denitrifying species *Ps. aeruginosa* (*nirSMCFDLGHJEN*) (Arai *et al.*, 1994), *Pa. denitrificans* (*nirXISECFDLGHJN*) (De Boer *et al.*, 1994, and unpublished data), and *Ps. stutzeri* (*nirSTBMCFDLGH* and *nirJEN*, the two clusters being separated by a part of the *nor*-gene cluster that encodes nitric oxide reductase) (Palmedo *et al.*, 1995). The *nirSCFDLGHJEN* genes are conserved in the three clusters although not always positionally. The *nirC* gene encodes a probable *c*-type cytochrome with a signal sequence for membrane translocation. Its function is unknown.

The *nirFDLGH* genes are proposed to encode a multimeric and multifunctional enzyme complex involved in the maturation and insertion of haem d_1 (Kawasaki *et al.*, 1997; Palmedo *et al.*, 1995). Notable sequence relatedness was found between NirF and NirS, but NirF lacks the N-terminal haem-*c*-binding domain of NirS. The amino-acid sequences deduced from the *nirDLGH* genes are homologous to each other and do not have motifs diagnostic for their precise function (De Boer *et al.*, 1994; Kawasaki *et al.*, 1997; Palmedo *et al.*, 1995). The *nirE* gene encodes a protein of 279 amino acids homologous to S-adenosyl-L-methionine:uroporphyrinogen-III methyltransferase from other bacterial strains. The protein is essential for haem-d_1 biosynthesis because it dimethylates uroporhyrin III to precorrin 2, the precursor of haem d_1 (De Boer *et al.*, 1994). NirE of *Ps. aeruginosa* shows 21% identity with NirF in the N-terminal 100-amino-acid residues. The *nirJ* gene encodes a protein of 387 amino acids, which shows partial identity with each of the *nirDLGH* genes and which is also required for biosynthesis of haem d_1 (Kawasaki *et al.*, 1997). The *nirN* gene encodes a protein of 493 amino acids with a conserved binding motif for haem *c* (CXXCH) and a typical N-terminal-signal sequence for membrane translocation. The derived NirN protein shows 23.9% identity with nitrite reductase (NirS) (Kawasaki *et al.*, 1997). The high degree of sequence relatedness between the *nirMC* genes, the *nirJDLGH* genes, and the *nirSFN* genes suggests that several gene duplication events have occurred in the *nir*-gene cluster.

Ps. stutzeri and *Ps. aeruginosa* have species-specific *nirTBM* genes and a *nirM* gene in between their *nirS* and *nirC* genes, respectively. Notable sequence relatedness was found between NirM and NirC (Palmedo *et al.*, 1995). The *nirT* gene encodes a presumed tetrahaem cytochrome that belongs to the NapC/NirT family of tetra-haem quinol dehydrogenases. In *Ps. stutzeri*, mutants deficient in NirT are unable to reduce nitrite (Jüngst *et al.*, 1991), which suggests that, in this organism in contrast to most other denitrifiers, electron transport proceeds to the nitrite reductase independently of the cytochrome bc_1 complex. Cytochrome bc_1 complex-dependent NirS systems, *e.g.*, that from *Pa. denitrificans*, do not have a *nirT* gene in the *nir* cluster. However, it has been shown that electron transfer to cytochrome cd_1 in *Ps. stutzeri* is as fully sensitive to the bc_1 inhibitors (Kucera *et al.*, 1988). Thus, it is possible that different routes of electron transfer to *Ps. stutzeri* NirS can operate under subtly different growth conditions.

Pa. denitrificans has a, so far, unique set of *nirIX* genes, which are located upstream of and divergently transcribed from the *nirS* gene (De Boer *et al.*, 1994; Murai *et al.*, 1998). NirI is a homologue of NosR and is essential for transcriptional regulation of the *nir* gene cluster. NirX is soluble and apparently exported to the periplasm *via* the Tat translocon, as judged by a twin-arginine motif present in its signal sequence. NirX is a homologue of NosX, both of which appear to have a role in copper insertion into nitrous oxide reductase in *Pa. denitrificans*. A single mutation of the *nirX* gene did not result in any apparent phenotype, but a double *nosX-nirX* mutant was unable to reduce nitrous oxide because nitrous oxide reductase in these cells lacked the Cu_A centre (Saunders *et al.*, 2000). The predicted NirX and NosX proteins share similarity with the YojL protein of *E. coli* and with the RnfF protein in some nitrogen-fixing organisms, where it has suggested roles both in electron transport to nitrogenase and in the regulation of the activity of a special sigma factor of RNA polymerase.

All of the species that have their cd_1-type nitrite reductase isolated and characterized are members of the alpha, beta, or gamma proteobacteria. Many other species that belong to these classes of bacteria also have the potential to express a cd_1-type nitrite reductase (see Table 1) as judged by inspection of their genome sequences and by Western analyses, gene probing, and *nirS*-specific PCR (Coyne *et al.*, 1989; Rosch *et al.*, 2002; Smith and Tiedje, 1992; Song and Ward, 2002). The only NirS-containing species that does not belong to these classes of bacteria is a *Flavobacterium* species (Coyne *et al.*, 1989; Smith and Tiedje, 1992), which belongs to the Bacteroidetes.

5.2. Copper-type Nitrite Reductases

The copper-type Nir has been isolated from species of diverse bacterial genera, like *Achromobacter cycloclastes* (Fenderson *et al.*, 1991; Inatomi, 1999), *Al. faecalis* S-

6 (Kukimoto *et al.*, 1994), *Alcaligenes xylosoxidans* (Abraham Zelda *et al.*, 1993), *Rh. sphaeroides sp. denitrificans* (Olesen *et al.*, 1998; Sawada and Satoh, 1980), *Pseudomonas aureofaciens* (Zumft *et al.*, 1987), the Gram-positive bacterium *Bacillus halodenitrificans* (Denariaz *et al.*, 1991), the actinomycete *Streptomyces thioluteus* (Shoun *et al.*, 1998), and the Archeaeon *Halobacterium denitrificans* (Coyne *et al.*, 1989).

5.2.1. Composition, Structure and Function of Copper-type Nitrite Reductases

Copper-type nitrite reductases are homotrimeric complexes of *ca.* 108 kDa (3 x 36 kDa), which contain three blue-green type I copper centres that are involved in electron transfer from donor to the active site, and three type-II copper centres, which form the active site. The type-II centres are located at the interfaces of the subunits (Adman *et al.*, 1995; Kukimoto *et al.*, 1994) and are ligated by three histidine residues; two being provided by one subunit and the third from the adjacent subunit (Godden *et al.*, 1991; Kukimoto *et al.*, 1994). A water (or hydroxide) molecule completes the tetrahedral coordination sphere of the Cu(II), but this ligand is absent in the reduced Cu(I) species. Spectroscopic studies have demonstrated that nitrite binds to the oxidised cupric type-II centre (Howes *et al.*, 1994; Strange *et al.*, 1995) and only binds poorly to the reduced cuprous type-II (Strange *et al.*, 1999). Thus, a mechanism in which nitrite binds to an oxidised type-II copper centre displacing the hydroxide ion has emerged (Murphy *et al.*, 1997). This binding may raise the redox potential of this centre and so promote transfer of an electron from the type-I centre for reduction of the nitrite to NO.

Some copper nitrite reductases appear blue, most others green (Abraham Zelda *et al.*, 1993). The crystal structure of the blue nitrite reductase from *Al. xylosoxidans* contains both type-1 and -II copper sites. Interestingly, the geometry of the type-I sites were tetrahedrally distorted. The superpositioning of the type-I copper sites in the blue enzyme and a green enzyme revealed that the orientation of the side chain containing methionine at position 150 differed between the two enzymes causing the difference in their colour (Inoue *et al.*, 1998). Mutant studies have revealed that the C-terminal part of the enzyme is essential for maintaining the quaternary structure as well as for full enzymatic activity (Chang *et al.*, 1998).

It is generally assumed that the type-I copper centre of a copper nitrite reductase receives electrons from a cupredoxin, either pseudoazurin or azurin depending upon the organism. Electron flow to the copper nitrite reductase is usually assumed to be *via* the cytochrome bc_1 complex although experimental evidence to support this view is often lacking. However, nitrite reduction is fully inhibited in intact cells of *Al. xylosoxidans* by the cytochrome bc_1 complex inhibitors, antimycin or mucidin (Kucera *et al.*, 1988).

5.2.2. Genetics of Copper-type Nitrite Reductases

The *nirK* genes that encode the copper-type nitrite reductase have been cloned and sequenced from *Al. xylosoxidans* (Prudencio *et al.*, 1999; Suzuki *et al.*, 1999), *Ac. cycloclastes* (Chen *et al.*, 1996), *Rh. sphaeroides* (Tosques Vvan *et al.*, 1997), *Ps. aureofaciens* (Glockner *et al.*, 1993), *Ps. G-179* (Ye *et al.*, 1992), and the gonococcal counterpart of *nirK*, the *aniA* gene from *Neisseria gonorrhoeae* (Hoehn and Clark, 1992; Mellies *et al.*, 1997). The genes are usually preceded by DNA-binding sites for FNR-like proteins (Householder *et al.*, 1999; Mellies *et al.*, 1997; Tosques Vvan *et al.*, 1997). In some species, the *nirK* gene is clustered with a downstream-located *nirV* gene (Bedzyk *et al.*, 1999; Jain and Shapleigh, 2001). The *nirV*-gene product is periplasmic as judged by its signal sequence. The role of the protein is unknown. Inactivation of *nirV* from *Rh. sphaeroides* 2.4.3 had no effect on cell growth and no effect on either nitrite reductase expression or activity (Jain and Shapleigh, 2001). Many other species have the potential to express a copper-type nitrite reductase (see Table 1; Braker *et al.*, 1998; Coyne *et al.*, 1989; Hallin and Lindgren, 1999; Smith and Tiedje, 1992). This potential makes the copper-type nitrite reductase a widely distributed enzyme within many phyla of the bacterial kingdom.

6. NITRIC OXIDE REDUCTASES

Because NO is toxic and highly reactive, it is effectively utilized by NO reductase ensuring that the free NO-concentration in denitrifying organisms is kept in the nanomolar range. Two types of nitric oxide reductase (NO reductase, Nor) have been identified in bacterial denitrifying species. One type receives electrons from either cytochrome *c* or pseudoazurin and is referred to as cNor; the second type receives electrons from quinol and is referred to as qNor. Both types are structurally-related integral-membrane proteins, which catalyse the two-electron reduction of two nitric oxide molecules to nitrous oxide.

6.1. cNor, the Cytochrome c/Pseudoazurin-dependent Nitric Oxide Reductase

Nitric oxide reductase of the cNor-type is a membrane-bound enzyme, which has been purified and characterized from *Ps. stutzeri* (Heis *et al.*, 1989), *Pa. denitrificans* (Carr and Ferguson, 1990), *Pa. pantotrophus* (Fujiwara and Fukumori, 1996), *Ac. cycloclastes* (Jones and Hollocher, 1993), *Ha. halodenitrificans* (Sakurai and Sakurai, 1997), and *Rh. sphaeroides* (Mitchell *et al.*, 1998).

6.1.1. Composition, Structure and Function of cNor

Purified cNor from *Ps. stutzeri* contained cardiolipin, phosphatidylglycerol, and phosphatidylethanolamine, which were required for high catalytic activity by the isolated enzyme (Kastrau *et al.*, 1994). All preparations, thus far, show that cNor contains two subunits, a small one (17 kDa) containing haem *c* and a larger one (53 kDa) with haems *b* and a non-haem iron (Hendriks *et al.*, 2000; Hendriks *et al.*, 1998). The small subunit has a soluble periplasmic haem *c*-binding domain, anchored to the cytoplasmic membrane *via* a single trans-membrane α-helix. A histidine and a methionine residue axially coordinate the haem group. The large subunit contains two haems *b* and a non-haem iron (Hendriks *et al.*, 1998). It has twelve membrane-spanning α-helices with the N- and C-termini located in the cytoplasm. Helices #two, #six, #seven and #ten contain six conserved histidine residues (Kastrau *et al.*, 1994; Zumft *et al.*, 1994).

The topological arrangement of the helices of subunit-I of cNor as well as the location of invariant histidine residues within them is similar to those of subunits-I of haem copper oxidases, suggesting that cNor is an ancient member of the family of haem copper oxidases (Saraste and Castresana, 1994; Van der Oost *et al.*, 1994). This family includes the different terminal oxidase complexes, which have been described for aerobic prokaryotes. Well-described members that share the characteristic features of the family are the aa_3-type cytochrome *c* oxidase (which is related to the mitochondrial counterpart), the bo_3-type and ba_3-type quinol oxidases, and the cbb_3-type cytochrome *c* oxidase. These oxidases all have twelve trans-membrane helices, which contain six invariant histidine residues that ligate a low-spin haem, a high-spin haem, and a copper atom, CuB; the latter two comoponents comprise the active site where O_2 is reduced to water.

Images from two-dimensional crystals obtained of cNor from *Pa. denitrificans* revealed that the NorB subunit is indeed homologous to the main catalytic subunit of cytochrome oxidase and is predicted to contain an active site similar to the haem copper oxidases (Hendriks *et al.*, 1998). The higher-order model structures constructed from the amino-acid sequences of NorC and NorB of *Ha. halodenitrificans* confirmed the topology of the helical segments and the locations of the metal centres positioned by the conserved histidine residues (Sakurai *et al.*, 1998). The main feature by which cNor is distinguished from the haem-copper oxidases is the elemental composition of the active site, which contains a non-haem iron (Fe_B) rather than a copper atom (Cu_B) as found in oxidases (Cheesman *et al.*, 1998; Hendriks *et al.*, 1998). Also, from the viewpoint of functionality, it appears that cNor is related to the haem-copper oxidases because aa_3-type cytochrome *c* oxidase is able to reduce nitric oxide (Brudvig *et al.*, 1980; Giuffre *et al.*, 1999), whereas cNor is able to reduce O_2 (Fujiwara and Fukumori, 1996). The question of whether not nitric oxide reductase or an ancient oxidase is the true ancestor of the superfamily of haem copper oxidases is still a matter of debate. Using the assumption that respiratory complexity and efficiency progressively increased throughout the evolutionary process, a detailed phylogenetic analysis suggested that oxygenic respiration is the oldest process, which pre-dates the denitrification mode of respiration (Musser and Chan, 1998).

Electron transfer from donor molecules to cNor is mediated by the cytochrome bc_1 complex and either a soluble cytochrome *c* or pseudoazurin. Electrons are donated to the haem *c* in subunit-II and travel *via* the low-spin haem *b* to the active site. There, two molecules of nitric oxide bind sequentially to the reduced enzyme under turnover conditions (Girsch and De Vries, 1997; Hendriks *et al.*, 1998), after which nitric oxide reduction to nitrous oxide occurs by interplay between the high-spin haem b_3 and the non-haem iron (Girsch and De Vries, 1997; Gronberg *et al.*, 1999; Sakurai *et al.*, 1998). Redox potentiometry has revealed that the non-haem iron of the dinuclear centre has an E_m (pH 7.6) of +320 mV, wheras the high-spin haem b_3 has a surprisingly low midpoint redox potential with an E_m (pH 7.6) = +60 mV) (Gronberg *et al.*, 1999). In the absence of bound substrate, this situation imposes a large thermodynamic barrier to reduction by the low-spin electron-transferring haem *c* (E_m (pH 7.6) = +310 mV) and haem *b* (E_m (pH 7.6) = +345 mV). Thus, it is possible that a one-electron reduced (mixed valence) active site (Fe^{2+}-haemFe^{3+}) is the relevant substrate-binding state for the catalytic cycle. The steady-state concentration of nitric oxide in denitrifying cells is very low and estimated to vary between 1-100 nM (Goretski *et al.*, 1990), which indicates that cNor has an extremely high affinity for nitric oxide. The enzyme is subject to substrate inhibition, possibly by binding of NO to the oxidized enzyme (Girsch and De Vries, 1997; Koutny and Kucera, 1999).

In contrast to the haem copper oxidases (De Gier *et al.*, 1996; Papa *et al.*, 1994; Raitio and Wikstrom, 1994; Saraste *et al.*, 1991), cNor is not a proton pump (Bell *et al.*, 1992; Carr and Ferguson, 1990; Hendriks *et al.*, 2000; Shapleigh and Payne, 1985). Furthermore, the catalytic site of cNor is at the periplasmic side, rather then on the cytoplasmic side as is found in oxidases. Electrons and protons required for NO-reduction are taken up at the same side of the membrane (the periplasm), which makes the enzymes non-electrogenic. Consistent with its non-electroogenic status, the amino-acid residues that form the so-called D- and K-channels, which are important in the delivery of protons from the cytoplasm to the dinuclear centre (Vygodina *et al.*, 1997), are absent from Nor. There are, however, a number of conserved glutamic-acid residues in periplasmic-loop regions and putative trans-membrane helices in Nor that are absent in the oxidases. These may play a role in proton movements, forming an 'E' channel that connects the dinuclear active site with the periplasm. Investigations into this possibility have shown that both E-125, which lies at periplasmic side of helix-IV, and E-198, which lies in the middle of helix-VI, are essential for activity, but not assembly, of *Pa. denitrificans* NorCB (Butland *et al.*, 2001). Further support for a route of proton uptake from the periplasm has recently come from electrometric studies on NorCB (Hendriks *et al.*, 2002). Little is currently known about the mechanism of proton output from cytochrome aa_3 and cytochrome bo_3, but it is conceivable that the proton-input channel used in NorB evolved into the proton-output channel in haem copper oxidases. Comparative studies on proton input in NOR and proton output in cytochrome aa_3 oxidase may prove illuminating.

6.1.2. Genetics of cNor

The minimum genetic potential to express cNor appears to be a *norCBQD* operon, such as found in *Nitrosomonas europaea* and *Rh. sphaeroides* (Bartnikas *et al.*, 1997). Most denitrifiers, however, possess additional *norEF* genes in their *nor*-gene clusters. Somewhat confusingly, the counterparts of the *norQEF* genes are referred to as *nirQOP* in *Ps. stutzeri* (Jüngst and Zumft, 1992), *Ps. aeruginosa* (Arai *et al.*, 1994) and *Ps. fluorescens* (Philippot *et al.*, 2001). The *norC* and *norB* genes encode subunit-II and subunit-I of cNor, respectively (Arai *et al.*, 1995; De Boer *et al.*, 1996).

The *norQ* gene encodes a 30-kDa cytoplasmic protein with an ATP-binding motif, which has high similarity to NirQ from *Ps. stutzeri* and *Ps. aeruginosa* (De

Boer *et al.*, 1996). A homologue of the *norQ* gene, called *cbbQ*, has been found downstream from the *cbbLS* genes that encode ribulose 1,5-bisphosphate carboxylase/oxygenase (RubisCO) in the thermophilic H_2-oxidizing bacterium, *Pseudomonas hydrogenothermophila* strain TH-1 (Yokoyama *et al.*, 1995). The *cbbQ* gene product also has a putative ATP-binding domain and is required for activation of the RubisCO. Remarkably, the *nirQ*-gene product from *Ps. aeruginosa* was able to activate RubisCO from *Ps. hydrogenothermophila* (Hayashi Nobuhiro *et al.*, 1998). The *norD* gene encodes a protein of unknown structure and function and with no similarity to other proteins. The NorQ and NorD proteins are essential for activation of cNor (Arai *et al.*, 1994; De Boer *et al.*, 1996; Jüngst and Zumft, 1992).

The *norB*- and *norC*-gene products from *Al. faecalis* expressed in *E. coli* showed no enzyme activity, probably due to lack of NorQ and NorD, which appear to perform an essential function in the activation of cNor in the recombinant *E. coli* (Kukimoto *et al.*, 2000). In this respect, it was notable that both genes were co-expressed with *norCB* (as part of the entire *norCBQDEF* operon) in the successful expression of active *Pa. denitrificans* NorCB in *E. coli* (Butland *et al.*, 2001).

The *norE* gene encodes a protein with five putative trans-membrane α-helices and has similarity to CoxIII, the third subunit of the aa_3-type cytochrome *c* oxidases. The *norF* gene encodes a small protein with two putative trans-membrane α-helices (Arai *et al.*, 1994; De Boer *et al.*, 1996). Both the *Pa. denitrificans norE* and *norF* mutant strains had specific growth rates and Nor contents similar to those of the wild-type strain, but had reduced Nor and Nir activities, indicating that their gene products are involved in maturation and/or stability of Nor activity (De Boer *et al.*, 1996).

The organization of the *nor* genes of different species is not always conserved as judged by the presence or absence of the *norEF* genes as well as their location within the different *nor*-gene clusters. Also, the position of the *norQ* gene varies. *Ps. stutzeri* (Jüngst and Zumft, 1992; Zumft *et al.*, 1994) and possibly also *Ps. fluorescens* (Philippot *et al.*, 2001) have their *nor*-gene cluster apparently split into *norQEF* and *norCBD* parts, which are located up- and down-stream of their respective *nir*-gene clusters.

6.2. qNor, the Quinol-dependent Nitric Oxide Reductase

A distinct member of the Nor family, designated qNor, was predicted from the genome sequence of the cyanobacterium *Synechocystis* sp. strain PCC6803 (Kaneko *et al.*, 1996). This enzyme, encoded by a single *norB* gene, has fourteen putative trans-membrane helices of which the region encompassing helices #3 to #14 shows high similarity to the NorB subunit of Nor, including the positions of the six invariant histidine residues. The region at the amino terminus is likely to fold into two additional helices connected by a hydrophilic loop of about 200 amino acids, which shows a weak resemblance to the soluble domains of the NorC subunits (Hendriks *et al.*, 2000). The enzyme has been purified from *Ra. eutropha* and consists of a single subunit of 75 kDa, which contains both high-spin and low-spin haem *b*, plus a non-haem ferric ion. Haem *c* is not detected in the active qNor enzyme and quinol analogues rather than cytochromes *c* are the electron donors for qNor (Kukimoto *et al.*, 2000).

Ra. eutropha contains two copies of the *norB* gene; these are *norB1* (formerly designated *norB* and residing on a megaplasmid) and *norB2* (formerly designated *norZ* and located on the genome), encoding two independent nitric oxide reductases with sequence identity of 90% (Cramm *et al.*, 1997). No *norC* homologues occur in the vicinity of the *norB* genes. Mutants with single-site deletions in either of the *norB* genes had no apparent phenotype but inactivation of both gene copies was lethal to the cells under anaerobic growth conditions, so indicating that both qNor enzymes perform similar roles during denitrification. The *norB* genes cluster with *norA*, but the function of NorA is not clear. It has homology with the *fnrN*-gene product from *Ps. stutzeri*. This gene is located upstream from the *dnrD* gene in *Ps. stutzeri* and expressed by its product DnrD in response to NO. DnrN affects the kinetics of synthesis and stability of *nirSTB* mRNA (Vollack and Zumft, 2001). A *norR* gene, which encodes an NO-responsive regulator, is located upstream and is divergently transcribed from both *norAB*-gene clusters (Pohlmann *et al.*, 2000).

Genes encoding quinol-dependent nitric oxide reductases are encountered in species from diverse phyla: several *Neisseria* and *Ralstonia* species (beta proteobacteria), *Geobacter metallireducens* (delta proteobacteria), several *Bacilli* (Gram positives), *Synechocystis* (cyanobacteria), *Cytophaga hutchinsonii* (Bacteroidetes), and both *Pyrobaculum aerophilum* and *Sulfolobus solfataricus* (Crenarchaeota). The *Neisseria* strains studied thus far all contain *nirK/aniA*-like genes that encode a copper-type nitrite reductase (Householder *et al.*, 2000; Parkhill *et al.*, 2000; Tettelin *et al.*, 2000) located close to their corresponding qNor genes. The qNor-encoding genes in *Ne. gonorrhoeae* are required for anaerobic growth, but their absence did not dramatically decrease anaerobic survival. The potential to express these genes in anaerobic niches of eucaryotic hosts may be essential for the adaptation of these pathogenic species and an important strategy in immune evasion.

A search of the unfinished genome-sequence databases revealed that genes encoding qNor are also present in the genomes of many pathogens, like *Corynebacterium diphteriae*, *Staphylococcus aureus*, *Legionella pneumophila*, and *Mycobacterium avium*. Just like *Synechocystis* and *Py. aerophilum*, these species do not contain genes that encode either NirS or NirK/AniA. Because NO reduction by qNor does not contribute to the generation of an electrochemical gradient across the membrane, it may well be that qNor in these species is recruited for a relatively rapid removal of toxic NO produced either in their environment or by killing cells from the host rather than being involved in free-energy transduction during anoxic conditions.

The Gram-positive bacterium *Bacillus azotoformans* has a special type of qNor that is composed of two subunits; one subunit is a structural and functional orthologue of other qNor enzymes, whereas the other contains a binuclear Cu_A centre similar to those found both in subunit-II of aa_3-type oxidases and in nitrous oxide reductase. The enzyme is, therefore, termed qCu_ANor. It appears to be a hybrid between Cu_A-containing cytochrome oxidases and the NO reductases present in Gram-negative bacteria and so may be an ancient progenitor of the family of haem-copper oxidases. It uses menaquinol as electron donor but not cytochrome *c*, which is normally an electron donor to Cu_A centres of oxidases and nitrous oxide reductase (Suharti *et al.*, 2001).

7. NITROUS OXIDE REDUCTASE

The last step in denitrification is the two-electron reduction of nitrous oxide to N_2 gas. This reaction is carried out by nitrous oxide reductase (Nos), which is a homodimeric soluble protein located in the periplasmic space. The enzyme has been purified from a large number of denitrifying strains, including *Pa. denitrificans* (Snyder and Hollocher, 1987), *Pa. pantotrophus* (Berks *et al.*, 1993), and *Ps. stutzeri* (Coyle *et al.*, 1985). Nos is a homo-dimer of a 65-kDa copper-containing subunit that binds a mixed valence di-nuclear Cu_A electron-entry site and a tetra-nuclear Cu_Z catalytic centre. Each monomer is made up of two domains; the 'Cu_A domain' has a cupredoxin fold and the 'Cu_z domain' consists of a seven-bladed propeller of β-sheets. The dimer organisation is such that inter-dimer electron transfer must take place between the Cu_A centre of one monomer and the Cu_Z centre of the second monomer (Brown *et al.*, 2000; Brown *et al.*, 2000). Electron input into Cu_A is usually *via* either mono-haem *c*-type cytochromes or cupredoxins (with E_m values around +250 mV).

The Cu_A site, or 'purple copper centre', is similar to the Cu_A site found in subunit-II of aa_3-type cytochrome *c* oxidases (Farrar *et al.*, 1991). Many lines of evidence have shown that the Cu_A site of both oxidases and nitrous oxide reductase is a binuclear site with two histidine and two bridging cysteine residues as ligands and a copper-copper bond. An unpaired electron is equally distributed between the two Cu atoms, making the Cu_A centre effectively a Cu(1.5)-Cu(1.5) centre (Farrar *et al.*, 1995; Kroneck *et al.*, 1990). The Cu_A site functions in the one-electron transfer from the soluble physiological electron donors (*i.e.*, cytochrome *c* and/or pseudoazurin) to the catalytic Cu_Z site. The Cu_Z domain has the active Cu_Z site in the centre of the β-propeller. The Cu_Z site consists of four copper atoms, three of which are ligated by two histidine residues each and the fourth by a histidine and an oxygen atom. The copper atoms are symmetrically arranged and bridged by a single sulfide ion; all copper-sulphide bonds are 2.3 Å. N_2O may bind to the Cu_Z centre via a single copper ion, with the remaining copper ions acting as an electron reservoir (Brown *et al.*, 2000; Brown *et al.*, 2000). This situation would then allow for fast electron transfer, so avoiding the formation of dead-end products. The [Cu_4S] centre can, in principle, exist in five different oxidation states and this could allow gating of electron transfer from the Cu_A (which is a one-electron donor) to facilitate coordinated two-electron reduction of N_2O to N_2.

The assembly of the novel [Cu_4S] is intriguing as Nos is a periplasmic enzyme. The signal sequence of Nos contains a twin-arginine motif diagnostic for proteins that are translocated to the periplasm in a folded or partially folded state *via* a special Tat (twin arginine translocon) system (Berks, 1996; Berks *et al.*, 2000). Tat substrates are often redox proteins for which the redox centre(s) is assembled in the cytoplasm prior to export. However, a mutant of *Ps. stutzeri* with a defective Tat system accumulated the NosZ apoprotein in the cytoplasm, suggesting that the copper centres are incorporated in the periplasm (Heikkila *et al.*, 2001). However, mutation of the Cu_A site in the C-terminal region of the protein was deleterious for protein export to the periplasm (Heikkila *et al.*, 2001), suggesting that, in this case, the Tat system may accommodate a substantially folded apoprotein rather than a holoprotein containing a redox cofactor as is generally thought to be the case for this protein-transport system.

The *nos*-gene clusters consist of at least seven genes located in the same transcriptional direction and in the order *nosRZDFYLX*. The putative promoter regions contain DNA sequences that resemble the binding site for FNR-like proteins. The *nosZ* gene encodes the monomers of nitrous oxide reductase. The *nosDFYL* genes encode proteins that are apparently required for copper incorporation into Nos, although their specific role(s) remains to be established. Copper ions are mostly in the insoluble cuprous state under reducing conditions at low O_2 concentrations, which demands special proteins for their uptake and oxidation to the biologically active cupric state. The NosD protein has some resemblance to the ATP-binding subunit of ABC transporters.

The proteins translated from the *nosR* and *nosX* genes showed remarkable similarity with the NirI and NirX proteins of *Pa. denitrificans*, respectively. Strains carrying a mutation in either *nirX* or *nosX* had growth characteristics virtually similar to that of the wild type regardless of the growth condition. The double *nirX.nosX* mutant was, in contrast, completely devoid of Nos activity. Although Nos was still expressed in the double mutant, it lacked Cu_A, suggesting that NosX is required for processing of one or both of the copper centres of Nos and that NirX can assume that role in its absence. Like NirI, NosR has six putative transmembrane helices, a large periplasmic domain, and a cytoplasmic domain harbouring cysteine motifs diagnostic of the presence of two [4Fe-4S] clusters. NosR is required for the transcriptional activation of the *nos* promoter in *Ps. stutzeri*. Lack of NOS activity also resulted from a *nosX* mutation in *Si. meliloti*. Surprisingly, complementation of the latter mutation succeeded only when a plasmid-borne copy of the *nosX* gene was supplied together with the *nosR* gene (Chan *et al.*, 1997). This result suggests that the NosR and NosX proteins are functionally coupled. A *nosC* gene is present in the *nosCRZDFYLX* gene cluster of *Pa. denitrificans* and *Rh. capsulatus*, but absent in *Ps. stutzeri*. NosC contains a conserved cysteine motif of unknown function.

8. LINKAGE OF THE DENITRIFICATION GENE CLUSTERS

In virtually all denitrifiers, Nir is co-ordinately expressed with Nor. This makes sense because NO produced by Nir is extremely cytotoxic as judged by Nor mutants being unable to denitrify, apparently as a consequence of the accumulation of cytotoxic NO (De Boer *et al.*, 1996). The only exception to date is *Rhizobium sullae* HCNT1, which is unusual in that it produces large quantities of NO due to the activity of a copper-type nitrite reductase but contains no NO-reductase (Toffanin *et al.*, 1996). The functional clustering of both nitrite reductase and nitric oxide reductase activities (Zumft, 1993) is reflected in the genetic clustering of the *nir* and *nor* genes in many denitrifying species (Figure 4).

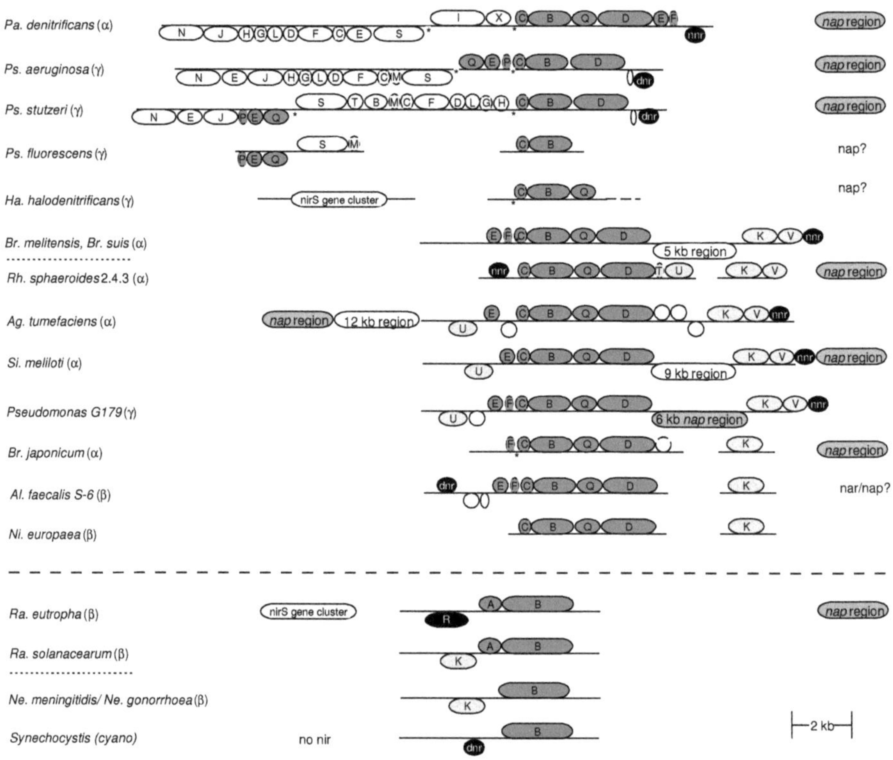

Figure 4. Linkage of the nir- *and* nor-*gene clusters in different denitrifiers. Where known, the position of the* nap-*gene clusters is indicated as well. The transcription direction is from left-to-right for genes above the lines and from right-to-left for genes below the lines. Open bars represent* nir *genes;* nirKV *and* nap *genes are in lighter gray; darker gray bars represent* nor *genes; and regulatory genes are in black. Species above and below the dashed line have cNor and qNor, respectively. Species above the dotted lines (at left) have Nar.*

In some but not all species, *nos-* or *nap*-gene clusters are also found in close proximity to their *nir* and *nor* loci, indicative of organizationed as genetic islands of denitrification genes (Figure 5). Four groups of denitrifying species, whose *nir* and *nor* loci are known, may be distinguished on the basis of the type of Nir and Nor that they express.

The first group of organisms expresses a cd_1-type nitrite reductase together with a cNor. This group includes *Pa. denitrificans*, *Ps. aeruginosa*, *Ps. stutzeri*, *Ps.*

fluorescens, and *Ha. halodenitrificans*. The genetic information on denitrification in *Ps. fluorescens* and *Ha. halodenitrificans* is incomplete. The other species have their *nir*- and *nor*-gene clusters linked to one another. In *Ps. stutzeri*, the *nir*-gene cluster is even sandwiched between its *nor* genes (see above). It is tempting to speculate that this organization is the consequence of a recombination event, which resulted in the inverse integration of the original *nirHGLDFCMBTS-norQEF* DNA-region. This suggestion is based on comparisons of the organization of the *nir* and *nor* loci from *Ps. aeruginosa* and *Pa. denitrificans* and of the *nor* loci from other species, where in all cases *norQ* is located close to the *norCB* genes. The *Ps. stutzeri* organization of *nir* and *nor* genes is apparently similar in *Ps. fluorescens* (Philippot *et al.*, 2001), suggesting that these two species are closely related and diverged after inversion of the *nir* region in the ancestor.

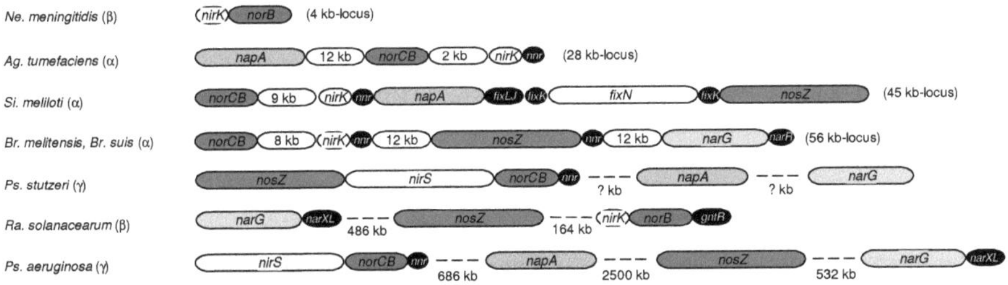

Figure 5. Linkage of denitrification gene clusters in denitrifying species.

Pa. denitrificans, Ps. aeruginosa, and *Ps. stutzeri* have an *fnr*-like gene downstream and divergently transcribed from their *nor*-gene clusters. This gene, designated *dnrD* (*Ps. stutzeri*), *dnr* (*Ps. aeruginosa*), *nnrR* (*Rh. sphaeroides*), or *nnr* (*Pa. denitrificans*) encodes an Fnr-like transcriptional activator required for expression of the *nir*- and *nor*-gene clusters. Their DNA-binding sites (FNR-box) conform to the consensus sequence TTGAT-N4-ATCAA, which resembles the one for the *E. coli* Fnr protein. These FNR-boxes are found in all *nor* and *nir* promoters. As far as is known, the species from this group have *nar*-, *nap*- and *nos*-gene clusters. In *Ps. stutzeri*, the *nos* genes are in close proximity to the *nir* and *nor* genes on a 30-kb chromosomal fragment (Glockner and Zumft, 1996; Zumft and Koerner, 1997; Zumft *et al.*, 1994). In *Ps. aeruginosa*, the *nir/nor* locus and the *nos*-, *nar*- and *nap*-gene clusters are unlinked and dispersed over the chromosome.

The second group expresses a copper-type nitrite reductase together with a cNor. This group includes *Br. suis, Br. melitensis, Ag. tumefaciens, Si. meliloti, Ps. G179, Br. japonicum, Al. faecalis, Rh. sphaeroides*, and *Ni. europaea*. The first five species have their *nirK* gene located relatively close to the *nor*-gene cluster (maximum distance of 9 kb in S. meliloti). *Br. japonicum* has its *nir* and *nor* genes dispersed over the chromosome (Mesa *et al.*, 2001). For the other three, it is not yet known how far the genes are separated. Except for *Ni. europaea*, all species of this group have an orthologue of the gene encoding the Nnr-type transcriptional activator adjacent to their *nir* or *nor* genes. Most of the species from this group have a *nap*-gene cluster but lack the *nar* genes. The exceptions are *Br. suis* and *Br. melitensis*, which lack *nap* genes, and have their *nor*-, *nir*-, *nos*-, and *nar*-gene clusters localized at a 56-kb locus. The *nap*-gene clusters of *Rh. sphaeroides, Ag. tumefaciens, Si. meliloti, Ps. G179*, and *Br. japonicum* are located close to or within a distance of 12-kb from their *nir* and *nor* genes. In *Ps. G*-179, the 6-kb *nap*-gene cluster is even sandwiched between the *nirK* gene and the *nor*-gene cluster. *Si. meliloti* has its *nor, nir, nap*, and *nos* genes clustered on a 45-kb locus, which also includes the *fix* genes that encode the high-affinity cbb_3-type oxidase. In contrast to *Rh. sphaeroides* 2.4.3, *Rh. sphaeroides* 2.4.1 has no nitrite-reductase activity (Bell *et al.*, 1992) due to the absence of *nir* genes. It has been suggested that a common ancestor of both *Rh. sphaeroides* 2.4.3 and 2.4.1 had both nitrite and nitric oxide reductase activity but, as the strains diverged, strain 2.4.1 lost *nirK* and *nirV*, making it incapable of nitrite reduction (Jain and Shapleigh, 2001).

The third group, which so far has only one member, *Ra. eutropha* (Braker *et al.*, 1998; Hallin and Lindgren, 1999), expresses a cd_1-type nitrite reductase together with a qNor. Perhaps *Ba. azotoformans* also belongs to this group because it has a special type of qNor (Suharti *et al.*, 2001) and western analyses suggests that it has a cd_1-type nitrite reductase (Coyne *et al.*, 1989). The *Ra. eutropha norB* gene is preceded by *norA* and *norR* genes. The latter encodes an NO-responsive regulator of the *norAB* promoter. It is not yet clear either how far the *nir* and *nor* genes are separated from one another or where its *nar*-, *nap*- and *nos*-gene clusters are located.

The fourth group expresses a copper-type Nir together with qNor. This group includes the *Neisseria* species and *Ra. solanacearum*. Their *nirK/aniA* and *norB* genes are adjacent to one another. *Ra. solanacearum* has its *nar*-, *nos*- and *nir/nor*-gene clusters located on a megaplasmid, but the three clusters are not linked. Unlike *Ra. eutropha, Ra. solanacearum* does not have a *norR* gene close to its *norB* gene.

9. BIOENERGETICS OF DENITRIFICATION

During respiratory electron transfer, membrane-bound redox enzymes transduce the redox free energy into a proton electrochemical gradient across the membrane by moving positive charges out and/or negative charges in. The resulting proton motive force (pmf) is used for ATP synthesis and energy-demanding processes. Some enzymes possess haem *b* groups at opposite sites of the membrane, which allows trans-membrane electron transfer. During oxidation of their substrate, protons are released at the periplasmic site of the membrane while the electrons move inwards *via* haems *b* to an acceptor at the cytoplasmic site of the membrane. This classical Mitchellian redox-loop mechanism results in a net charge separation of one per electron transferred through the enzyme from donor to acceptor molecule (1 q/e^-) (Mitchell, 1961). Examples of these enzymes are HydABC-type hydrogenase, FdnGHI-type formate dehydrogenase (Simon, 2002), the cytochrome bc_1 complex (Trumpower and Gennis, 1994), and Nar (Berks *et al.*, 1995).

Other enzymes are true proton pumps, which may transfer up to two protons from cytoplasm to periplasm per electron transferred through the enzyme from donor to acceptor. Examples of these enzymes are NADH dehydrogenase (two proton translocations per electron, 2 q/e$^-$; Friedrich, 2001) and most members of the family of haem copper oxidases (one proton translocation per electron, 1 q/e$^-$). The latter enzymes employ yet another mechanism to build up the pmf. Reduction of O_2 at the catalytic site involves proton transfer from the cytoplasm and electron transfer from a periplasmic donor. As a consequence, these charge movements from opposite sides of the membrane give rise to a net charge separation of 1 per electron transferred, 1 q/e$^-$. Therefore, the stoichiometry of charge movements in the oxidases adds up to 2 q/e$^-$ transferred from donor to O_2 (Michel, 1999; Wikstrom, 2000a; 2000b). Either periplasmic or membrane-bound enzymes that are not designed to contribute to the generation of a pmf will dissipate the redox free energy as heat; examples include succinate dehydrogenase, Nap, Nir, Nor, and Nos.

A complete denitrification pathway involves the sequential reduction of two nitrate ions to one N_2 using five molecules of NADH. Two NADH supply the four electrons for nitrate reduction, resulting in a total of twelve charge separations *via* NADH dehydrogenase (8 q/4e$^-$) and Nar (4 q/4e$^-$). The other three NADH deliver the electrons to Nir, Nor and Nos for further reduction of nitrite to N_2. Because these last enzymes are neither electrogenic nor a proton pump, pmf is only generated at the level of NADH dehydrogenase (12 q/6e$^-$) and the cytochrome bc_1 complex (6 q/6e$^-$). The total for all reactions is 30 q/10e$^-$ transferred from NADH to nitrate and its products, such that N_2 is finally produced. The stoichiometry will change accordingly either when the cytochrome bc_1 complex is by-passed during electron transfer (qNor *vs.* cNor) or when non-electrogenic electron donors (succinate dehydrogenase *vs.* NADH dehydrogenase) and/or acceptors (Nap *vs.* Nar) are recruited (Table 4).

Table 4. Electron transfer-mediated charge separations

Electron acceptors	Terminal oxidoreductases		q/10e$^-$ Electron donor NADH	q/10e$^-$ Electron donor succinate
$2.5\ O_2 \rightarrow 5\ H_2O$	*aa*$_3$		50	30
	ba$_3$		40	20
$2\ NO_3^- \rightarrow N_2$	Nar, Nir, Nos	cNor	30	10
		qNor	28	8
	Nap, Nir, Nos	cNor	26	6
		qNor	24	4

q/10e$^-$, charge separations per 10 electrons transferred from donor to acceptor.
aa$_3$, aa$_3$-type cytochrome c oxidase; ba$_3$, ba$_3$-type quinol oxidase.

The efficiency of free-energy transduction during denitrification is only 60% of that obtained during aerobic respiration. During aerobic electron transfer from five molecules of NADH to O_2 *via* NADH dehydrogenase, the bc_1 complex, and an aa_3-type oxidase, the net stoichiometry of charge movements is 50q/10e$^-$ (Table 4).

10. REGULATION OF TRANSCRIPTION OF DENITRIFICATION GENES

Facultative aerobic chemoorganotrophic organisms tightly regulate their respiration according to an energy hierarchy, which ensures that the most efficient pathways operate under each given growth condition. Under aerobic conditions, their respiratory networks are designed such that the electrons flow to O_2 as it is the most efficient route with respect to free-energy transduction. Only when the O_2 concentration drops to low levels and when either nitrate or nitrite is available are the denitrification enzymes expressed. The sequential reduction steps during denitrification imply that the product of the one enzyme is a substrate for the next one. Finely tuned regulation of the concentration and activity of the denitrification enzymes is, therefore, required in order to keep the free concentrations of nitrite and NO below cytotoxic levels.

Nar is the first enzyme that is induced in response to these conditions, whose activity gives rise to an initial increase in both the intra- and extra-cellular nitrite concentration (Baumann *et al.*, 1996; Korner, 1993; Korner and Zumft, 1989). Then, in sequence, nitrite reductase and nitric oxide reductase are co-ordinately expressed (Philippot *et al.*, 2001) with their concentrations and activities tuned such that the free nitrite concentration drops to the μM range while NO is kept in the nM range. The induction profile of nitrous oxide reductase resembles that of the nitrite and nitric oxide reductases to a large extent, suggesting that the regulation of the three corresponding promoters involves the same transcriptional regulator. An intriguing difference is that Nos is expressed at a later stage in the growth phase than the other two reductases, indicating that specific regulators other than the shared one control temporally the sequential expression of the denitrifying enzymes.

The molecular basis of the regulatory networks in denitrifying species, which control the oxic-anoxic shift as well as the fine-tuning of transcriptional activation, is beginning to emerge. Not surprisingly, the key molecules that act as signals to these regulation pathways are O_2, nitrate, nitrite, and NO. Sensing of each of these molecules involves more than one type of sensor protein, which are used in different combinations in different denitrifying species. Despite the differences in make-up, the regulatory networks operate such that homeostasis of nitrite and NO is accomplished in all cases.

10.1. O_2 Sensors

There are two types of O_2 sensor involved in regulation of denitrification, FixL and Fnr.

FixL is a membrane-bound sensor found in *Br. japonicum* and related species. Together with its cognate response regulator FixJ, these proteins belong to the group of two-component regulatory systems. The N-terminal domain of FixL contains a PAS domain with an O_2- responsive haem group (Gilles-Gonzalez *et al.*, 1994; Zhulin *et al.*, 1997). The C-terminal domain has histidine-kinase activity for autophosphorylation and subsequent phosphoryl group transfer to FixJ. The binding of O_2 to the haem in FixL switches the iron from high to low spin, resulting in a conformational change (Gong *et al.*, 1998), which inactivates the kinase activity and terminates the cascade (Gilles-Gonzalez *et al.*, 1995). FixL is not inactivated by NO (Gong *et al.*, 2000).

Fnr is a single component, O_2-sensitive protein that belongs to an expanding family of Crp/Fnr-like transcriptional activators. All these regulators contain a signal-sensing domain, a dimerization domain, a helix-turn-helix DNA-binding domain, and up to three sites that are involved in contacting RNA polymerase. One such site is a surface-exposed loop that contacts the sigma factor of RNA-polymerase (Williams *et al.*, 1991; Williams *et al.*, 1996). Crp-type proteins have conserved residues in their sensing domain for reversible binding of cAMP. Once bound, the monomers dimerize and bind to conserved sequences (a stretch of nucleotides with dyad symmetry, each half-site of which binds one monomer) of their target promoters. In type II promoters, this site is at a position directly upstream of the site where RNA-polymerase has its binding site (Busby and Ebright, 1997). The physical contact between the dimer and RNA-polymerase initiates a change in the geometry of the ternary transcription initiation complex and transcription of the DNA gets underway. Structural studies of the purified complex have revealed the key amino-acid residues of Crp as well as of the nucleotides of the promoter sequence involved in protein-protein and protein-DNA interactions (McKay and Steitz, 1981). The structural features of Fnr-type regulators are quite similar to those of Crp except that the sensing domains differ (Shaw *et al.*, 1983; Spiro, 1994).

Fnr of *E. coli* contains a sensory domain with four conserved cysteine residues that ligate an O_2-sensitive [4Fe-4S] cluster. This cluster disintegrates by direct interaction with O_2, thus making Fnr inactive. Once the O_2 concentration drops below threshold levels, the cluster is re-assembled. This event triggers a conformational change, resulting in dimerization of Fnr. The dimer binds to a conserved target sequence, TTGAT-N4-ATCAA (the Fnr-box), directly upstream of the RNA-polymerase binding site to facilitate contact with RNA-polymerase. Just as for Crp, this contact stabilizes the transcription initiation complex and promotes transcription of the target promoters (Green *et al.*, 2001; Kiley and Beinert, 1998). Under anaerobic conditions, NO reversibly inactivates Fnr as the result of NO binding to the [4Fe-4S] cluster and formation of a dinitrosyl-iron-cysteine complex (Cruz-Ramos *et al.*, 2002).

Homologues of Fnr are widespread in Nature and have been encountered in a variety of prokaryotes that have to cope with changes in O_2 availability, including pathogens. Only some of these protein members have cysteine signatures diagnostic for binding an O_2-sensitive [4Fe-4S] cluster. Most of them lack the cluster, indicating that these Fnr-homologues respond to signals other than O_2.

10.2. Nitrate and Nitrite Sensors

Three types of nitrate/nitrite-sensing systems have been characterized in denitrifying species, NarXL, NarQP, and NarR.

NarXL and NarQP are members of two-component regulatory systems. The NarX and NarQ proteins are the signal sensors, both of which respond to nitrate and nitrite although with different affinities. NarX is more specific for nitrate and NarQ for nitrite (Rabin and Stewart, 1993). The NarL and NarP proteins are their cognate response regulators, respectively. In *E. coli*, they bind DNA to control induction of the *nar* and *nap* operons and repression of genes that encode alternate anaerobic respiratory enzymes (Stewart, 1993; Darwin *et al*, 1998). The NarX and NarQ proteins communicate with both the NarP and NarL proteins in *E. coli*. The *narXL* and *narQP* genes from this bacterium are found upstream of the *narGHJI* and *narZYWV* operons, respectively. Denitrifiers from both the beta and gamma proteobacteria have genes encoding orthologues of NarXL also located either up- or down-stream of their *narGHJI* operons. Their products regulate the nitrate-induced expression of Nar. Expression of the *Neisseria nir* and *nor* genes are subject to a nitrite response mediated by a NarQP couple, the genes of which are separate from the *nir* and *nor* genes (Lissenden *et al.*, 2000).

NarR is a member of the Fnr family of transcriptional activators, but it lacks the cysteines to incorporate a [4Fe-4S] cluster. NarR of both *Pa. pantotrophus* and *Pa. denitrificans* is specifically required for transcription of the *narKGHJI* genes in response to increasing concentrations of nitrite. The mechanism of the nitrite response is not clear, but it is notable that NarR can also be activated by azide, which normally binds to metal centres, raising the possibility that NarR is a metalloprotein (Wood *et al.*, 2001; Wood 2002). Genes encoding NarR are found in the alpha proteobacteria *Br. suis, Br. melitensis, Pa. denitrificans*, and *Pa. pantotrophus* upstream of their *narGHJI*-gene clusters. There ise no indication that they have counterparts of *narXL*. It, therefore, seems that NarR substitutes for the NarXL system in the alpha proteobacteria.

10.3. Nitric Oxide Sensors

The first clue to the identification of specific NO sensors came from the observation that mutants disturbed in nitrite reduction were unable to activate the transcription of *nir* and *nor* promoters, apparently due to the lack of NO formation (Zumft *et al.*, 1994). NO was suggested to be required as an inducer for its own reductase. This view turned out to be correct as judged by the identification and characterization of NO-responsive transcriptional activators. As yet, two different types of such activators have been characterized in denitrifying species, Nnr and NorR.

Nnr (nitrite and nitric oxide gene regulator) is also a member of the Fnr family of transcriptional activators but, just like NarR, it lacks the cysteines to incorporate a [4Fe-4S] cluster. Nnr orthologues are responsive to NO, but there are no indications of the mechanism. An alignment of Nnr-like proteins does not reveal protein motifs diagnostic for ligating an NO-binding metal centre. Apart from the glycine residues involved in folding and the serine and arginine residues required for DNA binding, strictly conserved residues, possibly diagnostic for their functional importance, are Phe-42, Phe-82, Ala-97, Leu-135, Ile-142, and Arg-153 (numbering of *Pa. denitrificans* Nnr). Studies on Nnr derivatives, which were heterologously expressed in *E. coli* and in which some of these conserved amino acids were replaced, did not reveal likely candidates for NO sensing either. Except for *Ni. europaea,* and both *Neisseria* and *Ralstonia* species, Nnr orthologues are found in virtually all species that have Nir and Nor regardless of the types. These orthologues, sometimes referred to as DnrD, have a dedicated task in the coordinate transcription of the *nir*- and *nor*-gene clusters in response to NO. Without exception, the *nnr* gene in these species is located in close proximity of its target operons, which is suggestive for local specificity. The promoters of these operons contain Nnr-binding sites that resemble the consensus Fnr-box to a large extent.

The Nnr orthologues split in two phylogenetically distinct subgroups. One subgroup comprises the proteins from *Ps. aeruginosa, Ps. stutzeri, Pa. denitrificans*, and *Synechocystis* and the other group those from *Ps. G179, Al. faecalis, Rh. sphaeroides, Br. melitensis, Br. suis, Ag. tumefaciens*, and *Si. meliloti*. Not including the *Synechocystis* protein, the most striking difference between the two subgroups concerns the DNA-binding domain. Subgroup 1 and 2 proteins have conserved GHLxxQPExxSR and ExTxxTLHxxSR sequences within that domain, respectively. The corresponding sequence in *E. coli* is NYLxxTVExxSR, in which the glutamate and serine residues make contact with the cytosine and adenine of the Fnr-box half-site, 5'-ATCAA-3', at position 3 and 5, respectively. Species from the first group possess genes encoding the cd_1-type nitrite reductase, whereas those from the second group possess a gene encoding a copper-type nitrite reductase. *Br. suis* has two copies of the *nnr* gene, one located close to the *nir/nor* locus, and one next to the *nos* locus. The proteins are 60% identical and have the ExTxxTLHxxSR sequence in their DNA-binding domain. It has been suggested that labels be adopted for the Nnr proteins to reflect their phylogenetic placement (Zumft, 2002).

A likely evolutionary scenario might be that an ancestor of denitrifying species had a *nor*-gene cluster together with a gene encoding a Nnr-type regulator, perhaps recruited for NO scavenging at that time. A descendant of that species might then have obtained genes, which encode one of the two types of Nir, that were inserted close to the *nor/nnr* locus and the system was regulated and optimised for denitrifying purposes. As the two types of Nnr diverged in the ancestor and the descendant, the *nir-nor-nnr* locus of the latter was then spread among many other species without a clear phylogenetic pattern, indicative of lateral gene transfer. Another descendant of the ancestor received the genetic potential to express the other type of Nir. The locus that resulted from integration of these genes close to the *nor/nnr* gene cluster was laterally transferred to yet other phylogenetically unrelated species.

NorR is another protein involved in NO-responsive transcriptional regulation, but so far unique to *Ra. eutropha* (Pohlmann *et al.*, 2000). This bacterium has two copies of the *norR* gene, both of which are located upstream of their *norAB* gene clusters. NorR is a member of the NtrC family of response regulators. The absence of possible phosphorylation sites as well as the presence of a conserved GAF domain indicative for signal perception suggests that the protein belongs to a sub-family of response regulators, which sense their signal themselves rather than *via* a cognate signal sensor (Shingler, 1996). The mechanism of NO sensing is not clear because the GAF-like domain does not contain obvious metal cofactor-binding motifs.

10.4. NirI and NosR

The NosR sequence corresponds to that of a membrane-bound protein with six trans-membrane helices, a large periplasmic domain, and cysteine-rich cytoplasmic domains that resemble the binding sites of [4Fe-4S] clusters in many ferredoxin-like proteins. The first two helices connect a large periplasmic domain. The cytoplasmic domains that are held by the other four helices contain two conserved cysteine signatures (Cys-(N3)-Cys-Pro) as well as two [4Fe-4S] clusters. The protein is essential for expression of the *nos*-gene cluster encoding nitrous oxide reductase in *Ps. stutzeri* (Cuypers *et al.*, 1992), *Si. meliloti* (Chan *et al.*, 1997), and *Pa. denitrificans* (unpublished data).

The latter species also expresses a homologue of NosR, NirI, which has the same structural features but which has a specific role in transcription of the *nir*-gene cluster encoding nitrite reductase in response to O_2 limitation and the presence of N-oxides (Saunders *et al.*, 1999). The *nirI* gene is in a two-gene operon together with *nirX* just upstream the *nirS*-gene cluster. A genetic organization with two adjacent genes where the stop codon of the one gene overlaps with the start codon of the other is often suggestive for genes encoding proteins that interact with one another (Dandekar *et al.*, 1998). Transcription of the *nirIX*-gene cluster is controlled by Nnr. A single Nnr-binding sequence is located in the middle of the intergenic region between the *nirI* and *nirS* genes with its centre located at position -41.5 relative to the transcription start sites of both genes. NirI mutants are unable to express Nir. Remarkably, attempts to complement a NirI mutation were successful only when the *nirI* gene was reintegrated into the chromosome of the NirI-deficient mutant *via* homologous recombination in such a way that the wild-type *nirI* gene was present directly upstream of the *nir* operon. This result suggests a transcriptional and translational coupling of NirI and its target promoter (Saunders *et al.*, 1999). Interestingly, downstream-located mutations in NirI do not affect transcription but do yield inactive Nir, which suggests that the N-terminal domain of NirI is involved in the regulatory pathway while the C-terminal domain has a role either in maturation of or metal insertion into Nir.

10.5. Regulatory Networks in Denitrifiers

Despite the variations in make up of the regulatory networks, homeostasis of nitrite and NO is, in all denitrifiers, realized according to a more or less general concept. In this concept, the extent of transcriptional regulation depends on the concentration and activity of each of the regulators (Figure 6). Little is known about their concentrations, but at least some of the regulators are subject to autoregulation.

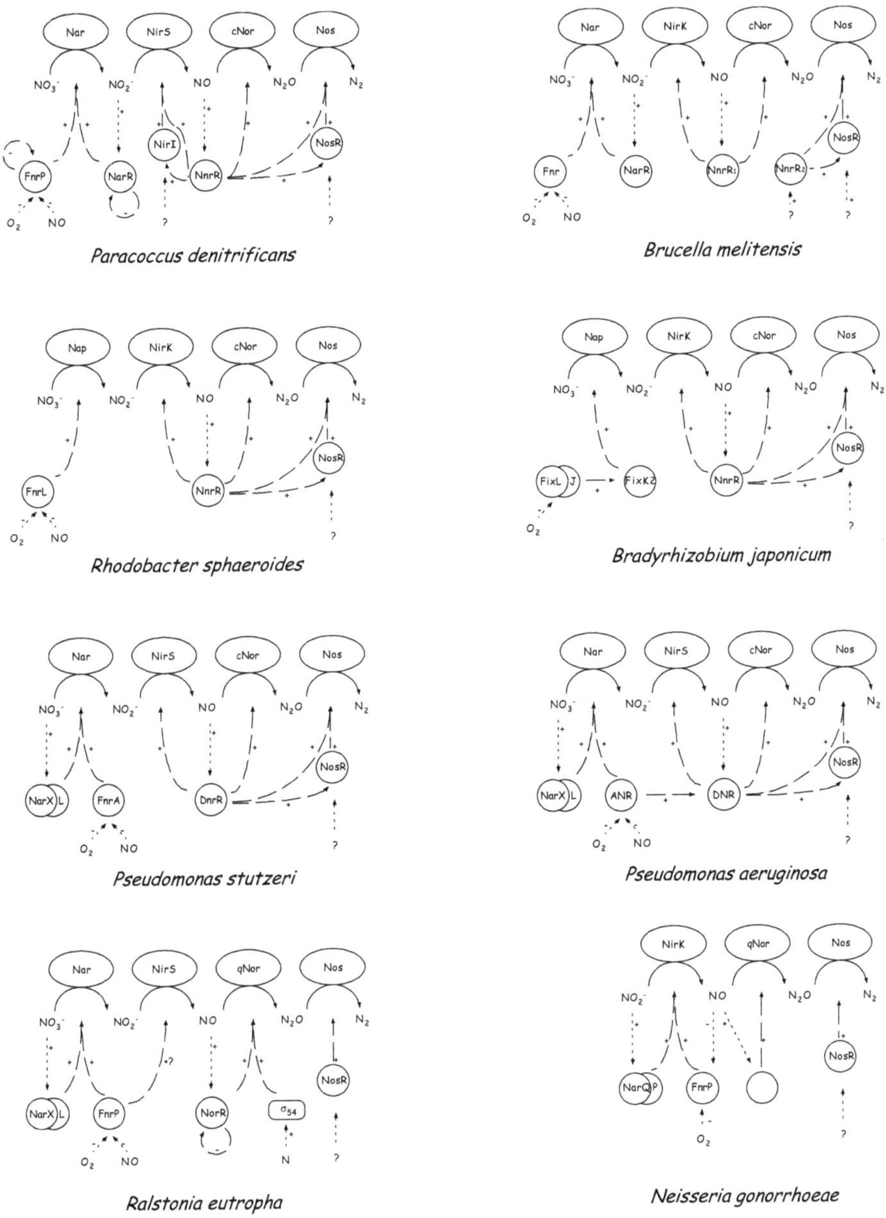

Figure 6. Schemes for the regulation of denitrification in some denitrifiers. Nar, nitrate reductase; NirS or NirK, cd_1-type or copper-type nitrite reductases, respectively; cNor or qNor, bc-type or quinol-dependent NO reductases, respectively; Nos, nitrous oxide reductase. Dotted lines, activating (+) or inactivating (-) signals; dashed lines, positive (-) or negative (-) transcription regulation of the corresponding genes by the regulators.

Their activities depend on the free concentrations of their signalling molecules, which differ at the different phases of denitrification.

10.5.1. Paracoccus denitrificans

Three Fnr-type transcriptional regulators, FnrP (an O_2 sensor), NarR (a nitrite sensor), and Nnr (a nitric oxide sensor) orchestrate the transcriptional changes during the shift from aerobic respiration to denitrification in *Pa. denitrificans*. Upon O_2 depletion, FnrP is activated and induces transcription of the *nar* operon in an unknown interplay with NarR. Nitrite required for activation of NarR may initially become available from Nap activity and subsequent passive diffusion over the membrane. Once Nar is active, nitrite comes available in the cytoplasm and a progressive increase in *nar*-gene expression, nitrate consumption, and nitrite accumulation is expected due to the product-induction mechanism. Both FnrP and NarR are subject to negative autoregulation so as to balance their concentrations.

Some NO may be formed during the first steps of nitrate reduction giving rise to activation of Nnr and initial expression of Nir and Nor. Because Nir expression is also subject to a product-induction mechanism, the concentration of the enzyme progressively increases when more NO molecules come available from its nitrite-reducing activities. A finely tuned regulation of the concentration of Nir is accomplished by a mechanism involving NirI, whose expression is also under control of Nnr. It is not yet known which signals are fed into the NirI pathway. As a result of Nir and Nor expression, nitrite and NO are consumed. Perhaps increasing concentrations of NO may also result in inactivation of FnrP (Cruz-Ramos *et al.*, 2002) and subsequent lowering of *nar*-gene expression. All together, these systems realize homeostasis of nitrite and NO. At a later state of denitrification, Nos is expressed under control of Nnr and NosR, resulting in nitrous oxide reduction. As for NirI, the mechanism of Nos activation by NosR is unknown.

Although the three Fnr orthologues are highly homologous to one another, they display a remarkable degree of specificity towards their target promoters. In spite of these specificities, their binding sites on the DNA resemble each other to a large extent and are sometimes even identical. These sites are correctly positioned relative to the RNA-polymerase-binding site as judged from transcript start analyses. Perhaps subtle differences in the binding strength between the regulator and the DNA underlie the observed specificity. Alternatively, the Fnr family members interact with specific, as yet unknown, transcription factors (possibly σ-factors of RNA-polymerase). Dedicated σ-factors would bind to unique sequences downstream of the FNR-box. Indeed, the surface-exposed loops of any of the three types of Fnr are composed of charged amino-acid residues. In addition, the sequences downstream of the FNR-boxes, where the σ-factor-binding sites are located in FnrP-, Nnr-, and NarR-regulated promoters, differ.

10.5.2. Brucella melitensis

The regulation of denitrification in *Br. melitensis* is unknown, but this bacterium has regulatory genes encoding most of the *Pa. denitrificans* counterparts. The differences are: (i) it lacks the *nirI* gene for fine-tuned expression of the *nir*-gene cluster; and (ii) it has two copies of the *nnr* gene, one close to the *nir-nor* locus and the other adjacent to the *nos*-gene cluster. Both Nnr proteins share a high degree of identity. It may well be that they are specifically involved in expression of the *nir*/*nor* genes and *nos*-gene cluster, respectively.

10.5.3. Rhodobacter sphaeroides

Nap is the enzyme responsible for denitrification in *Rh. sphaeroides* because strains with mutations in the *nap*-gene cluster lose nitrate-reductase activity as well as the ability to grow with nitrate under anaerobic-dark conditions (Liu *et al.*, 1999). This conclusion agrees well with the preliminary finding that *nar*-gene sequences are absent from the partial genome sequence of *Rh. sphaeroides*. The onset of nitrate reduction by Nap requires O_2 depletion and nitrate/nitrite availability. The O_2 response is regulated by FnrL, but the proteins involved in nitrate/nitrite sensing are not known. The *Rh. sphaeroides* genome does not contain sequences that translate into either NarXL or NarR-like proteins. Just as in *Pa. denitrificans*, NnrR regulates expression of both the nitrite and nitric oxide reductases. It is tempting to speculate that *nos*-gene expression proceeds in a way similar to that in *Pa. denitrificans*.

10.5.4. Bradyrhizobium japonicum.

Nap is responsible for anaerobic growth of *B. japonicum* under nitrate-respiring conditions (Delgado *et al.*, 2002). Expression of genes involved in anaerobic metabolism is controlled by a regulatory cascade initiated by the O_2-responsive FixLJ two-component regulator (Fischer, 1994). FixJ activates the expression of *fixK*$_2$, which is a member of the Fnr family but lacks the O_2-sensing [4Fe-4S] cluster. The FixK$_2$ protein then activates a number of target genes or operons, including the *fixNOQP* and *fixGHIS* operons that encode the high-affinity *cbb*$_3$–type terminal oxidase for respiration under micro-aerobic conditions. *B. japonicum fixLJ* and *fixK*$_2$ mutants are defective for anaerobic growth with nitrate as the terminal electron acceptor, suggesting that some genes either for nitrate reduction or for the entire denitrification pathway are subjected to control by FixLJ/K$_2$. In fact, it has been shown that both *nirK* and *nor* of *B. japonicum* are induced *via* the FixLJ/FixK$_2$ regulatory system (Velasco *et al.*, 2001; Mesa *et al.*, 2002a). It is not known if nitrate or nitrite signals are fed into this cascade. Recent data indicate that *B. japonicum* also has the genetic potential to encode an NnrR-homologue (Mesa et al. 2002b). Expression of Nir, Nor, and Nos may, therefore, proceed similarly as in *Pa. denitrificans*, *Br. suis* and *Rh. sphaeroides*.

10.5.5. Pseudomonades

Denitrification in *Pseudomonades* is initiated with transcription of the *nar*-gene cluster under control of both an O_2-responsive Fnr orthologue (FnrA or Anr) and the nitrate-responsive NarXL two-component regulatory proteins (Hartig *et al.*, 1999), just like regulation of *narGHJI* transcription in *E. coli*. Nar expression is, therefore, subject to a substrate-induction mechanism rather than a product-induction mechanism that operates in the alpha proteobacteria. In *Ps. aeruginosa*, Anr also controls the expression of Dnr in a cascade like fashion (Arai *et al.*, 1995). Dnr is the NO-sensing orthologue of Nnr, which regulates the expression of Nir and Nor, and perhaps also of Nos. Nos regulation also requires NosR. The regulation of denitrification is likely to be similar in *Ps. stutzeri*, although it is not clear if FnrA controls the expression of Dnr as well. It is not yet clear either how or if the Fnr and NarL proteins interact to mediate dual positive control of transcription initiation.

10.5.6. Ralstonia eutropha

Regulation of *nar* expression in *Ra. eutropha* is similar to that in *Pseudomonades* in that it requires an O_2-responsive Fnr orthologue and the nitrate responsive NarXL two-component regulatory proteins. It is not known which proteins regulate the expression of Nir and Nos. Expression of qNor requires NO, the NorR protein, and σ-54. NorR represses its own synthesis *via* negative autoregulation. The protein has no role in expression of the *nir*-, *nos*-, and *nar*-gene clusters in *Ra. eutropha* (Pohlmann *et al.*, 2000).

10.5.7. Neisseria gonorrhoeae

This pathogenic species has genes encoding NirK and qNor that make up an incomplete denitrification pathway, perhaps to cope with nitrite and NO challenges in the host and/or, in part, for bioenergetic purposes. Both an O_2-responsive Fnr orthologue and the nitrite-responsive NarQP two-component regulatory proteins control expression of NirK. The expression of the *norB* gene was not regulated by either Fnr or NarP but required anaerobic conditions and the presence of nitrite, as well as a functional nitrite reductase encoded by the gonococcal *aniA* gene. Evidence NO is likely the inducer for qNor expression in this bacterium (Householder *et al.*, 2000).

11. REGULATION OF DENITRIFICATION BY ENVIRONMENTAL FACTORS

The major factors regulating denitrification are the availability of nitrogen oxides and a decreased O_2 concentration. These two factors are, in turn, governed by many other factors, such as the availability of reductant (mostly organic carbon compounds but inorganic compounds may be used), temperature, water content, pH, porosity, and the presence of inhibitory compounds, which may act to cause accumulation of the intermediates of denitrification. The prediction of rates of denitrification and release of these intermediates in particular environments is, therefore, difficult (Knowles, 1996).

O_2 is one of the environmental factors that control the activity of denitrification. Denitrifying cells of *Pa. denitrificans* immediately stop nitrate reduction once they are exposed to O_2, apparently through the inactivation of the nitrate transporter, NarK, because Nar was still active under these conditions. O_2 also affects the activities of Nir, Nor, and Nos in *Ps. fluorescens* as judged by the observation that the reduction of nitrite, NO and nitrous oxide were each inhibited to increasingly greater extents by O_2 (McKenney *et al.*, 1994). The O_2 supply to soil microbes is largely determined by the water content of the soil. Water blocks the soil pores and lowers the diffusion rate of O_2 by 3-4 orders of magnitude, resulting in low free O_2 concentrations in niches below the water films. These situations support the growth of denitrifying bacteria.

The pH of an environment is another effector of denitrification activity. *Pa. denitrificans* was not able to build up a functional denitrification pathway after a transition from aerobic to anaerobic conditions at the sub-optimal pH of 6.8. Nitrite accumulated in the medium as the predominant denitrification product. Although the nitrite reductase gene was induced properly, the enzyme could not be detected at sufficient amounts in the culture. These observations indicate that either translation was somehow inhibited or once synthesized, nitrite reductase was inactivated possibly by the relatively high concentrations of nitrous acid (HNO_2). Interestingly, a *Pa. denitrificans* culture, which was grown to steady state anaerobically and then exposed to sub-optimal pH conditions, exhibited a reduced overall denitrification activity, but neither nitrite nor any other denitrification intermediate accumulated (Baumann *et al.*, 1997). The amount of nitrite generated during the first enzymatic stage of the process is also a most important factor in the regulation of denitrification by *Ps. stutzeri* under nutrient-rich conditions. The generated nitrite appeared to cause inhibition of the denitrification pathway above 5 mM and growth restriction at higher concentrations (15 mM) (Firth and Edwards, 1999). Net NO production and consumption by *Flexibacter canadensis* cells under anaerobic conditions is also dependent on the pH, with NO consumption increasing at pH values higher than or equal to 6.5 (Wu *et al.*, 1995).

12. DIVERSITY OF DENITRIFICATION

Zumft has made a list of genera of denitrifying species grouped according to either their principal growth mode (photo- or chemo-trophic, organo- or litho-trophic, hetero- or auto-trophic) or otherwise dominant physiological features, like hyperthermophily and alkaliphily (Zumft, 1997). Together with the available genome sequences, it follows that denitrification is carried out by many species of the alpha, beta, and gamma proteobacteria and by some species of the epsilon proteobacteria, Gram-positive bacteria, Bacteroidetes, and Euryarchaeota (Tables 1 and 2).

Enterobacteriaceae from the gamma proteobacteria and most of the species that belong to the delta and epsilon proteobacteria do not have the potential to denitrify (Table 3). Rather they combine nitrate reduction with respiratory nitrite ammonification by Nap and Nrf, respectively. Some mostly pathogenic species that are unable to denitrify have qNor, but this enzyme may be required for detoxification of NO in the host rather than for bioenergetic purposes.

Below, we describe the current knowledge on species that combine denitrification with other processes in the nitrogen cycle, *i.e.*, with either nitrogen fixation or ammonia oxidation.

12.1. Denitrification in Nitrogen-fixing Species

Nitrogen fixation is the conversion of N_2 into ammonia catalyzed by nitrogenases. This enzyme is found in many types of bacteria among which are rhizobia species. Rhizobia are Gram-negative nitrogen-fixing bacteria, which belong to the alpha proteobacteria. They fix N_2 in symbiosis with legume plants in so-called root nodules on legume roots and on the stems of some aquatic legumes. As such, they make an important contribution to the fixed-nitrogen input into agriculture (Amarger, 2001).

Following invasion of the plant cells *via* a complex signalling pathway between bacteria and plant, rhizobia stop dividing and undergo differentiation into nitrogen-fixing bacteroids. In the nodule, maintenance of nitrogenase activity is subject to a delicate equilibrium. A high rate of O_2 respiration is necessary to supply the energy demands of the nitrogen-reduction process but, on the other hand, O_2 also irreversibly inactivates the nitrogenase complex. A delicate balance in O_2 supply and demand solves this apparent conflict in order to keep the steady-state concentration of free O_2 low. The cortex of nodules acts as a diffusion barrier, which greatly limits permeability to O_2 (Hunt and Layzell, 1993). O_2 is delivered to the bacteroids by the plant O_2 carrier, leghaemoglobin, which reversibly binds O_2 and which is present exclusively in the infected nodule tissue (Appleby, 1984). To cope with the low ambient O_2 concentration in the nodule (10-50 nM O_2), nitrogen-fixing bacteroids use a high-affinity cytochrome *cbb*$_3$-type oxidase that is encoded by the *fixNOQP* operon (Preisig *et al.*, 1993).

The critical balance of O_2 input and consumption may easily be disturbed in the natural environment of the symbionts, for example, by flooding, which results in transient anoxic conditions close to the roots. These conditions may even differ spatially in the nodules. The intriguing question then arises of how the symbionts cope with these transient situations. Early studies indicated that many rhizobial species are able to denitrify in their free-living form or in legume root nodules or as isolated bacteroids (O'Hara and Daniel, 1985). In recent years, genes encoding periplasmic Nap, the copper-type Nir (NirK), cNor, and Nos have also been identified in several rhizobial species, such as *Rh. sullae* (Toffanin *et al.*, 1996), *Si. meliloti* (Barnett *et al.*, 2001), and *Br. japonicum* (Mesa *et al.*, 2001; Velasco *et al.*, 2001), indicating that these bacteria indeed have the full potential to denitrify. However, there is very little information available about the function of those genes in symbiosis. Perhaps the denitrifying potential is recruited in bacteroids to proceed with respiration coupled to free-energy transduction for nitrogen fixation under transient anoxic conditions. *Br. japonicum* can couple nitrate reduction to ATP generation when cultured anaerobically (Bandhari *et al.*, 1984) and can both assimilate nitrate and denitrify simultaneously (Vairinhos *et al.*, 1989).

The presence of a complete denitrification system has also been demonstrated in soybean nodules and isolated bacteroids of *Br. japonicum* (Arrese *et al.*, 1997; Delgado *et al*, 1991). The significance of denitrification coupled to ATP production can be appreciated when viewed from the point that, under the low O_2 conditions prevalent in the nodules, nitrate can function as an alternate electron acceptor to O_2, thereby providing the cellular ATP needed for nitrogen fixation in the nodules of legumes. This potential might come in very useful if the O_2 concentration drops to virtually zero as during the flooding of roots. Moreover, the responses of soybean to hypoxia (1.5 kPa pO_2) wee found to be consistent with responses to flooding stress (Bacanamwo and Purcell, 1999). Under these conditions, denitrifying activity could work as a mechanism to generate ATP to maintain nodule functioning. An associated role of denitrification in nodules would be the detoxification of cytotoxic compounds produced either as intermediates of denitrification reactions or emerging from the host plant. Apart from the denitrification process, NO may also be formed by a nitric oxide-synthase activity present in the roots and nodules (Cueto *et al.*, 1996).

Nitrate, when present in soils at high levels, has dramatic effects on both nodulation and nitrogenase activity (Streeter, 1988). The mechanisms responsible for the inhibition of nitrogenase activity in legume nodules by nitrate are still unclear. One suggested hypothesis is that inhibition is the result of toxic effects of nitrite, the first product of nitrate reduction in the nodule by the cytosolic nitrate reductase (Hunt and Layzell, 1993). Nitrite and nitric oxide significantly inhibit nitrogenase activity in bacteroids (Casella *et al.*, 1988). They also affect O_2 binding to leghemoglobin (Kanayama and Yamamoto, 1991) and interfere with haem-based sensors, such as FixL (Winkler *et al.*, 1996). In addition to the pleiotropic effects of nitrite and NO on cell integrity and enzyme activity, nitrous oxide is also a potent inhibitor of nitrogenase-catalyzed N_2 reduction (Christiansen *et al.*, 2000).

12.2. Denitrification in Nitrifiers

Species from the genus, *Nitrosomonas*, are obligate chemolithoautotrophic bacteria belonging to the beta-proteobacteria. They are found in soil, sewage, and marine environments, most preferentially in places with a sufficient supply of O_2, but depleted of organic carbon- and free-energy sources and rich in ammonium. They fix carbon *via* the Calvin-cycle and their energy demands are met by the oxidation of ammonia *via* hydroxylamine to nitrite. The initial step is a reductive one carried out by ammonia monooxygenase (AMO), which converts ammonia to hydroxylamine (Whittaker *et al.*, 2000). Hydroxylamine is subsequently oxidized to nitrite by hydroxylamine oxidoreductase (HAO). The majority of the electrons that are released during this step are transferred to O_2 *via* a cytochrome bc_1 complex and an aa_3-type oxidase, whose activities are coupled to free-energy transduction. Ammonia oxidizers thus utilize O_2 not only for the monooxygenase reaction but also as the preferred terminal electron acceptor of respiration. Therefore, nitrifying ammonia-oxidizing bacteria are traditionally considered to be obligatory aerobes with an inorganic nitrogen metabolism that involves only the oxidation of ammonia to nitrite. Evidence accumulated over the last three decades supporting a more versatile N-metabolism has challenged this view.

The production of nitric oxide, nitrous oxide, and N_2 by pure cultures of ammonia-oxidizing bacteria (Bock *et al.*, 1995; Kester *et al.*, 1997; Poth, 1986) suggests activity by nitrite, nitric oxide and nitrous oxide reductases. Enzymes with nitrite-reductase activity have been isolated from *Ni. europaea* (Beaumont *et al.*, 2002; Hooper, 1968; Miller and Nicholas, 1985). These findings, together with reports of the survival of ammonia-oxidizing bacteria under anoxic conditions (Abeliovich and Vonshak, 1992; Schmidt and Bock, 1997), indicate that these bacteria may combine aerobic nitrification with denitrification (nitrifier denitrification).

Nitrogen dioxide has been suggested as an alternative oxygen atom donor for the ammonia monooxygenase reaction under anaerobic conditions. *Nitrosomonas eutropha* was able to nitrify and denitrify simultaneously under anoxic conditions when gaseous nitrogen dioxide was supplemented to the atmosphere, whereas NO was shown to inhibit anaerobic ammonia oxidation (Schmidt and Bock, 1997). In contrast to the anaerobic situation, NO seems to be required for a high rate of aerobic denitrification (Zart and Bock, 1998). Removal of NO from the culture medium by means of intensive aeration and turbulence lowered the rate of aerobic denitrification (Zart *et al.*, 2000).

A search of the complete, but preliminary, genome sequence of *Ni. europaea* identified several genes and gene clusters that may be involved in denitrification. *Ni. europaea* expresses a copper-type nitrite reductase (NirK) and potentially a nitric oxide reductase (Nor) (Beaumont *et al.*, 2002). The *nirK* gene is the last in a cluster of four genes, the first of which encodes a blue copper protein, and the second and third of which encode di- and mono-haem *c*-type cytochromes, respectively (Beaumont *et al.*, 2002; Whittaker *et al.*, 2000). All four proteins are predicted to have a leader peptide indicating that they reside in the periplasm. Because their genes are clustered, it is likely that these proteins are functionally coupled (Dandekar *et al.*, 1998).

The blue copper protein has homology to the copper-resistance protein of *Pseudomonas syringae*, which resides in the periplasm and is involved in the sequestering of copper (Cha and Cooksey, 1991). This *Ni. europaea* protein has been isolated and its EPR spectrum showed signatures diagnostic for type-I and type-II copper centres (DiSpirito *et al.*, 1985). Accordingly, the residues involved in the ligation of these copper atoms are present. It can reduce O_2 with a range of artificial and natural electron donors and also shows some nitrite-reductase activity. The cytochromes *c* have been isolated and co-purified during all purification steps (Whittaker *et al.*, 2000). Neither of the cytochromes shared an overall homology with known cytochromes *c*. The promoter region does not contain sequences that resemble an FNR-box.

N. europaea also possesses a *norCBQD*-gene cluster, which resembles those present in heterotrophic denitrifiers. The *nor*-gene cluster is neither accompanied by an *nnr* gene encoding a homologue of the Fnr-family nor has in its promoter region a potential binding site for an Fnr-homologue. Expression of Nor in *Ni. europaea* is not apparently under the control of either O_2 or NO. *Ni. europaea* has a gene encoding an Fnr-homologue. The cysteines in the predicted protein sequence align with those from the *E. coli* protein. Genes encoding a nitrous oxide reductase are absent.

The induction profile of nitrite reductase in *Nitrosomonas* is controversial, with both repression and expression reported under aerobic conditions (Miller and Nicholas, 1985; Whittaker *et al.*, 2000). If this variation is the result of an O_2 response, then it is probably not mediated by Fnr because the *nirK*-gene cluster is not preceded by an Fnr-binding site. Additionally, a mutation in the *fnr* gene did not abolish the expression of nitrite reductase (Beaumont *et al.*, 2002). Aerobic induction of nitrite reductase is uncommon because the enzymes involved in denitrification are usually expressed in denitrifiers under low O_2 concentrations or anaerobiosis (Zumft, 1997). Regarding the possible aerobic expression of nitrite reductase, it has been suggested that the activity of this protein may be regulated at the metabolic level (Whittaker *et al.*, 2000). There are no reports on the expression and/or activity of nitric oxide reductase in *Ni. europaea*, but it is likely that it is coordinately expressed with NirK in order to manage the concentration of NO released by that enzyme.

13. YEAST AND FUNGAL DENITRIFICATION

NO production and consumption are processes integrated in denitrification pathways of many bacterial species but some fungi and yeasts have that potential as well. These eucaryotic microorganisms may contain any or all of nitrate, nitrite and nitric oxide reductases, but nitrous oxide reductases have not been described yet in these species. Both the nitrate and nitrite reductases are counterparts of the bacterial dissimilatory nitrate reductases and copper-type nitrite reductases, respectively. Nitric oxide reductase is the third enzyme in fungal and yeast denitrification. This unique enzyme, referred to as P450nor, resembles bacterial P450 cytochromes and it has been suggested that they are evolutionarily related.

Leptosphaeria maculans has the potential to express nitrate and nitrite reductases. Its nitrite reductase-encoding gene (*niiA*) is located close to the nitrate reductase-encoding gene (*niaD*) and the two genes are transcribed in the same direction (Williams *et al.*, 1995). Similar gene organizations are found in *Aspergillus parasiticus, Aspergillus oryzae, Aspergillus nidulans* and *Neurospora crassa* (Chang *et al.*, 1996).

Fusarium oxysporum contains dissimilatory nitrate and nitrite reductases in its mitochondria, which could be supplied with electrons from respiratory substrates, such as succinate, formate or malate combined with pyruvate. These electron-transfer pathways are coupled to the synthesis of ATP (Kobayashi *et al.*, 1996). The membrane-bound Nar is distinct from the soluble, assimilatory nitrate reductase. Further, the spectral and other properties of the fungal Nar were similar to Nar enzymes of *E. coli* and denitrifying bacteria. Formate-nitrate oxidoreductase activity was also detected in the mitochondrial fraction, which was shown to arise from the coupling of formate dehydrogenase (Fdh), Nar, and an ubiquinone/ubiquinol pool. The coupling with Fdh showed that the fungal Nar system is more similar to the system involved in nitrate respiration by *E. coli* than the system in the bacterial denitrifying pathway. Analyses of mutant species of *Fu. Oxysporum*, which were defective in Nar and/or assimilatory nitrate reductase, conclusively showed that Nar is essential for fungal denitrification (Uchimura *et al.*, 2002).

P450nor from *Fu. oxysporum* exhibited a potent nitric oxide-reductase activity to form nitrous oxide with NADH as the sole and direct electron donor. The apparent maximum turnover rate of NO was exceptionally high (31,500 min^{-1}) and the decomposition of the ferrous-P450(NO) complex was the rate-limiting step during turnover (Nakahara *et al.*, 1993). Redox potential measurements revealed a midpoint redox potential of –307 mV, which is apparently favourable for the reductive activation of the NO ligand. Electron paramagnetic resonance (EPR) spectra under different turnover conditions indicated that the NO ligand bound to the ferric iron is reduced by two electrons from NADH and then reacts with another NO molecule to yield nitrous oxide and water (Shiro *et al.*, 1995). Following crystallization, the structures of the enzyme in the ferric resting and the ferrous-CO states have been solved at 2.0-Å resolution (Nakahara *et al.*, 1994; Park *et al.*, 1997). These structures showed that the haem distal pocket is open to solvent, implicating this region as a possible NADH-binding site. The analyses suggested a hydrogen-bonding network possibly involved in the delivery of protons required for NO reduction at the catalytic site (Park *et al.*, 1997).

The gene that encodes P450nor, called CYP55, of *Fu. oxysporum* contains seven introns suggesting that horizontal transfer of the gene from bacteria to the fungus, if it occurred, was an early event in evolution. Transcription of the gene required both anoxic conditions and the presence of nitrate or nitrite (Tomura *et al.*, 1994). It has been suggested that a Rox1p-type protein, which represses anoxic genes under aerobic conditions in yeast, and a NirA-type protein, which induces expression of the nitrate-assimilatory genes in fungi, are likely candidates for the O_2 and nitrate response, respectively (Takaya *et al.*, 2002).

The fungus *Cylindrocarpon tonkinense* expresses a nitrite reductase under restricted aeration but could not reduce nitrate by dissimilatory metabolism. Nitrite reduction during anaerobic growth was suppressed by respiration inhibitors, suggesting that denitrification plays a physiological role in respiration in this fungus (Usuda *et al.*, 1995). *Cy. tonkinense* expresses two isoenzymes of cytochrome P450nor, which differed in their specificity for NADH and NADPH, suggesting that they play different roles in anaerobic metabolism (Usuda *et al.*, 1995). The cDNA for P450nor1 contained a targeting-like presequence upstream the N-terminus of mature protein, whereas that for P450nor2 did not, suggesting different intra-cellular localisation for them (Kudo *et al.*, 1996).

P450-type nitric oxide reductases, with close relatedness to the fungal P450nor, are also expressed in some yeast strains of the genus *Trichosporon* (Stundl *et al.*, 2000). The yeast *Trichosporon* SBUG 752 isolated from soil produced two isoforms of this cytochrome during the stationary phase of growth on glucose. These cytochromes had an apparent Mr of 43,000, pI values of 5.9 and 6.2, respectively, and could use both NADH and NADPH as reductant (Stuendl *et al.*, 1998).

14. CONCLUDING REMARKS

Specialized denitrifiers, such as *Paracoccus denitrificans* and the denitrifying *Pseudomonades*, contain more than 40 genes, which encode the proteins that make up a full denitrification pathway. They include the structural genes for the enzymes and electron donors, their regulators as well as many accessory genes required for assembly, and cofactor synthesis and insertion into the enzymes. In contrast, some denitrifiers only carry out the two central reactions of the pathway, which they use to support growth. The cost of maintaining this capability is a very small amount of genome space. Yet other, mostly pathogenic, organisms have the potential to express a special nitric oxide reductase apparently to detoxify NO released in their environment.

Recent years have seen the emergence of a great deal of genome information that is allowing a much greater insight into the phylogenetic distribution, evolution, and the propensity of the systems for lateral gene transfer of the denitrification genes and gene clusters. In addition, it is providing insight into the regulation of gene expression and the way in which some denitrification enzymes play different roles in different bacteria. As these genome data have emerged, so also have great strides been made in understanding the biochemistry of denitrification. High-resolution crystal structures of many of the terminal oxidoreductases and closely related enzymes are now available. Coupling this structural information to the genetic information will, in conjunction with biochemical, physiological and spectroscopic analyses, place researchers in an excellent position to achieve a full understanding of denitrification at the level of genetic cluster organisation, regulation of gene expression, biosynthesis, and enzyme structure-mechanism in a wide range of environmentally, industrially, and medically important bacteria.

ACKNOWLEDGMENTS

Maria J. Delgado acknowledges support by Marie Curie 40 REF: MCFI-2001-01944 and by the Junta de Andalucia CVI-275.

REFERENCES

Abeliovich, A., and Vonshak, A. (1992). Anaerobic metabolism of *Nitrosomonas europaea*. *Arch. Microbiol. 158*, 267-270.

Abraham, Z. H. L., Lowe, D. J., and Smith, B. E. (1993). Purification and characterization of the dissimilatory nitrite reductase fom *Alcaligenes xylosoxidans* subsp. *xylosoxidans* (N.C.I.M.B. 11015): Evidence for the presence of both type 1 and type 2 copper centres. *Biochem. J., 295*, 587-593.

Adman, E. T., Godden, J. W., and Turley, S. (1995). The structure of copper-nitrite reductase from *Achromobacter cycloclastes* at five pH values, with NO_2- bound and with type II copper depleted. *J. Biol. Chem., 270*, 27458-27474.
Alefounder, P. R., and Ferguson, S. J. (1980). The location of dissimilatory nitrite reductase and the control of dissimilatory nitrate reductase by oxygen in *Paracoccus denitrificans. Biochem. J., 192*, 231-240.
Allen, J. W., Watmough, N. J., and Ferguson, S. J. (2000). A switch in heme axial ligation prepares *Paracoccus pantotrophus* cytochrome *cd*1 for catalysis. *Nat. Struct. Biol., 7*, 885-888.
Amarger, N. (2001). Rhizobia in the field. *Adv. Agron., 73*, 109-168.
Appleby, C. A. (1984). Leghemoglobin and *Rhizobium* respiration. *Ann. Rev. Plant Physiol., 35*, 443-478.
Arai, H., Igarashi, Y., and Kodama, T. (1994). Structure and ANR-dependent transcription of the *nir* genes for denitrification from *Pseudomonas aeruginosa. Biosci. Biotechnol. Biochem., 58*, 1286-1291.
Arai, H., Igarashi, Y., and Kodama, T. (1995a). The structural genes for nitric oxide reductase from *Pseudomonas aeruginosa. BBA-Gene Struct. Express, 1261*, 279-284.
Arai, H., Igarashi, Y., and Kodama, T. (1995b). Expression of the *nir* and *nor* genes for denitrification of *Pseudomonas aeruginosa* requires a novel CRP/FNR-related transcriptional regulator, DNR, in addition to ANR. *FEBS Lett., 371*, 73-76.
Arrese, I. C., Minchin, F. R., Gordon, A. J., and Nath, A. K. (1997). Possible causes of the physiological decline in soybean nitrogen fixation in the presence of nitrate. *J. Exp. Bot., 309*, 905-913.
Bacanamwo, M., and Purcell, L. C. (1999). Soybean dry matter and N accumulation responses to flooding stress and hypoxia. *J. Exp. Bot., 50*, 689-696.
Baker, S. C., Ferguson, S. J., Ludwig, B., Page, M. D., Richter, O. M. H., and Van Spanning, R. J. M. (1998). Molecular genetics of the genus *Paracoccus*: Metabolically versatile bacteria with bioenergetic flexibility. *Microbiol. Mol. Biol. Rev., 62*, 1046-1078.
Baker, S. C., Saunders, N. F. W., Willis, A. C., Ferguson, S. J., Hajdu, J., and Fulop, V. (1997). Cytochrome cd_1 structure: Unusual haem environments in a nitrite reductase and analysis of factors contributing to beta-propellor folds. *J. Mol. Biol., 269*, 440-455.
Ballard, A. L., and Ferguson, S. J. (1988). Respiratory nitrate reductase from *Paracoccus denitrificans*. Evidence for two *b*-type heams in the gamma subunit and properties of a water-soluble active enzyme containing alpha and beta subunits. *Eur. J. Biochem., 174*, 207-212.
Bandhari, B., Naik, M. S., and Nicholas, D. J. D. (1984). ATP production coupled to denitrification of nitrate in *Rhizobium japonicum* grown in culture and in the bacteroids from *Glycine meliloti*. *Science, 293*, 668-672.
Barnett, M. J., Fisher, R. F., Jones, T., Komp, C., Abola, A. P., Barloy-Hubler, *et al.* (2001) Nucleotide sequence and predicted functions of the entire *Sinorhizobium meliloti* pSymA megaplasmid. *Proc. Natl. Acad. Sci. USA, 98*, 9883-9888.
Bartnikas, T. B., Tosques, I. E., Laratta, W. P., Shi, J. R., and Shapleigh, J. P. (1997). Characterization of the nitric oxide reductase-encoding region in *Rhodobacter sphaeroides* 2.4.3. *J. Bacteriol., 179*, 3534-3540.
Baumann, B., Snozzi, M., Zehnder, A. J. B., and Vandermeer, J. R. (1996). Dynamics of denitrification activity of *Paracoccus denitrificans* in continuous culture during aerobic-anaerobic changes. *J. Bacteriol., 178*, 4367-4374.
Baumann, B., Vandermeer, J. R., Snozzi, M., and Zehnder, A. J. B. (1997). Inhibition of denitrification activity but not of mRNA induction in *Paracoccus denitrificans* by nitrite at a suboptimal pH. *Antonie van Leeuwenhoek, 72*, 183-189.
Beaumont, H. J., Hommes, N. G., Sayavedra-Soto, L. A., Arp, D. J., Arciero, D. M., Hooper, A. B., *et al.* (2002). Nitrite reductase of *Nitrosomonas europaea* is not essential for production of gaseous nitrogen oxides and confers tolerance to nitrite. *J. Bacteriol., 184*, 2557-2560.
Bedzyk, L., Wang, T., and Ye, R. W. (1999). The periplasmic nitrate reductase in *Pseudomonas* sp. strain G-179 catalyzes the first step of denitrification. *J. Bacteriol., 181*, 2802-2806.
Bell, L. C., Page, M. D., Berks, B. C., Richardson, D. J., and Ferguson, S. J. (1993). Insertion of transposon Tn5 into a structural gene of the membrane-bound nitrate reductase of *Thiosphaera pantotropha* results in anaerobic overexpression of periplasmic nitrate reductase activity. *J. Gen. Microbiol., 139*, 3205-3214.

Bell, L. C., Richardson, D. J., and Ferguson, S. J. (1992). Identification of nitric oxide reductase activity in *Rhodobacter capsulatus* - The electron transport pathway can either use or bypass both cytochrome-*c*2 and the cytochrome-*bc*1 complex. *J. Gen. Microbiol., 138*, 437-443.

Bergmann, D. J., Arciero, D. M., and Hooper, A. B. (1994). Organization of the *hao* gene cluster of *Nitrosomonas europaea*: Genes for two tetraheme *c* cytochromes. *J. Bacteriol., 176*, 3148-3153.

Berks, B. C. (1996). A common export pathway for proteins binding complex redox cofactors? *Mol. Microbiol., 22*, 393-404.

Berks, B. C., Baratta, D., Richardson, D. J., and Ferguson, S. J. (1993). Purification and characterization of a nitrous oxide reductase from *Thiosphaera pantotropha*. Implications for the mechanism of aerobic nitrous oxide reduction. *Eur. J. Biochem., 212*, 467-476.

Berks, B. C., Ferguson, S. J., Moir, J. W. B., and Richardson, D. J. (1995). Enzymes and associated electron transport systems that catalyse the respiratory reduction of nitrogen oxides and oxyanions. *BBA-Bioenergetics, 1232*, 97-173.

Berks, B. C., Richardson, D. J., Reilly, A., Willis, A. C., and Ferguson, S. J. (1995). The *napEDABC* gene cluster encoding the periplasmic nitrate reductase system of *Thiosphaera pantotropha*. *Biochem. J., 309*, 983-992.

Berks, B. C., Richardson, D. J., Robinson, C., Reilly, A., Aplin, R. T., and Ferguson, S. J. (1994). Purification and characterization of the periplasmic nitrate reductase from *Thiosphaera pantotropha*. *Eur. J. Biochem., 220*, 117-124.

Berks, B. C., Sargent, F., and Palmer, T. (2000). The Tat protein export pathway. *Mol. Microbiol., 35*, 260-274.

Besson, S., Carneiro, C., Moura, J. J. G., Moura, I., and Fauque, G. (1995). A cytochrome *cd*(1)-type nitrite reductase isolated from the marine denitrifier *Pseudomonas nautica* 617: Purification and characterization. *Anaerobe, 1*, 219-226.

Blasco, F., Guigliarelli, B., Magalon, A., Asso, M., Giordano, G., and Rothery, R. A. (2001). The coordination and function of the redox centres of the membrane-bound nitrate reductases. *Cell Mol. Life Sci., 58*, 179-193.

Blasco, F., Iobbi, C., Ratouchniak, J., Bonnefoy, V., and Chippaux, M. (1990). Nitrate reductases of *Escherichia coli*: Sequence of the second nitrate reductase and comparison with that encoded by the *narGHJI* operon. *Mol. Gen. Genet., 222*, 104-111.

Blasco, F., Pommier, J., Augier, V., Chippaux, M., and Giordano, G. (1992). Involvement of the *narJ* or *narW* gene product in the formation of active nitrate reductase in *Escherichia coli*. *Mol. Microbiol., 6*, 221-230.

Blasco, R., Castillo, F., and Martinez-Luque, M. (1997). The assimilatory nitrate reductase from the phototrophic bacterium, *Rhodobacter capsulatus* E1F1, is a flavoprotein. *FEBS Lett., 414*, 45-49.

Bock, E., Schmidt, I., Stueven, R., and Zart, D. (1995). Nitrogen loss caused by denitrifying *Nitrosomonas* cells using ammonium or hydrogen as electron donors and nitrite as electron acceptor. *Arch. Microbiol., 163*, 16-20.

Boogerd, F. C., Van Verseveld, H. W., and Stouthamer, A. H. (1983). Dissimilatory nitrate uptake in *Paracoccus denitrificans* via a DmH$^+$-dependent system and a nitrate-nitrite antiport system. *Biochim. Biophys. Acta, 723*, 415-427.

Braker, G., Fesefeldt, A., and Witzel, K. P. (1998). Development of PCR primer systems for amplification of nitrite reductase genes (*nirK* and *nirS*) to detect denitrifying bacteria in environmental samples. *Appl. Environ. Microbiol., 64*, 3769-3775.

Breton, J., Berks, B. C., Reilly, A., Thomson, A. J., Ferguson, S. J., and Richardson, D. J. (1994). Characterization of the paramagnetic iron-containing redox centres of *Thiosphaera pantotropha* periplasmic nitrate reductase. *FEBS Lett., 345*, 76-80.

Brondijk, T. H., Fiegen, D., Richardson, D. J., and Cole, J. A. (2002). Roles of NapF, NapG and NapH, subunits of the *Escherichia coli* periplasmic nitrate reductase, in ubiquinol oxidation. *Mol. Microbiol., 44*, 245-255.

Brown, K., Djinovic-Carugo, K., Haltia, T., Cabrito, I., Saraste, M., Moura, J. J., Moura, I., Tegoni, M., and Cambillau, C. (2000). Revisiting the catalytic Cu_Z cluster of nitrous oxide (N_2O) reductase. Evidence of a bridging inorganic sulfur. *J. Biol. Chem., 275*, 41133-41136.

Brown, K., Tegoni, M., Prudencio, M., Pereira, A. S., Besson, S., Moura, J. J., Moura, I., and Cambillau, C. (2000). A novel type of catalytic copper cluster in nitrous oxide reductase. *Nat. Struct. Biol. 7*, 191-195.

Brudvig, G. W., Stevens, T. H., and Chan, S. I. (1980). Reactions of nitric oxide with cytochrome *c* oxidase. *Biochemistry, 19*, 5275-5285.

Bursakov, S. A., Carneiro, C., Almendra, M. J., Duarte, R. O., Caldeira, J., Moura, I., and Moura, J. J. (1997). Enzymatic properties and effect of ionic strength on periplasmic nitrate reductase (NAP) from *Desulfovibrio desulfuricans* ATCC 27774. *Biochem. Biophys. Res. Commun., 239*, 816-822.

Busby, S., and Ebright, R. H. (1997). Transcription activation at class II CAP-dependent promoters. *Mol. Microbiol., 23*, 853-859.

Butland, G., Spiro, S., Watmough, N. J., and Richardson, D. J. (2001). Two conserved glutamates in the bacterial nitric oxide reductase are essential for activity but not assembly of the enzyme. *J. Bacteriol., 183*, 189-199.

Butler, C. S., Charnock, J. M., Bennett, B., Sears, H. J., Reilly, A. J., Ferguson, S. J., Garner, C. D., Lowe, D. J., Thomson, A. J., Berks, B. C., and Richardson, D. J. (1999). Models for molybdenum coordination during the catalytic cycle of periplasmic nitrate reductase from *Paracoccus denitrificans* derived from EPR and EXAFS spectroscopy. *Biochemistry, 38*, 9000-9012.

Carr, G. J.and Ferguson, S. J. (1990). Nitric oxide formed by nitrite reductase of *Paracoccus denitrificans* is sufficiently stable to inhibit cytochrome oxidase activity and is reduced by its reductase under aerobic conditions. *Biochim. Biophys. Acta, 1017*, 57-62.

Carr, G. J., and Ferguson, S. J. (1990). The nitric oxide reductase of *Paracoccus denitrificans. Biochem. J., 269*, 423-429.

Carter, J. P., Richardson, D. J., and Spiro, S. (1995). Isolation and characterisation of a strain of *Pseudomonas putida* that can express a periplasmic nitrate reductase. *Arch. Microbiol., 163*, 159-166.

Cartron, M. L., Roldan, M. D., Ferguson, S. J., Berks, B. C., and Richardson, D. J. (2002). Identification of two domains and distal histidine ligands to the four haems in the bacterial *c*-type cytochrome NapC; the prototype connector between quinol/quinone and periplasmic oxido-reductases. *Biochem. J., 368*, 425-432.

Casella, S., Shapleigh, J. P., Lupi, F., and Payne, W. J. (1988). Nitrite reduction in bacteroids of *Rhizobium hedysari* strain HCNT1. *Arch. Microbiol., 149*, 384-388.

Cha, J. S., and Cooksey, D. A. (1991). Copper resistance in *Pseudomonas syringae* mediated by periplasmic and outer membrane proteins. *Proc. Natl. Acad. Sci. USA, 88*, 8915-8919.

Chan, Y., Mccormick, W., and Watson, R. (1997). A new *nos* gene downstream from *nosDFY* is essential for dissimilatory reduction of nitrous oxide by *Rhizobium (Sinorhizobium) meliloti*. *Microbiology, 143*, 2817-2824.

Chang, C. K., Timkovich, R., and Wu, W. (1986). Evidence that heme d_1 is a 1,3-porphyrindione. *Biochemistry, 25*, 8447-8453.

Chang, C. K., and Wu, W. (1986). The porhyrindione structure of heme d_1. *J. Biol. Chem., 261*, 8593-8596.

Chang, P. K., Ehrlich, K. C., Linz, J. E., Bhatnagar, D., Cleveland, T. E., and Bennett, J. W. (1996). Characterization of the *Aspergillus parasiticus niaD* and *niiA* gene cluster. *Curr. Genet., 30*, 68-75.

Chang, W.-C., Chen, J.-Y., Chang, T., Liu, M.-Y., Payne, W. J., Legall, J., and Chang, W.-C. (1998). The C-terminal segment is essential for maintaining the quaternary structure and enzyme activity of the nitro oxide forming nitrite reductase from *Achromobacter cycloclastes. Biochem. Biophys. Res. Commu., 250*, 782-785.

Cheesman, M. R., Ferguson, S. J., Moir, J. W. B., Richardson, D. J., Zumft, W. G., and Thomson, A. J. (1997). Two enzymes with a common function but different heme ligands in the forms as isolated. Optical and magnetic properties of the heme groups in the oxidized forms of nitrite reductase, cytochrome cd_1, from *Pseudomonas stutzeri* and *Thiosphaera pantotropha. Biochemistry, 36*, 16267-16276.

Cheesman, M. R., Zumft, W. G., and Thomson, A. J. (1998). The MCD and EPR of the heme centers of nitric oxide reductase from *Pseudomonas stutzeri* - evidence that the enzyme is structurally related to the heme-copper oxidases. *Biochemistry, 37*, 3994-4000.

Chen, J.-Y., Chang, W.-C., Chang, T., Chang, W.-C., Liu, M.-Y., Payne, W. J., and Legall, J. (1996). Cloning, characterization, and expression of the nitric oxide-generating nitrite reductase and of the blue copper protein genes of *Achromobacter cycloclastes. Biochem. Biophys. Res. Commun., 219*, 423-428.

Christiansen, J., Seefeldt, L. C., and Dean, D. R. (2000). Competitive substrate and inhibitor interactions at the physiologically relevant active site of nitrogenase. *J. Biol. Chem., 275*, 36104-36107.

Clegg, S., Yu, F., Griffiths, L., and Cole, J. A. (2002). The roles of the polytopic membrane proteins NarK, NarU and NirC in *Escherichia coli* K-12: Two nitrate and three nitrite transporters. *Mol. Microbiol., 44*, 143-155.

Cole, J. (1996). Nitrate reduction to ammonia by enteric bacteria: Redundancy, or a strategy for survival during oxygen starvation? *FEMS Microbiol. Lett., 136*, 1-11.

Coyle, C. L., Zumft, W. G., Kroneck, P. M. H., Körner, H., and Jakob, W. (1985). Nitrous oxide reductase from denitrifying *Pseudomonas perfectomarina*. Purification and properties of a novel multicopper enzyme. *Eur. J. Biochem., 153*, 459-467.

Coyne, M. S., Arunakumari, A., Averill, B. A., and Tiedje, J. A. (1989). Immunological identification and distribution of dissimilatory heme cd_1 and nonheme copper nitrite reductases in denitrifying bacteria. *Appl. Environ. Microbiol., 55*, 2924-2931.

Cramm, R., Siddiqui, R. A., and Friedrich, B. (1997). Two isofunctional nitric oxide reductases in *Alcaligenes eutrophus* h16. *J. Bacteriol., 179*, 6769-6777.

Cruz-Ramos, H., Crack, J., Wu, G., Hughes, M. N., Scott, C., Thomson, A. J., Green, J., and Poole, R. K. (2002). NO sensing by FNR: Regulation of the *Escherichia coli* NO-detoxifying flavohaemoglobin, Hmp. *EMBO J., 21*, 3235-3244.

Cueto, M., Hernandez-Perera, O., Martin, R., Bentura, M. L., Rodrigo, J., Lamas, S., and Golvano, M. P. (1996). Presence of nitric oxide synthase activity in roots and nodules of *Lupinus albus*. *FEBS Lett., 398*, 159-164.

Cutruzzola, F. (1999). Bacterial nitric oxide synthesis. *Biochim. Biophys. Acta, 1411*, 231-249.

Cuypers, H., Viebrock-Sambale, A., and Zumft, W. G. (1992). *NosR*, a membrane-bound regulatory component necessary for expression of nitrous oxide reductase in denitrifying *Pseudomonas stutzeri*. *J. Bacteriol., 174*, 5332-5339.

Dandekar, T., Snel, B., Huynen, M., and Bork, P. (1998). Conservation of gene order: a fingerprint of proteins that physically interact. *Trends Biochem. Sci., 23*, 324-328.

Danneberg, G., Zimmer, W., and Bothe, H. (1986). Aspects of nitrogen fixation and denitrification by *Azospirillum*. *Plant Soil, 90*, 193-202.

Darwin, A. J., Ziegelhoffer, E. C., Kiley, P. J., and Stewart, V. (1998). Fnr, NarP, and NarL regulation of *Escherichia coli* K-12 *napF* (periplasmic nitrate reductase) operon transcription in vitro. *J. Bacteriol., 180*, 4192-4198.

De Boer, A. P. N., Reijnders, W. N. M., Kuenen, J. G., Stouthamer, A. H., and Van Spanning, R. J. M. (1994). Isolation, sequencing and mutational analysis of a gene cluster involved in nitrite reduction in *Paracoccus denitrificans*. *Ant. van Leeuwenhoek, 66*, 111-127.

De Boer, A. P. N., Van Der Oost, J., Reijnders, W. N. M., Westerhoff, H. V., Stouthamer, A. H., and Van Spanning, R. J. M. (1996). Mutational analysis of the *nor* gene cluster which encodes nitric-oxide reductase from *Paracoccus denitrificans*. *Eur. J. Biochem., 242*, 592-600.

De Gier, J. W. L., Schepper, M., Reijnders, W. N. M., Van Dyck, S. J., Slotboom, D. J., Warne, *et al.* (1996). Structural and functional analysis of aa_3-type and cbb_3-type cytochrome *c* oxidases of *Paracoccus denitrificans* reveals significant differences in proton-pump design. *Mol. Microbiol., 20*, 1247-1260.

Delgado, M. J., Drevon, J. J., and Bedmar, E. J. (1991). Denitrificacion by bacteroids of an uptake hydrogenase negative (Hup^-) and its isogenic Hup^+ parental strain of *Bradyrhizobium japonicum*. *FEMS Microbiol. Lett., 77*, 157-162.

Delgado, M. J., Bonnard, N., Bedmar, E. J., and Müller, P. (2002). The *Bradyrhizobium japonicum napEDABC* genes encoding the periplasmic nitrate reductase are essential for nitrate respiration. *The Fifth European Nitrogen Fixation Conference, 6-10 September, Norwich. Abstract.*

DelVecchio, V. G., Kapatral, V., Redkar, R. J., Patra, G., Mujer, C., Los, T., *et al.* (2002). The genome sequence of the facultative intracellular pathogen *Brucella melitensis*. *Proc. Natl. Acad. Sci. USA, 99*, 443-448.

Demoss, J. A., and Hsu, P. Y. (1991). NarK enhances nitrate uptake and nitrite excretion in *Escherichia coli*. *J. Bacteriol., 173*, 3303-3310.

Denariaz, G., Payne, W. J., and LeGall, J. (1991). The denitrifying nitrite reductase of *Bacillus halodenitrificans*. *Biochim. Biophys. Acta, 1056*, 225-232.

Dias, J. M., Than, M. E., Humm, A., Huber, R., Bourenkov, G. P., Bartunik, H. D., *et al.* (1999). Crystal structure of the first dissimilatory nitrate reductase at 1.9 A solved by MAD methods. *Structure Fold Des., 7*, 65-79.

DiSpirito, A. A., Taaffe, L. R., Lipscomb, J. D., and Hooper, A. B. (1985). A blue copper oxidase from *Nitrosomonas europaea*. *Biochim. Biophys. Acta, 827*, 320-326.

Doi, M. Y., Shioi, Y., Morita, K., and Takamiya, K. (1989). Two types of cytochrome cd_1 in the aerobic photosynthetic bacterium *Erythrobacter* sp. Och 114. *Eur. J. Biochem., 184*, 521-527.

Einsle, O., Stach, P., Messerschmidt, A., Klimmek, O., Simon, J., Kroger, A., and Kroneck, P. M. (2002). Crystallization and preliminary X-ray analysis of the membrane-bound cytochrome *c* nitrite reductase complex (NrfHA) from *Wolinella succinogenes*. *Acta Crystallogr. D. Biol. Crystallogr., 58*, 341-342.

Farrar, J. A., Lappalainen, P., Zumft, W. G., Saraste, M., and Thomson, A. J. (1995). Spectroscopic and mutagenesis studies on the CuA centre from the cytochrome-*c* oxidase complex of *Paracoccus denitrificans*. *Eur. J. Biochem., 232*, 294-303.

Farrar, J. A., Thomson, A. J., Cheesman, M. R., Dooley, D. M., and Zumft, W. G. (1991). A model of the copper centres of nitrous oxide reductase (*Pseudomonas stutzeri*) - Evidence from optical, EPR and MCD spectroscopy. *FEBS Lett., 294*, 11-15.

Fenderson, F. F., Kumar, S., Adman, E. T., Liu, M. Y., Payne, W. J., and Legall, J. (1991). Amino acid sequence of nitrite reductase - A copper protein from *Achromobacter cycloclastes*. *Biochemistry, 30*, 7180-7185.

Ferguson, S. J. (1994). Denitrification and its control. *Anton van Leeuwenhoek Int. J. Gen. M., 66*, 89-110.

Ferguson, S. J. (1998). Nitrogen cycle enzymology. *Curr. Opin. Chem. Biol., 2*, 182-93.

Firth, J. R., and Edwards, C. (1999). Effects of cultural conditions on denitrification by *Pseudomonas stutzeri* measured by membrane inlet mass spectrometry. *J. Appl. Microbiol., 87*, 353-358.

Fischer, H. M. (1994). Genetic regulation of nitrogen fixation in rhizobia. *Microbiol. Rev., 58*, 352-386.

Friedrich, T. (2001). Complex I: A chimaera of a redox and conformation-driven proton pump? *J. Bioenerg. Biomembr., 33*, 169-177.

Fujiwara, T., and Fukumori, Y. (1996). Cytochrome *cb*-type nitric oxide reductase with cytochrome *c* oxidase activity from *Paracoccus denitrificans* ATCC 35512. *J. Bacteriol., 178*, 1866-1871.

Fulop, V., Moir James, W. B., Ferguson S. J., and Hajdu, J. (1995). The anatomy of a bifunctional enzyme: Structural basis for reduction of oxygen to water and synthesis of nitric oxide by cytochrome *cd*-1. *Cell, 81*, 369-377.

Gangeswaran, R., Lowe, D. J., and Eady, R. R. (1993). Purification and characterization of the assimilatory nitrate reductase of *Azotobacter vinelandii*. *Biochem. J., 289*, 335-342.

George, S. J., Allen, J. W., Ferguson, S. J., and Thorneley, R. N. F. (2000). Time-resolved infrared spectroscopy reveals a stable ferric heme-NO intermediate in the reaction of *Paracoccus pantotrophus* cytochrome *cd*1 nitrite reductase with nitrite. *J. Biol. Chem., 275*, 33231-33237.

Gilles-Gonzalez, M. A., Gonzalez, G., and Perutz, M. F. (1995). Kinase activity of oxygen sensor FixL depends on the spin state of its heme iron. *Biochemistry, 34*, 232-236.

Gilles-Gonzalez, M. A., Gonzalez, G., Perutz, M. F., Kiger, L., Marden, M. C., and Poyart, C. (1994). Heme-based sensors, exemplified by the kinase FixL, are a new class of heme protein with distinctive ligand binding and autoxidation. *Biochemistry, 33*, 8067-8073.

Girsch, P., and De Vries, S. (1997). Purification and initial kinetic and spectroscopic characterization of NO reductase from *Paracoccus denitrificans*. *Biochim. Biophys. Acta, 1318*, 202-216.

Giuffre, A., Stubauer, G., Sarti, P., Brunori, M., Zumft, W. G., Buse, G., and Soulimane, T. (1999). The heme-copper oxidases of *Thermus thermophilus* catalyze the reduction of nitric oxide: Evolutionary implications. *Proc. Natl. Acad. Sci. USA, 96*, 14718-14723.

Glockner, A. B., Jüngst, A., and Zumft, W. G. (1993). Copper-containing nitrite reductase from *Pseudomonas aureofaciens* is functional in a cytochrome cd_1-free background (nirS$^-$) of *Pseudomonas stutzeri*. *Arch. Microbiol., 160*, 18-26.

Glockner, A. B., and Zumft, W. G. (1996). Sequence analysis of an internal 9.72-kb segment from the 30-kb denitrification gene cluster of *Pseudomonas stutzeri*. *Biochim. Biophys. Acta, 1277*, 6-12.

Godden, J. W., Turley, S., Teller, D. C., Adman, E. T., Liu, M. Y., Payne, W. J., and LeGall, J. (1991). The 2.3 angstrom X-ray structure of nitrite reductase from *Achromobacter cycloclastes*. *Science, 253*, 438-442.

Goldman, B. S., and Roth, J. R. (1993). Genetic structure and regulation of the *cysG* gene in *Salmonella typhimurium*. *J. Bacteriol., 175*, 1457-1466.

Gong, W., Hao, B., and Chan, M. K. (2000). New mechanistic insights from structural studies of the oxygen-sensing domain of *Bradyrhizobium japonicum* FixL. *Biochemistry, 39*, 3955-3962.

Gong, W., Hao, B., Mansy, S. S., Gonzalez, G., Gilles-Gonzalez, M. A., and Chan, M. K. (1998). Structure of a biological oxygen sensor: A new mechanism for heme-driven signal transduction. *Proc. Natl. Acad. Sci. USA, 95*, 15177-15182.

Goretski, J., Zarifou, O. C., and Hollocher, T. C. (1990). Steady-state nitric oxide concentrations during denitrification. *J. Biol. Chem., 265*, 11535-11538.

Green, J., Scott, C., and Guest, J. R. (2001). Functional versatility in the CRP-FNR superfamily of transcription factors: FNR and FLP. *Adv. Microb. Physiol., 44*, 1-34.

Grönberg, K. L. C., Roldan, M. D., Prior, L., Butland, G., Cheesman, M. R., Richardson, D. J., *et al.* (1999). A low-redox potential heme in the dinuclear center of bacterial nitric oxide reductase: Implications for the evolution of energy-conserving heme-copper oxidases. *Biochemistry, 38*, 13780-13786.

Grove, J., Tanapongpipat, S., Thomas, G., Griffiths, L., Crooke, H., and Cole, J. (1996). *Escherichia coli* K-12 genes essential for the synthesis of *c*-type cytochromes and a third nitrate reductase located in the periplasm. *Mol. Microbiol., 19*, 467-481.

Hallin, S.and Lindgren, P. E. (1999). PCR detection of genes encoding nitrite reductase in denitrifying bacteria. *Appl. Environ. Microbiol., 65*, 1652-1657.

Hartig, E., Schiek, U., Vollack, K. U., and Zumft, W. G. (1999). Nitrate and nitrite control of respiratory nitrate reduction in denitrifying *Pseudomonas stutzeri* by a two-component regulatory system homologous to NarXL of *Escherichia coli. J. Bacteriol., 181*, 3658-3665.

Hayashi, N. R., Arai, H., Kodama, T., and Igarashi, Y. (1998). The *nirQ* gene, which is required for denitrification of *Pseudomonas aeruginosa*, can activate the RubisCO from *Pseudomonas hydrogenothermophila. Biochim. Biophys. Acta, 1381*, 347-350.

Heikkila, M. P., Honisch, U., Wunsch, P., and Zumft, W. G. (2001). Role of the Tat transport system in nitrous oxide reductase translocation and cytochrome *cd*1 biosynthesis in *Pseudomonas stutzeri. J. Bacteriol., 183*, 1663-1671.

Heis, B., Frunzke, K., and Zumft, W. G. (1989). Formation of the N-N bond from nitric oxide by a membrane-bound cytochrome *bc* complex of nitrate-respiring (denitrifying) *Pseudomonas stutzeri. J. Bacteriol., 171*, 3288-3297.

Hendriks, J., Oubrie, A., Castresana, J., Urbani, A., Gemeinhardt, S., and Saraste, M. (2000). Nitric oxide reductases in bacteria. *Biochim. Biophys. Acta, 1459*, 266-273.

Hendriks, J., Warne, A., Gohlke, U., Haltia, T., Ludovici, C., Lubben, M., and Saraste, M. (1998). The active site of the bacterial nitric oxide reductase is a dinuclear iron center. *Biochemistry, 37*, 13102-13109.

Hendriks, J. H., Jasaitis, A., Saraste, M., and Verkhovsky, M. I. (2002). Proton and electron pathways in the bacterial nitric oxide reductase. *Biochemistry, 41*, 2331-2340.

Hernandez, D., and Rowe, J. J. (1988). Oxygen inhibition of nitrate uptake is a general regulatory mechanism in nitrate respiration. *J. Biol. Chem, 263*, 7937-7939.

Hille, R. (1996). The mononuclear molybdenum enzymes. *Chem. Rev., 96*, 2757-2816.

Hilton, J. C., and Rajagopalan, K. V. (1996). Identification of the molybdenum cofactor of dimethyl sulfoxide reductase from *Rhodobacter sphaeroides* f. sp. denitrificans as bis(molybdopterin guanine dinucleotide)molybdenum. *Arch. Biochem. Biophys., 325*, 139-143.

Hoehn, G. T., and Clark, V. L. (1992). Isolation and nucleotide sequence of the gene (*aniA*) encoding the major anaerobically induced outer membrane protein of *Neisseria gonorrhoeae. Infect Immun., 60*, 4695-4703.

Hooper, A. B. (1968). A nitrite reducing enzyme from *Nitrosomonas europaea. Biochim. Biophys. Acta, 162*, 49-65.

Householder, T. C., Fozo, E. M., Cardinale, J. A., and Clark, V. L. (2000). Gonococcal nitric oxide reductase is encoded by a single gene, *norB*, which is required for anaerobic growth and is induced by nitric oxide. *Infect. Immun., 68*, 5241-5246.

Householder, T. C., Belli, W. A., Lissenden, S., Cole, J. A., and Clark, V. L. (1999) cis- and trans-acting elements involved in regulation of *aniA*, the gene encoding the major anaerobically induced outer membrane protein in *Neisseria gonorrhoeae. J. Bacteriol., 181*, 541-551.

Howes, B. D., Abraham, Z. H., Lowe, D. J., Bruser, T., Eady, R. R., and Smith, B. E. (1994). EPR and electron nuclear double resonance (ENDOR) studies show nitrite binding to the type 2 copper centers of the dissimilatory nitrite reductase of *Alcaligenes xylosoxidans* (NCIMB 11015). *Biochemistry, 33*, 3171-3177.

Hunt, S., and Layzell, D. B. (1993). Gas exchange of legume nodules and the regulation of nitrogenase. *Annu. Rev. Plant Physiol. Plant Mol. Biol., 44*, 483-511.

Inatomi, K.-I. (1999). The subunit structure of nitrite reductase purified from the denitrifier *Achromobacter cycloclastes. Biosci. Biotechnol. Biochem., 63*, 2020-2022.

Inoue, T., Gotowda, M., Deligeer, A., Kataoka, K., Yamaguchi, K., Suzuki, S., Watanabe, H., Gohow, M., and Kai, Y. (1998). Type 1 Cu structure of blue nitrite reductase from *Alcaligenes xylosoxidans* GIFU 1051 at 2.05 A resolution: Comparison of blue and green nitrite reductases. *J. Biochem., 124*, 876-879.

Iwasaki, H., and Matsubara, T. (1971). Cytochrome c_{557} (551) and cytochrome *cd* of *Alcaligenes faecalis. J. Biochem., 69*, 847-857.

Jain, R., and Shapleigh, J. P. (2001). Characterization of *nirV* and a gene encoding a novel pseudoazurin in *Rhodobacter sphaeroides* 2.4.3. *Microbiology, 147*, 2505-2515.

Jones, A. M., and Hollocher, T. C. (1993). Nitric oxide reductase of *Achromobacter cycloclastes. Biochim. Biophys. Acta, 1144*, 359-366.

Jormakka, M., Tornroth, S., Abramson, J., Byrne, B., and Iwata, S. (2002). Purification and crystallization of the respiratory complex formate dehydrogenase-N from *Escherichia coli. Acta Crystallogr. D. Biol. Crystallogr., 58*, 160-162.

Jüngst, A., Wakabayashi, S., Matsubara, H., and Zumft, W. G. (1991). The *nirSTBM* region coding for cytochrome cd_1-dependent nitrite respiration of *Pseudomonas stutzeri* consist of a cluster of mono-, di- and tetraheme proteins. *FEBS Lett., 279*, 205-209.

Jüngst, A., and Zumft, W. G. (1992). Interdependence of respiratory NO reduction and nitrite reduction revealed by mutagenesis of *nirQ*, a novel gene in the denitrification gene cluster of *Pseudomonas stutzeri. FEBS Lett., 314*, 308-314.

Kanayama, Y., and Yamamoto, Y. (1991). Formation of nitrosylleghemoglobin in nodules of nitrate-treated cowpea and pea plants. *Plant Cell Physiol., 32*, 19-24.

Kaneko, T., Sato, S., Kotani, H., Tanaka, A., Asamizu, E., Nakamura, Y., *et al.* (1996). Sequence analysis of the genome of the unicellular cyanobacterium *Synechocystis* sp. strain PCC6803. II. Sequence determination of the entire genome and assignment of potential protein-coding regions. *DNA Res., 3*, 109-136.

Kastrau, D. H. W., Heiss, B., Kroneck, P. M. H., and Zumft, W. G. (1994). Nitric oxide reductase from *Pseudomonas stutzeri*, a novel cytochrome *bc* complex - Phospholipid requirement, electron paramagnetic resonance and redox properties. *Eur. J. Biochem., 222*, 293-303.

Kawasaki, S., Arai, H., Kodama, T., and Igarashi, Y. (1997). Gene cluster for dissimilatory nitrite reductase (*nir*) from *Pseudomonas aeruginosa*: Sequencing and identification of a locus for heme d_1 biosynthesis. *J. Bacteriol., 179*, 235-242.

Kester, R. A., De Boer, W., and Laanbroek, H. J. (1997). Production of NO and N_2O by pure cultures of nitrifying and denitrifying bacteria during changes in aeration. *Appl. Environ. Microbiol., 63*, 3872-3877.

Kiley, P. J., and Beinert, H. (1998). Oxygen sensing by the global regulator, FNR: The role of the iron-sulfur cluster. *FEMS Microbiol. Rev., 22*, 341-352.

Knowles, R. (1996). Denitrification: Microbiology and ecology. *Life Support Biosph. Sci., 3*, 31-34.

Kobayashi, M., Matsuo, Y., Takimoto, A., Suzuki, S., Maruo, F., and Shoun, H. (1996). Denitrification, a novel type of respiratory metabolism in fungal mitochondrion. *J. Biol. Chem., 271*, 16263-16267.

Koppenhofer, A., Turner, K. L., Allen, J. W., Chapman, S. K., and Ferguson, S. J. (2000). Cytochrome cd_1 from *Paracoccus pantotrophus* exhibits kinetically gated, conformationally dependent, highly cooperative two-electron redox behavior. *Biochemistry, 39*, 4243-4249.

Korner, H. (1993). Anaerobic expression of nitric oxide reductase from denitrifying *Pseudomonas stutzeri. Arch. Microbiol., 159*, 410-416.

Korner, H., and Zumft, W. G. (1989). Expression of denitrification enzymes in response to the dissolved oxygen level and respiratory substrate in continuous culture of *Pseudomonas stutzeri. Appl. Environ. Microbiol., 55*, 1670-1676.

Koutny, M., and Kucera, I. (1999). Kinetic analysis of substrate inhibition in nitric oxide reductase of *Paracoccus denitrificans. Biochem. Biophys. Res. Commun., 262*, 562-564.

Koutny, M., Kucera, I., Tesarik, R., Turanek, J., and Van Spanning, R. J. M. (1999). Pseudoazurin mediates periplasmic electron flow in a mutant strain of *Paracoccus denitrificans* lacking cytochrome c_{550}. *FEBS Lett., 448*, 157-159.

Krause, B., and Nealson, K. H. (1997). Physiology and enzymology involved in denitrification by *Shewanella putrefaciens. Appl. Environ. Microbiol., 63*, 2613-2618.

Kroneck, P. M., Antholine, W. E., Kastrau, D. H., Buse, G., Steffens, G. C., and Zumft, W. G. (1990). Multifrequency EPR evidence for a bimetallic center at the CuA site in cytochrome *c* oxidase. *FEBS Lett., 268*, 274-276.

Kucera, I., Hedbavny, R., and Dadak, V. (1988). Separate binding sites for antimycin and mucidin in the respiratory chain of the bacterium *Paracoccus denitrificans* and their occurrence in other denitrificans bacteria. *Biochem. J., 252*, 905-908.

Kudo, T., Tomura, D., Liu, D. L., Dai, X. Q., and Shoun, H. (1996). Two isozymes of P450nor of *Cylindrocarpon tonkinense*: Molecular cloning of the cDNAs and genes, expressions in the yeast, and the putative NAD(P)H-binding site. *Biochimie, 78*, 792-799.

Kukimoto, M., Nishiyama, M., Murphy, M. E. P., Turley, S., Adman, E. T., Horinouchi, S., and Beppu, T. (1994). X-Ray structure and site-directed mutagenesis of a nitrite reductase from *Alcaligenes faecalis* S-6. Roles of two copper atoms in nitrite reduction. *Biochemistry, 33*, 5246-5252.

Kukimoto, M., Nishiyama, M., Tanokura, M., and Horinouchi, S. (2000). Gene organization for nitric oxide reduction in *Alcaligenes faecalis* S-6. *Biosci. Biotechnol. Biochem., 64*, 852-857.

LeGall, J., Payne, W. J., Morgan, V., and DerVartanian, D. (1979). On the purification of nitrite reductase from *Thiobacillus denitrificans* and its reaction with nitrite under reducing conditions. *Biochem. Biophys. Res. Commun., 87*, 355-362.

Lin J. T., Goldman B. S., and Stewart, V. (1994). The *nasFEDCBA* operon for nitrate and nitrite assimilation in *Klebsiella pneumoniae* M5a1. *J. Bacteriol., 176*, 2551-2559.

Lin, J. T., and Stewart, V. (1998). Nitrate assimilation by bacteria. *Adv. Microb. Physiol., 38*, 1-30.

Lindsay, M. R., Webb, R. I., Strous, M., Jetten, M. S., Butler, M. K., Forde, R. J., and Fuerst, J. A. (2001). Cell compartmentalisation in planctomycetes: Novel types of structural organisation for the bacterial cell. *Arch. Microbiol., 175*, 413-429.

Lissenden, S., Mohan, S., Overton, T., Regan, T., Crooke, H., Cardinale, J. A., Householder, T. C., Adams, P., O'Conner, C. D., Clark, V. L., Smith, H., and Cole, J. A. (2000). Identification of transcription activators that regulate gonococcal adaptation from aerobic to anaerobic or oxygen-limited growth. *Mol. Microbiol., 37*, 839-855.

Liu, H. P., Takio, S., Satoh, T., and Yamamoto, I. (1999). Involvement in denitrification of the *napKEFDABC* genes encoding the periplasmic nitrate reductase system in the denitrifying phototrophic bacterium *Rhodobacter sphaeroides f. sp. denitrificans. Biosci. Biotechnol. Biochem., 63*, 530-536.

Mancinelli, R. L., Cronin, S., and Hochstein, L. I. (1986). The purification and properties of a *cd*-type nitrite reductase from *Paracoccus halodenitrificans*. *Arch. Microbiol., 145*, 202-208.

Marger, M. D., and Saier, Jr., M. H. (1993). A major superfamily of transmembrane facilitators that catalyse uniport, symport and antiport. *Trends Biochem. Sci., 18*, 13-20.

McAlpine, A. S., McEwan, A. G., and Bailey, S. (1998). The high resolution crystal structure of DMSO reductase in complex with DMSO. *J. Mol. Biol., 275*, 613-623.

McKay, D. B., and Steitz, T. A. (1981). Structure of catabolite gene activator protein at 2.9 A resolution suggests binding to left-handed B-DNA. *Nature, 290*, 744-749.

McKenney, D. J., Drury, C. F., Findlay, W. I., Mutus, B., McDonnell, T., and Gajda, C. (1994). Kinetics of denitrification by *Pseudomonas fluorescens*: Oxygen effects. *Soil Biol. Biochem., 26*, 901-908.

Mellies, J., Jose, J., and Meyer, T. F. (1997). The *Neisseria gonorrhoeae* gene *aniA* encodes an inducible nitrite reductase. *Mol. Gen. Genet., 256*, 525-532.

Mesa, S., Gottfert, M., and Bedmar, E. J. (2001). The *nir, nor*, and *nos* denitrification genes are dispersed over the *Bradyrhizobium japonicum* chromosome. *Arch. Microbiol., 176*, 136-142.

Mesa, S., Velasco, L., Manzanera, M. E., Delgado, M. J., and Bedmar, E. J. (2002a). Characterization and regulation of the nitric oxide reductase-encoding region of *Bradyrhizobium japonicum. Microbiology, 148*, 3553-3560.

Mesa, S., Hennecke, H., Bedmar, E. J., and Fischer, H. M. (2002b). The role of *nnR* in the control of *Bradyrhizobium japonicum* denitrification genes. *The Fifth European Nitrogen Fixation Conference, 6-10 September, Norwich. Abstract.*

Michel, H. (1999). Cytochrome *c* oxidase: catalytic cycle and mechanisms of proton pumping - a discussion. *Biochemistry, 38*, 15129-15140.

Miller, D. J., and Nicholas, D. J. D. (1985). Characterization of a soluble cytochrome oxidase/nitrite reductase from *Nitrosomonas europaea*. *J. Gen. Microbiol.*, *131*, 2851-2854.

Mitchell, D. M., Wang, Y., Alben, J. O., and Shapleigh, J. P. (1998). FT-IR analysis of membranes of *Rhodobacter sphaeroides* 2.4.3 grown under microaerobic and denitrifying conditions. *Biochim Biophys. Acta, 1409*, 99-105.

Mitchell, P. (1961). Coupling of phosphorylation to electron and proton transfer by a chemi-osmotic type of mechanism. *Nature, 191*, 144-148.

Moir, J. W., and Wood, N. J. (2001). Nitrate and nitrite transport in bacteria. *Cell Mol. Life Sci., 58*, 215-224.

Moir, J. W. B., Baratta, D., Richardson, D. J., and Ferguson, S. J. (1993). The purification of a cd_1-type nitrite reductase from, and the absence of a copper-type nitrite reductase from, the aerobic denitrifier *Thiosphaera pantotropha*: The role of pseudoazurin as an electron donor. *Eur. J. Biochem., 212*, 377-385.

Moir, J. W. B., and Ferguson, S. J. (1994). Properties of a *Paracoccus denitrificans* mutant deleted in cytochrome c_{550} indicate that a copper protein can substitute for this cytochrome in electron transport to nitrite, nitric oxide and nitrous oxide. *Microbiology-UK, 140*, 389-397.

Moreno-Vivian, C., and Ferguson, S. J. (1998). Definition and distinction between assimilatory, dissimilatory and respiratory pathways. *Mol. Microbiol.*, 29, 664-666.

Moura, I., and Moura, J. J. (2001). Structural aspects of denitrifying enzymes. *Curr. Opin. Chem. Biol.*, *5*, 168-175.

Murai, K., Miyake, K., Andoh, J., and Iijima, S. (1998). Cloning and nucleotide sequence of the nitric oxide reductase locus in *Paracoccus denitrificans* IFO 12442. *J. Ferment. Bioeng., 86*, 494-499.

Murphy, M. E., Turley, S., and Adman, E. T. (1997). Structure of nitrite bound to copper-containing nitrite reductase from *Alcaligenes faecalis*. Mechanistic implications. *J. Biol. Chem., 272*, 28455-28460.

Musser, S. M., and Chan, S. I. (1998). Evolution of the cytochrome *c* oxidase proton pump. *J. Mol. Evol.*, *46*, 508-520.

Myers, C. R., and Myers, J. M. (1997). Cloning and sequence of *cymA*, a gene encoding a tetraheme cytochrome *c* required for reduction of iron(III), fumarate, and nitrate by *Shewanella putrefaciens* MR-1. *J. Bacteriol., 179*, 1143-1152.

Myers, J. M., and Myers, C. R. (2000). Role of the tetraheme cytochrome CymA in anaerobic electron transport in cells of *Shewanella putrefaciens* MR-1 with normal levels of menaquinone. *J. Bacteriol.*, *182*, 67-75.

Nakahara, K., Shoun, H., Adachi, S., Iizuka, T., and Shiro, Y. (1994). Crystallization and preliminary X-ray diffraction studies of nitric oxide reductase cytochrome P450nor from *Fusarium oxysporum*. *J. Mol. Biol., 239*, 158-159.

Nakahara, K., Tanimoto, T., Hatano, K., Usuda, K., and Shoun, H. (1993). Cytochrome-P-450-55A1 (P-450dNIR) acts as nitric oxide reductase employing NADH as the direct electron donor. *J. Biol. Chem., 268*, 8350-8355.

Noji, S., Nohno, T., Saito, T., and Taniguchi, S. (1989). The *narK* gene product participates in nitrate transport induced in *Escherichia coli* nitrate-respiring cells. *FEBS Lett., 252*, 139-143.

Nurizzo, D., Cutruzzola, F., Arese, M., Bourgeois, D., Brunori, M., Cambillau, C., and Tegoni, M. (1998). Conformational changes occurring upon reduction and NO binding in nitrite reductase from *Pseudomonas aeruginosa*. *Biochemistry, 37*, 13987-13996.

Nurizzo, D., Silvestrini, M.-C., Mathieu, M., Cutruzzola, F., Bourgeois, D., Fulop, V., Hajdu, J., Brunori, M., Tegoni, M., and Cambillau, C. (1997). N-terminal arm exchange is observed in the 2.15 A crystal structure of oxidized nitrite reductase from *Pseudomonas aeruginosa*. *Structure, 5*, 1157-1171.

O'Hara, G. M., and Daniel, R. M. (1985). Rhizobial denitrification. *Ann. Rev. Soil Biol. Biochem., 17*, 1-9.

Olesen, K., Veselov, A., Zhao, Y., Wang, Y., Danner, B., Scholes, C. P., and Shapleigh, J. P. (1998). Spectroscopic, kinetic, and electrochemical characterization of heterologously expressed wild-type and mutant forms of copper-containing nitrite reductase from *Rhodobacter sphaeroides* 2.4.3. *Biochemistry, 37*, 6086-6094.

Palmedo, G., Seither, P., Korner, H., Matthews, J. C., Burkhalter, R. S., Timkovich, R., and Zumft, W. G. (1995). Resolution of the *nirD* locus for heme d_1 synthesis of cytochrome cd_1 (respiratory nitrite reductase) from *Pseudomonas stutzeri*. *Eur. J. Biochem., 232*, 737-746.

Papa, S., Capitanio, N., Glaser, P., and Villani, G. (1994). The proton pump of heme-copper oxidases. *Cell Biol. Int., 18*, 345-355.

Park, S. Y., Shimizu, H., Adachi, S., Nakagawa, A., Tanaka, I., Nakahara, K., Shoun, H., Obayashi, E., Nakamura, H., Iizuka, T., and Shiro, Y. (1997). Crystal structure of nitric oxide reductase from denitrifying fungus *Fusarium oxysporum*. *Nature Struct. Biol., 4*, 827-832.

Parkhill, J., Achtman, M., James, K. D., Bentley, S. D., Churcher, C., Klee, *et al.* (2000). Complete DNA sequence of a serogroup A strain of *Neisseria meningitidis* Z2491. *Nature, 404*, 502-506.

Philippot, L., Mirleau, P., Mazurier, S., Siblot, S., Hartmann, A., Lemanceau, P., and Germon, J. C. (2001). Characterization and transcriptional analysis of *Pseudomonas fluorescens* denitrifying clusters containing the *nar, nir, nor* and *nos* genes. *Biochim. Biophys. Acta, 1517*, 436-440.

Pohlmann, A., Cramm, R., Schmelz, K., and Friedrich, B. (2000). A novel NO-responding regulator controls the reduction of nitric oxide in *Ralstonia eutropha*. *Mol. Microbiol., 38*, 626-638.

Poth, M. (1986). Dinitrogen production from nitrite by a *Nitrosomonas* isolate. *Appl. Environ. Microbiol., 52*, 957-959.

Potter, L., Angove, H., Richardson, D., and Cole, J. (2001). Nitrate reduction in the periplasm of gram-negative bacteria. *Adv. Microb. Physiol., 45*, 51-112.

Potter, L. C., and Cole, J. A. (1999). Essential roles for the products of the *napABCD* genes, but not *napFGH*, in periplasmic nitrate reduction by *Escherichia coli* K-12. *Biochem. J., 344*, 69-76.

Preisig, O., Anthamattan, D., and Hennecke, H. (1993). Genes for a microaerobically induced oxidase complex in *Bradyrhizobium japonicum* are essential for a nitrogen-fixing endosymbiosis. *Proc. Natl. Acad. Sci. USA, 90*, 3309-3313.

Prudencio, M., Eady, R. R., and Sawers, G. (1999). The blue copper-containing nitrite reductase from *Alcaligenes xylosoxidans*: Cloning of the *nirK* gene and characterization of the recombinant enzyme. *J. Bacteriol., 181*, 2323-2329.

Rabin, R. S., and Stewart, V. (1993). Dual response regulators (NarL and NarP) interact with dual sensors (NarX and NarQ) to control nitrate-regulated and nitrite-regulated gene expression in *Escherichia coli* K-12. *J. Bacteriol., 175*, 3259-3268.

Raitio, M., and Wikstrom, M. (1994). An alternative cytochrome oxidase of *Paracoccus denitrificans* functions as a proton pump. *BBA-Bioenergetics, 1186*, 100-106.

Reitzer, L. J. (1996). Ammonia assimilation and the biosynthesis of glutamine, glutamate, aspartate, asparagine, L-alanine, and D-alanine. In F. C. Neidhardt, *et al.* (Eds.), *Escherichia coli and Salmonella: Cellular and molecular biology* (pp. 391-407). Washington D.C.: ASM Press.

Richardson, D., and Sawers, G. (2002). Structural biology. PMF through the redox loop. *Science, 295*, 1842-1843.

Richardson, D. J. (2000). Bacterial respiration: A flexible process for a changing environment. *Microbiology, 146*, 551-571.

Richardson, D. J., Berks, B. C., Russell, D. A., Spiro, S., and Taylor, C. J. (2001). Functional, biochemical and genetic diversity of prokaryotic nitrate reductases. *Cell Mol. Life Sci., 58*, 165-178.

Richardson, D. J., McEwan, A. G., Page, M. D., Jackson, J. B., and Ferguson, S. J. (1990). The identification of cytochromes involved in the transfer of electrons to the periplasmic nitrate-reductase of *Rhodobacter capsulatus*. *Eur. J. Biochem., 194*, 263-270.

Richardson, D. J., and Watmough, N. J. (1999). Inorganic nitrogen metabolism in bacteria. *Curr. Opin. Chem. Biol., 3*, 207-219.

Roldan, M. D., Sears, H. J., Cheesman, M. R., Ferguson, S. J., Thomson, A. J., Berks, B. C., and Richardson, D. J. (1998). Spectroscopic characterization of a novel multiheme *c*-type cytochrome widely implicated in bacterial electron transport. *J. Biol. Chem., 273*, 28785-28790.

Rosch, C., Mergel, A., and Bothe, H. (2002). Biodiversity of denitrifying and dinitrogen-fixing bacteria in an acid forest soil. *Appl. Environ. Microbiol., 68*, 3818-3829.

Rothery, R. A., Blasco, F., Magalon, A., Asso, M., and Weiner, J. H. (1999). The hemes of *Escherichia coli* nitrate reductase A (NarGHI): Potentiometric effects of inhibitor binding to NarI. *Biochemistry, 38*, 12747-12757.

Rothery, R. A., Blasco, F., and Weiner, J. H. (2001). Electron transfer from heme bL to the [3Fe-4S] cluster of *Escherichia coli* nitrate reductase A (NarGHI). *Biochemistry, 40*, 5260-5268.

Rothery, R. A., Magalon, A., Giordano, G., Guigliarelli, B., Blasco, F., and Weiner, J. H. (1998). The molybdenum cofactor of *Escherichia coli* nitrate reductase A (NarGHI). Effect of a *mobAB* mutation and interactions with [Fe-S] clusters. *J. Biol. Chem., 273*, 7462-7469.

Rowe, J. J., Ubbinkkok, T., Molenaar, D., Konings, W. N., and Driessen, A. J. M. (1994). NarK is a nitrite-extrusion system involved in anaerobic nitrate respiration by *Escherichia coli*. *Mol. Microbiol., 12*, 579-586.

Sakurai, N., and Sakurai, T. (1997). Isolation and characterization of nitric oxide reductase from *Paracoccus halodenitrificans*. *Biochemistry, 36*, 13809-13815.

Sakurai, N., and Sakurai, T. (1998). Genomic DNA cloning of the region encoding nitric oxide reductase in *Paracoccus halodenitrificans* and a structure model relevant to cytochrome oxidase. *Biochem. Biophys. Res. Commun., 243*, 400-406.

Sakurai, T., Sakurai, N., Matsumoto, H., Hirota, S., and Yamauchi, O. (1998). Roles of four iron centers in *Paracoccus halodenitrificans* nitric oxide reductase. *Biochem. Biophys. Res. Commun., 251*, 248-251.

Samuelsson, M. O. (1985). Dissimilatory nitrate reduction to nitrate, nitrous oxide, and ammonium by *Pseudomonas putrefaciens*. *Appl. Environ. Microbiol., 50*, 812-815.

Sann, R., Kostka, S., and Friedrich, B. (1994). A cytochrome cd_1-type nitrite reductase mediates the first step of denitrification in *Alcaligenes eutrophus*. *Arch. Microbiol., 161*, 453-459.

Saraste, M. (1994). Structure and evolution of cytochrome oxidase. *Anton Leeuwenhoek Int. J. Gen. M., 65*, 285-287.

Saraste, M., and Castresana, J. (1994). Cytochrome oxidase evolved by tinkering with denitrification enzymes. *FEBS Lett., 341*, 1-4.

Saraste, M., Holm, L., Lemieux, L., Lubben, M., and Vanderoost, J. (1991). The happy family of cytochrome oxidases. *Biochem. Soc. Trans., 19*, 608-612.

Saunders, N. F., Hornberg, J. J., Reijnders, W. N., Westerhoff, H. V., de Vries, S., and van Spanning, R. J. M. (2000). The NosX and NirX proteins of *Paracoccus denitrificans* are functional homologues: Their role in maturation of nitrous oxide reductase. *J. Bacteriol., 182*, 5211-5217.

Saunders, N. F., Houben, E. N., Koefoed, S., de Weert, S., Reijnders, W. N., Westerhoff, H. V., *et al.* (1999). Transcription regulation of the *nir* gene cluster encoding nitrite reductase of *Paracoccus denitrificans* involves NNR and NirI, a novel type of membrane protein. *Mol. Microbiol., 34*, 24-36.

Sawada, E., and Satoh, T. (1980). Periplasmic location of dissimilatory nitrate and nitrite reductases in a denitrifying phototrophic bacterium, *Rhodopseudomonas sphaeroides* forma sp. *denitrificans*. *Plant Cell Physiol., 21*, 205-210.

Schindelin, H., Kisker, C., Hilton, J., Rajagopalan, K. V., and Rees, D. C. (1996). Crystal structure of DMSO reductase: Redox-linked changes in molybdopterin coordination. *Science, 272*, 1615-1621.

Schmidt, I., and Bock, E. (1997). Anaerobic ammonia oxidation with nitrogen dioxide by *Nitrosomonas eutropha*. *Arch. Microbiol., 167*, 106-111.

Sears, H. J., Bennett, B., Spiro, S., Thomson, A. J., and Richardson, D. J. (1995). Identification of periplasmic nitrate reductase Mo(V) EPR signals in intact cells of *Paracoccus denitrificans*. *Biochem. J., 310*, 311-314.

Sears, H. J., Sawers, G., Berks, B. C., Ferguson, S. J., and Richardson, D. J. (2000). Control of periplasmic nitrate reductase gene expression (*napEDABC*) from *Paracoccus pantotrophus* in response to oxygen and carbon substrates. *Microbiology, 146*, 2977-2985.

Sears, H. J., Spiro, S., and Richardson, D. J. (1997). Effect of carbon substrate and aeration on nitrate reduction and expression of the periplasmic and membrane-bound nitrate reductases in carbon-limited continuous cultures of *Paracoccus denitrificans* Pd1222. *Microbiology, 143*, 3767-3774.

Shapleigh, J. P, .and Payne, W. J. (1985). Nitric oxide-dependent proton translocation in various denitrifiers. *J. Bacteriol., 163*, 837-840.

Shaw, A. L., Leimkuhler, S., Klipp, W., Hanson, G. R., and McEwan, A. G. (1999). Mutational analysis of the dimethylsulfoxide respiratory (*dor*) operon of *Rhodobacter capsulatus*. *Microbiology, 145*, 1409-1420.

Shaw, D. J., Rice, D. W., and Guest, J. R. (1983). Homology between CAP and Fnr, a regulator of anaerobic respiration in *Escherichia coli*. *J. Mol. Biol., 166*, 241-247.

Shingler, V. (1996). Signal sensing by sigma(54)-dependent regulators: Derepression as a control mechanism. *Mol. Microbiol., 19*, 409-416.

Shiro, Y., Fujii, M., Iizuka, T., Adachi, S., Tsukamoto, K., Nakahara, K., and Shoun, H. (1995). Spectroscopic and kinetic studies on reaction of cytochrome p450nor with nitric oxide - Implication for its nitric oxide reduction mechanism. *J. Biol. Chem., 270*, 1617-1623.

Shoun, H., Kano, M., Baba, I., Takaya, N., and Matsuo, M. (1998). Denitrification of actinomycetes and purification of dissimilatory nitrite reductase and azurin from *Streptomyces thioluteus*. *J. Bacteriol., 180*, 4413-4415.

Siddiqui, R.A., Warneckeeberz, U., Hengsberger, A., Schneider, B., Kostka, S., and Friedrich, B. (1993). Structure and function of a periplasmic nitrate reductase in *Alcaligenes eutrophus* H16. *J. Bacteriol., 175*, 5867-5876.

Simon, J. (2002). Enzymology and bioenergetics of respiratory nitrite ammonification. *FEMS Microbiol. Rev., 26*, 285.

Smith, G. B., and Tiedje, J. M. (1992). Isolation and characterization of a nitrite reductase gene and its use as a probe for denitrifying bacteria. *Appl. Environ. Microbiol., 58*, 376-384.

Snyder, S. W., and Hollocher, T. C. (1987). Purification and some characteristics of nitrous oxide reductase from *Paracoccus denitrificans*. *J. Biol. Chem., 262*, 6515-6525.

Song, B., and Ward, B. B. (2002). Nitrite reductase genes in halobenzoate degrading denitrifying bacteria. *Unpublished data.*

Spencer, J. B., Stolowich, N. J., Roessner, C. A., and Scott, A. I. (1993). The *Escherichia coli cysG* gene encodes the multifunctional protein, siroheme synthase. *FEBS Lett., 335*, 57-60.

Spiro, S. (1994). The FNR family of transcriptional regulators. *Anton Leeuwenhoek Int. J. Gen. M., 66*, 23-36.

Stein, L. Y., and Arp, D. J. (1998). Loss of ammonia monooxygenase activity in *Nitrosomonas europaea* upon exposure to nitrite. *Appl. Environ. Microbiol., 64*, 4098-4102.

Stewart, V. (1993). Nitrate regulation of anaerobic respiratory gene expression in *Escherichia coli*. *Mol. Microbiol., 9*, 425-434.

Stiefel, E. I. (1996). Molding metallocenters for biology. *Chem. Biol., 3*, 643-644.

Stolz, J. F., and Basu, P. (2002). Evolution of nitrate reductase: Molecular and structural variations on a common function. *Chembiochem., 3*, 198-206.

Stouthamer, A. H. (1991). Metabolic regulation including anaerobic metabolism in *Paracoccus denitrificans*. *J. Bioenerg. Biomembr.*, 23, 163-185.

Stouthamer, A. H. (1992). Metabolic pathways in *Paracoccus denitrificans* and closely related bacteria in relation to the phylogeny of prokaryotes. *Anton Leeuwenhoek Int. J. Gen. M., 61*, 1-33.

Strange, R. W., Dodd, F. E., Abraham, Z. H., Grossmann, J. G., Bruser, T., Eady, R. R., *et al.* (1995). The substrate-binding site in Cu nitrite reductase and its similarity to Zn carbonic anhydrase. *Nat. Struct. Biol., 2*, 287-292.

Strange, R. W., Murphy, L. M., Dodd, F. E., Abraham, Z. H., Eady, R. R., Smith, B. E., and Hasnain, S. S. (1999). Structural and kinetic evidence for an ordered mechanism of copper nitrite reductase. *J. Mol. Biol., 287*, 1001-1009.

Streeter, J. (1988). Inhibition of legume nodule formation and nitrogen fixation by nitrate. *Crit. Rev. Plant Sci., 7*, 1-23.

Strous, M., Fuerst, J. A., Kramer, E. H., Logemann, S., Muyzer, G., van de Pas-Schoonen, K. T., *et al.* (1999). Missing lithotroph identified as new planctomycete. *Nature, 400*, 446-449.

Stuendl, U. M., Schmidt, I., Scheller, U., Schmid, R., Schunck, W. H., and Schauer, F. (1998). Purification and characterization of cytosolic cytochrome P450 forms from yeasts belonging to the genus *Trichosporon*. *Arch. Biochem. Biophys., 357*, 131-136.

Stundl, U. M., Patzak, D., and Schauer, F. (2000). Purification of a soluble cytochrome P450 from *Trichosporon montevideense*. *J. Basic Microbiol., 40*, 289-292.

Suharti, K., Strampraad, M. J., Schroder, I., and de Vries, S. (2001). A novel copper A containing menaquinol NO reductase from *Bacillus azotoformans*. *Biochemistry, 40*, 2632-2639.

Suzuki, E., Horikoshi, N., and Kohzuma, T. (1999). Cloning, sequencing, and transcriptional studies of the gene encoding copper-containing nitrite reductase from *Alcaligenes xylosoxidans* NCIMB 11015. *Biochem. Biophys. Res. Commun., 255*, 427-431.

Takaya, N., Uchimura, H., Lai, Y., and Shoun, H. (2002). Transcriptional control of nitric oxide reductase gene (CYP55) in the fungal denitrifier *Fusarium oxysporum*. *Biosc. Biotechnol. Biochem., 66*, 1039-1045.

Tettelin, H., Saunders, N. J., Heidelberg, J., Jeffries, A. C., Nelson, K. E., Eisen, J. A., *et al.*, (2000). Complete genome sequence of *Neisseria meningitidis* serogroup B strain MC58. *Science, 287*, 1809-1815.

Timkovich, R., Dhesi, R., Martinkus, K. J., Robinson, M. K., and Rea, T. M. (1982). Isolation of *Paracoccus denitrificans* cytochrome cd_1 : comparative kinetics with other nitrite reductases. *Arch. Biochem. Biophys., 215*, 47-58.

Toffanin, A., Wu, Q., Maskus, M., Casella, S., Abruna, H. D., and Shapleigh, J. P. (1996). Characterization of the gene encoding nitrite reductase and the physiological consequences of its expression in the nondenitrifying *Rhizobium* "hedysari" strain HCNT1. *Appl. Environ. Microbiol., 62*, 4019-4025.

Tomura, D., Obika, K., Fukamizu, A., and Shoun, H. (1994). Nitric oxide reductase cytochrome P-450 gene, CYP 55, of the fungus *Fusarium oxysporum* containing a potential binding-site for FNR, the transcription factor involved in the regulation of anaerobic growth of *Escherichia coli. J. Biochem. Tokyo, 116*, 88-94.

Tosques van, E., Kwiatkowski, A. V., Shi, J., and Shapleigh, J. P. (1997). Characterization and regulation of the gene encoding nitrite reductase in *Rhodobacter sphaeroides* 2.4.3. *J. Bacteriol., 179*, 1090-1095.

Trumpower, B. L., and Gennis, R. B. (1994). Energy transduction by cytochrome complexes in mitochondrial and bacterial respiration: The enzymology of coupling electron transfer reactions to transmembrane proton translocation. *Annu. Rev. Biochem., 63*, 675-716.

Tyson, K. L., Cole, J. A., and Busby, S. J. W. (1994). Nitrite and nitrate regulation at the promoters of two *Escherichia coli* operons encoding nitrite reductase: Identification of common target heptamers for both NarP- and NarL-dependent regulation. *Mol. Microbiol., 13*, 1045-1055.

Uchimura, H., Enjoji, H., Seki, T., Taguchi, A., Takaya, N., and Shoun, H. (2002). Nitrate reductase-formate dehydrogenase couple involved in the fungal denitrification by *Fusarium oxysporum. J. Biochem. (Tokyo), 131*, 579-586.

Usuda, K., Toritsuka, N., Matsuo, Y., Kim, D.H., and Shoun, H. (1995). Denitrification by the fungus *Cylindrocarpon tonkinense*: anaerobic cell growth and two isozyme forms of cytochrome P-450nor. *Appl. Environ. Microbiol., 61*, 883-889.

Vairinhos, F., Wallace, W., and Nicholas, D. J. D. (1989). Simultaneous assimilation and denitrification of nitrate by *Bradyrhizobium japonicum. J. Gen. Micro., 135*, 189-193.

Van de Graaf, A. A., Mulder, A., Debruijn, P., Jetten, M. S. M., Robertson, L. A., and Kuenen, J. G. (1995). Anaerobic oxidation of ammonium is a biologically mediated process. *Appl. Environ. Microbiol., 61*, 1246-1251.

Van der Oost, J., Deboer, A. P. N., Degier, J. W. L., Zumft, W. G., Stouthamer, A. H., and Van Spanning, R. J. M. (1994). The heme-copper oxidase family consists of three distinct types of terminal oxidases and is related to nitric oxide reductase. *FEMS Microbiol. Lett., 121*, 1-9.

Velasco, L., Mesa, S., Delgado, M. J., and Bedmar, E. J. (2001). Characterization of the *nirK* gene encoding the respiratory, Cu-containing nitrite reductase of *Bradyrhizobium japonicum. Biochim. Biophys. Acta, 1521*, 130-134.

Vollack, K. U., Xie, J., Hartig, E., Romling, U., and Zumft, W. G. (1998). Localization of denitrification genes on the chromosomal map of *Pseudomonas aeruginosa. Microbiology, 144*, 441-448.

Vollack, K. U., and Zumft, W. G. (2001). Nitric oxide signaling and transcriptional control of denitrification genes in *Pseudomonas stutzeri. J. Bacteriol., 183*, 2516-2526.

Vygodina, T. V., Capitanio, N., Papa, S., and Konstantinov, A. A. (1997). Proton pumping by cytochrome c oxidase is coupled to peroxidase half of its catalytic cycle. *FEBS Lett., 412*, 405-409.

Walsh, T. A., Johnson, M. K., Greenwood, C., Barber, D., Springall, J. P., and Thomson, A. J. (1979). Some magnetic properties of *Pseudomonas* cytochrome oxidase. *Biochem. J., 177*, 29-39.

Wang, H., and Gunsalus, R. P. (2000). The *nrfA* and *nirB* nitrite reductase operons in *Escherichia coli* are expressed differently in response to nitrate than to nitrite. *J. Bacteriol., 182*, 5813-5822.

Warnecke-Eberz, U., and Friedrich, B. (1993). Three nitrate reductase activities in *Alcaligenes eutrophus. Arch. Microbiol., 159*, 405-409.

Watmough, N. J., Butland, G., Cheesman, M. R., Moir, J. W. B., Richardson, D. J., and Spiro, S. (1999). Nitric oxide in bacteria: Synthesis and consumption. *Biochim. Biophys. Acta, 141*, 456-474.

Weeg-Aerssens, E., Wu, W. S., Ye, R. W., Tiedje, J. M., and Chang, C. K. (1991). Purification of cytochrome *cd*1 nitrite reductase from *Pseudomonas stutzeri* JM300 and reconstitution with native and synthetic heme d_1. *J. Biol. Chem., 266*, 7496-7502.

Whittaker, M., Bergmann, D., Arciero, D., and Hooper, A. B. (2000). Electron transfer during the oxidation of ammonia by the chemolithotrophic bacterium *Nitrosomonas europaea. Biochim. Biophys. Acta, 1459*, 346-355.

Wikstrom, M. (2000a). Mechanism of proton translocation by cytochrome *c* oxidase: A new four-stroke histidine cycle. *Biochim. Biophys. Acta, 1458*, 188-198.

Wikstrom, M. (2000b). Proton translocation by cytochrome *c* oxidase: A rejoinder to recent criticism. *Biochemistry, 39*, 3515-3519.

Williams, P. A., Fulop, V., Garman, E. F., Saunders, N. F. W., Ferguson, S. J., and Hajdu, J. (1997). Haem-ligand switching during catalysis in crystals of a nitrogen-cycle enzyme. *Nature, 389*, 406-412.

Williams, R., Bell, A., Sims, G., and Busby, S. (1991). The role of two surface exposed loops in transcription activation by the *Escherichia coli* CRP and FNR proteins. *Nucleic Acids Res., 19*, 6705-6712.

Williams, R. M., Rhodius, V. A., Bell, A. I., Kolb, A., and Bushy, S. J. W. (1996). Orientation of functional activating regions in the *Escherichia coli* CRP protein during transcription activation at class II promoters. *Nucleic Acids Res., 24*, 1112-1118.

Williams, R. S. B., Davis, M. A., and Howlett, B. J. (1995). The nitrate and nitrite reductase-encoding genes of *Leptosphaeria maculans* are closely linked and transcribed in the same direction. *Gene, 158*, 153-154.

Winkler, W. C., Gonzalez, G., Wittenberg, J. B., Hille, R., Dakappagari, N., Jacob, A., *et al.* (1996). Nonsteric factors dominate binding of nitric oxide, azide, imidazole, cyanide, and fluoride to the rhizobial heme-based oxygen sensor FixL. *Chem. Biol., 3*, 841-850.

Wood, N. J., Alizadeh, T., Bennett, S., Pearce, J., Ferguson, S. J., Richardson, D. J., and Moir, J. W. (2001). Maximal expression of membrane-bound nitrate reductase in *Paracoccus* is induced by nitrate via a third FNR-like regulator named NarR. *J. Bacteriol., 183*, 3606-3613.

Wood, N. J., Alizadeh, T., Richardson, D. J., Ferguson, S. J., and Moir, J. W. (2002). Two domains of a dual-function NarK protein are required for nitrate uptake, the first step of denitrification in *Paracoccus pantotrophus. Mol. Microbiol., 44*, 157-170.

Wood, P. M. (1978). Periplasmic location of the terminal reductase in nitrite respiration. *FEBS Lett., 92*, 214-218.

Wu, J.-Y., Siegel, L. M., and Kredich, N. M. (1991) High-level Eexpression of *Escherichia coli* NADPH-sulfite reductase: Requirement for a cloned *cys*G plasmid to overcome limiting siroheme cofactor. *J. Bacteriol., 173*, 325-333.

Wu, Q. T., Knowles, R., and Chan, Y. K. (1995). Production and consumption of nitric oxide by denitrifying *Flexibacter canadensis. Can. J. Microbiol., 41*, 585-591.

Wu, S. Q., Chai, W., Lin, J. T., and Stewart, V. (1999). General nitrogen regulation of nitrate assimilation regulatory gene *nasR* expression in *Klebsiella oxytoca* M5al. *J. Bacteriol., 181*, 7274-7284.

Yamazaki, T., Oyanagi, H., Fujiwara, T., and Fukumori, Y. (1995). Nitrite reductase from the magnetotactic bacterium *Magnetospirillum magnetotacticum.* A novel cytochrome *cd*1 with Fe(II):nitrite oxidoreductase activity. *Eur. J. Biochem., 233*, 665-671.

Ye, R. W., Arunakumari, A., Averill, B. A., and Tiedje, J. M. (1992). Mutants of *Pseudomonas fluorescens* deficient in dissimilatory nitrite reduction are also altered in nitric oxide reduction. *J. Bacteriol., 174*, 2560-2564.

Yokoyama, K., Hayashi, N. R., Arai, H., Chung, S. Y., Igarashi, Y., and Kodama, T. (1995). Genes encoding RubisCO in *Pseudomonas hydrogenothermophila* are followed by a novel *cbbQ* gene similar to *nirQ* of the denitrification gene cluster from *Pseudomonas* species. *Gene, 153*, 75-79.

Zart, D., and Bock, E. (1998). High rate of aerobic nitrification and denitrification by *Nitrosomonas eutropha* grown in a fermentor with complete biomass retention in the presence of gaseous NO_2 or NO. *Arch. Microbiol., 169*, 282-286.

Zart, D., Schmidt, I., and Bock, E. (2000). Significance of gaseous NO for ammonia oxidation by *Nitrosomonas eutropha. Antonie van Leeuwenhoek, 77*, 49-55.

Zhulin, I. B., Taylor, B. L., and Dixon, R. (1997). PAS domain S-boxes in Archaea, Bacteria and sensors for oxygen and redox. *Trends Biochem. Sci., 22*, 331-333.

Zumft, W. G., and Koerner, H. (1997). Enzyme diversity and mosaic gene organization in denitrification. *Antonie van Leeuwenhoek, 71*, 43-58.

Zumft, W. G. (1993). The biological role of nitric oxide in bacteria. *Arch. Microbiol., 160*, 253-264.

Zumft, W. G. (1997). Cell biology and molecular basis of denitrification. *Microbiol. Mol. Biol. Rev., 61*, 533-616.

Zumft, W. G. (2002). Nitric oxide signaling and NO dependent transcriptional control in bacterial denitrification by members of the FNR-CRP regulator family. *J Mol Microbiol. Biotechnol., 4,* 277-286.

Zumft, W. G., Braun, C., and Cuypers, H. (1994). Nitric oxide reductase from *Pseudomonas stutzeri* - Primary structure and gene organization of a novel bacterial cytochrome *bc* complex. *Eur. J. Biochem., 219*, 481-490.

Zumft, W. G., Gotzmann, D. J., and Kroneck, P. M. (1987). Type 1 blue copper proteins constitute a respiratory nitrite-reducing system in *Pseudomonas aureofaciens*. *Eur. J. Biochem., 168*, 301-307.

SUBJECT INDEX

Nitrogen Fixation: Origins, Applications, and Research Progress

1. B.E. Smith, R.L. Richards and W.E. Newton (eds.): *Catalysts for Nitrogen Fixation.* Nitrogenases, Relevant Chemical Models and Commercial Processes. 2004 ISBN 1-4020-2508-4
2. W. Klipp, B. Masepohl, J.R. Gallon and W.E. Newton (eds.): *Genetics and Regulation of Nitrogen Fixation in Free-Living Bacteria.* 2004 ISBN 1-4020-2178-X
3. R. Palacios and W.E. Newton (eds.): *Genomes and Genomics of Nitrogen-fixing Organisms.* 2005 ISBN 1-4020-3053-3
4. D. Werner and W.E. Newton (eds.): *Nitrogen Fixation in Agriculture, Forestry, Ecology, and the Environment.* 2005 ISBN 1-4020-3542-X

springeronline.com

MIX
Papier aus verantwortungsvollen Quellen
Paper from responsible sources
FSC® C105338

If you have any concerns about our products,
you can contact us on
ProductSafety@springernature.com

In case Publisher is established outside the EU,
the EU authorized representative is:
Springer Nature Customer Service Center GmbH
Europaplatz 3, 69115 Heidelberg, Germany

Printed by Libri Plureos GmbH
in Hamburg, Germany